BIBLIOGRAPHY ON THE FATIGUE OF MATERIALS, COMPONENTS AND STRUCTURES

BIBLIOGRAPHY ON THE FATIGUE OF MATERIALS, COMPONENTS AND STRUCTURES

1838 — 1950

compiled by

J. Y. MANN

M.E., B.Sc., M.I.E.Aust., A.F.R.Ae.S., F.I.M.
Senior Research Scientist, Structures Division
Aeronautical Research Laboratories, Melbourne

published for and on behalf of

THE ROYAL AERONAUTICAL SOCIETY

(Fatigue Committee)

by

PERGAMON PRESS

Oxford · New York · Toronto
Sydney · Braunschweig

Pergamon Press Ltd., Headington Hill Hall, Oxford
Pergamon Press Inc., Maxwell House, Fairview Park, Elmsford, New York 10523
Pergamon of Canada Ltd., 207 Queen's Quay West, Toronto 1
Pergamon Press (Aust.) Pty. Ltd., 19a Boundary Street,
Rushcutters Bay, N.S.W. 2011, Australia
Vieweg & Sohn GmbH, Burgplatz 1, Braunschweig

First edition 1970

Library of Congress Catalog Card No. 71–102401

Printed in Hungary

08 006754 9

TO MY WIFE AND FAMILY

whose patience has enabled me to complete this volume

Preface

INVESTIGATIONS relating to the failure of materials and structures under repeated or cyclic loads have been in progress for over one hundred and twenty-five years. During this period much experimental data have been accumulated on the behaviour and performance of materials, mechanical components and structural elements under fatigue loading conditions. However, experience has suggested that, because of the diverse nature of the problem and the range of publication sources, much of the published information will remain generally unnoticed until a reasonably complete and readily accessible bibliography on the subject is available.

In July 1945 a bibliography on The Fatigue of Materials was published by the Division of Aeronautics, Council for Scientific and Industrial Research, Australia. This was compiled for the assistance of authors preparing papers for a Symposium on the Failure of Metals by Fatigue which was held at the University of Melbourne in December 1946. It has become increasingly evident that this bibliography, which at the time appeared to be fairly comprehensive, listed only a small portion of the vast amount of information which had been published on fatigue during the preceding one hundred years.

Consequently, in 1952, Mr. H. A. Wills, then Superintendent of the Structures Division at the Aeronautical Research Laboratories, suggested that I prepare a comprehensive bibliography on the fatigue of materials, components and structures for the assistance of students, engineers, metallurgists and research workers. A limited number of copies of the first volume of this bibliography[a] was issued by the Aeronautical Research Laboratories in 1954, at which stage the Royal Aeronautical Society expressed interest in sponsoring a revised version with a wider circulation. The present volume is based on that published by the A.R.L., amended and enlarged and extended to the year 1950.

In compiling this bibliography, extensive use has been made of abstracting journals and lists of references which have appeared in the numerous papers and several books dealing with the subject. Although it is believed that this volume includes references to most of the published information on fatigue up to the year 1950, there are doubtless a number of omissions and inconsistencies. I would be indebted to any person who can provide other references, information or corrections to those appearing in this volume.

The task of checking the bibliographical data has involved a considerable amount of effort, particularly as the majority of the references included in this work were not available in Australia. Many individual persons, universities, institutes, societies, associations and other organisations in a number of countries have provided assistance in locating and checking these references. Although it is not possible to personally thank all those who have assisted in this way, I would like to record my deep appreciation to the following:

The Librarian and Library staff of the Aeronautical Research Laboratories, Melbourne, in particular Miss Aileen Moore and Miss Annette O'Donovan;

Mr. W. J. Crozier of the Department of Supply, Australia House, London;

Professor Takeo Yokobori, Department of Mechanical Engineering, Tohuko University, Japan;

Mr. A. G. Vannucci, Technical Information Officer, A.G.A.R.D., Paris;

The late Dr. R. Cazaud, I.R.S.I.D., Paris;

The Librarian, the Joint Library of the Iron and Steel Institute and Institute of Metals, London;

The Library of the Verein Deutscher Ingenieure, Düsseldorf, W. Germany;

---

[a] J. Y. Mann, *Bibliography on the Fatigue of Materials, Components and Structures, vol. 1, 1843–1938*. Aeronautical Research Laboratories, Australia, August 1954.

The Librarian, the Patent Office Library, London;

The Head, Reference Section, Science and Technology Division, Library of Congress, Washington;

The Librarian, Zentralstelle der Luftfahrtdokumentation und -Information, München, W. Germany;

The Librarian, Science Museum Library, London;

Superintendent and Librarian, Soprintendenza Bibliografica per il Piemonte, Torino, Italy;

The Librarian, National Lending Library for Science and Technology, Boston Spa, England;

The Editors and Publishers of various journals who have provided information and copies of their publications;

My colleagues at the Aeronautical Research Laboratories: Mr. J. Solvey for his assistance in the translation of information from the German and Russian languages, and Messrs. J. M. Finney and R. Simpson for their patience and thoroughness in checking the manuscript and proofs of this work.

This book is published with the permission of the Chief Scientist, Australian Defence Scientific Service, Department of Supply, Melbourne, Australia.

JOHN Y. MANN

Aeronautical Research Laboratories,
Melbourne

Notes on format and notation adopted

THE references in this bibliography are listed chronologically according to their year of publication and alphabetically in each year according to the name of the first author. Papers without specific authors are listed at the conclusion of the alphabetical section of each year. No attempt was made to segregate the references according to subject matter as it was thought that titles alone could be misleading, and in any case this would result in a much more bulky volume with many references duplicated or even triplicated. In order to provide easy access to particular subject matter the bibliography includes a comprehensive Subject Index (based, however, in many cases on the title of the paper alone), together with an Author Index.

Extensive use has been made of abbreviations for the titles of journals and other publications. Those adopted have been based on the forms suggested in the *World List of Scientific Periodicals*, Third Edition,[a] but reference has also been made to the Fourth Edition of this publication[b]. The order adopted for the bibliographical data in the various references is that suggested in British Standard 1629: 1950—*Bibliographical References*.

Where possible, the titles of the various papers are given in the language of publication, followed by a translation into English. However, in some instances, and particularly in the case of Russian and some Japanese references where the original characters require transliteration, the title is given in English only followed by a letter to indicate the language of publication. Examples are: Russian [R]; Japanese [J]; Chinese [C]; Polish [P].

In the case of the proceedings or transactions of conferences, symposia, etc., the year of entry mostly corresponds to the date of publication of the proceedings, when given, and not to the year in which the relevant conference was held. The number of pages quoted, in the case of books, reports, etc., refers to the numbered pages and does not include unnumbered pages occupied by diagrams or other illustrations. Columns, where numbered, have been regarded as pages.

[a] W. A. Smith, F. L. Kent and G. B. Stratton [Editors], *World List of Scientific Periodicals Published in the Years 1900–1950*, Third Edition. London, Butterworths Scientific Publications, 1952.

[b] P. Brown and G. B. Stratton [Editors], *World List of Scientific Periodicals Published in the Years 1900–1960*. Fourth Edition. London, Butterworths, 1963–1965.

1838

1. ALBERT, W. A. J. Über Treibseile am Harz. [Driving ropes in the Harz.] *Archiv für Mineralogie, Geognosie, Bergbau und Hüttenkunde*, vol. 10, 1838, pp. 215–234.

1842

2. HOOD, C. On some peculiar changes in the internal structure of iron, independent of, and subsequent to, the several processes of its manufacture. *Phil. Mag.*, ser. 3, vol. 21, no. 136, Aug. 1842, pp. 130–137.

1843

3. HODGKINSON, E. Experiments to prove that all bodies are to some degree inelastic and a proposed law for estimating the deficiency. *Rep. Brit. Ass.*, pt. 2, 1843, pp. 23–25.

4. RANKINE, W. J. M. On the causes of the unexpected breakage of the journals of railway axles; and on the means of preventing such accidents by observing the law of continuity in their construction. *Min. Proc. Instn civ. Engrs*, vol. 2, session 1843, 1842–1843, pp. 105–108.

5. YORK, J. O. Account of a series of experiments on the comparative strength of solid and hollow axles. *Min. Proc. Instn civ. Engrs*, vol. 2, session 1843, 1842–1843, pp. 89–94.

1844

. GLYNN, J. On the causes of fracture of the axles of railway carriages. *Min. Proc. Instn civ. Engrs*, vol. 3, session 1844, 1844, pp. 202–203.

1848

7. STOPFL, P. Achsenbrüche an Lokomotiven, Tender und Wagen, ihre Erklärung und Beseitigung. [Axle failures in locomotives, tenders and wagons—their causes and methods of avoiding.] *Org. Eisenbahnw.*, vol. 3, no. 2, 1848, pp. 55–67.

1849

8. THE LORD WROTTESLEY; WILLIS, R.; JAMES H.; RENNIE, G.; CUBITT, W. *and* HODGKINSON, E. *Report of the Commissioners appointed to inquire into the application of iron to railway structures.* London, Her Majesty's Stationery Office, Command Paper 1123, 1849, pp. 435.

9. McCONNELL, J. E. On railway axles. *Proc. Instn mech. Engrs, Lond., 1847–1849*, Oct. 1849, pp. 13–27.

1850

10. McCONNELL, J. E. On the deterioration of railway axles. *Proc. Instn mech. Engrs, Lond., 1850–1851*, Jan. 1850, pp. 5–19 and April 1850, pp. 3–14.

11. THORNEYCROFT, G. B. On the manufacture of malleable iron; with the results of experiments on the strength of railway axles. *Min. Proc. Instn civ. Engrs*, vol. 9, session 1849–1850, 1850, pp. 294–302.

12. THORNEYCROFT, T. On the form of shafts and axles. *Proc. Instn mech. Engrs, Lond., 1850–1851*, July 1850, pp. 35–41 and Oct. 1850, pp. 4–15.

1853

13. BRAITHWAITE, F. On the fatigue and consequent fracture of metals. *Min. Proc. Instn civ. Engrs*, vol. 13, 1853–1854, pp. 463–475.

14. MORIN, A. *Leçons de mécanique practique —résistance des matériaux. [Studies in practical mechanics—strength of materials.]* Paris, Librairie de L. Hachette et Cie, 1853, pp. 456. [See pages 348–354 for notes on the observations of M. Marcoux and M. C. Arnoux.]

1858

15. WÖHLER, A. Bericht über die Versuche, welche auf der Königl. Niederschleesisch-Märkischen Eisenbahn mit Apparaten zum Messen der Biegung und Verdrehung von Eisenbahnwagen-Achsen während de Fahrt, angestellt wurden. [Report on tests of the Königl. Niederschleesisch-Märkischen Eisenbahn made with apparatus for the measurement of the bending and torsion of railway axles in service.] *Z. Bauw.*, vol. 8, 1858, pp. 642–651.

1860

16. WÖHLER, A. Versuche zur Ermittelung der auf die Eisenbahnwagen-Achsen einwirkenden Kräfte und der Widerstandsfähigkeit der Wagen-Achsen. [Tests to determine the forces acting on railway wagon axles, and the capacity of resistance of the axles.] *Z. Bauw.*, vol. 10, 1860, pp. 583–616.

1861

17. FAIRBAIRN, W. On the effects of vibratory action and long continued changes of load upon wrought iron bridges and girders. *Civil Engineer and Architects Journal*, vol. 24, Nov. 1, 1861, pp. 327–329.

1863

18. WÖHLER, A. Über die Versuche zur Ermittelung der Festigkeit von Achsen, welche in den Werkstätten der Niederschlesisch-Märkischen Eisenbahn zu Frankfurt a.d.O. angestellt sind. [On tests for determining the strength of axles made in the workshops of the Niederschlesisch-Märkischen Eisenbahn at Frankfurt a.d.O.] *Z. Bauw.*, vol. 13, 1863, pp. 233–258.

1864

19. FAIRBAIRN, W. Experiments to determine the effect of impact, vibratory action and long-continued changes of load on wrought iron girders. *Phil. Trans.*, vol. 154, 1864, pp. 311–326.

20. FAIRBAIRN, W. The strength of iron structures. *Engineer, Lond.*, vol. 18, Nov. 11, 1864, pp. 293–294.

1866

21. WÖHLER, A. Resultate der in der Central-Werkstatt der Niederschlesisch-Märkischen Eisenbahn zu Frankfurt a.d.O. angestellten Versuche über die relative Festigkeit von Eisen, Stahl und Kupfer. [Results of tests made in the central workshop of the Niederschlesisch-Märkischen Eisenbahn at Frankfurt a.d.O., on the relative strength of iron, steel and copper.] *Z. Bauw.*, vol. 16, 1866, pp. 67–83.

1867

22. Wöhler's experiments on the strength of metals. *Engineering, Lond.*, vol. 4, Aug. 23, 1867, pp. 160–161.

23. Experiments on the "Fatigue" of Bessemer steel. *Engineering, Lond.*, vol. 4, Nov. 29, 1867, pp. 497–498.

1870

24. PONCELET, J. V. *Introduction à la mécanique industrielle, physique ou expérimentale. [Intro*

duction to industrial, physical or experimental mechanics.] Troisième édition. Paris, Imprimerie de Gauthier-Villars, 1870, pp. 757. [See particularly pp. 317–318.] [Deuxième edition, 1839.]

25. WÖHLER, A. Über die Festigkeits-Versuche mit Eisen und Stahl. [On strength tests of iron and steel.] *Z. Bauw.*, vol. 20, 1870, pp. 74–106.

26. WÖHLER, A. *Über die Festigkeitsversuchen mit Eisen und Stahl. Auf Anordnung des Ministers für Handel, Gerwerbe u. öffentl. Arbeiten, Grafen Itzenblitz, angestellt. [On strength tests with iron and steel. Carried out on the instruction of the Minister for Trade, Industry and Public Works, Count Itzenblitz.]* Berlin, Ernst und Korn, 1870, pp. 17.

1871

27. Wöhler's experiments on the "Fatigue" of metals. *Engineering, Lond.*, vol. 11, March 24, 1871, pp. 199–200; March 31, 1871, p. 221; April 7, 1871, pp. 243–244; April 14, 1871, p. 261; April 28, 1871, pp. 299–300; May 5, 1871, pp. 326–327; May 19, 1871, pp. 349–350; June 9, 1871, p. 397; and June 23, 1871, pp. 439–441.

1873

28. LAUNHARDT, W. Die Inanspruchnahme des Eisens. [The stressing of iron.] *Zeitschrift des Architekten- und Ingenieur-Vereins zu Hannover*, vol. 19, no. 1, 1873, cols. 139–144.

29. MÜLLER, G. Zulässige Inanspruchnahme des Schmiedeeisens bei Brückenconstructionen. [Allowable strains in wrought iron bridge structures.] *Zeitschrift des Österreichischen Ingenieur- und Architekten Vereins, Wien*, vol. 25, 1873, pp. 197–202.

1874

30. GERBER, H. Bestimmung der zulässigen Spannungen in Eisen-Constructionen. [Cal-culation of the allowable stresses in iron structures.] *Zeitschrift des Bayerischen Architekten und Ingenieur-Vereins*, vol. 6, no. 6, 1874, pp. 101–110.

31. SCHÄFFER, Bestimmung der zulässigen Spannung für Eisenconstructionen. [Determination of the allowable stress in iron structures.] *Z. Bauw.*, vol. 24, 1874, pp. 398–408.

32. SPANGENBERG, L. Über das Verhalten der Metalle bei wiederholten Anstrengungen. [On the behaviour of metals under repeated strains.] *Z. Bauw.*, vol. 24, 1874, pp. 473–495 and vol. 25, 1875, pp. 78–98.

1875

33. SPANGENBERG, L. *Über das Verhalten der Metalle bei wiederholten Anstrengungen. (Fortsetzung der Wöhler'schen Festigkeitsversuche). [On the behaviour of metals under repeated stress—continuation of the Wöhler strength experiments.]* Berlin, Ernst und Korn, 1875, pp. 32.

1876

34. WEYRAUCH, J. J. *Festigkeit und Dimensionenberechnung der Eisen- und Stahlconstructionen mit Rücksicht auf die neueren Versuche. Ein elementaren Anhang zu allen Lehrbüchern über Eisen- und Stahlconstructionen. [Strength and dimension calculations of iron and steel structures taking into consideration new tests—an elementary appendix to all text books on iron and steel structures.]* Leipzig, B. G. Teubner, 1876, pp. 116.

1877

35. METCALF, W. On the resistance of steel to vibration. *The Metallurgical Review*, vol. 1, no. 4, Dec. 1877, pp. 399–404.

36. WEYRAUCH, J. J. (Translation into English by A. J. DU BOIS) *Strength and determination*

of the dimensions of structures of iron and steel with reference to the latest investigations. New York, John Wiley and Sons, 1877, pp. 206.

1879

37. EGLESTON, T. The law of fatigue and refreshment of metals. *Trans. Amer. Inst. min. Engrs*, vol. 8, 1879–1880, pp. 398–404.

38. KENT, W. On an apparatus for testing the resistance of metals to repeated shocks. *Trans. Amer. Inst. min. Engrs*, vol. 8, 1879–1880, pp. 76–80.

39. LIPPOLD, H. Die Inanspruchnahme von Eisen und Stahl mit Rücksicht auf bewegte Last. — Im Anschluss an die Wöhler'schen Versuche, und im Vergleich mit den Resultaten der Methoden von Gerber, Launhardt, Weyrauch und Winkler. [The stressing of iron and steel with consideration of moving loads —Wöhler's tests compared with those of Gerber, Launhardt, Weyrauch and Winkler.] *Org. Eisenbahnw.*, vol. 16, no. 1, 1879, pp. 22–34.

40. SPANGENBERG, L. Über Festigkeits-Versuche mit Eisen und Stahl. [On strength tests with iron and steel.] *Glasers Ann. Gew.*, vol. 5, July 1, 1879, pp. 6–15. English abstr.: *Min. Proc. Instn civ. Engrs*, vol. 60, 1879–1880, pp. 415–416.

1880

41. BAUSCHINGER, J. Über das Kristallinischwerden und die Festigkeitsverminderung des Eisens durch den Gebrauch. [The crystallisation and the reduction of strength of iron in service.] *Dinglers J.*, vol. 235, no. 3, 1880, pp. 169–173.

42. WEYRAUCH, J. J. On the calculation of dimensions as depending on the ultimate working strength of materials. *Min. Proc. Instn civ. Engrs*, vol. 63, 1880–1881, pp. 275–296.

1881

43. MOHR, O. C. Über die Verwerthung der Wöhler'schen Versuche für die Dimensionierung der Eisenconstructionen, insbesondere der eisernen Brücken. [On the use of Wöhler's tests for the design of iron structures, particularly of iron bridges.] *Der Civil Ingenieur, Leipzig*, vol. 27, 1881, pp. 1–12.

1882

44. BAKER, B. Steel for tires and axles. *Min. Proc. Instn civ. Engrs*, vol. 67, 1882, pp. 353–357.

1884

45. KENNEDY, A. B. W. Frequent repetitions of load. *Professional Papers of the Corps of Royal Engineers, Chatam; Royal Engineers Institute Occasional Papers*, vol. 10, 1884, pp. 187–212.

1885

46. CONSIDÈRE, M. L'Emploi du Fer et de l'Acier dans les Constructions. [The application of iron and steel in structures.] *Ann. Ponts Chauss.*, ser. 6, vol. 9, 1885, pp. 574–775. [See particularly pp. 684–714.]

47. HILL, W. Crystallization of wrought iron. *Trans. Amer. Soc. mech. Engrs*, vol. 7, 1885–1886, pp. 241–272.

1886

48. BAUSCHINGER, J. Über die Veränderung der Elasticitätsgrenze und der Festigkeit des Eisens und Stahls durch Strecken und Quetschen, durch Erwärmen und Abkühlen und durch oftmal wiederholte Beanspruchung

[On the change of the elastic limit and strength of iron and steel by tension and compression, by heating and cooling and by often repeated loading.] *Mitt. mech.-tech. Lab. Münch.*, vol. 13, no. 8, 1886, pp. 1–115. English abstr.: *Min. Proc. Instn civ. Engrs*, vol. 87, 1886–1887, pp. 463–465.

49. UNWIN, W. C. A new view of the nature of the resistance of materials to repeated loads. *Engineer, Lond.*, vol. 62, Dec. 10, 1886, p. 457.

1887

50. BAKER, B. Some notes on the working stress of iron and steel. *Trans. Amer. Soc. mech. Engrs*, vol. 8, 1887, pp. 157–181.

51. BARLOW, W. H.; BRAMWELL, F. J.; THOMSON, J.; GALTON, J.; BAKER, B.; UNWIN, W. C.; KENNEDY, A. B. W.; BARLOW, C.; HELE-SHAW, H. S.; ROBERTS-AUSTEN, W. C. *and* ATCHISON, A. T. Report on the endurance of metals under repeated and varying stresses, and the proper working stresses on railway bridges and other structures subject to varying loads. *Rep. Brit. Ass.*, 1887, pp. 424–438.

52. MARTENS, A. Über das Kleingefüge des schmiedbaren Eisens, besonders des Stahles. [On the microstructure of wrought irons, and in particular steels.] *Stahl u. Eisen*, vol. 7, no. 4, April 1887, pp. 235–242.

53. UNWIN, W. C. A new view of the resistance of materials. *Engineer, Lond.*, vol. 63, Jan. 7, 1887, pp. 2–3.

1889

54. EWING, J. A. On hysteresis in the relation of strain to stress. *Rep. Brit. Ass.*, 1889, pp. 502–504.

55. WEYRAUCH, J. J. *Die Festigkeitseigenschaften und die Methoden der Dimensionenberechnung von Eisen- und Stahlkonstruktionen. Ein Anhang zu allen über die statische Berechnung der Eisen- u. Stahlkonstruktionen. [Strength properties and methods of calculating the dimensions of iron and steel structures. Appendix to all books on static calculations of iron and steel structures.]* Leipzig, Verlag B. G. Teubner, 1889, pp. 223.

1890

56. PEARSON, K. On Wöhler's experiments on alternating stress. *Messeng. Math.*, vol. 20, 1890, pp. 21–37.

1892

57. SONDERICKER, J. A description of some repeated stress experiments. *Technol. Quart.*, vol. 5, 1892, pp. 70–80.

1893

58. RUDELOFF, M. Berichte über die im Auftrage des Herrn Ministers für Handel und Gewerbe ausgeführten vergleichenden Untersuchungen von Seilverbindungen für Fahrstuhlbetrieb. Theil II. Ergebnisse der Untersuchungen bei stossweiser Inanspruchnahme. [Report on comparative investigation of rope connections for elevator drives carried out on the instructions of the Minister for Commerce and Trade. No. 2. Results of investigation under impact loading.] *Mitt. tech. VersAnst. Berl.*, vol. 11, 1893, pp. 177–199.

1894

59. KREUZPOINTNER, P. Do iron and steel crystallize in service? *Iron Age*, vol. 54, July 5, 1894, pp. 9–12 and Sept. 27, 1894, pp. 516–517.

60. Does the vibration of stamp-stems change their molecular structure? *Trans. Amer. Inst. min. Engrs*, vol. 24, 1894, pp. 809–846.

[Continuation of discussion to paper by T. A. RICKARD, *ibid.*, vol. 23, 1893, pp. 137–147. pp. 557, 560–561, 573–577 are relevant.]

1895

61. ANDREWS, T. The effect of strain on railway axles, and a minimum flexion resistance-point in axles. *Trans. Soc. Engrs, Lond.*, 1895, pp. 181–234.

62. KREUZPOINTNER, P. Kristallisieren Eisen und Stahl im Betriebe? [Does iron and steel crystallize in service?] *Stahl u. Eisen*, vol. 15. no. 10, May 15, 1895, pp. 474–483.

63. KREUZPOINTNER, P. Die Übermüdung der Metalle. [The overtiredness of metals.] *Stahl u. Eisen*, vol. 15, no. 18, Sept. 15, 1895, pp. 865–867.

64. MARTENS, A. Einfluss der Gewindeform auf die Festigkeit der Schraubenbolzen. [Influence of thread form on the strength of threaded bolts.] *Z. Ver. dtsch. Ing.*, vol. 39, no. 17, April 27, 1895, pp. 505–508.

65. The influence of vibration on steel. *Railr. Gaz.*, vol. 27, Oct. 4, 1895, p. 655.

1896

66. ANDREWS, T. Microscopic internal flaws in steel rails and propeller shafts. *Engineering, Lond.*, vol. 61, Jan. 17, 1896, pp. 91–92.

67. ANDREWS, T. Microscopic internal flaws inducing fracture in steel. *Engineering, Lond.*, vol. 62, July 10, 1896, pp. 35–37; July 17, 1896, pp. 68–69 and July 24, 1896, pp. 118–120.

68. HOPPE, O. Alberts Versuche und Erfindungen. Zugleich Beiträge zur Frage der Gefüge-veränderung von Eisen durch wiederholte Stösse und zur Erfindung des Drahtseiles und der Förderung mit Ketten ohne Ende. [Alberts experiments and inventions. Simultaneously, contributions to the problem of structural changes of iron through repeated impact; on the invention of a wire rope; and on transmission through endless chains.] *Stahl u. Eisen*, vol. 16, no. 12, June 15, 1896, pp. 437–441 and no. 13, July 1, 1896, pp. 496–500.

1897

69. AISBITT, M. W. On certain defects found in propeller shafts. *Trans. Inst. Mar. Engrs*, vol. 9, paper no. 72, 1897–1898, pp. 82.

70. ANDREWS, T. Microscopic observations on the deterioration by fatigue in iron and steel. *Engineering, Lond.*, vol. 63, Feb. 26, 1897, pp. 265–266.

71. ANDREWS, T. Microscopic observations on the deterioration by fatigue in steel rails. *Engineering, Lond.*, vol. 63, April 16, 1897, pp. 499–502, 504; June 25, 1897, pp. 840–842; vol. 64, July 23, 1897, pp. 99–102; Aug. 27, 1897, pp. 249–251; Sept. 3, 1897, pp. 298–299; Oct. 15, 1897, pp. 455–456; Dec. 3, 1897, pp. 676–678; vol. 65, Jan. 7, 1898, pp. 7–10; Feb. 18, 1898, pp. 201–204; April 15, 1898, pp. 451–453 and May 20, 1898, pp. 617–619.

72. FÖPPL, A. Dauerversuche von Bauschinger, ausgeführt in den Jahren 1886–1893. [Endurance tests of Bauschinger between the years 1886 and 1893.] *Mitt. mech.-tech. Lab. Münch.*, no. 25, 1897, pp. 3–19.

1898

73. COKER, E. G. Note on the endurance of steel bars subjected to repetitions of torsional stress. *Min. Proc. Instn civ. Engrs*, vol. 135, 1898–1899, pp. 294–299.

74. PORTER, H. F. J. Fatigue of metal in wrought iron and steel forgings. *J. Franklin Inst.*, vol. 145, no. 4, April 1898, pp. 241–261 and no. 5, May 1898, pp. 321–355.

75. WARREN, W. H. A testing machine for equal alternating stresses. *J. roy. Soc. N.S.W.*, vol. 32, 1898, pp. cxxix–cxxxi.

1899

76. BERGER, K. Beitrag zum Studium des elastischen Verhaltens von Gusseisen bei wiederholter Beanspruchung auf Zug und Druck. [Contribution to the study of the elastic properties of cast iron under repeated tensile and compressive stresses.] *Mitt. technol. GewMus. Wien*, vol. 9, 1899, pp. 13–26. English abstr.: *Min. Proc. Instn civ. Engrs*, vol. 136, 1898–1899, pp. 370–371.

77. SONDERICKER, J. Repeated stresses. *Technol. Quart.*, vol. 12, 1899, pp. 5–22.

78. SONDERICKER, J. Repeated stresses. *Mech. World*, vol. 26, no. 658, Aug. 11, 1899, pp. 67–68; no. 662, Sept. 8, 1899, pp. 115–116; no. 667, Oct. 13, 1899, pp. 174–175 and no. 668, Oct. 20, 1899, p. 185.

1900

79. GILCHRIST, J. On Wöhler's laws. *Engineer, Lond.*, vol. 90, Aug. 31, 1900, pp. 203–204.

80. SCHANZER, R. On mysterious fractures of steel shafts. *Engineering, Lond.*, vol. 69, April 27, 1900, pp. 563–567. *Metallographist*, vol. 3, no. 4, Oct. 1900, pp. 320–329.

1901

81. HOWARD, J. E. On the effects of repeated strains in structures as exemplified in the Brooklyn Bridge fractures. *Iron Age*, vol. 68, Aug. 8, 1901, p. 38.

82. MARTENS, A. Dauerversuche mit nahtlosen Stahlflaschen zur Aufbewahrung von Kohlensäure. [Endurance tests with seamless steel cylinders for carbon dioxide.] *Mitt. tech. VersAnst. Berl.*, vol. 19, 1901, pp. 217–258.

1902

83. ARNOLD, J. O. The dangerous crystallisation of mild steel and wrought iron. *Min. Proc. Instn civ. Engrs*, vol. 154, 1902–1903, pp. 122–124.

84. EWING, J. A. *and* HUMFREY, J. C. W. The fracture of metals under repeated alternations of stress. *Phil. Trans. (A)*, vol. 200, 1902, pp. 241–250. *Metallographist*, vol. 6, no. 2, April 1903, pp. 96–110.

85. FOSTER, F. A possible explanation of the phenomena caused by repetition of stress. *Mech. Engr*, vol. 10, no. 252, Nov. 22, 1902, pp. 704–706 and no. 253, Nov. 29, 1902, pp. 740–742. Abstr.: *Sci. Abstr.*, vol. 6, 1903, no. 866.

86. REYNOLDS, O. *and* SMITH, J. H. On a throw testing machine for reversals of mean stress. *Phil. Trans. (A)*, vol. 199, 1902, pp. 265–297. Abstr.: *Proc. roy Soc.*, vol. 70, 1902, pp. 44–46.

87. TURNER, C. A. P. Thermoelectric determination of stress. *Trans. Amer. Soc. civ. Engrs*, vol. 48, 1902, pp. 140–179.

88. Prime movers and their ailments. *Engineering, Lond.*, vol. 74, Oct. 17, 1902, pp. 511–512.

1903

89. FOSTER, F. On phenomena due to repetitions of stress, and on a new testing machine. *Mem. Manchr lit. phil. Soc.*, vol. 48, no. 7, 1903–1904, pp. 1–20.

90. LINDENTHAL, G. Alternate stresses in bridge members. *Proc. Amer. Soc. Test. Mater.*, vol. 3, 1903, pp. 169–174.

91. VAN ORNUM, J. L. The fatigue of cement products. *Trans. Amer. Soc. civ. Engrs*, vol. 51, 1903, pp. 443–451.

92. STEAD, J. E. *and* RICHARDS, A. W. The restoration of dangerously crystallised steel by heat treatment. *J. Iron St. Inst.*, vol. 64, no. 2, 1903, pp. 119–140.

1904

93. ARNOLD, J. O. The fracture of structural steels under alternating stress. *Engineer, Lond.*, vol. 98, Sept. 2, 1904, p. 227. *Engineering, Lond.*, vol. 78, Sept. 2, 1904, pp. 307–308. *Mech. Engr*, vol. 14, no. 348, Sept. 24, 1904, pp. 446–448. *Iron Steel Mag.*, vol. 8, no. 5, Nov. 1904, pp. 433–438.

94. DUDLEY, C. B. Alternate bending stresses. *Iron Steel Mag.*, vol. 7, no. 2, Feb. 1904, pp. 134–139.

95. OSMOND, F.; FRÉMONT, CH. *and* CARTAUD, G. Les modes de déformation et de rupture des fers et des aciers doux. [The manner of deformation and of rupture of iron and two steels.] *Rev. Métall.*, vol. 1, 1904, pp. 11–45.

96. SEATON, A. E. *and* JUDE, A. Impact tests on the wrought steels of commerce. *Proc. Instn mech. Engrs, Lond.*, pts. 3–4, 1904, pp. 1135–1230. *Iron Steel Mag.*, vol. 9, no. 2, Feb. 1905, pp. 135–150.

97. Alternating stress testing machine. *Engineer, Lond.*, vol. 98, Sept. 23, 1904, p. 308.

1905

98. ARNOLD, J. O. Note on the crystallisation of wrought iron. *Engineer, Lond.*, vol. 100, Aug. 18, 1905, p. 158.

99. GARDNER, J. C. Effects caused by the reversal of stresses in steel. *J. Iron St. Inst.*, vol. 67, no. 1, 1905, pp. 481–483.

100. HOUGHTON, S. A. Note on the failure of an iron plate through "fatigue". *J. Iron St. Inst.*, vol. 67, no. 1, 1905, pp. 383–389. *Engineering, Lond.*, vol. 79, June 9, 1905, p. 756. *Iron Steel Mag.*, vol. 10, no. 1, July 1905, pp. 11–17.

101. MILTON, J. T. Fractures in large steel boiler plates. *Trans. Instn nav. Archit., Lond.*, vol. 47, pt. II, July 1905, pp. 359–391. *Engineering, Lond.*, vol. 80, Aug. 4, 1905, pp. 164–166 and Aug. 11, 1905, pp. 195–197.

102. RICHARDS, A. W. *and* STEAD, J. E. Overheated steel. *J. Iron St. Inst.*, vol. 68, no. 2, 1905, pp. 84–117. *Iron Steel Mag.*, vol. 10, no. 5, Nov. 1905, pp. 386–404.

103. ROGERS, F. (a) Troostite; and (b) Heat treatment and fatigue of steel. *J. Iron St. Inst.*, vol. 67, no. , 1905, pp. 484–494.

104. SMITH, J. H. A new testing machine for reversals of stress. *Engineering, Lond.*, vol. 79, March 10, 1905, pp. 307–308.

105. STANTON, T. E. Alternating stress testing machine at the National Physical Laboratory. *Engineering, Lond.*, vol. 79, Feb. 17, 1905, pp. 201–203.

106. STANTON, T. E. *and* BAIRSTOW, L. On the resistance of iron and steel to reversals of

direct stress. *Min. Proc. Instn civ. Engrs*, vol. 166, pt. 4, 1905–1906, pp. 78–134.

107. WAZAU, G. Neuere Dauerversuchmaschinen. [New fatigue testing machines.] *Dinglers J.*, vol. 320, no. 31, Aug. 5, 1905, pp. 481–486.

108. Sankey's vibratory testing machine. *Mech. Engr*, vol. 16, no. 407, Nov. 11, 1905, p. 690.

1906

109. BERLINER, S. Über das Verhalten des Gusseisens bei langsamen Belastungswechseln. [The behaviour of cast iron under slow reversals of stress.] *Ann. Phys., Lpz.*, ser. 4, vol. 20, no. 8, 1906, pp. 527–562. English translation: *Brit. cast Iron Res. Ass. Trans.*, No. 1.

110. EDEN, E. M. The endurance of metals under alternating stresses and effect of rate of alternation on endurance. *Proc. Univ. Durham phil. Soc.*, vol. 3–4, 1906–1910, pp. 251–266. Abstr.: *Sci. Abstr.*, vol. 13, 1910, no. 1384.

111. FINLEY, W. H. A case of failure of iron from "fatigue". *Engng News*, vol. 55, no. 18, May 3, 1906, p. 487. Abstr.: *Sci. Abstr.*, vol. 9, 1906, no. 1200.

112. HOWARD, J. E. Methods of testing metals by alternate strains and thermic treatment of steels to increase their resistance. *Engng Rec.*, vol. 54, no. 12, Sept. 22, 1906, pp. 334–336. Abstr.: *Sci. Abstr.*, vol. 9, 1906, no. 1808.

113. PREUSS, E. Zur Geschichte der Dauerversuche mit Metallen. [History of fatigue tests of metals.] *Baumaterialienkunde*, vol. 11, no. 16, Aug. 15, 1906, pp. 245–249.

114. ROGERS, F. Sur quelques effets microscopiques produits sur les métaux par l'action des efforts. [On the microscopic effects produced by the action of stresses on metals.] *Rev. Métall.*, vol. 3, no. 9, Sept. 1906, pp. 518–527.

115. STANTON, T. E. Repeated impact testing machine at the National Physical Laboratory. *Engineering, Lond.*, vol. 82, July 13, 1906, pp. 33–34.

1907

116. DE FREMINVILLE, CH. Charactères des vibrations accompagnant le choc. Déduits de l'examen des cassures. [Nature of vibrations accompanied by impact. Inferences from examination of the cracks.] *Rev. Métall.*, vol. 4, no. 9, Sept. 1907, pp. 833–884.

117. HOWARD, J. E. Notes on the endurance of steels under repeated alternate stresses. *Proc. Amer. Soc. Test. Mater.*, vol. 7, 1907. pp. 252–257.

118. VAN ORNUM, J. L. The fatigue of concrete. *Trans. Amer. Soc. civ. Engrs*, vol. 58, 1907, pp. 294–320.

119. PREUSS, E. Ergebnisse neuere Dauerversuche an Metallen. [Results of new endurance tests on metals.] *Dinglers J.*, vol. 322, no. 7, Feb. 16, 1907, pp. 100–102.

120. ROGERS, F. Microscopic studies of strain in metals. *J. R. micr. Soc.*, Feb. 1907, pp. 14–18. Abstr.: *Engineering, Lond.*, vol. 82, Dec. 21, 1906, p. 842.

121. SOUTHER, H. White–Souther endurance test specimen. *Proc. Amer. Soc. Test. Mater.*, vol. 7, 1907, pp. 616–623.

122. STANTON, T. E. A factor in the design of machine details. *Engineering, Lond.*, vol. 83, April 19, 1907, p. 505.

1908

123. ALLAN, J. M. Fatigue of copper pipes. *Trans. N.-E. Cst Instn Engrs Shipb.*, vol. 25, pts. 3–4, 1908–1909, pp. 135–154.

124. ARNOLD, J. O. Factors of safety in marine engineering. *Trans. Instn nav. Archit., Lond.*, vol. 50, 1908, pp. 260–289. *Engineering, Lond.*, vol. 85, April 24, 1908, pp. 565–566 and May 1, 1908, pp. 598–601.

125. BERRY, H. C. Some tests of reinforced concrete beams under oft-repeated loading. *Proc. Amer. Soc. Test. Mater.*, vol. 8, 1908, pp. 454–468.

126. HANCOCK, E. L. Tests on staybolts. *Proc. Amer. Soc. Test. Mater.*, vol. 8, 1908, pp. 369–372.

127. ROSENHAIN, W. The study of breakages. *Engineering, Lond.*, vol. 86, Sept. 11, 1908, pp. 340–343.

128. SCHUCHART, A. Untersuchung der Biegbarkeit von Drähten. [Investigation of the flexibility of wires.] *Stahl u. Eisen*, vol. 28, no. 27, July 1, 1908, pp. 945–949 and no. 28, July 8, 1908, pp. 988–993. Abstr.: *Sci. Abstr.*, vol. 11, 1908, no. 1362.

129. SOUTHER, H. Characteristic results of endurance tests on wrought iron, steel and alloys. *Proc. Amer. Soc. Test. Mater.*, vol. 8, 1908, pp. 379–385.

130. STANTON, T. E. A new fatigue test for iron and steel. *J. Iron St. Inst.*, vol. 76, no. 1, 1908, pp. 54–70. Abstr.: *Engineering, Lond.*, vol. 85, May 22, 1908, pp. 696–697.

131. STANTON, T. E. *and* BAIRSTOW, L. The resistance of materials to impact. *Proc. Instn mech. Engrs, Lond.*, pts. 3–4, 1908, pp. 889–919.

132. Motor driven vibratory testing machine. *Elect. Rev., Lond.*, vol. 62, no. 1586, April 17, 1908, pp. 656–657.

133. Working stresses. *Engineering, Lond.*, vol. 85, May 15, 1908, pp. 653–654.

1909

134. ARNOLD, J. O. The mysteries of metals. *Engineering, Lond.*, vol. 87, Feb. 5, 1909, pp. 170–173.

135. BAIRSTOW, L. The elastic limits of iron and steel under cyclical variations of stress. *Phil. Trans. (A)*, vol. 210, 1909–1910, pp. 35–55.

136. LE CHATELIER, H. Sur l'essai des métaux par amortissement des mouvements vibratoires. [On the testing of metals by damping of vibratory movements.] *Rev. Métall.*, vol. 6, no. 7, July 1909, pp. 887–889.

137. LE CHATELIER, H. L'essai des métaux aux efforts alternatifs. [Tests of metals under alternating loads.] *Rev. Métall.*, vol. 6, no. 11, Nov. 1909, pp. 1156–1160.

138. FÖPPL, A. Dauerversuche mit eingekerbten Stäben. [Fatigue tests with notched bars.] *Mitt. mech.-tech. Lab. Münch.*, no. 31, 1909. pp. 1–42. German abstr.: *Stahl u. Eisen*, vol. 29, no. 11, March 17, 1909, pp. 409–410.

139. GUILLET, A. Intervention de l'amortissement dans l'essai des fers. [Utilization of damping in tests on iron.] *Rev. Métall.*,

vol. 6, no. 7, July 1909, pp. 885–887. *Stahl u. Eisen*, vol. 29, no. 25, June 23, 1909, pp. 956–957.

140. GULLIVER, G. M. Internal friction in loaded materials. *Congr. int. Ass. Test. Mat.*, Fifth Congr., Copenhagen, 1909, paper VIII$_{(9)}$.

141. HOWARD, J. E. The resistance of steels to repeated alternate stresses. *Congr. int. Ass. Test. Mat.*, Fifth Congr., Copenhagen, 1909, paper IV$_{(1)}$. *Mech. Engr*, vol. 24, no. 623, Dec. 31, 1909, pp. 828–829. Abstr.: *Sci. Abstr.*, vol. 13, 1910, no. 218.

142. ROSENHAIN, W. The fatigue and crystallisation of metals. *J. W. Scot. Iron St. Inst.*, vol. 16, 1909, pp. 129–149.

143. SCHÜLE, F. and BRUNNER, E. Quality tests and endurance tests of copper wires. *Congr. int. Ass. Test. Mat.*, Fifth Congr., Copenhagen, 1909, paper IV$_{(2)}$.

144. SMITH, J. H. A fatigue testing machine. *Engineering, Lond.*, vol. 88, July 23, 1909, pp. 105–107.

145. WITHEY, M. O. Tests of bond between concrete and steel in reinforced concrete beams. *Bull. Univ. Wis. Engng Ser.*, no. 321, Oct. 1909, pp. 64.

146. The Landgraf–Turner alternating impact machine. *Iron Steel Times*, vol. 1, June 24, 1909, pp. 340–343. *Amer. Mach., N.Y.*, vol. 32, pt. 1, 1909, pp. 641–642. Abstr.: *Sci. Abstr.*, vol. 12, 1909, no. 1348.

147. Machine à essayer les poutres en béton arné à la flexion alternative. [Machine for testing beams of reinforced concrete in alternating bending.] *Génie civ.*, vol. 55, no. 24, Oct. 9, 1909, p. 446.

1910

148. BASQUIN, O. H. The exponential law of endurance tests. *Proc. Amer. Soc. Test. Mater.*, vol. 10, 1910, pp. 625–630.

149. BERRY, H. C. Apparatus for repeated loads on concrete cylinders and a typical result. *Proc. Amer. Soc. Test. Mater.*, vol. 10, 1910, pp. 581–587.

150. BLOUNT, B.; KIRKALDY, W. G. and SANKEY, H. R. Comparison of the tensile, impact-tensile and repeated bending methods of testing steel. *Proc. Instn mech. Engrs, Lond.*, pts. 1–2, 1910, pp. 715–772.

151. FRÉMONT, CH. La fatigue des métaux et les nouvelles méthodes d'essais. [The fatigue of metals and new methods of testing.] *Génie civ.*, vol. 57, no. 25, Oct. 22, 1910, pp. 460–464; no. 26, Oct. 29, 1910, pp. 493–495 and vol. 58, no. 3, Nov. 19, 1910, pp. 53–55. German abstr.: *Stahl u. Eisen*, vol. 31, no. 19, May 11, 1911, p. 776.

152. MESNAGER, A. Sur un signe exterieur d'altération des métaux. [On a surface indication of changes in metals.] *Tech. mod.*, vol. 2, no. 9, Sept. 1910, pp. 514–515.

153. MESNAGER, A.; BREUIL, P.; SCHÜLE, F.; Do, Capitaine; REJTŐ, A.; GRENET, L.; CELLERIER, F. and GUILLET, L. Enquête sur la "Fatigue des Métaux". [Inquiry on the fatigue of metals.] *Tech. mod.*, vol. 2, no. 1, Jan. 1910, pp. 19–21; no. 2, Feb. 1910, pp. 83–84; no. 3, March 1910, pp. 151–154; no. 4, April 1910, pp. 210–214; no. 5, May 1910, pp. 280–284 and no. 6, June 1910, pp. 345–347.

154. MORLEY, A. Strength of materials under combined stresses. *Engineering, Lond.*, vol. 89, April 29, 1910, p. 555.

155. RITCHIE, J. B. An apparatus for inducing fatigue in wires by means of repeated extensional and rotational strains, with the effects produced by such fatigue in the laws of torsional oscillation. *Proc. roy. Soc. Edinb.*, vol. 31, 1910–1911, pp. 440–447. Abstr.: *Sci. Abstr.*, vol. 14, 1911, no. 1311.

156. SMITH, J. H. Some experiments on fatigue of metals. *J. Iron St. Inst.*, vol. 82, no. 2, 1910, pp. 246–318. Abstr.: *Engineering, Lond.*, vol. 90, Oct. 7, 1910, pp. 494–495.

157. WAWRZINIOK, O. *Die Ermüdung des Eisenbahnschienenmaterials.* [*The fatigue of railway rail materials.*] Berlin, Springer, 1910, pp. 47.

158. WITHEY, M. O. Tests of reinforced concrete columns subjected to repeated and eccentric loads. *Proc. Amer. Soc. Test. Mater.*, vol. 10, 1910, pp. 361–375.

159. Impact testing machine. *Engineering, Lond.*, vol. 89, May 6, 1910, p. 572.

160. Stresses and strains. *Engineering, Lond.*, vol. 90, Nov. 11, 1910, pp. 669–670.

1911

161. EDEN, E. M.; ROSE, W. N. and CUNNINGHAM, F. L. The endurance of metals—experiments on rotating beams at University College, London. *Proc. Instn mech. Engrs, Lond.*, pts. 3–4, 1911, pp. 839–974. *Engineering, Lond.*, vol. 92, Oct. 27, 1911, pp. 575–580 and Nov. 3, 1911, pp. 612–613.

162. HOPKINSON, B. *and* WILLIAMS, G. T. The elastic hysteresis of steel. *Proc. roy. Soc.*, (A), vol. 87, 1911–1912, pp. 502–511. *Engineering, Lond.*, vol. 94, Dec. 13, 1912, pp. 827–828.

163. LILLY, W. E. The elastic limits and strength of materials. *Trans. Instn civ. Engrs Ire.*, vol. 38, 1911–1912, pp. 27–52.

164. MACGREGOR, J. S. *and* STOUGHTON, B. A new method of testing the endurance of case-hardened gears and pinions. *Proc. Amer. Soc. Test. Mater.*, vol. 11, 1911, pp. 822–832.

165. RÉSAL, J. Théorie des vibrations transversales d'une barre élastique. [Theory of transverse vibrations in an elastic bar.] *Rev. Métall.*, vol. 8, no. 5, May 1911, pp. 346–366.

166. STANTON, T. E. *and* PANNELL, J. R. Experiments on the strength and fatigue properties of welded joints in iron and steel. *Min. Proc. Instn civ. Engrs*, vol 188, pt. 2, 1911–1912, pp. 1–77. Abstr.: *Engineering, Lond.*, vol. 92, Dec. 15, 1911, p. 814.

167. TURNER, L. B. The strength of steel in compound stress and endurance under repetition of stress. *Engineering, Lond.*, vol. 92, July 28, 1911, pp. 115–117; Aug. 11, 1911, pp. 183–185; Aug. 25, 1911, pp. 246–250 and Sept. 8, 1911, pp. 305–307.

1912

168. BOUDOUARD, O. Breakdown tests of metals. *Congr. int. Ass. Test. Mater.*, Sixth Congr., New York, 1912, paper V_3.

169. FEA, L. Mechanical tests of special steels for naval construction. *Congr. int. Ass. Test. Mater.*, Sixth Congr., New York, 1912, paper I_1.

170. HAIGH, B. P. A new machine for alternating load tests. *Engineering, Lond.*, vol. 94, Nov. 22, 1912, pp. 721–723. Abstr.: *Rep. Brit. Ass.*, sect. G, 1912, pp. 569–570.

171. HOPKINSON, B. A high-speed fatigue tester and the endurance of metals under alternating stress of high frequency. *Proc. roy. Soc. (A)*, vol. 86, 1912, pp. 131–149. Abstr.: *Engineer, Lond.*, vol. 113, Feb. 2, 1912, pp. 113–114.

172. JOB, R. Types of rail failures and some methods for their detection. *Congr. int. Ass. Test. Mater.*, Sixth Congr., New York, 1912, pp. 78–90.

173. KOMMERS, J. B. Repeated stress testing. I—An investigation of a commercial endurance test. II—A proposed quality factor. *Congr. int. Ass. Test. Mater.*, Sixth Congr., New York, 1912, papers V_{4a} and V_{4b}.

174. ROOS-HJELMSÄTER, J. O. On endurance tests of machine steel. *Congr. int. Ass. Test. Mater.*, Sixth Congr., New York, 1912, paper V_{2a}.

175. ROOS-HJELMSÄTER, J. O. Some static and dynamic endurance tests. *Congr. int. Ass. Test. Mater.*, Sixth Congr., New York, 1912, paper V_{2b}.

176. STANTON, T. E. Recent researches made at the National Physical Laboratory, Teddington, England, on the resistance of metals to alternating stresses. *Congr. int. Ass. Test. Mater.*, Sixth Congr., New York, 1912, paper V_1.

177. UPTON, G. B. *and* LEWIS, G. W. The fatigue failure of metals. *Amer. Mach., N. Y.*, vol. 37, Oct. 17, 1912, pp. 633–635 and Oct. 24, 1912, pp. 678–683.

178. The Witton-Kramer fatigue tester. *Engineering, Lond.*, vol. 94, Dec. 13, 1912, pp. 805–806.

1913

179. BOUDOUARD, O. Fatigue des métaux. [Fatigue of metals.] *Rev. Métall.*, vol. 10, no. 1, Jan. 1913, pp. 70–78.

180. KOMMERS, J. B. Essais de durée I—Recherche sur un essai commercial d'endurance. [Endurance tests I—Research on a commercial endurance test.] *Rev. Métall.*, vol. 10, no. 1, Jan. 1913, pp. 79–87.

181. KOMMERS, J. B. Essais aux efforts alternatifs II—Proposition d'un facteur de qualité et d'une forme normale d'essai, avec des résultats pour divers matériaux. [Tests under alternating loads II—Specification for quality of manufacture and the normal type of test, with results for various materials.] *Rev. Métall.*, vol. 10, no. 1, Jan. 1913, pp. 88–92.

182. LUDWIK, P. Ursprungsfestigkeit und statische Festigkeit, eine Studie über Ermüdungserscheinungen. [Original strength and static strength, a study of fatigue phenomenon.] *Z. Ver. dtsch. Ing.*, vol. 57, no. 6, Feb. 8, 1913, pp. 209–213. Abstr.: *Sci. Abstr.*, vol. 16, 1913, no. 542.

183. MASON, W.; ROGERS, F. *and* EDEN, E. M. Report on alternating stress. *Rep. Brit. Ass.*, 1913, pp. 183–213.

184. POPPLEWELL, W. C. The connection between the elastic phenomena exhibited during slow reversals of stress and the ultimate endurance of steel. *Min. Proc. Instn civ. Engrs*, vol. 197, pt. 3, 1913–1914, pp. 264–290.

185. ROGERS, F. So-called "Crystallisation through Fatigue". *J. Iron St. Inst.*, vol. 88,

no. 2, 1913, pp. 392–398. Abstr.: *Mech. Engr*, vol. 32, no. 815, Sept. 5, 1913, pp. 213–214. *Iron Age*, vol. 92, no. 11, Sept. 11, 1913, pp. 554–555.

186. ROOS-HJELMSÄTER, J. O. Essais d'endurance de l'acier pour machines. [Fatigue tests on steels for machines.] *Rev. Métall.*, vol. 10, no. 1, Jan. 1913, pp. 58–62.

187. ROOS-HJELMSÄTER, J. O. Quelques essais statiques et dynamiques d'endurance. [Some static and dynamic endurance tests.] *Rev. Métall.*, vol. 10, no. 1, Jan. 1913, pp. 63–69. Abstr.: *Sci. Abstr.*, vol. 16, 1913, no. 747.

188. STANTON, T. E. Récentes recherches faites au National Physical Laboratory de Teddington (Angleterre) sur la résistance des métaux aux efforts alternatifs. [Recent research at the National Physical Laboratory at Teddington, England, on the resistance of metals to alternating loads.] *Rev. Métall.*, vol. 10, no. 1, Jan. 1913, pp. 53–57.

189. STANTON, T. E. *and* PANNELL, J. R. Experiments on the strength and fatigue properties of welded joints in iron and steel. *Collected Researches*, National Physical Laboratory, Teddington, vol. 9, 1913, pp. 191–217.

190. THEARLE, S. J. P. Note on some cases of "fatigue" in the steel materials of steamers. *Trans. Instn nav. Archit., Lond.*, vol. 55, pt. 2, 1913, pp. 28–36. *Engineering, Lond.*, vol. 95, June 27, 1913, pp. 891–893.

191. UPTON, G. B. A recently installed fatigue testing machine—description of machine and records of some tests. *Sci. Amer. Suppl.*, vol. 76, no. 1980, Dec. 13, 1913, pp. 372–373.

1914

192. BOEKE, C. *Over breuk na herhaalde belasting; een onderzoek naar de oorzaken van breuk en de optredende spanningen bij duurzaamheidsproeven van metalen. [Failure after repeated loading; an investigation on the causes of failure and the stresses in endurance tests of metals.]* Rotterdam, W. L. and J. Brusse, 1914, pp. 107.

193. HEYN, E. Die Kerbwirkung und ihre Bedeutung für den Konstrukteur. [The notch effect and its importance to the designer.] *Z. Ver. dtsch. Ing.*, vol. 58, no. 10, March 7, 1914, pp. 383–391.

194. HUNNINGS, S. V. A new vibratory testing machine. *Iron Age*, vol. 94, no. 2, July 9, 1914, pp. 84–86.

195. KREUZPOINTNER, P. Fatigue of rails. *Rly Age, N. Y.*, vol. 57, no. 17, Oct. 23, 1914, pp. 755–756.

196. LUFTSCHITZ, V. Neuere Materialprüfungsmethoden und Apparate zu ihrer Durchführung. [New materials testing methods and testing apparatus.] *Mitt. tech. VersAmt., Wien*, vol. 3, no. 2, 1914, pp. 28–33.

197. MARTENS, A. Über die in den Jahren 1892 bis 1912 im Königlichen Materialprüfungsamt in Berlin-Lichterfelde ausgeführten Dauerversuche mit Flusseisen. [Fatigue tests with mild steel made between 1892 and 1912 in the Königlichen Materialprüfungsamt in Berlin-Lichterfelde.] *Mitt. MatPrüfAmt Berl.*, vol. 32, no. 1, 1914, pp. 51–85. English abstr.: *Sci. Abstr.*, vol. 17, 1914, no. 1371.

198. NUSBAUMER, E. Rotary bend tests, alternating bend tests and repeated shock tests. *Carnegie Schol. Mem.*, vol. 6, 1914, pp. 94–169.

199. NUSBAUMER, E. Étude comparative sur les essais au choc simple, les essais aux chocs

répétés les essais de flexion rotative et les essais de flexion alternée. [A comparative study of simple impact tests, repeated impact tests, rotating bending tests and alternating bending tests.] *Rev. Métall.*, vol. 11, no. 11, Nov. 1914, pp. 1133–1190.

200. PREUSS, E. Kerbwirkung bei Dauerschlag-beanspruchung. [Notch effect under repeated impact stress.] *Z. Ver. dtsch. Ing.*, vol. 58, no. 18, May 2, 1914, pp. 701–703.

201. PREUSS, E. Die Festigkeit von Schweisseisen gegenüber Stossbeanspruchung. [The strength of wrought iron under impact stresses.] *Stahl u. Eisen*, vol. 34, no. 29, July 16, 1914, pp. 1207–1208.

202. READ, A. A. *and* GREAVES, R. H. The influence of nickel on some copper-aluminium alloys. *J. Inst. Met.*, vol. 11, no. 1, 1914, pp. 169–213. *Engineering, Lond.*, vol. 97, March 20, 1914, pp. 399–404.

203. STROMEYER, C. E. The determination of fatigue limits under alternating stress conditions. *Proc. roy. Soc. (A)*, vol. 90, 1914, pp. 411–425.

204. STROMEYER, C. E. The elasticity and endurance of steam pipes. *Trans. Instn nav. Archit., Lond.*, vol. 56, 1914, pp. 144–157. Abstr.: *Engineering, Lond.*, vol. 97, June 19, 1914, pp. 856–858.

205. STROMEYER, C. E. Fatigue of metals. *Proc. Sheffield Soc. Engrs Metall.*, Oct.—Dec. 1914, pp. 3–14.

206. STROMEYER, C. E. Practical method of determining the fatigue limit of metals. *Power*, vol. 40, Dec. 1, 1914, pp. 793–794.

207. A method of determining fatigue by calorimetry. *Engineer, Lond.*, vol. 118, Sept. 18, 1914, pp. 281–282. *Iron Age*, vol. 94, no. 19, Nov. 5, 1914, p. 1066.

208. Repeated stresses. *Engineering, Lond.*, vol. 98, Oct. 2, 1914, pp. 420–421.

1915

209. BROWN, W. The fatigue of nickel and iron wires when subjected to the influence of alternating magnetic fields at frequency 50 per second. *Sci. Proc. R. Dublin Soc.*, vol. 14, Jan. 1915, pp. 336–344.

210. BROWN, W. The subsidence of torsional oscillations and the fatigue of nickel wires when subjected to the influence of alternating magnetic fields of frequencies up to 250 per second. *Sci. Proc. R. Dublin Soc.*, vol. 14, March 1915, pp. 521–528.

211. HAIGH, B. P. Report on alternating stress tests of a sample of mild steel received from the British Association Stress Committee. *Rep. Brit. Ass.*, 1915, pp. 163–170. Abstr.: *Engineering, Lond.*, vol. 100, Oct. 8, 1915, pp. 379–380.

212. HAIGH, B. P. The endurance of metals under alternating stresses. *J. W. Scot. Iron St. Inst.*, vol. 23, 1915–1916, pp. 17–56.

213. HUNTINGTON, A. K. The effect of heat and work on the mechanical properties of metals. *J. Inst. Met.*, vol. 13, no. 1, 1915, pp. 23–79.

214. JONSON, E. Fatigue of copper alloys. *Proc. Amer. Soc. Test. Mater.*, vol. 15, pt. 2, 1915, pp. 101–111.

215. KOMMERS, J. B. Repeated stress tests of steel. *Amer. Mach., N. Y.*, vol. 42, April 1, 1915, pp. 551–553.

216. KREUZPOINTNER, P. The fatigue and disease of metals. *Iron Age*, vol. 95, no. 17, April 29, 1915, pp. 950–951.

217. MASON, W. On speed effect and recovery in slow-speed alternating stress tests. *Proc. roy. Soc. (A)*, vol. 92, 1915–1916, pp. 373–376.

218. MITINSKI, A. N. The fatigue of metals, axles and tires. [R] *Zh. russk. metall. Obshch.*, vol. 1, no. 2, 1915, pp. 191–228. English abstr.: *Chem. Abstr.*, vol. 11_1, 1917, p. 1126.

219. MOORE, H. F. *and* SEELY, F. B. The failure of materials under repeated stress. *Proc. Amer. Soc. Test. Mater.*, vol. 15, pt. 2, 1915, pp. 437–466.

220. RUDELOFF, M. Erfahrungen über das Unbrauchbarwerden von Drahtseilen. [Experiences in connection with the deterioration of wire cables.] *Mitt. MatPrüfAmt Berl.*, vol. 33, nos. 3–4, 1915, pp. 198–209.

221. SMITH, J. H. *and* WEDGWOOD, G. A. Stress-strain loops for steel in the cyclic state. *J. Iron St. Inst.*, vol. 91, no. 1, 1915, pp. 365–397.

222. STROMEYER, C. E. A machine for determining fatigue limits calorimetrically. *Rep. Brit. Ass.*, 1915, p. 638.

223. STROMEYER, C. E. The law of fatigue applied to crankshaft failures. *Trans. Instn nav. Archit., Lond.*, vol. 57, 1915, pp. 174–184. Abstr.: *Engineering, Lond.*, vol. 99, April 9, 1915, p. 400.

224. UHLER, J. L. Dynamic properties of steel castings; vibratory results on carbon, vanadium and nickel-chrome steels compared; the historical steps in fatigue testing. *Iron Age*, vol. 96, no. 14, Sept. 30, 1915, pp. 754–756.

225. The endurance of metals under repeated stresses, some new facts, and a new method of testing. *Locomotive*, vol. 30, no. 5, Jan. 1915, pp. 130–142.

1916

226. CORSE, W. M. *and* COMSTOCK, G. F. Aluminium bronze, some recent tests and their significance. *Proc. Amer. Soc. Test. Mater.*, vol. 16, pt. 2, 1916, pp. 118–150.

227. GUEST, J. J. *and* LEA, F. C. Torsional hysteresis of mild steel. *Proc. roy. Soc. (A)*, vol. 93, 1916–1917, pp. 313–332.

228. HAIGH, B. P. The endurance of metals under alternating stresses. *Mech. World*, vol. 59, no. 1528, April 14, 1916, pp. 177–178.

229. LUDWIK, P. Über die Ermüdung der Metalle. [On the fatigue of metals.] *Z. öst. Ing.- u. ArchitVer.*, vol. 68, no. 42, 1916, pp. 795–798.

230. McADAM, D. J. Endurance and impact tests of metals. *Proc. Amer. Soc. Test. Mater.*, vol. 16, pt. 2, 1916, pp. 292–308. *Iron Tr. Rev.*, vol. 59, no. 24, Dec. 21, 1916, pp. 1257–1260.

231. MASON, W. The hysteresis of steel under repeated torsion. *Rep. Brit. Ass.*, sect. G, 1916, pp. 285–288. Abstr.: *Engineering, Lond.*, vol. 102, Sept. 15, 1916, p. 269.

232. MITINSKI, A. N. Fatigue des métaux, service des essieux et des bandages. [Fatigue of metal axles and tires in service.] *Rev. Métall.*, vol. 13, 1916, pp. 67–71.

233. MOORE, H. F. *and* SEELY, F. B. Constants and diagrams for repeated stress calculations. *Proc. Amer. Soc. Test. Mater.*, vol. 16, pt. 2, 1916, pp. 470–475.

234. POPPLEWELL, W. C. The resistance of iron and steel to complete reversals of stress. *Engng Rev., Lond.*, vol. 30, no. 4, Oct. 16, 1916, pp. 120–122.

235. POPPLEWELL, W. C. The influence of speed on endurance tests. *Engineer, Lond.*, vol. 122, Oct. 20, 1916, pp. 339–340.

236. PRICHARD, H. S. The effects of straining structural steel and wrought iron. *Trans. Amer. Soc. civ. Engrs*, vol. 80, 1916, pp. 1429–1542.

237. RUDELOFF, M. Der heutige Stand der Dauerversuche mit Metallen. [The present position regarding the fatigue of metals.] *Verh. Ver. GewFleiss., Berl.*, vol. 95, no. 7, 1916, pp. 343–369. German abstr.: *Stahl u. Eisen*, vol. 37, no. 14, April 5, 1917, pp. 334–338.

238. STANTON, T. E. Resistance of wood to stress reversals. [abstract.] *Engineering, Lond.*, vol. 101, June 23, 1916, p. 605.

239. STANTON, T. E. *and* BATSON, R. G. On the fatigue resistance of mild steel under various conditions of stress distribution. *Rep. Brit. Ass.*, sect. G, 1916, pp. 288–291. *Engineering, Lond.*, vol. 102, Sept. 15, 1916, pp. 269–270.

240. STROMEYER, C. E. Fatigue des matériaux métalliques soumis à des efforts répétés. [Fatigue of metallic materials under repeated loads.] *Rev. Métall.*, vol. 13, 1916, pp. 71–73.

241. THOMPSON, F. C. Surface tension effects in the intercrystalline cement in metals and the elastic limit. *J. Iron St. Inst.*, vol. 93, no. 1, 1916, pp. 155–210. *Engineering, Lond.*, vol. 101, May 19, 1916, pp. 472–474; May 26, 1916, pp. 513–514 and June 2, 1916, pp. 540–542.

1917

242. BAILEY, R. W. Ductile materials under variable shear stress. *Engineering, Lond.*, vol. 104, July 27, 1917, pp. 81–83.

243. BROWN, W. The fatigue of nickel and iron wires when subjected to the influence of transverse alternating magnetic fields. *Sci. Proc. R. Dublin Soc.*, vol. 15, Jan. 1917, pp. 163–170. Abstr.: *Elect. World, N.Y.*, vol. 69, June 9, 1917, p. 1119.

244. HAIGH, B. P. Experiments on the fatigue of brasses. *J. Inst. Met.*, vol. 18, no. 2, 1917, pp. 55–86. *Engineering, Lond.*, vol. 104, Sept. 21, 1917, pp. 315–319.

245. HANKINS, G. A. Alternating stress tests of aluminium alloys. (i) Preliminary report on the relative resistance to reversals of bending stress of various aluminium alloys when tested at normal temperature and at 180°C. (ii) On the relative resistance to reversals of bending stress of various rolled aluminium alloys when tested at normal temperature and at 150°C. *Advisory Committee for Aeronautics (Great Britain)*, Light Alloys Sub-Committee Report 12, July 1917, pp. 8.

246. HOWARD, J. E. Transverse fissures in steel rails. *Trans. Amer. Inst. min. (metall.) Engrs*, vol. 58, 1917–1918, pp. 597–649.

247. LEON, A. Über die Ermüdung von Maschinenteilen. [On the fatigue of machine parts.] *Z. Ver. dtsch. Ing.*, vol. 61, no. 9, March 3, 1917, pp. 192–196 and no. 10, March 10, 1917, pp. 214–218. English abstr.: *J. Amer. Soc. mech. Engrs*, vol. 39, Oct. 1917, p. 885.

248. McADAM, D. J. An alternating torsion testing machine. *Proc. Amer. Soc. Test. Mater.*, vol. 17, pt. 2, 1917, pp. 599–602. Abstr.: *Iron Age*, vol. 100, July 19, 1917, p. 125.

249. McADAM, D. J. An impact-endurance testing machine. *J. Amer. Soc. nav. Engrs*, vol. 29, no. 24, Nov. 1917, pp. 663–672.

250. MASON, W. Alternating stress experiments. *Proc. Instn mech. Engrs, Lond.*, Jan.–May 1917, pp. 121–196. *Engineering, Lond.*, vol. 103, Feb. 23, 1917, pp. 187–190, and March 2, 1917, pp. 211–214. Abstr.: *Iron Age*, vol. 100, Oct. 11, 1917, pp. 874–875.

251. STANTON, T. E. *and* BATSON, R. G. C. The effect on the resistance to fatigue of crankshafts of a variation on the radius of curvature of fillets. *Advisory Committee for Aeronautics (Great Britain)*, Internal Combustion Engine Sub-Committee Report 15, Oct. 1917, pp. 6.

252. WOLFF, E. B. The failure of boiler plates in service, and investigation of the stresses that occur in riveted joints. *Engineering, Lond.*, vol. 104, Sept. 28, 1917, pp. 326–330. Abstr.: *Engineering, Lond.*, vol. 124, Nov. 23, 1917, pp. 456–457. *Power*, vol. 47, no. 5, Jan. 29, 1918, pp. 166–167.

253. Rail failure attributed to track service—transverse fissures are fatigue fractures according to report on Texas accident—experiments at making fissures at will. *Iron Age*, vol. 99, June 7, 1917, pp. 1382–1385.

1918

254. ABELL, W. S. Experiments on the application of electric welding to large structures. *Min. Proc. Instn civ. Engrs*, vol. 208, pt. 2, 1918–1919, pp. 104–126, 181–182.

255. COMSTOCK, G. F. Metallographic investigation of transverse fissures in rails with special reference to high-phosphorous streaks. *Trans. Amer. Inst. Min. (metall.) Engrs*, vol. 62, 1918–1919, pp. 703–753.

256. LUDWIK, P. Über Dauerversuche. [On fatigue tests.] *Mitt. tech. VersAmt., Wien*, vol. 7, no. 2, 1918, pp. 22–38.

257. MATSUMURA, T. Tension, impact and repeated impact tests of mild and hard steels. *Mem. Coll. Engng Kyoto*, vol. 2, no. 2, 1918, pp. 63–69.

258. MOORE, H. F. *and* PUTMAN, W. J. Effect of cold working and of rest on resistance of steel to fatigue under reversed stress. *Trans. Amer. Inst. Min. (metall.) Engrs*, vol. 62, 1918–1919, pp. 397–419.

259. PRICHARD, H. S. Overstrain and fatigue failure of steel as related to grain structure. *Engng News Rec.*, vol. 80, no. 23, June 6, 1918, pp. 1086–1090.

260. SCHLINK, F. J. Study of mechanical hysteresis will advance our knowledge of materials. *Engng News Rec.*, vol. 80, no. 22, May 30, 1918, pp. 1035–1037.

261. STODOLA, A. *and* SCHÜLE, F. Hohlkehlenschärfung und Dauerbiegung. [Fillet sharpness and bending fatigue.] *Schweiz. Bauztg*, vol. 71, no. 13, March 30, 1918, pp. 145–14'

262. WEBB, H. A. *and* BARLING, W. H. Failure of ductile materials under fluctuating stresses. *Aeronautics, Lond.*, vol. 14, no. 235, April 1, 1918, pp. 331–333. Abstr.: *J. Amer. Soc. mec Engrs*, vol. 40, July 1918, pp. 584–585.

263. Tests on tie bars from the Menai suspension bridge. *Engineer, Lond.*, vol. 126, July 1, 1918, pp. 47–49.

264. Ironing out cracks in iron: Bardeen electrical method of eliminating fatigue of metals. *Sci. Amer. Suppl.*, vol. 86, no. 2220, July 20, 1918, p. 48.

1919

265. FARMER, F. M. Fatigue testing machine. *Proc. Amer. Soc. Test. Mater.*, vol. 19, pt. 2, 1919, pp. 709–719. Abstr.: *Amer. Mach.*, *N.Y.*, vol. 51, no. 6, Aug. 7, 1919, pp. 271–273.

266. FRÉMONT, C. Sur la rupture prématurée des pièces d'acier soumises à des efforts répétés. [On the premature rupture of pieces of steel subjected to repeated loads.] *C. R. Acad. Sci.*, *Paris*, vol. 168, Jan. 6, 1919, pp. 54–56. English abstr.: *J. Inst. Met.*, vol. 21, no. 1, 1919, pp. 469–470.

267. FULTON, A. R. Experiments on the effect of alternations of tensile stress at low frequencies on the elastic properties of mild steel. *Rep. Brit. Ass.*, 1919, pp. 484–485. Abstr.: *Engineering*, *Lond.*, vol. 109, Jan. 9, 1920, pp. 65–66.

268. HATFIELD, W. H. The mechanical properties of steel with some considerations of the question of brittleness. *Proc. Instn mech. Engrs*, *Lond.*, Jan.–May 1919, pp. 347–533. *Engineering*, *Lond.*, vol. 107, May 9, 1919, pp. 615–618; May 16, 1919, pp. 634–636 and May 23, 1919, pp. 686–688.

269. KOMMERS, J. B. Broader use of Johnson's Formula for repeated stress. *Engng News Rec.*, vol. 83, no. 20, Nov. 27—Dec. 4, 1919, 942–944.

270. LUDWIK, P. Über die Änderung der Festigkeitseigenschaften der Metalle bei Wechselnder Beanspruchung. [On the change in strength properties of metals under repeated straining.] *Z. Metallk.*, vol. 11, 1919, pp. 157–168.

271. MERICA, P. D. *and* KARR, C. P. Some tests of light aluminium casting alloys: the effect of heat treatment. *Proc. Amer. Soc. Test. Mater.*, vol. 19, pt. 2, 1919, pp. 297–327.

272. MOORE, H. F. The fatigue of metals. *J. Engrs' Cl. Philad.*, vol. 36, no. 173, April 1919, pp. 138–143.

273. MOORE, H. F. Fatigue of metals under repeated stress. *J. West. Soc. Engrs*, vol. 24, no. 6, June 1919, pp. 331–340.

274. MOORE, H. F. *and* GEHRIG, A. G. Some fatigue tests of nickel steel and chrome-nickel steel. *Proc. Amer. Soc. Test. Mater.*, vol. 19, pt. 2, 1919, pp. 206–223. Abstr.: *Iron Age*, vol. 104, July 3, 1919, p. 42.

275. ROY, L. Sur la résistance dynamique de l'acier. [On the dynamic resistance of steels.] *C.R. Acad. Sci.*, *Paris*, vol. 168, Feb. 10, 1919, pp. 304–307.

276. Fatigue phenomena in metals—a report summarising the available facts and theories relating to fatigue failure and a discussion of some of the unsolved problems. *J. Amer. Soc. mech. Engrs*, vol. 41, no. 9, Sept. 1919, pp. 731–738.

277. Fatigue tests of nickel and chrome-nickel steel. *Engng World*, vol. 15, no. 8, Oct. 15, 1919, pp. 31–34.

1920

278. DALBY, W. E. Researches on the elastic properties and the plastic extension of metals. *Phil. Trans. (A)*, vol. 221, 1920–1921, pp. 117–138.

279. FRÉMONT, C. Genèse des fissurations de certains essieux. [Origin of cracks in axles.]

C.R. Acad. Sci., Paris, vol. 170, May 1920, pp. 1161–1164.

280. FULTON, A. R. Experiments on the effect of alternations of tensile stress at low frequencies on the elastic properties of mild steel. *Engineering, Lond.*, vol. 109, Jan. 9, 1920, pp. 65–66.

281. GIBSON, W. A. Fatigue and impact fatigue tests of aluminium alloys. *Proc. Amer. Soc. Test. Mater.*, vol. 20, pt. 2, 1920, pp. 115–136.

282. GRIFFITH, A. A. The phenomena of rupture and flow in solids. *Phil. Trans. (A)*, vol. 221, 1920–1921, pp. 163–198.

283. GUTHRIE, R. G. Stress results within the elastic limit. *Iron Age*, vol. 106, no. 11, Sept 9, 1920, pp. 649–651.

284. GUTHRIE, R. G. Recoverance. *Chem. metall. Engng*, vol. 23, no. 14, Oct. 6, 1920, pp. 671–672.

285. HATFIELD, W. H. Further notes on automobile steels. *Proc. Instn Auto. Engrs*, vol. 15, 1920–1921, pp. 465–508.

286. HATFIELD, W. H. Steel, from the standpoint of marine engineering. *J. W. Scot. Iron St. Inst.*, vol. 28, 1920–1921, pp. 52–78.

287. JENKIN, C. F. The strength and suitability of engineering materials—address to Engineering Section of the British Association. *Rep. Brit. Ass.*, sect. G, 1920, pp. 125–134. Abstr.: *Engineering, Lond.*, vol. 110, Aug. 27, 1920, pp. 290–292.

288. KOMMERS, J. B. Repeated stress safety factors quickly determined. *Engng News Rec.*, vol. 85, no. 9, Aug. 26, 1920, pp. 393–394.

289. LANGDON, S. C. *and* GROSSMAN, M. A. The embrittling effects of cleaning and pickling upon carbon steels. *Trans. Amer. electrochem. Soc.*, vol. 37, 1920, pp. 543–578.

290. MCADAM D. J. A high-speed alternating torsion testing machine. *Proc. Amer. Soc. Test. Mater.*, vol. 20, pt. 2, 1920, pp. 366–371.

291. MOORE, H. F. *and* KOMMERS, J. B. Fatigue of metals under repeated stresses. *Iron Age*, vol. 105, June 3, 1920, pp. 1595–1598. *Blast. Furn.*, vol. 8, June 1920, pp. 368–372.

292. MÜLLENHOFF, K. A. Belastungshöhe bei Dauerversuchen. [Load levels in fatigue tests.] *Stahl u. Eisen*, vol. 40, no. 8, Jan. 15, 1920, pp. 91–92. French abstr.: *Rev. Métall.*, Extraits, vol. 17, no. 8, Aug. 1920, pp. 365–366.

293. RITTERSHAUSEN, FR. *and* FISCHER, FR. P. Dauerbrüche an Konstruktionsstahlen und die Krupp'sche Dauerschlagprobe. [Fatigue failure in constructional steels and the Krupp repeated impact test.] *Krupp. Mh.*, vol. 1, no. 6, June 1920, pp. 93–104.

294. ROSENHAIN, W. The engineering relations of shock and fatigue. *Automot. Industr. N.Y.*, vol. 42, no. 24, June 10, 1920, pp. 1293–1296.

295. SLATER, W. A.; SMITH, G. A. *and* MUELLER, H. P. Effects of repeated reversals of stress on double reinforced concrete beams. *Tech. Pap. U.S. Bur. Stand.*, no. 182, Dec. 20, 1920, pp. 51.

296. WILLIAMS, G. M. Some determinations of the stress-deformation relations for concretes under repeated and continuous loadings. *Proc. Amer. Soc. Test. Mater.*, vol. 20, pt. 2, 1920, pp. 233–265.

297. Shock and fatigue. *Engineer, Lond.*, vol. 129, April 30, 1920, p. 451.

298. Fatigue phenomena in metals—summary of available facts and theories relating to fatigue failure and a discussion of some unsolved problems. *Sci. Amer. Mon.*, vol. 1, no. 3, March 1920, pp. 221–228.

299. Fatigue limit in metals. *Iron Coal Tr. Rev.*, vol. 101, Aug. 27, 1920, p. 265.

1921

300. DACY, G. H. Why weary metal fails under light loads: the causes of steel fatigue, and the ticklish problem of testing against it. *Sci. Amer.*, vol. 125, Dec. 1921, pp. 109–110.

301. DESCH, C. H. Brittleness and fatigue in metals. *Trans. Instn Engrs Shipb. Scot.*, vol. 65, 1921–1922, pp. 584–598.

302. ELOY, F. L'influence de chocs répétés a la compression sur les aciers. [Influence of repeated compression shocks on steels.] *Rev. Industr. min.*, no. 19, 1921, pp. 603–606.

303. FARMER, F. M. The desirability of standardisation in the testing of welds. *Engineering, Lond.*, vol. 111, Feb. 25, 1921, pp. 239–242.

304. GOUGH, H. J. Improvements in methods of fatigue testing. *Engineer, Lond.*, vol. 132, Aug. 12, 1921, pp. 159–162.

305. GOUGH, H. J. Some experiments on the fatigue of materials under alternating torsion. *Rep. Memor. aero. Res. Comm., Lond.*, no. 743, April 1921, pp. 14.

306. GRIFFITH, A. A. Stress concentrations in theory and practice. *Rep. Brit. Ass.*, 1921, pp. 316–324.

307. GRIMME, J. Merkwürdige Brucherscheinungen bei Eisenstäben. [Peculiar fracture phenomena in iron bars.] *Z. Ver. dtsch. Ing.*, vol. 65, no. 23, June 4, 1921, pp. 603–604.

308. GUILLET, L. Quelques Essais aux Chocs répétés. [Some repeated impact tests.] *Rev. Métall.*, Mem., vol. 18, no. 2, Feb. 1921, pp. 96–100.

309. GUILLET, L. Nouvelles expériences de chocs répétés. [Recent experiments on repeated impact.] *Rev. Métall.*, Mem., vol. 18, no. 12, Dec. 1921, pp. 755–757.

310. GUTHRIE, R. G. The relation of recoverance to the fatigue of metals. *J. Soc. automot. Engrs, N.Y.*, vol. 8, no. 1, Jan. 1921, pp. 65–68.

311. HAIGH, B. P. The strain energy function and the elastic limit. *Rep. Brit. Ass.*, sect. G, 1921, pp. 324–329.

312. HANKINS, G. A. Properties of commercially pure nickel as a standard material for fatigue investigations. Preliminary report. *Rep. Memor. aero. Res. Comm., Lond.*, no. 789, Nov. 1921, pp. 8.

313. HOWARD, J. E. Relations between the physical properties of steels and their endurance of service stresses. *Trans. Amer. Soc. Steel Treat.*, vol. 1, no. 11, Aug. 1921, pp. 673–682.

314. HOWARD, J. E. Internal service strains in steel. *Trans. Faraday Soc.*, vol. 17, pt. 1, Dec. 1921, pp. 117–122.

315. KNERR, H. C. Influence of surface flaws on strength of metals. *Automot. Industr., N.Y.*, vol. 45, no. 25, Dec. 22, 1921, pp. 1216–1217.

316. KREUZPOINTER, P. The fatigue of metals. *Engng Min. J.*, vol. 112, Aug. 6, 1921, pp. 216–217.

317. MCADAM, D. J. The endurance of steel under repeated stresses. *Chem. metall. Engng*, vol. 25, Dec. 14, 1921, pp. 1081–1087.

318. MASON, W. *and* DELANEY, W. J. Alternating combined stress experiments. *Rep. Brit. Ass.*, sect. G, 1921, pp. 329–341.

319. MERICA, P. D.; WALTENBERG, R. G. *and* MCCABE, A. S. Some mechanical properties of hot-rolled Monel metal. *Proc. Amer. Soc. Test. Mater.*, vol. 21, 1921, pp. 922–939.

320. MILLER, J. Fatigue breakdown in automobile steels. *Trans. Amer. Soc. Steel Treat.*, vol. 1, no. 6, March 1921, pp. 321–325.

321. MOORE, H. F. Tests support theory of fatigue. *Iron Tr. Rev.*, vol. 68, no. 13, March 31, 1921, pp. 895–897.

322. MOORE, H. F. Progressive failure or fatigue of metals under repeated stress. *Trans. Amer. Soc. Steel Treat.*, vol. 1, no. 6, March 1921, pp. 327–330.

323. MOORE, H. F. Investigation of fatigue of metals under stress. *Trans. Amer. Soc. Steel Treat.*, vol. 2, no. 1, Oct. 1921, pp. 70–71. *Min. and Metall. N.Y.*, no. 174, June 1921, p. 74.

324. MOORE, H. F. *and* KOMMERS, J. B. An investigation of the fatigue of metals. *Uni. Ill. Engng Exp. Sta. Bull.*, 124, 1921, pp. 178, Abstr.: *Chem. metall. Engng*, vol. 25, no. 25, Dec. 21, 1921, pp. 1141–1144. *Engng News Rec.*, vol. 88, no. 2, Jan. 12, 1922, pp. 76–78.

Power, vol. 54, no. 24, Dec. 13, 1921, p. 922. *Automot. Industr. N.Y.*, vol. 46, no. 5, Feb. 2, 1922, p. 229.

325. MÜLLER, W. *and* LEBER, H. Querschnittsübergänge und Biegefestigkeit bei Dauerbeanspruchung durch Stösse. [Changes of cross-section and bending strength under repeated impact stresses.] *Z. Ver. dtsch. Ing.*, vol. 65, no. 42, Oct. 13, 1921, pp. 1087–1093. English abstr.: *Engineering, Lond.*, vol. 113, Feb. 24, 1922, p. 246.

326. NASH, C. W. The fatigue of metals. *Pract. Engr, Lond.*, vol. 64, no. 1804, Sept. 22, 1921, pp. 183–184. *Commonw. Engr*, vol. 8, no. 12, July 1, 1921, pp. 350–353.

327. ONO, A. Fatigue resistance of steel under repeated combined stresses. [J] *J. Soc. mech. Engrs, Japan*, vol. 23, no. 62, Sept. 1921, pp. 201–216.

328. ONO, A. Fatigue of steel under combined bending and torsion. *Mem. Coll. Engng Kyushu*, vol. 2, no. 2, 1921, pp. 117–142. Abstr. *Sci. Abstr.*, vol. 24, 1921, no. 1191.

329. RAWDON, H. S. The presence of internal fractures in steel rails and their relation to the behaviour of the material under service stresses. *Trans. Faraday Soc.*, vol. 17, pt. 1 Dec. 1921, pp. 110–116.

330. RITTERSHAUSEN, FR. *and* FISCHER, FR. F Dauerbrüche an Konstruktionsstählen un die Kruppsche Dauerschlagprobe. [Fatigue failures in structural steels and the Krup impact fatigue test.] *Stahl u. Eisen*, vol. 4 no. 47, Nov. 24, 1921, pp. 1681–1690.

331. ROSENHAIN, W. Hardness and its relatio to ductility and fatigue range. *Automo Industr., N.Y.*, vol. 44, no. 11, March 1 1921, pp. 604–607.

332. SHAPIRA, G. Endurance testing machines. *Found. Tr. J.*, vol. 24, no. 276, Dec. 1, 1921, pp. 446–447.

333. SHIMER, W. R. Manufacture of steel from raw materials to finished product: Remarks on heat treatment and fatigue failures. *Trans. Amer. Soc. Steel Treat.*, vol. 1, no. 8, May 1921, pp. 423–435.

334. STENGER, E. P. *and* STENGER, B. H. Effect of heat-treatment on the fatigue strength of steel. *Trans. Amer. Soc. Steel Treat.*, vol. 1, no. 11, Aug. 1921, pp. 617–638. Abstr.: *Chem. metall. Engng*, vol. 23, no. 13, Sept. 29, 1920, p. 635.

335. Vermoeiingsproeven algemeen. [The fatigue resistance of duralumin.] *Nat. Luchtvaartlab.*, *Amsterdam*, Rep. M.17A, 1921, pp. 13.

336. The fatigue of steel. *Iron Age*, vol. 107, no. 25, June 23, 1921, p. 1687.

337. Discussion on drill steel involves fatigue tests. *Iron Tr. Rev.*, vol. 68, Feb. 24, 1921, pp. 544–545.

338. Alternating stress testing machine. *Engineer, Lond.*, vol. 131, May 20, 1921, p. 550.

339. The Haigh alternating stress machine. *Engineer, Lond.*, vol. 132, no. 3422, July 29, 1921, pp. 116–117.

340. Information on fatigue failures. *Chem. metall. Engng*, vol. 24, no. 9, March 2, 1921, pp. 370–371.

341. Fatigue. *Engineer, Lond.*, vol. 131, March 18, 1921, p. 296.

1922

342. ALBERT, C. D. Factors of safety and allowable stress. *Amer. Mach., N.Y.*, vol. 57, no. 2, Sept. 2, 1922, pp. 54–57.

343. BAIRSTOW, L. The fatigue of metals. *Beama J.*, vol. 11, Oct. 1922, pp. 652–657; Nov. 1922, pp. 731–737, and Dec. 1922, pp. 811–818.

344. CLEMMER, H. F. Fatigue of concrete. *Proc. Amer. Soc. Test. Mater.*, vol. 22, pt. 2, 1922, pp. 408–419.

345. DESCH, C. H. Brittleness and fatigue of metals. *Mech. World*, vol. 72, no. 1868, Oct. 20, 1922, pp. 269–270 and no. 1869, Oct. 27, 1922, pp. 289–291.

346. ENSSLIN, M. Brüche an gekröpften Kurbeln und Vorbeugungsmassnahmen. [Fractures in crankshafts and methods of prevention.] *Maschinenbau*, vol. 2, 1922, pp. 107–108.

347. GOUGH, H. J. On the elastic limits of copper under cyclic stress variation. *Engineering, Lond.*, vol. 114, Sept. 8, 1922, pp. 291–293.

348. GRINDLEY, J. H. Repeated stress testing and its results. *Proc. Staffs. Iron St. Inst.*, vol. 38, 1922–1923, pp. 80–96.

349. HAIGH, B. P. Elastic and fatigue limits in metals. *Bgham metall. Soc. J.*, vol. 8, no. 9, Feb. 1922, pp. 412–422. Abstr.: *Metal Ind., Lond.*, vol. 21, no. 20, Nov. 17, 1922, pp. 466–467.

350. HATT, W. K. Note on the fatigue of mortar. *Proc. Amer. Concr. Inst.*, vol. 18, 1922, pp. 167–173.

351. HEATHECOTE, H. L. *and* WHINFREY, C. G. Tearing tests of metals. *Chem. metall. Engng*, vol. 27, no. 7, Aug. 16, 1922, pp. 310–311.

352. HORSBURGH, E. M. The alternating torsion of rope-wire. *Engineering, Lond.*, vol. 114, Dec. 22, 1922, pp. 759–760.

353. JENKIN, C. F. A mechanical model illustrating the behaviour of metals under static and alternating loads. *Engineering, Lond.*, vol. 114, Nov. 17, 1922, p. 603.

354. JENKIN, C. F. Fatigue in metals. *Engineer, Lond.*, vol. 134, no. 3493, Dec. 8, 1922, pp. 612–614.

355. JOHNSON, J. B. *and* DANIELS, S. Study of some failures in aircraft plane and engine parts. *Trans. Amer. Soc. Steel Treat.*, vol. 2, no. 12, Sept. 1922, pp. 1167–1176, 1212.

356. KNERR, H. C. Remarks on fatigue failure of metal parts, their cause and prevention. *Forg. Heat. Treat.*, vol. 8, no. 1, Jan. 1922, pp. 40–42.

357. LEES, S. On a simple model to illustrate elastic hysteresis. *Phil. Mag.*, ser. 6, vol. 44, no. 261, Sept. 1922, pp. 511–537.

358. MASON, W. The mechanics of the Wöhler rotating bar fatigue test. *Rep. Memor. aero. Res. Comm., Lond.*, no. 838, Sept. 1922, pp. 8. *Engineering, Lond.*, vol. 115, no. 2996, June 1, 1923, pp. 698–699.

359. MAUKSCH, W. Der Arbeitsverbrauch bei oftmals wiederholter Zugbeanspruchung von Eisen und Kupfer bei verschiedenen Temperaturen. [The energy consumed by iron and copper under repeated tension loading at different temperatures.] *Mitt. K.-Wilh.-Inst. Metallforsch.*, vol. 1, 1922, pp. 41–57.

360. MOORE, H. F. *and* JASPER, T. M. Recent developments in fatigue of metals. *Iron Age*, vol. 110, no. 13, Sept. 28, 1922, pp. 779–784.

361. MOORE, H. F. *and* KOMMERS, J. B. Fatigue of metals under repeated stress. *Trans. Amer. Soc. Steel Treat.*, vol. 2, no. 4, Jan. 1922, pp. 305–319 and no. 9, June 1922, pp. 812–818.

362. MOORE, H. F.; KOMMERS, J. B. *and* JASPER, T. M. Fatigue or progressive failure of metals under repeated stress. *Proc. Amer. Soc. Test. Mater.*, vol. 22, pt. 2, 1922, pp. 266–311.

363. MÜLLER, W. Schlagbiegefestigkeit und Schlaghärte legierter Konstruktions-stähle. [On the bending impact endurance strength and impact hardness of alloy constructional steels.] *ForschArb. IngWes.*, no. 247, 1922, pp. 38.

364. MÜLLER, W. *and* LEBER, H. Über die Ermüdung geglühter und vergüteter Kohlenstoffstähle. [On the fatigue of normalised and heat treated carbon steels.] *Z. Ver. dtsch. Ing.*, vol. 66, no. 22, June 3, 1922, pp. 543–546.

365. SCHULZ, E. H. *and* PÜNGEL, W. Beiträge zur Ermüdungsprobe von Stahl auf dem Kruppschen Dauerschlagwerk. [Contribution to the fatigue testing of steel using the Krupp impact endurance machine.] *Mitt. VersAnst. dtsch-luxemb. Bergw. Hütten. Dortmund*, vol. 1, no. 2, 1922, pp. 43–51.

366. STROMEYER, C. E. Fatigue of metals. *Proc. S. Wales Inst. Engrs*, vol. 38, no. 3, May 1922,

pp. 285–331. Abstr.: *Iron Coal Tr. Rev.*, vol. 104, no. 2831, June 2, 1922, pp. 822–824. *Mech. World*, vol. 72, no. 1682, Sept. 8, 1922, pp. 167–168. *Colliery Guard.*, vol. 123, no. 3206, June 9, 1922, pp. 1422–1423. *Power*, vol. 56, no. 1, July 4, 1922, p. 33.

367. TAUBERT, R. Über die Entstehung von Dauerbrüchen. [On the origin of fatigue fractures]. *Maschinenbau*, vol. 2, 1922, pp. 261–264.

368. TEMPLIN, R. L. Non-ferrous metal fatigue. *Iron Age*, vol. 110, Aug. 10, 1922, p. 356. *Metal Ind., Lond.*, vol. 21, no. 11, Sept. 15, 1922, pp. 247–248.

369. THOMPSON, F. C. *and* WHITEHEAD, E. Some mechanical properties of the nickel-silvers. *J. Inst. Met.*, vol. 27, no. 1, 1922, pp. 227–266.

370. WILSON, J. S. *and* HAIGH, B. P. The influence of rivet holes on steel structures. *Engineering, Lond.*, vol. 114, Sept. 8, 1922, pp. 309–312. Abstr.: *Rep. Brit. Ass.*, 1922, p. 382.

371. Failure by fatigue. *Engineering, Lond.*, vol. 113, no. 2939, April 28, 1922, pp. 525–526.

372. New tests in fatigue of metals. *Automot. Industr., N.Y.*, vol. 46, no. 23, June 8, 1922, pp. 1270–1271.

373. Fatigue resistance of duralumin. *Tech. Memor. nat. Adv. Comm. Aero., Wash.*, no. 135, Sept. 1922, pp. 15.

74. Fatigue. *Engineer, Lond.*, vol. 134, Dec. 15, 1922, pp. 637–638.

1923

375. AITCHISON, L. The low apparent elastic limit of quenched or work-hardened steels. *Carnegie Schol. Mem.*, vol. 12, 1923, pp. 113–217.

376. BYRNE, B. R. Fatigue of iron and steel. *J. Instn Loco. Engrs, Lewes*, vol. 13, no. 62, Sept.–Dec. 1923, pp. 766–818.

377. CREPPS, R. B. Fatigue of mortar. *Proc. Amer. Soc. Test. Mater.*, vol. 23, pt. 2, 1923, pp. 329–340.

378. DALBY, W. E. Further researches on the strength of materials. *Proc. roy. Soc., (A)*, vol. 103, 1923, pp. 8–25.

379. FÖPPL, O. Schwingungsbeanspruchung und Rissbildung insbesondere von Konstruktionsstählen. [Fatigue stress and crack formation, especially in structural steel.] *Schweiz. Bauztg.*, vol. 81, no. 8, Feb. 24, 1923, pp. 87–91.

380. FÖPPL, O. Versuchsanordnungen zur Bestimmung der Schwingungsfestigkeit von Materialien. [Experimental methods for finding the fatigue strength of materials.] *Maschinenbau*, vol. 2, 1923, pp. 1002–1004.

381. FORCELLA, P. Le rotture accidentali dei materiali metallici in opera ed il modo di prevenirle. [The premature failure of metallic materials in service and methods of prevention.] *Elettrotecnica*, vol. 10, no. 28, Oct. 5, 1923, pp. 672–676. *Rivista tecnica delle Ferrovie Italiane*, vol. 24, no. 6, Dec. 15, 1923, pp. 261–264. French abstr.: *Bulletin technique de la Suisse Romande, Lausanne*, vol. 50, no. 6, March 15, 1924, p. 75.

382. GOUGH, H. J. *and* HANSON, D. The behaviour of metals subjected to repeated stresses.

Proc. roy. Soc., (A), vol. 104, July 1923, pp. 538–565.

383. HAIGH, B. P. Thermodynamic theory of mechanical fatigue and hysteresis in metals. *Rep. Brit. Ass.*, 1923, pp. 358–368. Abstr.: *Iron Coal Tr. Rev.*, vol. 105, 1923, p. 464.

384. HARSCH, J. W. Short time tests for long-time endurance. How the testing of metal for fatigue limits is being put upon a production basis. *Sci. Amer.*, vol. 128, no. 4, April 1923, pp. 264 and 288.

385. HARSCH, J. W. Heat treatment and the strength of steel under repeated stress. *Forg. Heat Treat.*, vol. 9, Jan. 1923, p. 57.

386. HEINDLHOFER, K. *and* SJÖVALL, H. Endurance test data and their interpretation. *Trans. Amer. Soc. mech. Engrs*, vol. 45, 1923, pp. 141–150. Abstr.: *Mech. Engng, N.Y.*, vol. 45, Oct. 1923, pp. 579–580.

387. JASPER, T. M. The value of the energy relation in the testing of ferrous metals at varying ranges of stress and at intermediate and high temperatures. *Phil. Mag.*, ser. 6, vol. 46, Oct. 1923, pp. 609–627.

388. JENKIN, C. F. Fatigue failure of metals. *Proc. roy. Soc., (A)*, vol. 103, April 3, 1923, pp. 121–138.

389. JENKIN, C. F. Fatigue in metals. *J. R. aero. Soc.*, vol. 27, no. 147, March 1923, pp. 89–104. *Chem. metall. Engng*, vol. 28, May 7, 1923, pp. 811–815.

390. LEA, F. C. Tensile tests of materials at high temperatures. *Engineer, Lond.*, vol. 135, Feb. 16, 1923, pp. 182–183.

391. LEA, F. C. The effect of repetition stresses on materials. *Engineering, Lond.*, vol. 115, Feb. 16, 1923, pp. 217–219 and Feb. 23, 1923, pp. 252–254.

392. LESSELLS, J. M. Static and dynamic tests for steel. *Trans. Amer. Soc. Steel Treat.*, vol. 4, no. 4, Oct. 1923, pp. 536–545.

393. LESSELLS, J. M. Notes on the fatigue of metals. *Mech. Engng, N.Y.*, vol. 45, no. 12, Dec. 1923, pp. 695–696.

394. LEWTON, R. E. Some endurance tests of spring steels. *Trans. Amer. Soc. Steel Treat.*, vol. 3, no. 9, June 1923, pp. 944–953.

395. LUDWIK, P. Die Veränderung der Metalle bei Wiederholter Beanspruchung. [The changes in metals under repeated stress.] *Z. Metallk.*, vol. 15, no. 3, March 1923, pp. 68–73.

396. LUDWIK, P. *and* SCHEU, R. Das Verhalten der Metalle bei Wiederholterbeanspruchung. [The behaviour of metals under repeated stress.] *Z. Ver. dtsch. Ing.*, vol. 67, no. 6, Feb. 10, 1923, pp. 122–126.

397. MCADAM, D. J. Endurance of steel in tension, torsion and impact. *Engng News Rec.*, vol. 91, no. 8, Aug. 23, 1923, pp. 298–301.

398. MCADAM, D. J. Endurance properties of steel: their relation to other physical properties and to chemical composition. *Proc. Amer. Soc. Test. Mater.*, vol. 23, pt. 2, 1923, pp. 56–105.

399. MASON, W. The distribution of stress in round mild steel bars under alternating torsion or bending. *Rep. Brit. Ass.*, 1923, pp. 386–409.

400. MASON, W. The effect of a temperature of 212°F on steel submitted to alternating torsion. *Rep. Memor. aero. Res. Comm., Lond.*, no. 863, Feb. 1923, pp. 12.

401. MASON, W. The mechanics of the Wöhler rotating bar fatigue test. *Engineering, Lond.*, vol. 115, no. 2996, June 1, 1923, pp. 698–699.

402. MOORE, H. F. Fatigue of metals and the basic assumptions of mechanics of materials. *Mich. Technic.*, vol. 37, no. 1, Nov. 1923, pp. 17–18, 30.

403. MOORE, H. F. *and* JASPER, T. M. An investigation of the fatigue of metals: series of 1922. *Uni. Ill. Engng Exp. Sta. Bull.* no. 136, 1923, pp. 97.

404. MOORE, R. R. Resistance of manganese bronze, duralumin and Electron metal to alternating stresses. *Proc. Amer. Soc. Test. Mater.*, vol. 23, pt. 2, 1923, pp. 106–129.

405. MÜLLER, W. *and* LEBER, H. Beanspruchungshöhe, Korngrösse und Temperatur bei Ermüdungserscheinungen. [Stress magnitude, grain size and temperature in the fatigue phenomenon.] *Z. Ver. dtsch. Ing.*, vol. 67, no. 15, April 14, 1923, pp. 357–363.

406. ROBSON, T. Determination of the fatigue resisting capacity of steel under alternating stress. *Engineering, Lond.*, vol. 115, no. 2977, Jan. 19, 1923, pp. 67–68.

407. ROSENHAIN, W. Strain and fracture in metals. *Chem. metall. Engng*, vol. 28, no. 23, June 11, 1923, pp. 1026–1030.

408. ROSENHAIN, W.; ARCHBUTT, S. L. *and* WELLS, S. A. E. Production and heat treatment of chill castings in an aluminium alloy. *J. Inst. Met.*, vol. 29, pt. 1, 1923, pp. 191–216.

409. SCOBLE, W. A. The repeated bending of steel wire. *Rep. Brit. Ass.*, 1923, pp. 409–411.

410. STRIBECK, R. Dauerfestigkeit von Eisen und Stahl bei wechselnder Biegung, verglichen mit den Ergebnissen des Zugversuchs. [Fatigue strength of iron and steel under alternating bending compared with tensile test results.] *Z. Ver. dtsch. Ing.*, vol. 67, no. 26, June 30, 1923, pp. 631–636.

411. THOMAS, W. N. The effect of scratches and of various workshop finishes upon the fatigue strength of steel. *Engineering, Lond.*, vol. 116, Oct. 12, 1923, pp. 449–454 and Oct. 19, 1923, pp. 483–485. *Rep. Memor. aero. Res. Comm., Lond.*, no. 860, March, 1923, pp. 28.

412. WILSON, J. S. *and* HAIGH, B. P. Stresses in bridges. *Rep. Brit. Ass.*, 1923, pp. 368–385. *Engineering, Lond.*, vol. 116, Sept. 28, 1923, pp. 411–413 and Oct. 12, 1923, pp. 446–448.

413. WOOD, J. K. Oscillations and fatigue of springs. *Amer. Mach., N.Y.*, vol. 58, no. 2, March 3, 1923, pp. 67–70.

414. The fatigue limit and the proportionality of Monel metal. *Engineering, Lond.*, vol. 116, July 20, 1923, p. 88.

415. The testing of materials. *Engineering, Lond.*, vol. 116, Nov. 16, 1923, pp. 633–634.

416. Fatigue failures cause heavy losses. *Iron Tr. Rev.*, vol. 72, no. 21, May 24, 1923, p. 1530.

417. Investigations of fatigue of metals. *Engng and Contr.*, vol. 59, May 23, 1923, pp. 1200–1201.

418. Notches and scratches. *Engineer, Lond.*, vol. 136, Oct. 12, 1923, p. 397.

1924

419. AITCHISON, L. Cold work and fatigue. *Rep. Memor. aero. Res. Comm.*, *Lond.*, no. 923, May 1924, pp. 6.

420. AITCHISON, L. Materials in aircraft construction. *Engineering, Lond.*, vol. 117, Jan. 18, 1924, pp. 89–91.

421. BATSON, R. G. Testing wires and wire ropes, with special reference to fatigue. *Testing, N.Y.*, vol. 1, Jan. 1924, pp. 7–22.

422. GILLETT, H. W. *and* MACK, E. L. Notes on some endurance tests of metals. *Proc. Amer. Soc. Test. Mater.*, vol. 24, pt. 2, 1924, pp. 476–546.

423. GOUGH, H. J. *The fatigue of metals.* London, Scott, Greenwood and Son, 1924, pp. 304.

424. GOUGH, H. J. Fatigue in metals. *Found. Tr. J.*, vol. 30, Dec. 18, 1924, pp. 527–528. *Engineer, Lond.*, vol. 138, Dec. 12, 1924, p. 664.

425. GOUGH, H. J.; HANSON, D. *and* WRIGHT, S. J. The behaviour of single crystals of aluminium under static and repeated stresses. *Rep. Memor. aero. Res. Comm.*, *Lond.*, no. 995, Nov. 1924, pp. 54.

426. GRIFFITHS, A. A. The impressed conditions of fatigue tests. *Rep. Brit. Ass.*, 1924, pp. 325–326.

427. HAIGH, B. P. Theory of rupture in fatigue. *Proc. Int. Congr. app. Mech.*, First Congr., Delft, 1924, pp. 326–332.

428. HAIGH, B. P. Slag inclusions in relation to fatigue. *Trans. Faraday Soc.*, vol. 20, 1924–1925, pp. 153–158.

429. HAIGH, B. P. *and* BEALE, A. The influence of circular holes on the fatigue strength of hard steel plates. *Rep. Brit. Ass.*, 1924, pp. 326–331.

430. HATT, W. K. Fatigue of concrete. *Proc. Highw. Res. Bd, Wash.*, vol. 4, 1924, pp. 47–60.

431. HORT, W. Ermüdungsfestigkeit bei hohen Beanspruchungsfrequenzen. [Fatigue strength with high frequency loadings.] *Z. tech. Phys.* vol. 5, no. 10, 1924, pp. 433–436. *Maschinenbau*, vol. 3, no. 27, Nov. 13, 1924, pp. 1038–1040.

432. JANNIN, L. Quelques résultats d'essai des matériaux aux vibrations. [Some results of vibrational tests on materials.] *Rev. Métall.*, vol. 21, no. 12, Dec. 1924, pp. 742–749.

433. JENKIN, C. F. The work of the fatigue panel of the Aeronautical Research Committee. *Engineering, Lond.*, vol. 118, no. 3059, Aug. 15, 1924, p. 245. Abstr.: *Rep. Brit. Ass.*, 1924, pp. 414–415.

434. KÄNDLER, H. Neue Wege zur Herabsetzung der Kerbwirkung. [New ways to reduce the notch effect.] *Z. tech. Phys.*, vol. 5, no. 4, 1924, pp. 151–154.

435. LEA, F. C. The effect of low and high temperatures on materials. *Proc. Instn mech.*

Engrs., Lond., vol. 2, July–Dec., 1924, pp. 1053–1096. *Engineering, Lond.,* vol. 118, Dec. 12, 1924, pp. 816–817 and Dec. 19, 1924, pp. 843–845.

436. LEA, F. C. *and* BUDGEN, H. P. The failure of a nickel chrome steel under repeated stresses of various ranges. *Rep. Memor. aero. Res. Comm., Lond.,* no. 920, June 1924, pp. 4.

437. LEA, F. C. *and* BUDGEN, H. P. The effect of high temperature on the range of repetition stresses on steel. *Engineering, Lond.,* vol. 118, Oct. 3, 1924, p. 500 and Oct. 10, 1924, pp. 532–534. Abstr.: *Rep. Brit. Ass.,* 1924, p. 415.

438. LESSELLS, J. M. The elastic limit in tension, and its influence on the breakdown by fatigue. *Proc. Instn mech. Engrs, Lond.,* vol. 2, July–Dec. 1924, pp. 1097–1114. *Engineering, Lond.,* vol. 118, Dec. 12, 1924, pp. 813–814.

439. MCADAM, D. J. Endurance properties of corrosion-resistant steels. *Proc. Amer. Soc. Test. Mater.,* vol. 24, pt. 2, 1924, pp. 273–303.

440. MCADAM, D. J. Accelerated fatigue tests and some endurance properties of metals. *Proc. Amer. Soc. Test. Mater.,* vol. 24, pt. 2, 1924, pp. 454–475.

441. MCADAM, D. J. The endurance range of steel. *Proc. Amer. Soc. Test. Mater.,* vol. 24, pt. 2, 1924, pp. 574–600.

442. MCADAM, D. J. Correlation of endurance properties of metals. *Trans. Amer. Soc. Steel Treat.,* vol. 6, no. 3, Sept. 1924, pp. 393–395.

443. MÄILANDER, R. Ermüdungserscheinungen und Dauerversuche. [Fatigue phenomena and endurance testing.] *Stahl u. Eisen,* vol. 44, no. 21, May 22, 1924, pp. 585–589; no. 22,

May 29, 1924, pp. 624–629; no. 23, June 5, 1924, pp. 657–661; no. 24, June 12, 1924, pp. 684–691 and no. 25, June 19, 1924, pp. 719–725.

444. MASON, W. Note on the distribution of stress in fatigue test specimens (torsion and bending). *Rep. Brit. Ass.,* 1924, pp. 331–332.

445. MERRILS, F. S. Studies in the fatigue of metals. *Carnegie Schol. Mem.,* vol. 13, 1924, pp. 83–128.

446. MILLER, J. Case-hardening and fatigue resistance. *Iron Age,* vol. 113, May 1, 1924, pp. 1269–1271.

447. MILLINGTON, W. E. W. *and* THOMPSON, F. C. The investigation of a fatigue failure of brass tubes in a feed-water heater with a consideration of the nature of "fatigue". *J. Inst. Met.,* vol. 31, no. 1, 1924, pp. 81–120. Abstr.: *Engineering, Lond.,* vol. 117, March 14, 1924, pp. 344–345.

448. MOORE, H. F. *and* JASPER, T. M. Evidence for fatigue limits. *Engineering, Lond.,* vol. 118, Oct. 24, 1924, pp. 580–582 and Nov. 7, 1924, pp. 658–660. Abstr.: *Rep. Brit. Ass.,* 1924, p. 414.

449. MOORE, H. F. *and* JASPER, T. M. The evidence and determination of a fatigue limit in metals. *Technograph,* vol. 37, no. 1, Nov. 1924, pp. 12–16, 42.

450. MOORE, H. F. *and* JASPER, T. M. An investigation of the fatigue of metals: series of 1923. *Uni. Ill. Engng Exp. Sta. Bull.,* no. 142, 1924, pp. 86.

451. MOORE, R. R. Resistance of metals to repeated static and impact stresses. *Proc. Amer. Soc. Test. Mater.,* vol. 24, pt. 2, 1924, pp. 547–573.

452. Ono, A.. Experiments on the fatigue of steel. *Mem. Coll. Engng Kyushu*, vol. 3, no. 2, 1924, pp. 51–85.

453. Palmgren, A. Die Lebensdauer von Kugellagern. [The life of ball bearings.] *Z. Ver dtsch. Ing.*, vol. 68, no. 14, April 5, 1924, pp. 339–341.

454. Riede, W. Rekristallisationserscheinungen an dauerbeanspruchten Stählen. [Recrystallisation phenomena in the fatigue stressing of steels.] *Stahl u. Eisen*, vol. 44, no. 30, July 24, 1924, pp. 880–883.

455. Rosenhain, W. When metals get tired—modern endurance tests. *Conquest*, vol. 6, Nov. 1924, pp. 15–20.

456. Rowlinson, F. A. The failure of metals by fatigue—why steel and iron give way under repetition of relatively slight stress. *Sci. Amer.*, vol. 131, no. 4, Oct. 1924, pp. 240, 292.

457. Scoble, W. A. Wire ropes research. *Engineering, Lond.*, vol. 118, Dec. 26, 1924, pp. 856–860.

458. Shiba, C. *and* Yuasa, K. On the effect of fillet on the endurance of metals. *J. Soc. mech. Engrs, Japan*, vol. 27, no. 84, 1924, pp. 209–246.

459. Tuckerman, L. B. *and* Aitchison, C. S. Design of specimens for short-time "fatigue" tests. *Tech. Pap. U.S. Bur. Stand.*, vol. 19, no. 275, Dec. 22, 1924, pp. 47–55.

460. Welter, G. Die Statische und dynamische Elastizitätsgrenze in Material Prüfungs- und Konstruktionswesen. [The static and dynamic elastic limits in materials testing and methods of construction.] *Z. Ver. dtsch. Ing.*, vol. 68, no. 1, Jan. 5, 1924, pp. 9–11.

461. Woodvine, J. G. R. The behaviour of case-hardened parts under fatigue stresses. *Carnegie Schol. Mem.*, vol. 13, 1924, pp. 197–237.

462. Modifications of the fatigue of metals theory formulated. *Automot. Industr. N.Y.*, vol. 51, no. 3, July 17, 1924, pp. 161–162.

463. Endurance tests of molybdenum steels. *Amer. Mach., N.Y.*, vol. 59, no. 21, Jan. 12, 1924, p. 760.

464. Effect of sulfur on endurance properties of rivet steel. Third preliminary report of the Joint Committee on investigation of phosphorus and sulfur in steel. *Proc. Amer. Soc. Test. Mater.*, vol. 24, pt. 1, 1924, pp. 96–107.

465. Endurance and failure. *Engng News Rec.*, vol. 93, no. 7, Aug. 14, 1924, p. 251.

466. Recent English endurance tests of steel and other metals. *Engng News Rec.*, vol. 93, no. 18, Oct. 30, 1924, pp. 709–710.

467. Fatigue properties of steel. *Circ. U.S. Bur. Stand.*, no. 101, April 23, 1924, pp. 139–145.

468. New fatigue testing machine. *Iron Age*, vol. 114, no. 23, Dec. 4, 1924, p. 1482.

469. Selecting the steel for the job. *S.A.E. Jl.*, vol. 15, no. 6, Dec. 1924, pp. 483–484.

1925

470. AITCHISON, L. *and* JOHNSON, L. W. The effect of grain upon the fatigue strength of steels. *J. Iron St. Inst.*, vol. 111, no. 1, 1925, pp. 351–378. Abstr.: *Engineering, Lond.*, vol. 119, May 8, 1925, p. 585.

471. BAMFORD, T. G. Comparative tests on some varieties of commercial copper rod. *J. Inst. Met.*, vol. 33, no. 1, 1925, pp. 167–189.

472. BILLET, P. *and* WANTZ, H. Essai sur la fatigue du métaux das les bandages de roues de chemins de fer. [Tests on the fatigue of metal in the tires of railway wheels.] *Rev. Métall.*, vol. 22, no. 3, March 1925, pp. 154–169 and no. 4, April 1925, pp. 207–217.

473. BREUIL, P. *Les essais de fatigue des métaux et les machines Amsler pour leur exécution. [Amsler machines for the fatigue testing of metals.]* Paris, Dunod, 1925, pp. 69. French abstr.: *Génie civ.*, vol. 87, no. 3, (no. 2240), July 18, 1925, pp. 64–66.

474. DEWS, H. C. The growth of modern theories of fatigue failure. *Metal Ind., Lond.*, vol. 26, no. 23, June 5, 1925, p. 551–553.

475. GOUGH, H. J. The effect of keyways upon the strength and stiffness of shafts subjected to torsional stresses. *Rep. Memor. aero. Res. Comm., Lond.*, no. 864, April 1925, pp. 20.

476. GOUGH, H. J.; WRIGHT, S. J. *and* HANSON, D. An experiment to determine if slip can be detected during the unloading portion of a cycle of repeated tensile stresses. *Rep. Memor. aero. Res. Comm., Lond.*, no. 1022, Dec. 1925, pp. 6.

477. HAHNEMANN, W.; HECHT, H. *and* WILCKENS, E, Eine neue Materialprüfmaschine für Dauerbeanspruchungen. [A new material fatigue testing machine.] *Z. tech. Phys.*, vol. 6, no. 9, 1925, pp. 465–468.

478. HAIGH, B. P. Fatigue in non-ferrous alloys. *Bull. Brit. non-ferr. Met. Ass.*, no. 15, July 1925, pp. 11–16.

479. HUGENARD, E.; MAGNAN, A. *and* PLANIOL, A. Sur un appareil mesurant les déformations des ailes en vue de l'étude de la fatigue et du vieillissement des avions. [On an apparatus for measuring the deformation of wings in relation to the study of the fatigue and the ageing of aircraft.] *Bull. tech. Serv. Aérotech., Paris*, no. 24, Feb. 1925, pp. 14. Italian abstr.: *Aerotecnica, Roma*, vol. 5, no. 4, July–Aug. 1925, pp. 257–258.

480. IRWIN, P. L. Fatigue of metals by direct stress. *Proc. Amer. Soc. Test. Mater.*, vol. 25, pt. 2, 1925, pp. 53–65. Abstr.: *Engng News Rec.*, vol. 95, no. 8, Aug. 20, 1925, pp. 311–312.

481. JASPER, T. M. Typical static and fatigue test on steel at elevated temperatures. *Proc. Amer. Soc. Test. Mater.*, vol. 25, pt. 2, 1925, pp. 27–52. Abstr.: *Engng News Rec.*, vol. 95, no. 8, Aug. 20, 1925, pp. 309–311.

482. JASPER, T. M. An outline for the application of fatigue and elastic results to metal spring design. *Trans. Amer. Soc. mech. Engrs*, vol. 47, 1925, pp. 731–745.

483. JENKIN, C. F. High frequency fatigue tests. *Rep. Memor. aero. Res. Comm., Lond.*, 982, Oct. 1925, pp. 24. *Proc. roy. Soc., (A)*, vol. 109, Sept. 1925, pp. 119–143. Abstr.: *Metallurgist*, vol. 1, Oct. 30, 1925, pp. 145–146.

484. KÄNDLER, H. *and* SCHULZ, E. H. Ein neuer Weg zur Verminderung der Dauerbruchge-

fahr. [A new method of reducing the danger of fatigue failure.] *Stahl u. Eisen*, vol. 45, no. 38, Sept. 17, 1925, pp. 1589–1596.

485. KURAISHI, B. A new repeated torsion tester [J]. *J. Soc. mech. Engrs, Japan*, vol. 28, no. 102, 1925, pp. 797–817.

486. MCADAM, D. J. Endurance properties of alloys of nickel and of copper. *Trans. Amer. Soc. Steel Treat.*, vol. 7, no. 1, Jan. 1925, pp. 54–81; no. 2, Feb. 1925, pp. 217–236 and no. 5, May 1925, pp. 581–617.

487. MCADAM, D. J. Endurance properties of metals. *Mech. Engng, N.Y.*, vol. 47, no. 7, July 1925, pp. 566–572.

488. MCADAM, D. J. Effect of cold-working on endurance and other properties of metals. *Trans. Amer. Soc. Steel Treat.*, vol. 8, Dec. 1925, pp. 782–836.

489. MCINTOSH, F. F. *and* COCKRELL, W. L. The effect of phosphorus on the resistance of low carbon steel to repeated alternating stresses. *Bull. Carneg. Inst. Tech.*, no. 25, 1925, pp. 33.

490. MAILÄNDER, R. Die Ermüdung der Metalle. [The fatigue of metals.] *Stahl u. Eisen*, vol. 45, no. 41, Oct. 8, 1925, pp. 1713–1715.

491. MOORE, H. F. Fatigue tests of metals and the theory of elasticity. *Engng News Rec.*, vol. 94, no. 6, Feb. 5, 1925, pp. 225–226.

492. MOORE, H. F. Studying the fatigue of metals. *Amer. Mach., N.Y.*, vol. 62, no. 15, May 30, 1925, pp. 563–565.

493. MOORE, H. F. Notes on the fatigue of non-ferrous metals. *Min. and Metall. N.Y.*, vol. 6, no. 225, Sept. 1925, pp. 465–467.

494. MOORE, H. F. Stress repetition and fatigue in steel structures. *Engng News Rec.*, vol. 95, no. 10, Sept. 3, 1925, pp. 376–377.

495. MOORE, H. F. What happens when metal fails by fatigue? *Amer. Mach., N.Y.*, vol. 63, no. 13, Sept. 24, 1925, pp. 502–503.

496. MOORE, H. F. *and* JASPER, T. M. An investigation of the fatigue of metals—series of 1925. *Uni. Ill. Engng Exp. Sta. Bull.*, no. 152, 1925, pp. 89.

497. MOORE, R. R. Some fatigue tests on non-ferrous metals. *Proc. Amer. Soc. Test. Mater.*, vol. 25, pt. 2, 1925, pp. 66–96. Abstr.: *Aviation, N.Y.*, vol. 19, no. 1, July 6, 1925, p. 10.

498. MOORE, R. R. Fatigue of welds. *Mech. Engng, N.Y.*, vol. 47, Oct. 1925, pp. 794–795.

499. QUACK, W. Kesselexplosion in Westdeutschland infolge Ermüdungsumbruches scharfgebogener Bodenkrempung. [Boiler explosion in West Germany as a result of fatigue failure in sharp bottom flange.] *Wärme*, vol. 48, no. 29, 1925, pp. 369–372.

500. SHIBA, C. *and* YUASA, K. On the effect of fillets on the endurance of metals. *J. Fac. Engng Tokyo*, vol. 15, no. 10, Feb. 1925, pp. 317–338.

501. STONEY, G. The failure of materials in engineering. *Engineering, Lond.*, vol. 120, no. 3112, Aug. 21, 1925, p. 224.

502. TAPSELL, H. J. *and* BRADLEY, J. Mechanical tests at high temperatures on a non-ferrous alloy of nickel and chromium. *Engineering, Lond.*, vol. 120, Nov. 13, 1925, pp. 614–615; Nov. 20, 1925, pp. 648–649 and Dec. 11, 1925, pp. 746–747.

503. TIMOSHENKO, S. *and* DIETZ, W. Stress concentration produced by holes and fillets. *Trans. Amer. Soc. mech. Engrs*, vol. 47, 1925, pp. 199–237.

504. VOORHEES, R. R. Fatigue of metals in airplane parts. *Iron Age*, vol. 115, no. 21, May 21, 1925, p. 1498.

505. Spring testing machine for weighing and repetition tests. *Engineering, Lond.*, vol. 119, no. 3089, March 13, 1925, pp. 316–317.

506. Endurance properties of alloys of nickel and of copper. *Metallurgist*, vol. 1, April 24, 1925, pp. 55–60.

507. Universal endurance impact testing machine. *Engineering, Lond.*, vol. 119, May 15, 1925, pp. 604, 608.

508. American researches on the fatigue of steel. *Metallurgist*, vol. 1, Aug. 28, 1925, pp. 118–121.

509. Scratches and the fatigue strength of steel. *Metallurgist*, vol. 1, Oct. 30, 1925, pp. 150–151.

1926

510. BISSELL, A. G. Electric arc welding in the manufacture of structural steel. *J. Amer. Weld. Soc.*, vol. 5, no. 9, Sept. 1926, pp. 24–34.

11. BULLEID, C. H. Fatigue tests of cast iron. *Engineering, Lond.*, vol. 122, Oct. 1, 1926, pp. 429–430.

12. DUCHEMIN, E. Contribution à l'étude des essais de chocs répétés. [Contribution to the study of repeated impact tests.] *Rev. Métall.*, vol. 23, no. 12, Dec. 1926, pp. 718–722.

513. FÖPPL, O. Die Dämpfungsfähigkeit eines Baustahles bei Wechselbeanspruchungen. [The damping capacity of structural steel under alternating stresses.] *Z. Ver. dtsch. Ing.*, vol. 70, no. 39, Sept. 25, 1926, pp. 1291–1296.

514. FUJII, Y. The fatigue of steel and its recovery—part I. *Mem. Coll. Engng Kyoto*, vol. 4, no. 2, March 1926, pp. 37–62.

515. GOUGH, H. J. Some recent researches and their bearing on the theory of failure of metals. *Metallurgist*, vol. 2, Sept. 24, 1926, pp. 132–133.

516. GOUGH, H. J. *The fatigue of metals*. London, Ernest Benn Ltd., 1926, pp. 304.

517. GOUGH, H. J.; HANSON, D. *and* WRIGHT, S. J. The behaviour of single crystals of aluminium under static and repeated stresses. *Phil. Trans. (A)*, vol. 226, 1926–1927, pp. 1–30.

518. GOUGH, H. J. *and* TAPSELL, H. J. Some comparative fatigue tests in special relation to the impressed conditions of test. *Rep. Memor. aero. Res. Comm., Lond.*, no. 1012, April 1926, pp. 21.

519. GOUGH, H. J.; WRIGHT, S. J. *and* HANSON, D. Some further experiments on the behaviour of single crystals of aluminium under reversed torsional stresses. *Rep. Memor. aero. Res. Comm., Lond.*, no. 1023, Jan. 1926, pp. 13.

520. GOUGH, H. J.; WRIGHT, S. J. *and* HANSON, D. Some further experiments on single crystals of aluminium employing reversed direct

stresses. *Rep. Memor. aero. Res. Comm., Lond.*, no. 1024, Jan. 1926, pp. 14.

521. GOUGH, H. J.; WRIGHT, S. J. *and* HANSON, D. A test on a specimen consisting of three crystals under reversed torsional stresses. *Rep. Memor. aero. Res. Comm., Lond.*, no. 1025, Jan. 1926, pp. 5.

522. GOUGH, H. J.; WRIGHT, S. J. *and* HANSON, D. Further experiments on single crystals of aluminium under reversed torsional stresses. *J. Inst. Met.*, vol. 36, no. 2, 1926, pp. 173–190.

523. HAGUE, A. P. Corrosion fatigue. *Metallurgist*, vol. 2, Oct. 29, 1926, pp. 152–154.

524. HANKINS, G. A.; HANSON, D. *and* FORD, G. W. The mechanical properties of four heat treated spring steels. *J. Iron St. Inst.*, vol. 114, no. 2, 1926, pp. 265–294.

525. HANSON, D. The fatigue of metals. *Proc. Staffs. Iron St. Inst.*, vol. 42, 1926–1927, pp. 13–30. Abstr.: *Metal Ind., Lond.*, vol. 29, no. 19, Nov. 5, 1926, pp. 445–446.

526. HONDA, K. A comparison of static and dynamic tensile and notched bar tests. *J. Inst. Met.*, vol. 34, no. 2, 1926, pp. 27–37. *Engineering, Lond.*, vol. 122, Sept. 24, 1926, pp. 398–400.

527. HONEGGER, E. The fatigue of metals when stressed beyond the yield point. *Brown-Boveri Rev.*, vol. 13, no. 7, July 1926, pp. 169–174.

528. IRWIN, P. L. Fatigue of metals by direct stress. *Proc. Amer. Soc. Test. Mater.*, vol. 26, pt. 2, 1926, pp. 218–223.

529. JANNIN, L. Étude des phénomènes d'hysteresis élastique sur quelques alliages. [Study of the phenomena of elastic hysteresis in some alloys.] *Rev. Métall.*, vol. 23, 1926, pp. 709–717.

530. JASPER, T. M. Report on cause of failure in metal of A. O. Smith cylinder tested in repeated stress under repeated hydrogen static loading. *J. Amer. Weld. Soc.*, vol. 5, no. 1, Jan. 1926, pp. 11–14.

531. LEA, F. C. *and* BUDGEON, H. P. Combined torsional and repeated bending stresses. *Engineering, Lond.*, vol. 122, Aug. 20, 1926, pp. 242–245.

532. LECOEUVRE, R. Les essais d'endurance des matériaux. [The endurance testing of materials.] *Tech. aéro.*, vol. 17, no. 52, Feb. 15, 1926, pp. 34–47.

533. LEHMANN, G. D. The variation in the fatigue strength of metals when tested in the presence of different liquids. *Rep. Memor. aero. Res. Comm., Lond.*, 1054, Oct. 1926, pp. 13. *Engineering, Lond.*, vol. 122, Dec. 31, 1926, pp. 807–809.

534. LEHR, E. Die Dauerfestigkeit, ihre Bedeutung für die Praxis und ihre Kurzfristige Ermittlung mittels neuartiger Prüfmaschinen. [The endurance limit, its importance in practice and its rapid determination by means of new testing machines.] *Glasers Ann. Gew.*, vol. 99, no. 1184, Oct. 15, 1926, pp. 109–114; no. 1185, Nov. 1, 1926, pp. 117–122; no. 1188, Dec. 15, 1926, pp. 177–180 and vol. 100, no. 1191, Feb. 21, 1927, pp. 33–39.

535. LILJEBLAD, R. An investigation of the fatigue of metals due to locally concentrated stresses. *IngenVetenskAkad. Handl.*, no. 47, 1926, pp. 17. Abstr.: *Engineering, Lond.*, vol. 122, July 23, 1926, p. 93.

536. McADAM, D. J. Stress–strain-cycle relationship and corrosion fatigue of metals. *Proc. Amer. Soc. Test. Mater.*, vol. 26, pt. 2, 1926, pp. 224–254. Abstr.: *Metallurgist*, vol. 2, Sept. 24, 1926, pp. 130–131.

537. McINTOSH, F. F. Effect of phosphorus on the endurance limit of low-carbon steels. *Min. and Metall.*, *N.Y.*, vol. 7, no. 236, Aug. 1926, pp. 332–333.

538. MOORE, H. F. What happens when metal fails in fatigue? *Trans. Amer. Soc. Steel Treat.*, vol. 9, no. 4, April 1926, pp. 539–552.

539. MOORE, H. F. The mechanism of fatigue failure of metals. *J. Franklin Inst.*, vol. 202, no. 5, Nov. 1926, pp. 547–568.

540. MOORE, H. F. Tests on the fatigue strength of cast steel. *Uni. Ill. Engng Exp. Sta. Bull.*, no. 156, 1926, pp. 18.

541. MOORE, R. R. Effect of grooves, threads and corrosion upon the fatigue of metals. *Proc. Amer. Soc. Test. Mater.*, vol. 26, pt. 2, 1926, pp. 255–280.

542. ROWE, F. W. Fatigue failures in steel. *Metal Ind.*, *Lond.*, vol. 28, Feb. 5, 1926, pp. 133–135; Feb. 12, 1926, pp. 157–159 and Feb. 19, 1926, pp. 185–186.

543. SCHOTTKY, H. Schmelzschweissung und Dauerbruch. [Fusion welding and fatigue fracture.] *Krupp. Mh.*, vol. 7, no. 12, Dec. 1926, pp. 213–216.

544. SOUTHWELL, R. V. *and* GOUGH, H. J. On the concentration of stress in the neighbourhood of a small spherical flaw; and on the propagation of fatigue fractures in statistically isotropic materials. *Phil. Mag.*, ser. 7, vol. 1, no. 1, Jan. 1926, pp. 71–97. *Rep. Memor. aero. Res. Comm.*, *Lond.*, no. 1003, Jan. 1926, pp. 22.

545. TIMOSHENKO, S. Stress concentration produced by fillets and holes. *Proc. Int. Congr. app. Mech.*, *Second Congr.*, *Zurich*, 1926, pp. 419–426.

546. WELTER, G. Dauerschlagfestigkeit und dynamische Elastizitätsgrenze. [Repeated impact strength and dynamic elastic limit.] *Z. Ver. dtsch. Ing.*, vol. 70, no. 20, May 15, 1926, pp. 649–655 and no. 23, June 5, 1926, pp. 772–776. English abstr.: *Metallurgist*, vol. 2, July 30, 1926, pp. 108–110.

547. WELTER, G. Silumin: an improvement of its dynamic elastic qualities and endurance limit by the addition of copper. *J. Inst. Met.*, vol. 34, no. 2, 1926, pp. 325–358.

548. Fatigue of metals. *Machinery*, *Lond.*, vol. 27, no. 696, Jan. 28, 1926, p. 586.

549. Safe loads and endurance of steels under repeated bending stresses. *Engineer*, *Lond.*, vol. 141, Jan. 29, 1926, p. 130.

550. Fatigue of metals: its influence on mechanical design. *Chem. Age*, *Lond.*, vol. 14, Monthly Metallurgical Section, March 6, 1926, p. 18.

551. High frequency fatigue tests. *Engineer*, *Lond.*, vol. 141, April 16, 1926, pp. 434–435.

552. Machine pour les essais de chocs répétés. [A repeated impact testing machine.] *Chal. et Industr.*, vol. 7, June 1926, pp. 351–352.

553. The mechanism of fatigue. *Engineering, Lond.*, vol. 122, Oct. 1, 1926, pp. 422–423.

554. Tensile tests made with aluminium–silicon alloys. *Automot. Industr. N.Y.*, vol. 55, no. 20, Nov. 11, 1926, p. 825.

555. Endurance of annealed brass, hard-drawn brass, copper and duralumin. *Metal Ind., Lond.*, vol. 29, no. 21, Nov. 19, 1926, p. 490.

556. Fatigue machine is semi-automatic. *Abrasive Ind.*, vol. 7, Dec. 1926, pp. 389–390.

557. "Moore" fatigue machine. *Amer. Mach., N.Y.*, vol. 65, no. 18, Dec. 4, 1926, pp. 727–728.

558. Electro-magnetic alternating stress testing machine. *Engineering, Lond.*, vol. 122, no. 3178, Dec. 10, 1926, pp. 722–724.

1927

559. BROWN, R. M. An investigation into some effects of cold drawing on the strength and endurance of mild steel. *Trans. Instn Engrs Shipb. Scot.*, vol. 71, 1927–1928, pp. 495–579.

560. BULLEID, C. H. and ALMOND, A. R. The fatigue of cast iron. *Engineering, Lond.*, vol. 124, no. 3232, Dec. 23, 1927, p. 827.

561. BURNHAM, T. H. Torsional fatigue limits. *Engineering, Lond.*, vol. 124, July 8, 1927, pp. 33–34.

562. DRÄGER, K. Über Schwingungserscheinungen auf Freileitungen. [On vibration phenomena in overhead transmission lines.] *Elektrotech. u. Maschinenb.*, vol. 45, 1927, pp. 185–189.

563. FALKNER, V. M. Fluctuating bend tests of timber spars. *Roy. Aircr. Estab.*, Rep. M.T. 5236, June 1927, pp. 4 and Rep. M.T. 5236a, June 1927, pp. 8.

564. FORCELLA, P. La métallographie microscopique en relation avecs les essais de résilience et de durée. [Microscopic metallography in relation to resilience and fatigue tests.] *Congr. int. Ass. Test. Mat., Seventh Congr., Amsterdam*, 1927, vol. 1, pp. 229–252.

565. GOUGH, H. J. Some modern views on the fatigue of metals. *Struct. Engr*, vol. 5, no. 3, March 1927, pp. 70–83. *Engineer, Lond.*, vol. 143, no. 3720, April 29, 1927, pp. 474–477.

566. GOUGH, H. J. Note on some fatigue and density tests made on aluminium aggregate. *Rep. Memor. aero. Res. Comm., Lond.*, no. 1110, June 1927, pp. 8.

567. GOUGH, H. J. The behaviour of a single crystal of alpha iron subjected to alternating torsional stresses. *Rep. Memor. aero. Res. Comm., Lond.*, no. 1148, Oct. 1927, pp. 34.

568. GOUGH, H. J. Fatigue phenomenon with relation to cohesion problems. *Metal Ind., Lond.*, vol. 31, no. 24, Dec. 16, 1927, pp. 557–560.

569. GOUGH, H. J. Fatigue of metals and alloys. *International Critical Tables, New York*, vol. 2, First Edition, 1927, pp. 595–608.

570. HAIGH, B. P. Hysteresis in relation to cohesion and fatigue. *Metal Ind., Lond.*, vol. 31, no. 25, Dec. 23, 1927, pp. 584–585.

571. HULTGREN, A. The fatigue properties, impact resistance and hardness of certain nickel-

chromium steels for gears, case hardened, oil hardened and air hardened respectively. *Ingen-VetenskAkad. Handl.*, no. 59, 1927, pp. 1–55.

572. IKEDA, S. On the fatigue of steels for springs, axles and rails. [J] *Bull. Res. Off. Govt Rlys, Jap.*, vol. 15, no. 11, Nov. 1927, pp. 1744–1774.

573. INGLIS, N. P. Hysteresis and fatigue of the Wöhler rotating cantilever specimen. *Metallurgist*, vol. 3, Feb. 25, 1927, pp. 23–26.

574. INGLIS, N. P. Hysteresis in metals under alternating stress. *Metallurgist*, vol. 3, Sept. 30, 1927, pp. 138–140.

575. KÜHNEL, R. Die Gefahren der Schwingungsbeanspruchung für den Werkstoff. [The danger in fatigue loadings for materials.] *Z. Ver. dtsch. Ing.*, vol. 71, no. 17, April 23, 1927, pp. 557–561.

576. LEA, F. C. Some experiments on the effect of repeated stresses on materials. *J. Inst. aero. Engrs*, vol. 1, no. 5, May 1927, pp. 7–46.

577. LEA, F. C. The effect of heat-treatment on cold-drawn steel tubes. *Engineering, Lond.*, vol. 124, Dec. 23, 1927, pp. 797–800 and Dec. 30, 1927, pp. 831–834.

578. LEA, F. C. *and* HEYWOOD, F. The failure of some steel wires under repeated torsional stresses at various mean stresses determined from experiments on helical springs. *Proc. Instn mech. Engrs, Lond.*, pt. 1, April 1927, pp. 403–463. *Engineering, Lond.*, vol. 123, May 6, 1927, pp. 562–564 and May 20, 1927, pp. 621–623. Abstr.: *Engineer, Lond.*, vol. 143, May 6, 1927, pp. 487–488.

579. LEHR, E. Schwingungsfestigkeit und Ermüdungserscheinungen der Werkstoffe unter besonderer Berücksichtigung der für den Maschinenkonstrukteur massgebenden Gesichtspunkte. [Fatigue strength and fatigue phenomena in materials with particular regard to the viewpoint of the machine designer.] *Werkzeugmaschine*, vol. 31, 1927, pp. 400–408.

580. LESSELLS, J. M. Fatigue strength of hard steels and its relation to tensile strength. *Trans. Amer. Soc. Steel Treat.*, vol. 11, March 1927, pp. 413–424. Abstr.: *Amer. Mach., N.Y.*, vol. 65, no. 13, Oct. 30, 1926, pp. 525–526.

581. LUCAS, F. F. Observations on the microstructure of the path of fatigue failure in a specimen of Armco Iron. *Trans. Amer. Soc. Steel Treat.*, vol 11, April 1927, pp. 531–550. *Bell. Telep. Lab. Rep.*, B-257, June 1927.

582. MCADAM, D. J. Corrosion fatigue of metals as affected by chemical composition, heat treatment and cold working. *Trans. Amer. Soc. Steel Treat.*, vol. 11, no. 5, March 1927, pp. 355–390.

583. MCADAM, D. J. Corrosion fatigue of non-ferrous metals. *Proc. Amer. Soc. Test. Mater.*, vol. 27, pt. 2, 1927, pp. 102–127. Abstr.: *Engng News Rec.*, vol. 99, no. 3, July 21, 1927, pp. 95–96.

584. MCADAM, D. J. Fatigue and corrosion-fatigue of metals. *Congr. int. Ass. Test. Mat., Seventh Congr., Amsterdam*, 1927, vol. 1, pp. 305–358.

585. MASON, W. *and* INGLIS, N. P. The distribution of stress and strain in the Wöhler rotating cantilever fatigue test. *Rep. Memor. aero. Res. Comm. Lond.*, no. 1126, Oct. 1927, pp. 39.

586. MILLS, R. E. and DAWSON, R. F. Fatigue of concrete. *Proc. Highw. Res. Bd. Wash.*, vol. 7, pt. 1, 1927, pp. 160–172.

587. MOORE, H. F. The fatigue of metals. *Congr. int. Ass. Test. Mat., Seventh Congr., Amsterdam*, 1927, vol. 1, pp. 297–303.

588. MOORE, H. F. *Manual of the endurance of metals under repeated stress. A summary of views and test data with instructions for use. Compiled for designing, inspecting and testing engineers.* New York, Engineering Foundation Publication no. 13, 1927, pp. 63.

589. MOORE, H. F. A study of fatigue cracks in car axles. *Uni. Ill. Engng Exp. Sta. Bull.*, no. 165, June 1927, pp. 22. Abstr.: *Heat Treat. Forg.*, vol. 13, Nov., 1927, pp. 447–449.

590. MOORE, H. F. and KOMMERS, J. B. *The fatigue of metals.* New York, McGraw-Hill Book Co., Inc., 1927, pp. 326.

591. MOORE, H. F. and LYON, S. W. Tests on the endurance of gray cast iron under repeated stress. *Proc. Amer. Soc. Test. Mater.*, vol. 27, pt. 2, 1927, pp. 87–101.

592. MOORE, H. F. and LYON, S. W. Fatigue tests of cast iron. *Trans. Amer. Foundrym. Ass.*, vol. 35, 1927, pp. 410–426.

593. MOORE, H. F.; LYON, S. W. and INGLIS, N. P. Tests on the fatigue strength of cast iron. *Uni. Ill. Engng Exp. Sta. Bull.*, no. 164, June 1927, pp. 50.

594. MOORE, R. R. Fatigue resistance of welds. *J. Amer. Weld. Soc.*, vol. 6, no. 4, April 1927, pp. 11–32, Abstr.: *Engng News Rec.*, vol. 98, no. 24, June 16, 1927, p. 975. *Power*,

vol, 65, no. 21, May 24, 1927, pp. 796–797. German translation: Ermüdung von Schweissungen. *Z. Flugtech.*, vol. 19, no. 10, May 29, 1928, pp. 222–225.

595. MOORE, R. R. Effect of corrosion upon the fatigue resistance of thin duralumin. *Proc. Amer. Soc. Test. Mater.*, vol. 27, pt. 2, 1927, pp. 128–133. Abstr.: *Engng News Rec.*, vol. 99, no. 3, July 21, 1927, p. 97.

596. RABOZÉE, H. Influence du traitement thermique et du traitement mécanique sur la résistance de l'acier aux efforts répétés. [Influence of thermal and mechanical treatment on the resistance of steel to repeated stress.] *Congr. int. Ass. Test. Mat., Seventh Congr., Amsterdam*, 1927, vol. 1, pp. 291–296.

597. ROBINSON, T. L. Comparative tests on ball bearing steels. *Trans. Amer. Soc. Steel Treat.*, vol. 11, April 1927, pp. 607–618.

598. SHTEINBERG, S. S. On the fatigue of iron steel and other metals. [R] *Vest. metalloprom.* no. 11, 1927, pp. 80–92, and no. 12, 1927 pp. 6–13.

599. SIMINSKI, I. On the fatigue of timber i repeated loading. [R] *Vest. inzh. Tekh.*, no. 4 April 1927, pp. 137–140.

600. TAPSELL, H. J. and CLENSHAW, W. J. Properties of materials at high temperature *Dep. sci. industr. Res., Lond., Engng Res Spec. Rep.* no. 1, 1927, pp. 60, and no. 1927, pp. 16.

601. TOMLINSON, G. A. Rusting of steel surfac in contact. *Proc. roy. Soc., (A)*, vol. 1 1927, pp. 472–483.

602. TOWNSEND, J. R. Fatigue studies of te phone cable sheath alloys. *Proc. Amer. S*

Test. Mater., vol. 27, pt. 2, 1927, pp. 153–172. Abstr.: *Engng News Rec.*, vol. 99, no. 3, July 21, 1927, pp. 96–97.

603. WHITTEMORE, H. L. Suggested program for an investigation of the fatigue resistance of welds. *J. Amer. Weld. Soc.*, vol. 6, no. 1, Jan. 1927, pp. 21–24.

604. The fatigue of metals. *Machinery, Lond.*, vol. 29, no. 747, Feb. 3, 1927, pp. 596–597.

605. Repeated-bending testing machine. *Engineering, Lond.*, vol. 123, no. 3188, Feb. 18, 1927, pp. 212–214.

606. Fatigue fracture in welded parts. *Metallurgist*, vol. 3, April 29, 1927, p. 64.

607. Determination of the fatigue hardness of metals. *Engng Prog.*, vol. 8, no. 5, May 1927, pp. 131–132.

608. Elastic hysteresis. *Metallurgist*, vol. 3, July 29, 1927, pp. 105–107.

609. Fatigue failure of railway materials. *Metallurgist*, vol. 3, Aug. 26, 1927, pp. 123–124.

610. Corrosion fatigue. *Metallurgist*, vol. 3, Sept. 30, 1927, p. 131.

611. Fatigue cracks in axles. *Metallurgist*, vol. 3, Dec. 30, 1927, pp. 181–183.

1928

612. BATSON, R. G. *and* BRADLEY, J. Researches on springs. 5.—The effect of "nip" on the

mechanical properties of laminated springs. *Dep. sci. industr. Res., Lond., Engng Res., Spec. Rep.* no. 11, Sept. 1928, pp. 38.

613. BATSON, R. G. *and* BRADLEY, J. Researches on springs. 6.—Static and endurance tests of laminated springs made of carbon and alloy steels. *Dep. sci. industr. Res., Lond., Engng Res., Spec. Rep.* no. 13, Nov. 1928, pp. 33.

614. BECKER, E. *and* FÖPPL, O. Dauerversuche zur Bestimmung der Festigkeitseigenschaften, Beziehungen zwischen Baustoffdämpfung und Verformungsgeschwindigkeit. [Fatigue tests to determine strength properties and the relation between material damping properties and rate of deformation.] *ForschArb. IngWes.*, no. 304, 1928, pp. 28. English translation: *Roy. Aircr. Estab., Lib. Trans.*, no. 627, March 1957, pp. 55.

615. BECKINSALE, S. *and* WATERHOUSE, H. The deterioration of lead cable sheathing by cracking and its prevention. *J. Inst. Met.*, vol. 39, no. 1, 1928, pp. 375–406. *Engineering, Lond.*, vol. 125, March 9, 1928, pp. 299–300 and March 16, 1928, pp. 334–336.

616. BOHUSZEWICZ, O. *and* SPÄTH, W. Die Schnellbestimmung der Dauerwechselfestigkeit. [The rapid determination of fatigue strength.] *Arch. Eisenhüttenw.*, vol. 2, no. 4, Oct. 1928, p. 249–255. *Stahl u. Eisen*, vol. 48, no. 43, Oct. 25, 1928, pp. 1505–1506.

617. BRENNER, P. Über die dynamische Festigkeit von Flugzeug-Konstruktionsteilen. [On the dynamic strength of aircraft constructional materials.] *Z. Ver. dtsch. Ing.*, vol. 72, no. 40, Oct. 6, 1928, p. 1408. English abstr.: *Metallurgist*, vol. 5, April 26, 1929, pp. 52–53.

618. CZOCHRALSKI, J. Allgemeines zur Frage des Dauerbruches. [General survey of the

problem of fatigue fracture.] *Z. Metallk.*, vol. 20, no. 2, Feb. 1928, pp. 37–39.

619. CZOCHRALSKI, J. *and* HENKEL, E. Welche Veränderungen erleiden die mechanischen Eigenschaften durch Ermüdung? [What changes occur in mechanical properties under fatigue?] *Z. Metallk.*, vol. 20, no. 2, Feb. 1928, pp. 58–63.

620. DEUTSCH, W. *and* FIEK, G. Dauerprüfmaschinen. [Fatigue testing machines.] *Z. Ver. dtsch. Ing.*, vol. 72, no. 48, Dec. 1, 1928, pp. 1760–1764.

621. DIXON, S. M.; HOGAN, M. A. *and* ROBERTSON, J. M. The deterioration of colliery winding ropes in service, with descriptions of some typical failures. *Pap. Saf. Min. Res. Bd. Lond.*, no. 50, 1928, pp. 42.

622. FISCHER, M. F. Note on the effect of repeated stresses on the magnetic properties of steel. *J. Res. nat. Bur. Stand.*, vol. 1, no. 5. Nov. 1928, pp. 721–732.

623. FÖPPL, O. Bestimmung der Werkstoffdämpfung mittels der Verdrehungs-ausschwingmaschine. [Determination of material damping using the torsion fatigue machine.] *Z. Ver. dtsch. Ing.*, vol. 72, no. 37, Sept. 15, 1928, pp. 1293–1296.

624. FREEMAN, J. R. Fatigue resistance of rail steel. *Iron Age*, vol. 121, June 21, 1928, pp. 1743–1745.

625. FREEMAN, J. R.; DOWDELL, R. L. *and* BERRY, W. J. Endurance and other properties of rail steel. *Tech. Pap. U.S. Bur. Stand.*, vol. 22, no. 363, Jan. 12, 1928, pp. 269–365. Abstr.: *Metallurgist*, vol. 4, Aug. 31, 1928, pp. 122–123.

626. GOUGH, H. J. Note on some fatigue phenomena with special relation to cohesion problems. *Trans. Faraday Soc.*, vol. 24, March 1928, pp. 137–148.

627. GOUGH, H. J. The behaviour of a single crystal of alpha iron subjected to alternating torsional stresses. *Proc. roy. Soc.*, *(A)*, vol. 118, April, 2, 1928, pp. 498–534.

628. GOUGH, H. J. Fatigue phenomena. *Proc. Instn Auto. Engrs*, vol. 23, 1928–1929, pp. 341–352.

629. GOUGH, H. J. Fatigue phenomena, with special reference to single crystals. *J. R. Soc. Arts*, vol. 76, no. 3955, Sept. 7, 1928, pp. 1025–1044; no. 3956, Sept. 14, 1928, pp. 1045–1062; no. 3957, Sept. 21, 1928, pp. 1065–1081; no. 3958, Sept. 28, 1928, pp. 1085–1114 and no. 3959, Oct. 5, 1928, pp. 1117–1142. Abstr.: *Engineering, Lond.*, vol. 125, Feb. 17, 1928, pp. 200–201; Feb. 24, 1928, pp. 232–233 and March 2, 1928, p. 264.

630. GOUGH, H. J. *and* COX, H. L. The behaviour of a single crystal of zinc subjected to alternating torsional stresses. *Rep. Memor. aero. Res. Comm., Lond.*, no. 1183, July 1928, pp. 23.

631. GOUGH, H. J. *and* MURPHY, H. J. The causes of failure of wrought iron chains. *Dep. sci. industr. Res., Lond., Engng Res., Spec. Rep.*, no. 3, 1928, pp. 167.

632. HAIGH, B. P. Hysteresis in relation to cohesion and fatigue. *Trans. Faraday Soc.*, vol. 24, March 1928, pp. 125–137. Abstr.: *Engineering, Lond.*, vol. 125, March 9, 1928, p. 295.

633. HANKINS, G. A. The endurance of spring steel plates under repetition of reversed bend-

ing stress. *Dep. sci. industr. Res., Lond., Engng Res., Spec. Rep.*, no. 5, 1928, pp. 26. Abstr.: *Engineering, Lond.*, vol. 126, no. 3268, Aug. 31, 1928, p. 273.

634. HANKINS, G. A. Researches on springs. 3.—Torsional fatigue tests on spring steels. *Dep. sci. industr. Res., Lond., Engng Res., Spec. Rep.*, no. 9, April 1928, pp. 24.

635. HARDER, O. E. Why some drill rod steels fail. *Iron Age*, vol. 121, no. 8, Feb. 23, 1928, pp. 532–534.

636. HERBST, H. Ansprüche an Förderseile und ihre Prüfung. [Requirements and testing of mine ropes.] *Z. Ver. dtsch. Ing.*, vol. 72, no. 10, March 10, 1928, pp. 345–349. English translation: *Henry Brutcher Tech. Trans.*, no. 344.

637. HEROLD, W. Dauerbeanspruchung, Gefüge und Dämpfung. [Fatigue stress, structure and damping.] *Arch. Eisenhüttenw.*, vol. 2, no. 1, July 1928, pp. 23–39. Abstr.: *Stahl u. Eisen*, vol. 48, no. 31, Aug. 2, 1928, pp. 1051–1052.

638. HORT, W. Dauerbruch als dynamische und schwingungstechnische Erscheinung. [Fatigue failure as a dynamic and vibration phenomenon.] *Z. Metallk.*, vol. 20, no. 2, Feb. 1928, pp. 40–44.

639. IKEDA, S. A rapid method of determining endurance limit by means of measuring electrical resistance. *J. Soc. mech. Engrs, Japan*, vol. 31, no. 136, Aug. 1928, pp. 447–476. Abstr.: *Iron Age*, vol. 123, June 6, 1929, p. 1572.

640. JENKIN, C. F. *and* LEHMANN, G. D. High frequency fatigue. *Rep. Memor. aero. Res. Comm., Lond.*, no. 1222, Dec. 1928, pp. 34.

641. KAUL, L. Molekularstruktur des Eisens und Ermüdungsproben bei Dampfkesseln. [Molecular structure and fatigue tests of steam boilers.] *Metallbörse*, vol. 18, 1928, pp. 932–933.

642. KOMMERS, J. B. The static and fatigue properties of some cast irons. *Proc. Amer. Soc. Test. Mater.*, vol. 28, pt. 2, 1928, pp. 174–204.

643. KUNTZE, W. Statische Grundlagen zum Schwingungsbruch. [Static principles for fatigue failure.] *Z. Ver. dtsch. Ing.*, vol. 72, no. 42, Oct. 20, 1928, pp. 1488–1492.

644. KUNTZE, W.; SACHS, G. *and* SIEGLER-SCHMIDT, H. Elastizität, statische Versuche und Dauerprüfung. [Elasticity, static and fatigue tests.] *Z. Metallk.*, vol. 20, no. 2, Feb. 1928, pp. 64–68.

645. LÁSZLÓ, F. Die Kerbe. [The notch.] *Z. Ver. dtsch. Ing.*, vol. 72, no. 24, June 16, 1928, pp. 851–856.

646. LAUTE, K. *and* SACHS, G. Was ist Ermüdung? [What is fatigue?] *Z. Ver. dtsch. Ing.*, vol. 72, no. 34, Aug. 25, 1928, pp. 1188–1190. English abstr.: *Metallurgist*, vol. 4, Nov. 30, 1928, pp. 173–174.

647. LEA, F. C. *and* BATEY, R. A. The properties of cold-drawn wires, with particular reference to repeated torsional stresses. *Proc. Instn mech. Engrs, Lond.*, pt. 2, Dec. 1928, pp. 865–899.

648. LECOEUVRE, R. *and* CAZAUD, R. Application de la micrographie à l'étude des phénomènes de fatigue dans le duralumin. [Microscopic studies of the phenomenon of fatigue in duralumin.] *Bull. tech. Serv. Aérotech., Paris*, no. 52, 1928, pp. 34–41.

649. LEHR, E. Oberflächenempfindlichkeit und innere Arbeitsaufnahme der Werkstoffe bei Schwingungsbeanspruchung. [Surface sensitivity and internal absorption of energy of materials during the endurance test.] Z. Metallk., vol. 20, no. 2, Feb. 1928, pp. 78–90.

650. LJUNGBERG, K. Konstant Brottarbete som Förklaring till Brott Genom Utmattnings- och andra Belastningar. [Constant energy to rupture in relation to fracture by fatigue and other loadings.] Tekn. Tidskr., Stockh., vol. 58, no. 44, Nov. 3, 1928, pp. 409–411 and no. 45, Nov. 10, 1928, pp. 418–420.

651. LUDWIK, P. Bruchgefahr und Material-prüfung. [Dangers of fracture, and materials testing.] Ber. eidgenöss. MatPrüfAnst., no. 35, 1928, pp. 38.

652. LYON, A. J. Etching of aluminium alloy propellers. Aviation, N.Y., vol. 24, no. 21, May 21, 1928, pp. 1456–1457 and pp. 1475–1478.

653. McADAM, D. J. Some factors involved in corrosion and corrosion-fatigue of metals. Proc. Amer. Soc. Test. Mater., vol. 28, pt. 2, 1928, pp. 117–158.

54. McADAM, D. J. Corrosion of metals as affected by time and by cyclic stress. Trans. Amer. Inst. min. (metall.) Engrs, vol. 78, 1928, pp. 571–615.

655. MATSUSHITA, T.; NAGASAWA, K. and KOMATSU, J. On the fatigue of steels. [J] Tetsu to Hagane, vol. 14, 1928, pp. 985–996.

656. MELCHIOR, P. Dauerbruch. [Fatigue fracture.] Z. Ver. dtsch Ing., vol. 72, no. 16, April 21, 1928, pp. 537–540.

657. MOORE, H. F. Endurance limit. J. Engng Educ., vol. 18, no. 6, Feb. 1928, pp. 574–575.

658. MOORE, H. F. Repeated stress endurance of metals. S.A.E. Jl., vol. 22, no. 2, Feb. 1928, pp. 289–291.

659. MOORE, H. F. "Tired" steel. Engrs ana Engng, vol. 45, no. 8, Aug. 1928, pp. 171–178.

660. MOORE, H. F. Gleitung, Bruch und Er-müdung von Metallen. [Slip, fracture and fatigue of metals.] Metallwirtschaft, vol. 7, no. 47, Nov. 23, 1928, pp. 1272–1273.

661. MOORE, H. F. What happens in couplings; fatigue of materials; what it is and how it occurs in flexible couplings. Iron Steel Engr, vol. 5, Dec. 1928, pp. 513–517.

662. MOORE, H. F. and ALLEMAN, N. J. Fatigue tests of carburized steel. Trans. Amer. Soc. Steel Treat., vol. 13, no. 3, March 1928, pp. 405–419.

663. MOORE, H. F. and HOWARD, F. C. A metallographic study of the path of fatigue failure in copper. Uni. Ill. Engng Exp. Sta. Bull., no. 176, May 1928, pp. 27. Metal Ind., Lond., vol. 32, no. 24, June 15, 1928, pp. 589–592.

664. MOORE, H. F. and LYON, S. W. Fatigue tests of cast iron. Canad. Foundrym., vol. 19, no. 2, Feb. 1928, pp. 11–12, 36.

665. MOORE, H. F.; LYON, S. W. and ALLEMAN, N. J. Tests of the fatigue strength of steam turbine blade shapes. Uni. Ill. Engng Exp. Sta Bull., no. 183, Oct. 1928, pp. 36.

666. NAGASAWA, Y. A proposed method for determining the fatigue limit. [J] J. Soc. mech Engrs, Japan, vol. 31, no. 135, July 1928 pp. 259–274.

667. NEEDHAM, W. R. Fatigue of mild steel. *Machinery, Lond.*, vol. 32, no. 820, June 28, 1928, p. 405.

668. NEKRYTYI, S. S. *and* BOKOV, I. I. Fatigue of light aluminium casting alloys subjected to repeatedly reversed stresses. [R] *Vest. inzh. Tekh.* no. 5, May 1928, pp. 245–251 and no. 6, June 1928, pp. 283–289. German abstr.: *Z. Metallk.*, vol. 22, no. 7, July 1930, pp. 247–249. English abstr.: *Trans. Amer. Soc. Steel Treat.*, vol. 14, Dec. 1928, p. 941.

669. RUDELOFF, M. Die Prüfung der Festigkeitseigenschaften metallischer Baustoffe auf der Werkstoffschau. [The testing of the mechanical properties of metallic constructional materials.] *Giesserei*, vol. 15, no. 9, March 2, 1928, pp. 196–200; no. 10, March 9, 1928, pp. 217–225; no. 11, March 16, 1928, pp. 237–245; no. 12, March 23, 1928, pp. 263–272 and no. 13, March 30, 1928, pp. 289–297.

670. SCHMID, E. Ermüdung vom Standpunkt der Vorgänge im Einkristall. [Fatigue from the viewpoint of the behaviour of single crystals.] *Z. Metallk.*, vol. 20, no. 2, Feb. 1928, pp. 69–77.

671. SCOBLE, W. A. Third report of the wire ropes research committee. *Proc. Instn mech. Engrs, Lond.*, April 1928, pp. 353–404. Abstr.: *Engineering, Lond.*, vol. 125, April 27, 1928, pp. 522–525.

672. SPELLER, F. N.; McCORKLE, I. B. *and* MUMMA, P. F. Influence of corrosion accelerators and inhibitors on the fatigue of ferrous metals. *Proc. Amer. Soc. Test. Mater.*, vol. 28, pt. 2, 1928, pp. 159–173.

673. SPINDLER, H. Genietete oder geschweisste Kraftwagenrahmen. [Riveted or welded automobile structures.] *Motorwagen*, vol. 31, no. 6, Feb. 29, 1928, pp. 102–105.

674. SUZUKI, M. Breaking of metal by fatigue. [J] *Bull. Res. Off. Govt Rlys, Jap.*, vol. 16, no. 8, 1928, pp. 1177–1201.

675. TAPSELL, H. J. The fatigue resisting properties of 0.17 percent carbon steel at different temperatures and at different mean tensile stresses. *J. Iron. St. Inst.*, vol. 117, no. 1, 1928, pp. 275–294. Abstr.: *Engineering, Lond.*, vol. 125, May 4, 1928, pp. 557–558.

676. TAPSELL, H. J.; ARCHBUTT, S. L. *and* JENKIN, J. W. Mechanical properties of pure magnesium and certain magnesium alloys in the wrought condition. Mechanical properties of an "Electron" alloy. *Rep. Memor. aero. Res. Comm., Lond.*, no. 1285, Feb. 1928, pp. 9.

677. TAYLOR, G. I. *and* GOUGH, H. J. Note on experiment aimed at determining whether reversed slipping occurs on plane of slip of single crystal of aluminium under alternating stresses. *Rep. aero. Res. Comm., Lond.*, no. E. F. 206, Jan. 1928, pp. 16.

678. URQUHART, J. W. Safe stress limit for forged steel. *Heat Treat. Forg.*, vol. 14, no. 11, 1928, pp. 1282–1284.

679. WAGNER, R. Die Bestimmung der Dauerfestigkeit der knetbaren, veredelbaren Leichtmetallegierungen. [The determination of the fatigue strength of wrought, age-hardening light alloys.] Berlin, Verlag Julius Springer, 1928, pp. 64.

680. YOUNGER, J. E. Preliminary study of fatigue failure of metal propellers caused by engine impulses and vibration. *Inform. Circ. U.S. Air Corp.*, vol. 7, no. 618, Aug. 1928, pp. 17.

681. ZANDER, W. Der Einfluss von Oberflächen-beschädigungen auf die Biegungsschwingungs-festigkeit. [The influence of surface damage on the bending fatigue strength.] *Veröff. Wöhler-Inst.*, no. 1, 1928, pp. 65.

682. Corrosion fatigue. *Metallurgist*, vol. 4, April 27, 1928, pp. 62–63.

683. Strength and endurance of cold drawn mild steel. *Metallurgist*, vol. 4, May 25, 1928, pp. 75–76.

684. The endurance properties of metals. *Engineering, Lond.*, vol. 126, no. 3279, Nov. 16, 1928, pp. 636–637.

685. Fatigue failures of large engine shafts. *Iron Age*, vol. 122, no. 22, Nov. 29, 1928, p. 1362.

686. Determination of endurance limits. *Metallurgist*, vol. 4, Dec. 28, 1928, pp. 187–189.

1929

687. ADAM, A. T. Notes on wire for mining ropes. *J. Iron St. Inst.*, vol. 120, no. 2, 1929, pp. 27–67.

688. ADERS, K. Einfluss des Alterns auf das Verhalten weichen Stahles bei Schwingungs-beanspruchungen. [Influence of ageing on the fatigue strength of mild steel.] *Mitt. ForschInst. ver. Stahlw. A.-G.*, vol. 1, no. 9, 1929, pp. 201–221.

689. BERNHARD, R. Dauerversuche an geniete-ten und geschweissten Brücken. [Fatigue tests on riveted and welded bridges.] *Z. Ver. dtsch. Ing.*, vol. 73, no. 47, Nov. 23, 1929, pp. 1675–1680. English abstr.: *Metallurgist*, vol. 6, Feb. 28, 1930, pp. 23–24.

690. BINNIE, A. M. The influence of oxygen on corrosion fatigue. *Rep. Memor. aero. Res.*

Comm., Lond., no. 1244, March 1929, pp. 3. Abstr.: *Engineering, Lond.*, vol. 128, Aug. 16, 1929, pp. 190–191.

691. BRADLEY, J. A high-speed endurance testing machine for leaf springs. *Engineering, Lond.*, vol. 127, no. 3287, Jan. 11, 1929, pp. 36–37.

692. BRENNER, P. Dynamische Festigkeit von Flugzeugkonstruktionsteilen. [Dynamic strength of aircraft structural members.] *Jb. 1929 dtsch. VersAnst. Luftf.*, pp. 149–155. *Luftfahrtforsch.*, vol. 3, no. 3, 1929, pp. 59–65.

693. CAZAUD, R. Le procéde d'élaboration de l'acier et les caractéristiques aux efforts alter-nés. [The process of steel-making and the characteristics of resistance to alternating stresses.] *Aciers spéc.*, vol. 4, no. 41, Jan. 1929, pp. 14–22.

694. CAZAUD, R. La fatigue des métaux. [The fatigue of metals.] *Aciers spéc.*, vol. 4, no. 47, July 1929, pp. 322–331; no. 48, Aug. 1929, pp. 361–370 and no. 49, Sept. 1929, pp. 423–428.

695. DORGERLOH, E. Eine neue Prüfungsmaschine zur Untersuchung der Werkstoffe bei wech-selnden, oftmals wiederholten Biegebean-spruchungen. [A new testing machine for examining materials under alternating, fre-quently repeated bending stresses.] *Metall-wirtschaft*, vol. 8, no. 41, Oct. 7, 1929, pp. 986–990.

696. DURRER, . Konstante Brucharbeit als Erklärung für den Bruch durch Ermüdungs-und andere Belastungen. [Constant energy to fracture as an explanation of fracture by fatigue and other loadings.] *Z. Ver. dtsch. Ing.*, vol. 73, no. 24, June 15, 1929, pp. 830–832.

697. ECKARDT, H. Dauerbeanspruchung von Stahl bei erhöhter Temperatur. [Repeated stressing of steel at elevated temperature.] *Veröffentlichungen des Zentralverbandes der Preussischen Dampfkessel-Überwachungsvereine, Halle*, vol. 7, 1929, pp. 5–71.

698. EWING, D. D. What happens when steel gets tired. *Elect. Rly J.*, vol. 73, no. 17, Aug. 1929, pp. 775–777.

699. EWING, D. D. Fatigue in concrete; an element to be taken into account in design. *Elect. Rly J.*, vol. 73, no. 18, Sept. 1929, pp. 829–830.

700. FÖPPL, O. Das Drücken der Oberflache von Bauteilen aus Stahl. [Compression of the surface of steel structural members.] *Stahl u. Eisen*, vol. 49, no. 17, April 25, 1929, pp. 575–577. English abstr.: *Metallurgist*, vol. 5, May 31, 1929, pp. 70–71.

701. FÖPPL, O. Die Steigerung der Dauerhaltbarkeit durch Oberflächendrücken. [Increasing the fatigue strength by surface pressure.] *Maschinenbau*, vol. 8, no. 22, Nov. 21, 1929, pp. 752–755.

702. FÖPPL, O.; BECKER, E. *and* VON HEYDEKAMPF, G. S. *Die Dauerprüfung der Werkstoffe hinsichtlich ihrer Schwingungsfestigkeit und Dämpfungsfähigkeit. [The endurance testing of materials with regard to their fatigue strength and damping capacity.]* Berlin, Julius Springer, 1929, pp. 124.

703. FÖPPL, O. *and* VON HEYDEKAMPF, G. S. Dauerfestigkeit und Konstruktion. [Fatigue strength and construction.] *Metallwirtschaft*, vol. 8, no. 45, Nov. 8, 1929, pp. 1087–1091.

704. FRANKE, E. Dauerbrüche an Schiffswellen. [Fatigue fractures of the shafts of ships.] *Schiffbau*, vol. 30, no. 5, 1929, pp. 112–113.

705. FRANKE, E. Über Dauerbrüche an Maschinenteilen und geeignete Prüfverfahren zur Bestimmung der Dauerfestigkeit. [The fatigue failure of machine parts and testing methods for determining their fatigue strength.] *Auto.-tech. Z.*, vol. 32, no. 29, Oct. 20, 1929, pp. 643–645 and no. 30, Oct. 31, 1929, pp. 676–678.

706. FREEMAN, J. R. *and* SOLAKIAN, H. N. Effect of service on the endurance properties of rail steels. *J. Res. nat. Bur. Stand.*, vol. 3, no. 2, Aug. 1929, pp. 205–246.

707. FULLER, T. S. Some aspects of corrosion-fatigue. *Trans. Amer. Inst. min. (metall.) Engrs*, vol. 83, 1929, pp. 47–55.

708. GERDIEN, H. Über einen neuen Apparat zur Untersuchung von Dauerbiegeschwingungen. [A new apparatus for testing under repeated bending.] *Z. tech. Phys.*, vol. 10, no. 9, 1929, pp. 389–392.

709. GILLETT, H. W. So-called "accelerated" endurance testing of metal with a review of Ikeda's method. *Metals and Alloys*, vol. 1, no. 1, July 1929, pp. 19–21.

710. GILLETT, H. W. Have heat-treated aluminium alloys a true endurance limit? *Metals and Alloys*, vol. 1, no. 1, July 1929, pp. 21–22, 37.

711. GILLETT, H. W. The need for information on notch propagation. *Metals and Alloys*, vol. 1, no. 3, Sept. 1929, pp. 114–116.

712. GOUGH, H. J. *and* COX, H. L. The behaviour of a single crystal of zinc subjected to alternating torsional stresses. *Proc. roy. Soc.*, *(A)*, vol. 123, 1929, pp. 143–167.

713. GOUGH, H. J. and COX, H. L. Further experiments on the behaviour of single crystals of zinc subjected to alternating torsional stresses. *Rep. Memor. aero. Res. Comm., Lond.*, no. 1322, Aug. 1929, pp. 18.

714. GOUGH, H. J. and COX, H. L. The behaviour of a single crystal of antimony subjected to alternating torsional stresses. *Rep. Memor. aero. Res. Comm., Lond.*, no. 1323, Nov. 1929, pp. 18.

715. GRAF, O. *Die Dauerfestigkeit der Werkstoffe und Konstruktionselemente. [The fatigue strength of materials and structural parts.]* Berlin, Julius Springer, 1929, pp. 131.

716. GREAVES, R. H. Repeated blow impact tests. *Metallurgist*, vol. 5, May 31, 1929, pp. 67–68 and June 28, 1929, pp. 88–90.

717. GÜNTHER, K. Der Einfluss von Oberflächenbeschädigungen auf die Biegungsschwingungsfestigkeit. [The influence of surface damage on the bending fatigue strength.] *Veröff. Wöhler-Inst.*, no. 2, 1929, pp. 68. *Metallwirtschaft*, vol. 8, no. 50, Dec. 13, 1929, pp. 1220–1224; no. 52, Dec. 27, 1929, pp. 1267–1271 and vol. 9, no. 2, Jan. 10, 1930, pp. 35–39.

718. HAIGH, B. P. The relative safeties of mild and high tensile alloyed steels under alternating and pulsating stresses. *J. Soc. chem. Ind., Lond.*, vol. 48, no. 2, Jan. 11, 1929, pp. 23–30. Abstr.: *Mech. Engng, N.Y.*, vol. 51, no. 5, May 1929, pp. 374–375.

719. HAIGH, B. P. The relative safety of mild and high tensile alloy steels under alternating and pulsating stresses. *Proc. Instn Auto. Engrs.*, vol. 24, 1929–1930, pp. 320–362.

720. HAIGH, B. P. Chemical action in relation to fatigue in metals. *Trans. Instn chem. Engrs, Lond.*, vol. 7, 1929, pp. 29–48.

721. HANKINS, G. A. and FORD, G. W. The mechanical and metallurgical properties of spring steels as revealed by laboratory tests. *J. Iron St. Inst.*, vol. 119, no. 1, 1929, pp. 217–253.

722. HERBST, H. Dauerbiegeversuche mit Drahtseilen. [Bending fatigue tests on wire ropes.] *Z. Ver. dtsch. Ing.*, vol. 73, no. 45, Nov. 9, 1929, p. 1623. English translation: *Henry Brutcher Tech. Trans.*, no. 429.

723. HEROLD, W. Über die Beziehungen der Dauerbiegefestigkeit zu den statischen Festigkeitswerten. [Relation between bending fatigue strength and static strength.] *Z. Ver. dtsch. Ing.*, vol. 73, no. 36, Sept. 7, 1929, pp. 1261–1266.

724. VON HEYDEKAMPF, G. S. Eine Dauerbiegemaschine mit schwingendem, in der Messstrecke gleichmässig beanspruchtem Probestab. [A plane bending fatigue testing machine for determining the proportions of test pieces.] *Veröff. Wöhler-Inst.*, no. 3, 1929, pp. 40.

725. VON HOUDREMONT, E. and MAILÄNDER, R. Dauerbiegeversuche mit Stählen. [Bending fatigue tests on steels.] *Stahl u. Eisen*, vol. 49, no. 23, June 6, 1929, pp. 833–839. *Krupp. Mh.*, vol. 10, no. 4/5, April–May 1929, pp. 39–49.

726. IKEDA, S. Rapid determination of the endurance limit by measuring the electrical resistance. *Technol. Rep. Tohoku Univ.*, vol. 8, no. 2, 1929, pp. 41–70.

727. IKEDA, S. A method of determining endurance limit by means of measuring electrical resistance. *Mech. World*, vol. 86, no. 2225, Aug. 23, 1929, pp. 178–181 and no. 2226, Aug. 30, 1929, pp. 192–195.

728. IKEDA, S. A rapid method of determining endurance limit. [J] *Kinzoku no kenkyu*, vol. 6, no. 11, 1929, pp. 475–492.

729. JENKIN, C. F. *and* LEHMANN, G. D. High frequency fatigue. *Proc. roy. Soc.*, (*A*), vol. 125, no. 796, Aug. 1, 1929, pp. 83–119.

730. JOHNSON, J. B. *and* OBERG, T. T. Fatigue resistance of some aluminium alloys. *Proc. Amer. Soc. Test. Mater.*, vol. 29, pt. 2, 1929, pp. 339–352. Abstr.: *Metallurgist*, vol. 6, Sept. 26, 1930, pp. 137–139.

731. KÖHLER, W. Einfluss der Wärmebehandlung auf die Schwingungsfestigkeit untereutektoider Stähle. [Influence of heat treatment on the fatigue strength of hypoeutectoid steel.] *Walzwerk u. Hütte*, no. 1, 1929, pp. 9–15 and no. 4, 1929, pp. 50–52, 54–56.

732. KOMMERS, J. B. The relative safeties of mild and high strength steels under fatigue stresses. *Chem. and Ind.* (*Rev.*), vol. 48, no. 51, Dec. 20, 1929, pp. 1223–1226.

733. KOMMERS, J. B. The fatigue properties of cast iron. *Proc. Amer. Soc. Test. Mater.*, vol. 29, pt. 2, 1929, pp. 100–108.

734. KÜHLE, A. Einfluss des Alterns und Blaubruchs auf die Dauerschlagprobe. [Influence of ageing and blue fracture on the impact fatigue test.] *Mitt. ForschInst. ver. Stahlw. A.-G.*, vol. 1, no. 4, 1929, pp. 83–102.

735. LAUTE, K. Schwingungsfestigkeit von Metallen — nach Arbeiten von McAdam. [Fatigue strength of metals—review of the work of McAdam.] *Z. Metallk.*, vol. 21, no. 5, May 1929, pp. 174–178.

736. LAUTE, K. *and* SACHS, G. Was ist Ermüdung? [What is fatigue?] *Mitt. dtsch. Mat-PrüfAnst.*, no. 9, 1929, pp. 89–91.

737. LEA, F. C. The penetration of hydrogen into metal cathodes and its effect upon the tensile properties of metals and their resistance to repeated stresses; with a note on the effect of non-electrolytic baths and nickel plating on these properties. *Proc. roy. Soc.*, (*A*), vol. 123, 1929, pp. 171–185.

738. LUDWIK, P. Schwingungsfestigkeit. [Fatigue strength.] *Z. öst. Ing.- u. ArchitVer.*, vol. 81, no. 41/42, Oct. 11, 1929, pp. 403–406.

739. LUDWIK, P. Dauerversuche an Werkstoffen. [Fatigue testing of materials]. *Z. Ver. dtsch. Ing.*, vol. 73, no. 51, Dec. 21, 1929, pp. 1801–1810.

740. LUDWIK, P. *and* SCHEU, R. Dauerversuche mit Metallen. [Fatigue tests of metals.] *Metallwirtschaft*, vol. 8, no. 1, Jan. 4, 1929, pp. 1–5.

741. MCADAM, D. J. Fatigue and corrosion-fatigue of spring material. *Trans. Amer. Soc. mech. Engrs.*, vol. 51, pt. 1, APM-51-5, 1929, pp. 45–58. Abstr.: *Mech. Engng, N.Y.*, vol. 50, no. 11, Nov. 1928, p. 888. *Amer. Mach.*, *N.Y.*, vol. 69, no. 25, Jan. 26, 1929, pp. 966–967.

742. MCADAM, D. J. Corrosion of metals under cyclic stress. *Proc. Amer. Soc. Test. Mater.*, vol. 29, pt. 2, 1929, pp. 250–313.

743. MCADAM, D. J. Corrosion of metals as affected by stress, time, and number of cycles. *Trans. Amer. Inst. min.* (*metall.*) *Engrs*, vol. 83, 1929, pp. 56–110.

744. MACGREGOR, R. A. The failure of steel castings and forgings through fatigue. *Trans. N.-E. Cst Instn Engrs Shipb.*, vol. 46, 1929–1930, pp. 221–238.

745. MATTHAES, K. Prüfung des Festigkeitseigenschaften von Alclad-Blechen. [Testing the strength properties of Alclad sheet.] *Luftfahrtforsch.*, vol. 3, May 29, 1929, pp. 153–160. Abstr.: *Z. Metallk.*, vol. 21, no. 11, Nov. 1929, pp. 394–395. English abstr.: *Metals and Alloys*, vol. 1, no. 3, Sept. 1929, p. 122.

746. MEYER, FR. Werkstoffprüfung. [Materials testing.] *Siemens-Z.*, vol. 9, no. 3, March 1929, pp. 175–177.

747. MOORE, H. F. Elastic failure and fatigue failure of metals. *Mech. Engng, N.Y.*, vol. 51, no. 4, April 1929, pp. 290–294.

748. MOORE, H. F. Fatigue tests on drums and shells. *J. Amer. Weld. Soc.*, vol. 8, no. 10, Oct., 1929, pp. 42–52. *Iron Age*, vol. 124, no. 10, Sept. 5, 1929, pp. 607–611.

749. MOORE, H. F. Crystallisation is not cause of fatigue failure in metals. *Iron Tr. Rev.*, vol. 85, no. 18, Oct. 31, 1929, pp. 1098–1099, 1151.

750. MOORE, H. F. Fatigue of metals—a review. *Iron Age*, vol. 124, no. 18, Oct. 31, 1929, pp. 1169–1170.

751. MOORE, H. F. Studies on the fatigue of metals. *Heat Treat. Forg.*, vol. 15, no. 11, Nov. 1929, pp. 1443–1450.

752. MOORE, H. F. Fatigue characteristics of cold worked metals. *Metal Stamp.*, vol. 2, no. 11, Nov. 1929, pp. 875–876, 900.

753. MOORE, H. F. Study of fatigue of metals yields data for machine design. *Mach. Design*, vol. 1, no. 3, Nov. 1929, pp. 35–38.

754. MOORE, H. F. The fatigue of metals—a review of progress from 1920 to 1929. *Amer-ican Iron and Steel Institute, Year Book* 1929, pp. 304–336. German abstr.: *Stahl u. Eisen*, vol. 50, no. 4, Jan. 23, 1930, pp. 111–112.

755. MOORE, H. F.; LYON, S. W. *and* ALLEMAN, N. J. A study of fatigue cracks in car axles— part 2. *Uni. Ill. Engng Exp. Sta. Bull.*, no. 197, Nov. 1929, pp. 27.

756. MUNDT, R. Ermüdungsbruch und zulässige Belastung von Wälzquerlagern. [Fatigue fractures and allowable loads on roller bearings. *Z. Ver. dtsch. Ing.*, vol. 73, no. 2, Jan. 12 1929, pp. 53–59.

757. NEEDHAM, W. R. Fatigue tests of cast iron *Iron Steel Ind.*, vol. 2, no. 4, Jan. 1929 pp. 101–102.

758. ONO, A. Some results of fatigue tests of metals. *J. Soc. mech. Engrs, Japan*, vol. 3; no. 148, Aug. 1929, pp. 331–341.

759. PETERSON, R. E. Fatigue tests of large specimens. *Proc. Amer. Soc. Test. Mater.*, vol. 2 pt. 2, 1929, pp. 371–380. Abstr.: *Heat Trea Forg.*, vol. 15, Aug. 1929, pp. 1009–1010.

760. PROBST, E. Die Rissebildung bei Beto und Eisenbeton-Konstruktionen unter beso derer Berücksichtigung des Einflusses wiede holter Belastungen. [The formation of crac in plain and reinforced concrete structur with special reference to the effect of repeat loading.] *Int. Congr. Br. Struct. Engng, Pr Second Congr.*, Vienna, 1929, pp. 492–505.

761. RAWDON, H. S. Corrosion embrittleme of duralumin. The effect of corrosion acco panied by stress on the tensile properties sheet duralumin. *Tech. Note nat. Adv. Con Aero., Wash.*, no. 305, May 1929, pp. 26.

762. RAWDON, H. S. Effect of corrosion accompanied by stress, on the tensile properties of sheet duralumin. *Proc. Amer. Soc. Test. Mater.*, vol. 29, pt. 2, 1929, pp. 314–338.

763. REAVELL, W. The standardisation of keys and keyways. *Engineering, Lond.*, vol. 127, April 26, 1929, pp. 529–531.

764. ROWELL, H. S. A new high speed fatigue testing machine. *Engineering, Lond.*, vol. 128, no. 3315, July 26, 1929, pp. 97–98.

765. SACHS, G. Die Ermüdung veredelbarer Aluminium-Walzlegierungen. [The fatigue of improvable aluminium wrought alloys.] *Z. Metallk.*, vol. 21, no. 1, Jan. 1929, pp. 27–29.

766. SCHAECHTERLE, K. Beitrag zur Auswertung von Dauerversuchen. [Contribution to the evaluation of fatigue tests.] *Stahlbau*, vol. 2, no. 20, Oct. 4, 1929, pp. 238–240.

767. SCHWINNING, W. Untersuchungen über die Wechselfestigkeit von Freileitungsdrähten. [Investigations on the fatigue strength of wires for overhead transmission lines.] *AluminZ.*, vol. 1, nos. 1/2, April–May 1929, pp. 52–59. Abstr.: *Z. Metallk.*, vol. 21, no. 10, Oct. 1929, p. 347.

768. SMITH, J. H.; CONNOR, C. A. *and* ARMSTRONG, F. H. The correlation of fatigue and overstress. *J. Iron. St. Inst.*, vol. 120, no. 2, 1929, pp. 267–295. *Engineering, Lond.*, vol. 128, Nov. 8, 1929, pp. 605–606 and Nov. 15, 1929, pp. 661–662. Abstr.: *Mech. Engng, N.Y.*, vol. 52, no. 2, Feb., 1930, pp. 159–160.

769. SPÄTH, W. Dynamische Untersuchungen an technischen Gebilden. [Dynamic tests on engineering components.] *Z. Ver. dtsch. Ing.* vol. 73, no. 27, July 9, 1929, pp. 963–965.

770. SPELLER, F. N.; McCORKLE, I. B. *and* MUMMA, P. F. The influence of corrosion accelerators and inhibitors on fatigue of ferrous metals. *Proc. Amer. Soc. Test. Mater.*, vol. 29, pt. 2, 1929, pp. 238–249.

771. THUM, A. *and* UDE, H. Die Elastizität und die Schwingungsfestigkeit des Gusseisens. [The elasticity and the fatigue strength of cast irons.] *Giesserei*, vol. 16, no. 22, May 31, 1929, pp. 501–513 and no. 24, June 14, 1929, pp. 547–556. English abstr.: *Found. Tr. J.*, vol. 42, no. 702, Jan. 30, 1930, pp. 90–91.

772. THUM, A. *and* WISS, W. Über dynamische Verfestigung und Über-lastungsfähigkeit von Stählen. [On the dynamic hardening and overstraining capacity of steels.] *Z. Ver. dtsch. Ing.*, vol. 73, no. 50, Dec. 14, 1929, pp. 1787–1788.

773. TOWNSEND, J. R. *and* GREENALL, C. H. Fatigue studies of non-ferrous sheet metals. *Proc. Amer. Soc. Test. Mater.*, vol. 29, pt. 2, 1929, pp. 353–370. *Bell Syst. tech. J.*, vol. 8, July 1929, pp. 576–590. *Bell Telep. Lab. Reprint*, B-406, Aug. 1929, pp. 15. Abstr.: *Metal Stamp.*, vol. 2, no. 8, Aug. 1929, pp. 599–600.

774. WEWERKA, A. Die Kerbwirkung. [The notch effect.] *Maschinenbau*, vol. 8, no. 2, Jan. 17, 1929, pp. 33–37.

775. WIESENÄCKER, H. Eine neue Dauerbiege-Prüfmaschine für Flachfedern und Bleche. [A new bending fatigue testing machine for flat springs and plates.] *Z. Ver. dtsch. Ing.*, vol. 73, no. 38, Sept. 21, 1929, pp. 1367–1368.

776. ZANDER, W. Der Einfluss von Oberflächenbeschädigungen auf die Biegungsschwingungsfestigkeit. [The influence of surface damage on the bending fatigue strength.] *Metallwirtschaft*,

vol. 8, no. 2, Jan. 11, 1929, pp. 29–32; no. 3, Jan. 18, 1929, pp. 53–57 and no. 4, Jan. 25, 1929, pp. 77–80.

777. Corrosion-fatigue of metals. *Metallurgist*, vol. 5, Jan. 25, 1929, pp. 3–6 and March 29, 1929, pp. 44–48.

778. Electrical resistance measurement determines endurance of metals. *Automot. Industr., N.Y.*, vol. 60, no. 12, March 23, 1929, pp. 476–478.

779. Fatigue limit of steel found by rapid electrical test. *Iron Age*, vol. 123, no. 23, June 6, 1929, p. 1572.

780. Surfaces. *Metallurgist*, vol. 5, July 26, 1929, pp. 98–99.

781. The relation between endurance limits and tensile properties. *Metallurgist*, vol. 5, July 26, 1929, pp. 105–107.

782. Testing the product for endurance. *Amer. Mach., N.Y.*, vol. 70, no. 26, Aug. 3, 1929, p. 1017.

783. The work of rupture in relation to fatigue. *Metallurgist*, vol. 5, Aug. 30, 1929, pp. 124–125.

784. Airplane disaster caused by fatigue of steel. *Iron Age*, vol. 124, no. 10, Sept. 5, 1929, p. 600.

785. The fatigue of metals. *Engineer, Lond.*, vol. 148, no. 3855, Nov. 29, 1929, p. 584.

1930

786. ADERS, K. Über das Verhalten weicher Stähle bei Schwingungsbeanspruchungen. [On the behaviour of mild steels under fatigue stresses.] *Stahl u. Eisen*, vol. 50, no. 31, July 31, 1930, pp. 1095–1096.

787. BACON, F. An enquiry into the causes of the breakage of rolls. *Proc. Staffs. Iron. St. Inst.*, vol. 46, 1930–1931, pp. 115–143.

788. BARKLIE, R. H. D. *and* DAVIES, H. J. The influence of electrodeposited metallic coatings on the resistance of mild steel to fatigue. *Rep. aero. Res. Comm., Lond.*, no. E. F. 262, Jan. 1930, pp. 23.

789. BARKLIE, R. H. D. *and* DAVIES, H. J. The effect of surface conditions and electrodeposited metals on the resistance of materials to repeated stresses. *Proc. Instn mech. Engrs, Lond.*, vol. 1, Jan.–May 1930, pp. 731–750. Abstr.: *Engineer, Lond.*, vol. 150, Dec. 19, 1930, pp. 670–673.

790. BARTELS, W. B. Die Dauerfestigkeit ungeschweisster und geschweisster Guss- und Walzwerkstoffe. [The fatigue strength of unwelded and welded cast and rolled materials.] *Ber. Inst. mech. Technol. Berl.*, no. 3, 1930. pp. 31.

791. BARTELS, W. B. Die Dauerfestigkeit geschweisster und ungeschweisster Guss- und Walzwerkstoffe. [The fatigue strength of welded and unwelded cast and rolled materials.] *Giessereiztg*, vol. 27, no. 22, Nov. 15, 1930. pp. 607–616; no. 23, Dec. 1, 1930, pp. 637–645 and no. 24, Dec. 15, 1930, pp. 661–669 *Z. Ver. dtsch. Ing.*, vol. 74, no. 41, Oct. 11 1930, pp. 1423–1426. English summary *Metals and Alloys*, vol. 2, no. 5, Nov. 1931 pp. 313–314.

792. BEHRENS, P. Das Oberflächendrücken zu Erhöhung der Drehschwingungsfestigkei [Surface compression to increase the torsiona fatigue strength.] *Veröff. Wöhler-Inst.*, no. 6 1930, pp. 51.

793. BERG, S. Zur Frage der Beanspruchung beim Dauerschlagversuch. [The question of stressing in impact fatigue tests.] *ForschArb. IngWes.*, no. 331, 1930, pp. 28.

794. CAZAUD, R. Recherches sur la fatigue des métaux. [Researches on the fatigue of metals.] *Bull. tech. Serv. Aérotech.*, *Paris*, no. 68, June 1930, pp. 62.

795. DEUTSCH, W. Maschinen für Dauerversuche. [Machines for fatigue tests.] *Z. Metallk.*, vol. 22, no. 2, Feb. 1930, pp. 56–61.

796. DINGER, H. C. Corrosion fatigue. An explanation of boiler plate failure. *Power*, vol. 71, no. 4, Jan. 28, 1930, pp. 133–134.

797. DINGER, H. C. Fatigue cracks and boiler plate. *Power*, vol. 71, no. 22, June 3, 1930, pp. 872–876.

798. DÖHRING, H. Das Drücken der Oberfläche und der Einfluss von Querbohrungen auf die Biegeschwingungsfestigkeit. [Surface compression and the influence of transverse holes on the bending fatigue strength.] *Veröff. Wöhler-Inst.*, no. 5, 1930, pp. 60. English Summary: *Metals and Alloys*, vol. 2, no. 2, Feb. 1931, pp. 94–95.

799. DÖHRING, H. Das Drücken der Oberfläche. [The compression of surfaces.] *Metallwirtschaft*, vol. 9, no. 34, Aug. 22, 1930, pp. 702–706.

800. DÖHRING, H. Der Einfluss von Querbohrung auf die Biegungsschwingungsfestigkeit. [The influence of transverse holes on the bending fatigue strength.] *Metallwirtschaft*, vol. 9, no. 38, Sept. 19, 1930, pp. 781–786.

801. DORGELOH, E. Eine neue Prüfungsmaschine zur Untersuchung der Werkstoffe bei wechselnden, oftmals wiederholten Biegebeanspruchungen. [A new testing machine for investigating materials under alternating and repeated bending stresses.] *Metallwirtschaft*, vol. 9, no. 18, May 2, 1930, pp. 381–386.

802. DOUGLAS, W. D. *and* LANE, A. S. Fatigue limit of spruce under alternating bending. *Roy. Aircr. Estab.*, *Rep.* M.T. 5489, Jan. 1930, pp. 9.

803. ESAU, A. *and* VOIGT, E. Verbesserungen an der Materialprüfmaschine für Zugdruckbeanspruchung. Vorspannung und Selbsterregung. [Improvements in testing machines for tension-compression stresses. Initial stress and self excitation.] *Z. tech. Phys.*, vol. 11, no. 2, Feb. 1930, pp. 55–58.

804. ESAU, A. *and* VOIGT, E. Beiträge zum Verhalten von Werkstoffen bei dynamischer Beanspruchung. [Contribution to the behaviour of materials under dynamic stresses.] *Z. tech. Phys.*, vol. 11, no. 3, March 1930, pp. 78–81.

805. FÖPPL, O. Die Dämpfung der Werkstoffe bei wechselnden Normalspannungen und bei wechselnden Schubspannungen. [The damping of materials under alternating direct stresses and under alternating shearing stresses.] *Z. Ver. dtsch. Ing.*, vol. 74, no. 40, Oct. 4, 1930, pp. 1391–1394. English abstr.: *Metallurgist*, vol. 7, Jan. 30, 1931, pp. 13–14.

806. FÖPPL, O. *and* SCHAAF, G. Die Werkstoffdämpfung bei Dreh- und Biegeschwingungsbeanspruchung. [The damping of materials under torsional and bending alternating stresses.] *ForschArb. IngWes.*, no. 335, 1930, pp. 27.

807. FREEMAN, J. R. *and* FRANCE, R. D. Endurance properties of some special rail steels. *J. Res. nat. Bur. Stand.*, vol. 4, June 1930, pp. 851–874.

808. FULLER, T. S. Endurance properties of steel in steam. *Trans. Amer. Inst. min. (metall.) Engrs*, vol. 90, 1930, pp. 280–292.

809. GILLETT, H. W. Strengthening by understressing. *Proc. Amer. Soc. Test. Mater.*, vol. 30, pt. 1, 1930, pp. 295–297.

810. GILLETT, H. W. Effect of alloying and heat treatment on the endurance limit of steel. *Proc. Amer. Soc. Test. Mater.*, vol. 30, pt. 1, 1930, pp. 291–293.

811. GILLETT, H. W. More on fatigue and high frequency fatigue testing. *Metals and Alloys*, vol. 1, no. 7, Jan. 1930, pp. 332–333.

812. GILLETT, H. W. A study of the Ikeda short-time resistance test for fatigue. *Metals and Alloys*, vol. 1, no. 11, May 1930, p. 520.

813. GILLETT, H. W. Corrosion fatigue. *Metals and Alloys*, vol. 1, no. 12, June 1930, p. 562.

814. GOUGH, H. J. Fatigue of single crystals of pure metals. *Commun. new int. Ass. Test. Mat.*, *(A)*, 1930, pp. 133–144.

815. GOUGH, H. J. *and* COX, H. L. The behaviour of a single crystal of antimony subjected to alternating torsional stresses. *Proc. roy. Soc.*, *(A)*, vol. 127, 1930, pp. 431–453.

816. GOUGH, H. J. *and* COX, H. L. Further experiments on the behaviour of single crystals of zinc subjected to alternating torsional stresses. *Proc. roy. Soc.*, *(A)*, vol. 127, 1930, pp. 453–479.

817. GOUGH, H. J. *and* COX, H. L. Single crystals of bismuth subjected to alternating torsional stresses. *Rep. Memor. aero. Res. Comm.*, *Lond.*, no. 1432, Dec. 1930, pp. 14.

818. GOUGH, H. J. *and* SOPWITH, D. G. Corrosion fatigue test on aluminium crystal. *Rep. Memor. aero. Res. Comm.*, *Lond.*, no. 1433, Sept. 1930, pp. 30.

819. GÜNTHER, K. Der Einfluss der Oberflächenbearbeitung auf die Dauerfestigkeit. [The influence of surface working on fatigue strength.] *Oberflächentechnik*, vol. 7, no. 6, 1930, pp. 54–57.

820. GÜNTHER, K. Die Steigerung der Dauerhaltbarkeit durch zusätzliche Kaltverformung der Oberfläche von dauerbeanspruchten Konstruktionsteilen. [Increasing the fatigue strength by additional cold-deformation of the surface of constructional parts subjected to fatigue loadings.] *Oberflächentechnik*, vol. 7, April 15, 1930, pp. 71–74.

821. HAIGH, B. P. The relative safety of mild and high-tensile alloy steels. *Auto. Engr*, vol. 20, no. 264, Feb. 1930, pp. 70–78.

822. HAIGH, B. P. The yield point and the fatigue limit. *Engineering, Lond.*, vol. 129, no. 3344, Feb. 14, 1930, pp. 231–234. *Engineer, Lond.*, vol. 149, no. 3868, Feb. 28, 1930, pp. 238–239; no. 3869, March 7, 1930, pp. 262–263 and no. 3870, March 14, 1930, pp. 304–305.

823. HAIGH, B. P. Brittle fracture in metals. *Engineering, Lond.*, vol. 130, Nov. 28, 1930, pp. 685–687; Dec. 5, 1930, pp. 717–719 and Dec. 12, 1930, pp. 752–753.

824. HAIGH, B. P. *and* JONES, B. Atmospheric action in relation to fatigue in lead. *J. Inst. Met.*, vol. 43, no. 1, 1930, pp. 271–295. Abstr.: *Engineering, Lond.*, vol. 129, March 28, 1930, pp. 423–425.

825. HAIGH, B. P. *and* THORNE, F. W. Rupture by fatigue. *Proc. Int. Congr. app. Mech.*, Third Congr., Stockholm, 1930, vol. II, pp. 300–304

826. HARVEY, W. E. Zinc as a protective coating against corrosion fatigue of steel. *Metals and Alloys*, vol. 1, no. 10, April 1930, pp. 458–461.

827. HEIDEBROEK, E. Maschinenteile und Werkstoffkunde. [Machine components and the science of materials.] *Z. Ver. dtsch. Ing.*, vol. 74, no. 37, Sept. 13, 1930, pp. 1259–1265.

828. VON HENGSTENBERG, O. *and* MAILÄNDER, R. Biegeschwingungsfestigkeit von nitrierten Stählen. [Bending fatigue strength of nitrided steels.] *Krupp. Mh.*, vol. 11, Sept.–Oct. 1930, pp. 252–254. *Z. Ver. dtsch. Ing.*, vol. 74, no. 32, Aug. 9, 1930, pp. 1126–1128. English abstr.: *Iron Age*, vol. 127, no. 5, Jan. 29, 1931, pp. 400–401.

829. HEROLD, W. Vergleich der Messergebnisse verschiedener Versuchsarten wie Zugfestigkeit, Härte, Dehnung, Einschnürung, Falt-, Biege-, Verdrehungsziffern, Struktur, Textur u.s.w. [Comparison of the results of different tests, such as tensile strength, hardness, elongation, reduction of area, folding, bending and torsion, relation to structure, texture, etc.] *Commun. new int. Ass. Test. Mat.*, *(D)*, 1930, pp. 9–17.

830. VON HEYDEKAMPF, G. S. Neuere Dauerbiegemaschinen mit schwingendem Probestab (Biegungsschwingungsmaschinen).[New bending fatigue machines with vibrating test specimens (vibrating bending machines).] *Metallwirtschaft*, vol. 9, no. 15, April 11, 1930, pp. 321–326.

831. VON HEYDEKAMPF, G. S. Cold rolling raises fatigue or endurance limit. *Iron Age*, vol. 126, Sept. 18, 1930, pp. 775–777, 829 and Oct. 2, 1930, pp. 928–929.

832. HODGE, J. C. The application of fusion welding to pressure vessels. *J. Amer. Weld. Soc.*, vol. 9, no. 10, Oct. 1930, pp. 93–116.

833. HOFFMANN, W. Dauerfestigkeit geschweisster Stahlverbindungen. [Fatigue strength of welded steel joints.] *Z. Ver. dtsch. Ing.*, vol. 74, no. 46, Nov. 15, 1930, pp. 1561–1564.

834. ILLIES, H. Ermüdungsversuche an genieteten und geschweissten Stahltrommeln und Zylindern. [Fatigue tests on riveted and welded steel drums and cylinders.] *Schmelzschweissung*, vol. 10, May 1930, pp. 120–121.

835. JASPER, T. M. Overstressing in fatigue. *Proc. Amer. Soc. Test. Mater.*, vol. 30, pt. 1, 1930, pp. 293–294.

836. JASPER, T. M. Cold working. *Proc. Amer. Soc. Test. Mater.*, vol. 30, pt. 1, 1930, pp. 294–295.

837. JENKINS, C. H. M. The effect of surface conditions on fatigue test results. *Commun. new int. Ass. Test. Mat.*, *(A)*, 1930, pp. 145–148.

838. JENNINGS, C. H. Fatigue and impact tests for welds. *J. Amer. Weld. Soc.*, vol. 9, no. 9, Sept. 1930, pp. 90–104.

839. JÜNGER, A. Erfahrungen über die Prüfung der Dauerfestigkeit verschiedener Werkstoffe auf der MAN-Biegeschwingungsmaschine. [Experiments on the fatigue strength of different materials using the MAN bending fatigue machine.] *Mitt. ForschAnst. Gutehoffn., Nürnberg*, vol. 1, no. 1, Sept. 1930, pp. 8–18. English abstr.: *Metallurgist*, vol. 7, March 27, 1931, pp. 44–45.

840. KNACKSTEDT, W. Die Werkstoffdämpfung bei Drehschwingungen nach dem Dauerprüfverfahren und dem Ausschwingverfahren. [The damping of materials in torsional oscillation by fatigue tests and oscillation tests.] *Veröff. Wöhler-Inst.*, no. 4, 1930, pp. 60. *Z. Ver. dtsch. Ing.*, vol. 74, no. 34, Aug. 23, 1930, pp. 1182–1184.

841. KOMMERS, J. B. Effect of range of stress and kind of stress on fatigue. *Proc. Amer. Soc. Test. Mater.*, vol. 30, pt. 1, 1930, pp. 272–284.

842. KOMMERS, J. B. Mechanical hysteresis and fatigue. *Proc. Amer. Soc. Test. Mater.*, vol. 30, pt. 1, 1930, pp. 287–289.

843. KOMMERS, J. B. The effect of under-stressing of cast iron and open hearth iron. *Proc. Amer. Soc. Test. Mater.*, vol. 30, pt. 2, 1930, pp. 368–383.

844. KORTUM, H. Über die Materialdämpfung bei Dauerbeanspruchung durch Torsionsschwingungen. [On material damping under fatigue stressing in torsional vibration.] *ForschArb. IngWes.*, vol. 1, 1930, pp. 297–307.

845. KRÄMER, O. Dauerbiegeversuche mit Hölzern. [Bending fatigue tests on wood.] *Luftfahrtforsch.*, vol. 8, no. 2, July 15, 1930, pp. 39–48. *Jb. dtsch. VersAnst. Luftf.*, 1930, pp. 411–420.

846. KÜHNEL, R. Dauerbruch und Dauerfestigkeit. Erfahrungen aus dem Betrieb der Deutschen Reichsbahn. [Fatigue fracture and fatigue strength. Service experiences from German railways.] *Z. Ver. dtsch. Ing.*, vol. 74, no. 6, Feb. 8, 1930, pp. 181–184.

847. KULKA, H. Dynamische Probleme im Brückenbau. [Dynamic problems in bridge structures.] *Stahlbau*, vol. 3, no. 26, Dec. 24, 1930, pp. 301–305.

848. KUNTZE, W. Vergleich der Messergebnisse verschiedener Versuchsarten, wie Zugfestigkeit, Härte, Dehnung, Einschnürung, Falt-, Biege-, Verdrehungsziffern, Schwingungsfestigkeit, Kerbschlagprobe u.s.w. [Comparison of the results of different tests, such as tensile strength, hardness, elongation, reduction of area, folding, bending, torsion, fatigue, impact, etc.] *Commun. new int. Ass. Test. Mat.*, (D), 1930, pp. 1–8.

849. KUNTZE, W. Statische Grundlagen zum Schwingungsbruch. [Static principles for fatigue failure.] *Mitt. dtsch. MatPrüfAnst.*, no. 14, 1930, pp. 17–22.

850. KUNTZE, W. Berechnung der Schwingungsfestigkeit aus Zugfestigkeit und Trennfestigkeit. [Calculations of the fatigue strength from the tensile strength and the cleavage strength.] *Mitt. dtsch. MatPrüfAnst.*, no. 14, 1930, pp. 82–85. *Z. Ver. dtsch. Ing.*, vol. 74, no. 8, Feb. 22, 1930, pp. 231–234.

851. KUNTZE, W. *and* SACHS, G. Der Fliessbeginn bei Wechselnder Zug-Druckbeanspruchung. [The commencement of creep under alternating tension-compression stress.] *Mitt. dtsch. MatPrüfAnst.*, no. 14, 1930, pp. 77–82. *Metallwirtschaft*, vol. 9, no. 4, Jan. 24, 1930, pp. 85–89.

852. KUNTZE, W.; SACHS, G. *and* SIEGLER SCHMIDT, H. Elastizität, statische Versuch und Dauerprüfung. [Elasticity, static and fatigue tests.] *Mitt. dtsch. MatPrüfAnst.*, no. 10, 1930, pp. 53–58.

853. LAURENT-ATTHALIN, F. L'mesure de la fatigue tuyauteries der vapeur à aut haute pression. [Measurement of fatigue of high pressure steam pipe lines.] *Tech. mod.*, vol. 22, no. , April 15, 1930, pp. 274–278.

854. LAUTE, K. Normalisierung des Prüfverfahrens zur Ermüdung an Metallen. [Standardisation of methods for testing the fatigue resistance of metals.] *Commun. new int. Ass. Test. Mat.*, (A), 1930, pp. 111–118.

855. LJUNGBERG, K. Konstante Brucharbeit Erklärung für der Bruch durch Ermüdung und andere Belastungen. [Constant energy

fracture as an explanation of fracture by fatigue and other loadings.] *Commun. new int. Ass. Test. Mat., (A)*, 1930, pp. 149–154. *Proc. Int. Congr. app. Mech.*, Third Congr., Stockholm, 1930, vol. II, pp. 294–299.

856. LUDWIK, P. Schwingungsfestigkeit und Gleitwiderstand. [Fatigue strength and resistance to slip.] *Z. Metallk.*, vol. 22, no. 11, Nov. 1930, pp. 374–378.

857. LUDWIK, P. Ermüdung. I — Verhältnis der Ermüdungsgrenze zur Elastizitätsgrenze und anderen Mechanischen Eigenschaften (pp. 119–124). II — Einfluss der Oberflächenbeschaffenheit auf die Ergebnisse der Ermüdungsprüfung (pp. 124–132). [Fatigue. I — Relation of the fatigue limit to the elastic limit and other mechanical properties. II — Effect of surface condition on fatigue test results.] *Commun. new int. Ass. Test. Mat., (A)*, 1930, pp. 119–132.

858. MCADAM, D. J. Stress and corrosion. *Proc. Int. Congr. app. Mech.*, Third Congr., Stockholm, 1930, vol. II, pp. 269–293.

859. MCADAM, D. J. Influence of cyclic stress on corrosion. *Tech. Publ. Amer. Inst. Min. Engrs*, no. 329, April 1930, pp. 42. Abstr.: *Mech. Engng, N.Y.*, vol. 52, no. 6, June 1930, pp. 619–620.

860. MCADAM, D. J. The influence of stress range and cycle frequency on corrosion. *Proc. Amer. Soc. Test. Mater.*, vol. 30, pt. 2, 1930, pp. 411–447.

861. MCCULLOUGH, P. J. Fatigue testing sucker rods to destruction. *Petrol. Engr*, vol. 2, April 1930, pp. 50–58.

862. MATTHAES, K. Ermüdungseigenschaften von Kurbelwellenstahl. [Fatigue characteristics of crankshaft steel.] *Maschinenbau,* vol. 9, no. 4, Feb. 20, 1930, pp. 117–122.

863. MATTHAES, K. Kurbelwellenbrüche und Werkstoff-fragen. [Crankshaft fractures and the question of the material.] *Luftfahrtforsch.*, vol. 8, July 1930, pp. 91–120.

864. MEMMLER, K. *and* LAUTE, K. Dauerversuche an der Hochfrequenz-Zug-Druck-Maschine (Bauart Schenck). [Fatigue tests with the Schenck high frequency tension-compression machine.] *ForschArb. IngWes.*, no. 329, 1930, pp. 32. Abstr.: *Z. Ver. dtsch. Ing.*, vol. 74, no. 6, Feb. 8, 1930, pp. 189–190. *Z. Metallk.*, vol. 22, no. 7, July 1930, pp. 249–250.

865. MOCHEL, N. L. A note on fatigue tests of nitrided steel. *Proc. Amer. Soc. Test. Mater.*, vol. 30, pt. 2, 1930, pp. 406–410.

866. MOORE, H. F. Fracture under repeated stress. *Proc. Amer. Soc. Test. Mater.,* vol. 30, pt. 1, 1930, pp. 286–287.

867. MOORE, H. F. The significance of fatigue test results. *Proc. Amer. Soc. Test. Mater.*, vol. 30, pt. 1, 1930, pp. 304–305.

868. MOORE, H. F. *and* KONZO, S. A study of the Ikeda short-time (electrical resistance) test for fatigue strength of metals. *Uni. Ill. Engng Exp. Sta. Bull.*, no. 205, April 1930, pp. 31. Abstr.: *Metals and Alloys*, vol. 1, no. 2, Aug. 1929, p. 70.

869. MOORE, H. F. *and* VER, T. A study of slip lines, strain lines, and cracks in metals under repeated stress. *Uni. Ill. Engng Exp. Sta. Bull.*, no. 208, June 1930, pp. 60.

870. MOORE, R. R. Endurance limit, machines, specimens and accelerated tests. *Proc. Amer*

Soc. Test. Mater., vol. 30, pt. 1, 1930, pp. 261–270.

871. MORGAN, P. D. *and* WHITEHEAD, S. A critical study of mechanical vibrations on overhead transmission lines. *Rep. Brit. elect. Ind. Res. Ass.*, no. F/T39, 1930, pp. 23.

872. MÜLLER, A. Werkstoffermüdung und Biegungsbeanspruchung bei Transmissionswellen. [Material fatigue and bending stresses in transmission shafts.] *Maschinenbau*, vol. 9, no. 19, 1930, pp. 640–646.

873. MUSATTI, I. Proprietà dinamiche delle leghe ultra leggere. [Dynamic properties of ultra-light alloys.] *Metallurg. ital.*, vol. 22, no. 14, Dec. 1930, pp. 1052–1068.

874. NEKRYTY, S. S. The methods of mechanical testing of light aluminium alloys. *Commun. new int. Ass. Test. Mat.*, *(A)*, 1930, pp. 277–292.

875. OELSCHLÄGER, J. Schwingungsprüfung von Materialien. [Fatigue testing of materials.] *Auto.-tech. Z.*, vol. 33, no. 26, Sept. 20, 1930, pp. 629–632.

876. ONO, A. Zur Theorie der Schwingungsfestigkeit. [On the theory of the fatigue strength.] *Proc. Int. Congr. app. Mech.*, Third Congr., Stockholm, 1930, vol. II, pp. 305–310.

877. PAPE, H. M. Beanspruchung schwingender Drahtseile unter besonderer Berüchsichtigung der Beanspruchungen an den Tragklemmen von Freileitungen. [Fatigue strength of wire cable with particular regard to the stress at the supporting points of overhead transmission lines.] *Veröff. Wöhler-Inst.*, no. 7, 1930, pp. 86.

878. PETERSON, R. E. Stress concentration and fatigue strength. *Proc. Amer. Soc. Test. Mater.*, vol. 30, pt. 1, 1930, pp. 298–304.

879. PETERSON, R. E. Fatigue tests of small specimens with particular reference to size effect. *Trans. Amer. Soc. Steel Treat.*, vol. 18, 1930, pp. 1041–1056.

880. PETERSON, R. E. *and* JENNINGS, C. H. Fatigue tests of fillet welds. *Proc. Amer. Soc. Test. Mater.*, vol. 30, pt. 2, 1930, pp. 384–394.

881. RAWDEN, H. S. Fatigue of metals as affected by metallographic features. *Proc. Amer. Soc. Test. Mater.*, vol. 30, pt. 1, 1930, pp. 284–286.

882. REININGER, H. Bruch der Achse einer grossen Ziehpresse. [Fracture of axles in large extrusion presses.] *Maschinenbau*, vol. 9, no. 6, 1930, pp. 213–214.

883. REITER, M. Entwicklung und gegenwärtiger Stand der Schienenschweissung. [Development and present position in the welding of rails.] *Org. Eisenbahnw.*, vol. 85, 1930, pp. 398–405 and 419–432.

884. VON ROESSLER, L. Über das Verhalten von autogen geschnittenem Material bei dauernder Beanspruchung durch Schlag. [On the behaviour of flame cut material under impact fatigue stresses.] *Autogene Metallbearb.*, vol. 23, no. 3, 1930, pp. 34–41.

885. SCHAECHTERLE, K. Dauerversuche mit Nietverbindungen. [Fatigue tests on riveted joints.] *Stahlbau*, vol. 3, no. 24, Nov. 28, 1930, pp. 276–281 and no. 25, Dec. 5, 1930, pp. 289–295.

886. SCHEU, R. Beziehungen zwischen den Schwingungsfestigkeiten bei Biegung und bei Verdrehung. [Relation between the fatigue strength in bending and torsion.] *S. B. Akad. Wiss.*, *Wien*, vol. 139, nos. 9/10, 1930, pp. 535–554.

887. SCHWINNING, W. and DORGERLOH, E. Untersuchungen über die Festigkeitseigenschaften von Freileitungsdrähten aus Elektrolytkupfer, Bronze, Aluminium und Aldrey bei Zerreissversuchen und bei Schwingungsbeanspruchung. [Investigation of the strength properties of overhead transmission cable wire of electrolytic copper, bronze, aluminium and Aldrey by tension test and by fatigue test.] HausZ. V.A.W., vol. 2, no. 4/5, July–Aug. 1930, pp. 99–130.

888. SCHWINNING, W. and STROBEL, E. Verfestigung durch Wechselbeanspruchung. [Work hardening under alternating stressing.] Z. Metallk., vol. 22, no. 11, Nov. 1930, pp. 378–381 and no. 12, Dec. 1930, pp. 402–404.

889. SODERBERG, C. R. Factor of safety and working stress. Trans. Amer. Soc. mech. Engrs, vol. 52, 1930, APM pp. 13–28.

890. SÖRENSEN, S. Dynamische Erscheinungen in Maschinen zur Prüfung der Schwingungsfestigkeit. [Dynamic phenomena in fatigue testing machines.] Commun. new int. Ass. Test. Mat., (D), 1930, pp. 40–47.

891. SPÄTH, W. Dauerprüfmaschine für Torsionwechselbelastung. [Fatigue testing machine for alternating torsional loads.] Werkzeugmaschine, vol. 34, no. 3, Feb. 15, 1930, pp. 45–49.

892. SPÄTH, W. Zur Konstruktion von Dauerprüfmaschinen. [The construction of fatigue testing machines.] Z. tech. Phys., vol. 11, no. 4, 1930, pp. 115–118.

893. STANTON, T. E. The adhesion and fatigue of thin coatings of white metal deposited on mild steel surfaces. Rep. Memor. aero. Res. Comm., Lond., no. 1424, Dec. 1930, pp. 8. Abstr.: Engineer, Lond., vol. 153, Feb. 19, 1932, p. 205.

894. STONE, M. and RITTER, J. G. Electrically welded structures under dynamic stress. J. Amer. Inst. elect. Engrs, vol. 49, no. 3, March 1930, pp. 202–205.

895. TEMPLIN, R. L. Detection of incipient fatigue failures. Proc. Amer. Soc. Test. Mater., vol. 30, pt. 1, 1930, pp. 289–291.

896. THOMAS, H. A. Final report of the investigation on the effects of annealing on chains. National Safety Council Nineteenth Annual Safety Congress Transactions, vol. 1, 1930, pp. 281–332. Abstr.: Metals and Alloys, vol. 2, no. 3, Sept. 1931, p. 172.

897. THORNTON, G. E. A study of welded metals under fatigue tests. Engng Bull. Wash. St. Coll., vol. 11, no. 11, April 1930, pp. 33. J. Amer. Weld. Soc., vol. 9, no. 10, Oct. 1930, pp. 48–66.

898. THUM, A. Dauerfestigkeit Kerbwirkung und Konstruktion. [Fatigue strength, notch effect and design.] Auto.-tech. Z., vol. 33, no. 34, Dec. 10, 1930, pp. 818–821; no. 35, Dec. 20, 1930, pp. 852–854 and no. 36, Dec. 31, 1930, pp. 880–883.

899. THUM, A. and BERG, S. Zur Frage der Beanspruchung beim Dauerschlagversuch. [On the problem of loading in impact fatigue tests.] Z. Ver. dtsch. Ing., vol. 74, no. 7, Feb. 15, 1930, pp. 200–204.

900. THUM, A. and UDE, H. Die mechanischen Eigenschaften des Gusseisens. [The mechanical properties of cast iron.] Z. Ver. dtsch. Ing., vol. 74, no. 9, March 1, 1930, pp. 257–264.

901. THUM, A. and WISS, W. J. An investigation of the phenomena of the strengthening of steel and its behaviour under repeated overstress.

Trans. Amer. Soc. Steel Treat., vol. 18, July 1930, pp. 1035–1040.

902. TOWNSEND, J. R. Correlation of endurance limit with other physical properties. *Proc. Amer. Soc. Test. Mater.*, vol. 30, pt. 1, 1930, pp. 270–272.

903. TOWNSEND, J. R. *and* GREENALL, C. H. Fatigue studies of telephone cable-sheath alloys — II. *Proc. Amer. Soc. Test. Mater.*, vol. 30, pt. 2, 1930, pp. 395–405.

904. YUASA, K. On the process of failure of metals under stress. *J. Fac. Engng Tokyo*, vol. 18, no. 9, March 1930, pp. 271–345.

905. Summary of present-day knowledge of fatigue phenomena in metals. *Proc. Amer. Soc. Test. Mater.*, vol. 30, pt. 1, 1930, pp. 260–310.

906. Experiments on laminated springs. *Engineering, Lond.*, vol. 129, no. 3338, Jan. 3, 1930, pp. 26–28.

907. Endurance or fatigue testing of materials. *Machinery, Lond.*, vol. 35, no. 900, Jan. 9, 1930, pp. 473–478.

908. Effect of service on fatigue properties of rails. *Metallurgist*, vol. 6, Jan. 31, 1930, pp. 12–14.

909. Machines for determining torsional fatigue developed in Germany. *Automot. Industr., N.Y.*, vol. 62, no. 6, Feb. 8, 1930, pp. 193–194.

910. Tensile tests and the fatigue limit. *Metallurgist*, vol. 6, March 28, 1930, pp. 34–35.

911. High frequency fatigue. *Nature, Lond.*, vol. 125, April 19, 1930, p. 617.

912. Stress and strength. *Engineering, Lond.*, vol. 129, no. 3354, April 25, 1930, pp. 543–544.

913. Fatigue and fatigue testing. *Metallurgist*, vol. 6, April 25, 1930, p. 55 and May 30, 1930, pp. 75–76.

914. Grinding test specimens reduce cost and improve product. *Grits & Grinds*, vol. 21, no. 5, May 1930, pp. 10–12.

915. The fatigue of metals. *Machinery, Lond.*, vol. 36, no. 922, June 12, 1930, pp. 334–338.

916. Zinc coatings and the corrosion fatigue of steel. *Metallurgist*, vol. 6, June 27, 1930, pp. 86–87.

917. Fatigue strength and tensile and rupture strength. *Metallurgist*, vol. 6, June 27, 1930, pp. 94–95.

918. Fatigue failures: the danger of sharp corners in parts subjected to severe stress. *Auto. Engr*, vol. 20, no. 269, July, 1930, p. 254.

919. Pilon Amsler pour essais l'endurance rapide des métaux à la flexion, compression et traction. [The Amsler machine for the rapid endurance testing of metals in bending, compression and tension.] *Aciers spéc.*, vol. 5, July 1930, pp. 330–332.

920. The National Physical Laboratory—fatigue in large single metallic crystals. *Engineering, Lond.*, vol. 130, no. 3367, July 25, 1930, pp. 117–118.

921. Fatigue of aluminium alloys. *Metallurgist*, vol. 6, Sept. 26, 1930, pp. 137–139.

922. Some American researches on the fatigue of metals. *Metallurgist*, vol. 6, Nov. 28, 1930, pp. 170–172 and Dec. 26, 1930, pp. 183–186.

1931

923. ARMBRUSTER, E. *Einfluss der Oberflächenbeschaffenheit auf den Spannungsverlauf und die Schwingungsfestigkeit.* [*The influence of surface condition on stress distribution and on the fatigue strength.*] Berlin, VDI-Verlag, 1931, pp. 64.

924. ASAKAWA, Y. On the fatigue of metals. [J] *Kinzoku no kenkyu*, vol. 8, no. 4, April 1931, pp. 221–236.

925. ASAKAWA, Y. A method of measuring fatigue degree and its recovery. *Int. Engng Congr., Tokyo, 1929*, vol. III, 1931, pp. 63–72.

926. AUGHTIE, F. A source of mechanical vibration for experimental purposes. *Phil. Mag.*, ser. 7, vol. 11, 1931, pp. 517–522.

927. AUSTIN, C. R. The effect of surface decarburization on fatigue properties of steel. *Metals and Alloys*, vol. 2, no. 3, Sept. 1931, pp. 117–119.

928. BACON, F. Fatigue stresses with special reference to the breakage of rolls. *Proc. S. Wales Inst. Engrs*, vol. 47, no. 2, April 1931, pp. 141–335. *Engineering, Lond.*, vol. 131, no. 3397, Feb. 20, 1931, pp. 280–282 and no. 3399, March 6, 1931, pp. 341–344. Abstr.: *Metallurgist*, vol. 7, July 31, 1931, pp. 103–107. *Blast Furn.*, vol. 20, May 1932, p. 448; June 1932, p. 538 and July 1932, p. 608.

929. BARNER, G. *Der Einfluss von Bohrungen auf die Dauerzugfestigkeit von Stahlstäben.* [*The influence of drilled holes on the tension fatigue strength of steel bars.*] Berlin, VDI-Verlag, 1931, pp. 50.

930. BATSON, R. G. C. *and* BRADLEY, J. Fatigue strength of carbon and alloy steel plates as used for laminated springs. *Proc. Instn mech. Engrs, Lond.*, vol. 120, Feb. 1931, pp. 301–332. Abstr.: *Engineering, Lond.*, vol. 131, no. 3401, March 20, 1931, pp. 405–406.

931. BATSON, R. G. *and* HANKINS, G. A. Springs and spring materials. *Int. Engng Congr., Tokyo, 1929*, vol. III, 1931, pp. 201–220.

932. BATSON, R. G. *and* TAPSELL, H. J. Materials at high temperatures. *Congr. int. Ass. Test. Mat., Eighth Congr., Zurich*, 1931, vol. 1, pp. 160–167.

933. BAUMGÄRTEL, K. Über Dauerprüfungen von Azetylen-Schweissungen. [On the fatigue testing of acetylene welds.] *Autogene Metallbearb.*, vol. 24, no. 6, March 15, 1931, pp. 81–87 and no. 7, April 1, 1931, pp. 96–99.

934. BAUMGÄRTEL, K. *and* BRÜSER, K. Dauerfestigkeit von geschweissten und gelöteten Fahrrad- und Kraftrad-Rahmenrohren. [Fatigue strength of welded and soldered bicycle and motorcycle frame tubes.] *Schmelzschweissung*, vol. 10, no. 8, Aug. 1931, pp. 204–208 and no. 9, Sept. 1931, pp. 216–220.

935. BEHRENS, P. Das Oberflächendrücken zur Erhöhung der Drehungsschwingungsfestigkeit. [Surface compression for increasing the torsional fatigue strength.] *Metallwirtschaft*, vol. 10, no. 22, May 29, 1931, pp. 431–435.

936. BEISSNER, H. Einfluss der Gasschmelzschweissung auf die Biegungsschwingungsfestigkeit von Chrom-Molybdän-Stahlrohren. [Influence of gas welding on the alternating bending strength of chromium–molybdenum steel tubes.] *Z. Ver. dtsch. Ing.*, vol. 75, no. 30, July 25, 1931, pp. 954–956.

937. BOLKHOVITINOV, N. F. Fatigue tests and modern equipment for carrying them out. [R] *Vest. inzh. Tekh.*, vol. 17, no. 1, Jan. 1931, pp. 23–25.

938. BRENNER, P. Baustoffragen bei der Konstruktion von Flugzeugen. [Materials problems in the construction of aircraft.] *Z. Flugtech.*, vol. 22, no. 21, Nov. 14, 1931, pp. 637–648. English translation: *Tech. Memor. nat. Adv. Comm. Aero.*, *Wash.*, no. 658, Feb. 1932, pp. 25.

939. BUCHHOLTZ, H. *and* SCHULZ, E. H. Zur Frage der Dauerfestigkeit des hochwertigen Baustahles St.52. [The question of the fatigue strength of the high-strength structural steel St.52.] *Mitt. ForschInst. ver. Stahlw. A.-G.*, vol. 2, no. 6, 1931, pp. 97–112. *Stahl u. Eisen*, vol. 51, no. 31, July 30, 1931, pp. 957–961.

940. CAZAUD, R. Influence de la grosseur du grain micrographique sur la résistance à la fatigue de l'acier doux. Effets de l'écrouissage, du recuit et de la surchauffe. [Effect of grain size on the fatigue resistance of mild steel. The effects of hammer hardening, annealing and overheating.] *C.R. Acad. Sci.*, *Paris*, vol. 192, no. 24, June 15, 1931, pp. 1558–1560.

941. CHAPMAN, E. Maximum stress: its influence on cost and service life of a structure. *J. Amer. Weld. Soc.*, vol. 10, no. 9, Sept. 1931, pp. 19–22.

942. DAWIDENKOW, N. *and* SCHEWANDIN, E. Über den Ermüdungsriss. [On fatigue cracks.] *Metallwirtschaft*, vol. 10, no. 37, Sept. 11, 1931, pp. 710–714.

943. DEHLINGER, U. Gefügeveränderungen beim Dauerbruch. [Structural changes associated with fatigue fracture.] *Metallwirtschaft*, vol. 10, no. 2, Jan. 9, 1931, pp. 26–28.

944. DÖRING, H. Das Drücken der Oberfläche. [The compression of surfaces.] *Oberflächentechnik*, vol. 8, May 5, 1931, pp. 97–99.

945. FAHRENHORST, W. *and* SCHMID, E. Wechseltorsionsversuche an Zink-Kristallen. [Alternating torsion tests with zinc crystals.] *Z. Metallk.*, vol. 23, no. 12, Dec. 1931, pp. 323–328.

946. FORCELLA, P. L'essai aux chocs répétés a flexion alternée sur les rails des chemins de fer Italiens. [Repeated impact and alternating bending tests on rails used by Italian railways.] *Congr. int. Ass. Test. Mat.*, *Eighth Congr.*, *Zurich*, 1931, vol. 1, pp. 247–265.

947. FORCELLA, P. Essais de torsion alternée de 360° sur fils de cables métalliques pour téléférages et funiculaires. [Alternating torsion tests round 360° of wire used in metal cables and funicular railways.] *Congr. int. Ass. Test. Mat. Eighth Congr.*, *Zurich*, 1931, vol. 1, pp. 266–277.

948. FRANCE, R. D. Endurance testing of steel comparison of results obtained with rotating beam versus axially loaded specimens. *Proc Amer. Soc. Test. Mater.*, vol. 31, pt. 2, 1931 pp. 176–193.

949. FRANKE, E. Probenentnahme und Werk stoffprüfung an Rotorkörpern. [Selection o samples and testing of materials for rotors *Elektrotech. Z.*, vol. 52, no. 19, May 7, 1931 pp. 600–601.

950. FULLER, T. S. Influence of corrosion o fatigue of notched specimens. *Min. ar Metall.*, *N.Y.*, vol. 12, no. 298, Oct. 193 p. 446.

951. FULLER, T. S. Endurance properties some well known steels in steam. *Trans. Ame Soc. Steel Treat.*, vol. 19, no. 2, 193 pp. 97–111.

952. GARNER, E. F. Stresses in fillets. *Pr Engng*, vol. 2, June 1931, pp. 259–261.

953. GILLETT, H. W. What is this thing called fatigue? *Metals and Alloys*, vol. 2, no. 2, Feb. 1931, pp. 71–79.

954. GOUGH, H. J. The present state of knowledge of fatigue of metals. *Congr. int. Ass. Test. Mat., Eighth Congr., Zurich*, 1931, vol. 1, pp. 207–227.

955. GRAF, O. Einige Bemerkungen über die Ermittlung der Dauerfestigkeit und der Zulässigen Anstrengungen der Werkstoffe. [Some observations on the determination of the endurance limit and the permissible strain of materials.] *Congr. int. Ass. Test. Mat., Eighth Congr., Zurich*, 1931, vol. 1, pp. 328–333.

956. GRAF, O. *Dauerfestigkeit von Stählen mit Walzhaut, ohne und mit Bohrung, von Niet- und Schweissverbindungen. [Fatigue strength of steels with rolling surface, with and without holes, riveted and welded joints.]* Berlin, VDI-Verlag, 1931, pp. 42. Abstr. *Stahlbau*, vol. 4, no. 22, Oct. 30, 1931, pp. 258–260. *Stahl u. Eisen*, vol. 52, no. 11, March 17, 1932, pp. 268–269.

957. GÜNTHER, K. Die Steigerung der Dauerhaltbarkeit durch Oberflächendrücken. [The increase of fatigue strength by surface compression.] *Maschinenschaden*, vol. 8, no. 8, 1931, pp. 125–129.

958. HAIGH, B. P. *and* ROBERTSON, T. S. A seven-ton 50-cycle fatigue testing machine. *Proc. Amer. Soc. Test. Mater.*, vol. 31, pt. 2, 1931, pp. 221–235.

959. HAIGH, B. P. *and* ROBERTSON, T. S. Plastic strain in relation to fatigue in mild steel. *Engineering, Lond.*, vol. 132, no. 3427, Sept. 18, 1931, pp. 389–390.

960. HANKINS, G. A. *and* BECKER, M. L. The effect of surface conditions produced by heat treatment on the fatigue resistance of spring steels. *J. Iron St. Inst.*, vol. 124, no. 2, 1931, pp. 387–460. Abstr.: *Iron Coal Tr. Rev.*, vol. 123, no. 3318, Oct. 2, 1931, pp. 489–490 and no. 3323, Nov. 6, 1931, pp. 698–699.

961. HEMPEL, M. Das Verhalten einiger Werkstoffe bei dynamischer Biegungsbeanspruchung. [The behaviour of some materials under dynamic bending stress.] *ForschArb. IngWes.*, vol. 2, no. 9, 1931, pp. 327–334.

962. HEROLD, W. Die Drehwechselfestigkeit verschiedener Stähle bei gleichzeitiger elastischer Beanspruchung. [The torsional fatigue strength of various steels with simultaneous elastic stressing.] *Maschinenbau*, vol. 10, no. 20, Oct. 15, 1931, pp. 637–643.

963. HERTEL, H. Dynamische Bruchversuche mit Flugzeugbauteilen. [Dynamic fracture testing of structural members for aircraft.] *Z. Flugtech.*, vol. 22, no. 15, Aug. 14, 1931, pp. 465–474 and no. 16, Aug. 28, 1931, pp. 489–502. *Jb. 1931 dtsch. VersAnst. Luftf.*, pp. 142–164. English translation: *Tech. Memor. nat. Adv. Comm. Aero., Wash.*, no. 698, Jan. 1933, pp. 42.

964. VON HEYDEKAMPF, G. S. Damping capacity of materials. *Proc. Amer. Soc. Test. Mater.*, vol. 31, pt. 2, 1931, pp. 157–175.

965. HIRSCHFELD, F. Das Feststellen von Risbildung und des Beginnens von Dauerbrüchen. [The detection of the initiation and formation of fatigue cracks.] *Auto.-tech. Z.*, vol. 34, no. 28, Oct. 10, 1931, pp. 639–640.

966. HOFFMANN, W. Dauerfestigkeit der mittels Azetylen-Sauerstoff geschweissten Stahlverbindungen. [Fatigue strength of oxyacetylene welded steel joints.] *Autogene Metallbearb.*, vol. 24, no. 7, April 1, 1931, pp. 99–102.

967. HOLT, L. T. Failure of machine members. *Mach. Design*, vol. 3, no. 6, June 1931, pp. 41–43.

968. HORST, C. G. *and* WESCOTT, B. B. Sucker rods—why they break. *Petrol. Engr*, vol. 3, no. 2, Nov. 1931, pp. 26–30.

969. IKEDA, S. A rapid method of determining endurance limit by means of measuring electrical resistance in heat-treated steel. *Int. Engng Congr., Tokyo, 1929*, vol. III, 1931, pp. 491–519.

970. IKEDA, S. Endurance limit of nickel–copper alloys determined by the electrical resistance method. [J] *Kinzoku no kenkyu*, vol. 8, no. 1, Jan. 1931, pp. 32–42.

971. INGLIS, N. P. *and* LAKE, G. F. Corrosion fatigue tests of mild steel and chromium–nickel austenitic steel in River Tees Water. *Trans. Faraday Soc.*, vol. 27, Dec. 1931, pp. 803–808.

972. ISEMER, H. Die Steigerung der Schwingungsfestigkeit von Gewinden durch Oberflächendrücken. [Increasing the fatigue strength of screw threads by surface rolling.] *Mitt. Wöhler-Inst.*, no. 8, 1931, pp. 64. *Metallwirtschaft*, vol. 10, no. 37, Sept. 11, 1931, pp. 714–716.

973. JASPER, T. M. Fatigue and impact testing of welded products. "*Symposium on welding*", American Society for Testing Materials, Dec. 1931, pp. 130–140.

974. JENNINGS, C. H. Dynamic properties of arc-welded joints. *Iron Steel Engr*, vol. 8, no. 12, Dec. 1931, pp. 498–500.

975. JOHNSON, J. B. Fatigue properties of welds. *Welding, Pittsb.*, vol. 2, no. 3, March 1931, pp. 159–163, 165.

976. JOHNSON, J. B. Fatigue of aircraft parts. *Aviation, N.Y.*, vol. 30, no. 8, Aug. 1931, pp. 462–464 and no. 9, Sept. 1931, pp. 542 and 546.

977. JÜNGER, A. Weitere Erfahrungen in der Prüfung der Dauerfestigkeit verschiedener Werkstoffe auf der MAN-Biegeschwingungsmaschine. [Further results of fatigue tests on various materials using the MAN bending fatigue machine.] *Mitt. ForschAnst. Gutehoffn., Nürnberg*, vol. 1, no. 3, Feb. 1931, pp. 45–58.

978. KARLSON, K. G. Utmattningshållfasthet. [Fatigue strength.] *Industritidn. Nord.*, vol. 59, no. 41, 1931, pp. 326–329.

979. KAUFMANN, E. *Über die Dauerbiegefestigkeit einiger Eisenwerkstoffe und ihrer Beeinflussung durch Temperatur und Kerbwirkung.* [*On the bending fatigue strength of certain ferrous materials and the influence of temperature and notches.*] Berlin, Julius Springer, 1931, pp. 89. English abstr.: *Metals and Alloys*, vol. 2, no. 3, Sept. 1931, p. MA 173.

980. KIDANI, Y. Crystallographical investigation of some mechanical properties of metals. V—fatigue of metals under alternating torsion. *J. Fac. Engng Tokyo*, vol. 19, no. 7, 1931, pp. 177–190.

981. KOMMERS, J. B. The static and fatigue properties of brass. *Proc. Amer. Soc. Test. Mater.*, vol. 31, pt. 2, 1931, pp. 243–258.

982. KOMMERS, J. B. Steady torsion combined with alternating bending. *Engineering, Lond.*, vol. 132, no. 3424, Aug. 28, 1931, p. 249.

983. KÜHNEL, R. Herabsetzung der Schwingungsfestigkeit durch Korrosion. [Reduction of fatigue strength by corrosion.] *Maschinenbau*, vol. 10, no. 22, Nov. 19, 1931, pp. 700–702.

984. KUNTZE, W. Zur Problemstellung der Metallermüdung. [The problem of fatigue in metals.] *Metallwirtschaft*, vol. 10, no. 48, Nov. 27, 1931, pp. 895–897. English translation: *Light Metals Res.*, vol. 1, no. 22, Dec. 12, 1931, pp. 8–12.

985. LEA, F. C. Some points in connection with the failure of metals under repeated stresses. *Int. Engng Congr., Tokyo, 1929*, vol. III, 1931, pp. 541–551.

986. LEA, F. C. *and* DICK, J. Torsional fatigue tests of cold-drawn wire. *Proc. Instn mech. Engrs, Lond.*, vol. 120, May 1931, pp. 661–677.

987. LEHMANN, G. Schäden an Aluminiumleitungen durch Seilschwingungen und Mittel für ihre Behebung. [Damage to aluminium transmission lines by vibration and means of prevention.] *Elektrizitätswirtschaft*, vol. 30, Sept. 1931, pp. 530–533.

988. LEHR, E. Die Nutzbarmachung der Werkstofforschung für die Konstruktion. [The utilization of material research for construction.] *Congr. int. Ass. Test. Mat., Eighth Congr., Zurich*, 1931, vol. 1, pp. 333–337.

989. LEHR, E. Wie lassen sich die Ergebnisse der Schwingungsprüfung der Werkstoffe für den Konstrukteur nutzbar machen? [How can the results of fatigue tests of materials be utilized by the designer?] *Sparwirtschaft*, vol. 9, no. 7, 1931, pp. 271–279 and no. 8, 1931, pp. 313–320.

90. DE LEIRIS, H. Recherches expérimentales sur la fatigue a la dilation des tuyautages de vapeur. [Experimental researches on the fatigue, by expansion, of steam pipes.] *Bull. tech. Bur. Veritas*, vol. 13, no. 6, June 1931, pp. 116–120.

91. LJUNBERG, K. Konstante Brucharbeit als Erklärung für den Bruch Ermüdungs- und andere Belastungen. [Constant energy to fracture as an explanation of fracture by fatigue and other loadings.] *Congr. int. Ass. Test. Mat., Eighth Congr., Zurich*, 1931, vol. 1, pp. 337–339.

992. LUDWIK, P. Kerb- und Korrosiondauerfestigkeit. [Notch and corrosion fatigue strength.] *Metallwirtschaft*, vol. 10, no. 37, Sept. 11, 1931, pp. 705–710.

993. LUDWIK, P. Ermüdung. [Fatigue.] *Congr. int. Ass. Test. Mat., Eighth Congr., Zurich*, 1931, vol. 1, pp. 190–206.

994. LUNDGREN, A. Utmattningsprov pa Järn och Stal. I—Roterande Utmattning. [Fatigue tests on iron and steel. I — rotating fatigue.] *Jernkontor. Ann.*, vol. 115, no. 1, Jan. 15, 1931, pp. 1–70.

995. McADAM, D. J. Influence of stress on corrosion. *Tech. Publ. Amer. Inst. Min. Engrs*, no. 417, 1931, pp. 39.

996. McADAM, D. J. Stress corrosion of metals. *Congr. int. Ass. Test. Mat., Eighth Congr., Zurich*, 1931, vol. 1, pp. 228–246.

997. MATTHAES, K. Statische und dynamische Festigkeitseigenschaften einiger Leichmetalle. [Static and dynamic tensile properties of several light metals.] *Ber. dtsch. VersAnst. Luftf.*, no. 250, 1931. *Jb. 1931 dtsch. VersAnst. Luftf.*, pp. 439–484.

998. MEMMLER, K. *and* LAUTE, K. Untersuchung metallischer Baustoffe auf Schwingungsfestigkeit mit der Hochfrequenz-Zug-Druck-Maschine (Bauart Schenck). [Investigation of the fatigue strength of structural metallic materials with the Schenck high frequency tension-compression machine.] *Mitt. dtsch. MatPrüfAnst.*, no. 15, 1931, pp. 39–70.

999. MILLER, H. L. Corrosion fatigue tests on various staybolt materials. *Boiler Mkr, N.Y.*, vol. 31, Aug. 1931, pp. 206–208.

1000. MILLER, H. L. Fatigue tests on staybolts. *Metals and Alloys*, vol. 2, no. 3, Sept. 1931, pp. 109–110.

1001. MOORE, H. F. Tests on the resistance to repeated pressure of forged, riveted and welded boiler shells. *Trans. Amer. Soc. mech. Engrs*, Fuels and Steam Power Division, vol. 53, no. 1, Jan.–April 1931, pp. 55–60.

1002. MOORE, H. F. Crackless plasticity, a new property of metals. *Iron Age*, vol. 128, no. 11, Sept. 10, 1931, pp. 674–677, 721.

1003. MOORE, H. F. *and* ALLEMAN, N. J. Progress report on fatigue tests of low-carbon steel at elevated temperatures. *Proc. Amer. Soc. Test. Mater.*, vol. 31, pt. 1, 1931, pp. 114–121.

1004. MOORE, H. F. *and* LEWIS, R. E. Fatigue tests in shear of three non-ferrous metals. *Proc. Amer. Soc. Test. Mater.*, vol. 31, pt. 2, 1931, pp. 236–242.

1005. MUNDT, R. Über die Tragfähigkeit von Zylinderrollenlagern. [The loading capacity of bearings for cylinder rolls.] *Maschinenbau*, vol. 10, May 21, 1931, pp. 354–357.

1006. ODING, I. *and* EFREMOV, A. The effect of cold working on fatigue strength of metals. [R] *Vest. metalloprom.*, no. 10, Oct. 1931, pp. 69–75.

1007. ONO, A. Fatigue of metals under repeated stress. *Int. Engng Congr., Tokyo, 1929*, vol. III, 1931, pp. 477–489.

1008. PETERSON, R. E. Fatigue tests of model turbo-generator rotors. *Mech. Engng, N.Y.*, vol. 53, March 1931, pp. 211–215.

1009. PETERSON, R. E. *and* JENNINGS, C. H. Fatigue tests of weld metal. *Proc. Amer. Soc. Test. Mater.*, vol. 31, pt. 2, 1931, pp. 194–203.

1010. POMP, A. *and* DUCKWITZ, C. A. Dauer-prüfungen unter wechselnden Zugbeanspru-chungen an Stahldrähten. [Fatigue tests of steel wires under pulsating tensile stresses.] *Mitt. K.-Wilh.-Inst. Eisenforsch.*, vol. 13, no. 5, 1931, pp. 79–91. Abstr.: *Stahl u. Eisen*, vol. 51, no. 20, May 14, 1931, pp. 620–622.

1011. POMP, A.; DUCKWITZ, C. *and* LINDEBERG, A. Några utmattningsförsök med patenterad och dragen ståltråd. [Fatigue tests on patented and drawn steel wire.] *Jernkontor. Ann.*, vol. 115, no. 8, Aug. 1931, pp. 371–403.

1012. PROBST, E. The influence of rapidly alter-nating loading on concrete and reinforced concrete. *Struct. Engr*, vol. 9, no. 10, Oct. 1931, pp. 326–340. Also vol. 9, no. 12, Dec. 1931, pp. 410–432.

1013. RAIRDEN, A. S. Unequal loading decreases elevator rope life. *Power*, vol. 74, no. 19, Nov. 10, 1931, pp. 683–684.

1014. RATHKE, K. Universalprüfmaschine für Dauerwechselbelastung. [Universal testing machine for alternating fatigue loading.] *Z Ver. dtsch. Ing.*, vol. 75, no. 41, Oct. 10, 1931 p. 1289.

1015. ROSENHAIN, W. *and* MURPHY, A. J. Accel erated cracking of mild steel (boiler plate under repeated bending. *J. Iron St. Inst.* vol. 123, no. 1, 1931, pp. 259–284. *Iron Coa Tr. Rev.*, vol. 122, May 8, 1931, pp. 739–743

1016. RUSSELL, H. W. *and* WELCKER, W. A.. Endurance and other properties at low temperatures of some alloys for aircraft use. *Tech. Note nat. Adv. Comm. Aero., Wash.*, no. 381, June 1931, pp. 26.

1017. RUSSELL, H. W. *and* WELCKER, W. A. Apparatus for low temperature endurance testing. *Proc. Amer. Soc. Test. Mater.*, vol. 31, pt. 1, 1931, pp. 122–128.

1018. SARAN, W. *Die Dauerfestigkeit der Leichmetall-Sandguss-Legierungen. [The fatigue strength of light metal sand casting alloys.]* Darmstadt, Verlag Pfeffer und Bolzer, 1931, pp. 87.

1019. SARAN, W. Materialprüfung im Motorenbau. [The testing of materials in automobile construction.] *Werkzeugmaschine*, vol. 35, no. 15, Aug. 15, 1931, pp .301–304 and no. 16, Aug. 31, 1931, pp. 329–333.

1020. SCHAECHTERLE, K. Zur Wahl der zulässigen Anstrengungen bei Stahlbrücken. [The choice of allowable loads in steel bridges.] *Stahlbau*, vol. 4, no. 8, April 17, 1931, pp. 89–92.

1021. SCHMIDT, J. Die Dämpfungsfähigkeit von Eisen- und Nichteisenmetallen bei Dreh- und Biegeschwingungs-Beanspruchung. [The damping power of ferrous and non-ferrous metals under repeated torsional and repeated bending stresses.] *Veröff. Wöhler-Inst.*, no. 9, 1931, pp. 52.

1022. SCHMIDT, W. Die Bedeutung des Kristallaufbaues für die Beurteilung der Elastizitätsgrenze und Dauerfestigkeit von Elektronmetall. [The significance of crystal structure for the evaluation of the elastic limit and fatigue strength of Elektron metal.] *Z., Metallk.*, vol. 23, no. 2, Feb. 1931, pp. 54–57.

1023. SCHNEIDER, W. Beitrag zur Frage der Schwingungsfestigkeit. [Contribution to the problem of fatigue strength.] *Stahl u. Eisen*, vol. 51, no. 10, March 5, 1931, pp. 285–292.

1024. SCHULZ, E. H. *and* BUCHHOLTZ, H. Über die Entwicklung der Dauerprüfung in Deutschland. [The development of fatigue testing in Germany.] *Congr. int. Ass. Test. Mat., Eighth Congr., Zurich*, 1931, vol. 1, pp. 278–303.

1025. SCHWINNING, W. *and* DORGERLOH, E. Neue Prüfmaschinen zur Bestimmung der Wechselfestigkeit für umlaufende Biegung. [New testing machines for determining fatigue strength under rotating bending.] *Z. Metallk.*, vol. 23, no. 6, June 1931, pp. 186–188.

1026. SHELTON, S. M. Fatigue testing of wire. *Proc. Amer. Soc. Test. Mater.*, vol. 31, pt. 2, 1931, pp. 204–220.

1027. SPOONER, E. C. R. Corrosion-fatigue in practice. *Commonw. Engr*, vol. 18, no. 12, July 1, 1931, pp. 441–442.

1028. STONES, F. Inclusions cause majority of valve spring wire failures. *Iron Age*, vol. 128, no. 20, Nov. 12, 1931, pp. 1234–1237.

1029. STUART, J. A. G. Fatigue of metals. *Mech. World*, vol. 89, no. 2302, Feb. 13, 1931, pp. 155–158.

1030. SWAN, A.; SUTTON, H. *and* DOUGLAS, W. D. An investigation of steels for aircraft engine valve springs. *Proc. Instn mech. Engrs, Lond.*, vol. 120, Feb. 1931, pp. 261–299. *Engineering, Lond.*, vol. 131, no. 3398, Feb. 27, 1931, pp. 314–316 and no. 3400, March 13, 1931, pp. 374–376.

1031. TEMPLIN, R. L. Fittings control the life of metallic tubing subjected to vibration. *Automot. Industr.*, *N.Y.*, vol. 65, no. 2, July 11, 1931, pp. 50–51.

1032. THUM, A. Zur Steigerung der Dauerhaltbarkeit gekerbter Konstruktionen. [Increasing the fatigue strength of notched constructional members.] *Z. Ver. dtsch. Ing.*, vol. 75, no. 43, Oct. 24, 1931, pp. 1328–1330.

1033. THUM, A. *and* BERG, S. Die Enlastungskerbe. [Load relieving notches.] *ForschArb. IngWes.*, vol. 2, no. 10, Oct. 1931, pp. 345–351.

1034. VERNOTTE, P. Le phénomène de fatigue des métaux. Moyens propres à en déterminer la limité. [The phenomena of fatigue of metals. Appropriate means of determining the limit.] *Sci. et Industr.*, vol. 15, no. 208, May 1931, pp. 223–229.

1035. WEDEMEYER, E. A. Beurteilung von Kurbelwellenbrüchen. [Interpretation of crankshaft fractures.] *Auto.-tech. Z.*, vol. 34, no. 20/21, July 20–31, 1931, pp. 472–475.

1036. WEGELIUS, E. Väsytyslujuus, sen merkitys ja määrääminen. [Fatigue strength, its meaning and determination.] *Tekn. Aikl.*, vol. 21, no. 9, Sept. 1931, pp. 435–443.

1037. WEINMAN, R. A. Fatigue properties of welds. *J. Amer. Weld. Soc.*, vol. 10, no. 10, Oct. 1931, pp. 12–18.

1038. WILLIAMS, F. H. The practical application of the microscope in railway service. *Engng J. Montreal*, vol. 14, Nov. 1931, pp. 558–563.

1039. Versuche mit Automobil-Tragfedern im National Physical Laboratory. Nach Berichten des N.P.L. bearbeitet von E. Lehr. [Experiments with automobile springs at the National Physical Laboratory. Recent reports of the N.P.L. summarised by E. Lehr.] *ForschArb. IngWes.*, vol. 2, 1931, pp. 287–290.

1040. Begriffliche und Prüfmethodische Beziehung zwischen Elastizität und Plastizität, Zähigkeit und Sprödigkeit. [Conceptions and test methods relating elasticity and plasticity, toughness and brittleness.] *Congr. int. Ass. Test. Mat.*, *Eighth Congr.*, *Zurich*, 1931, vol. 2, pp. 530–584.

1041. Fatigue of mild steel including steel castings and forgings. *Trans. N.-E. Cst Instn Engrs Shipb.*, vol. 48, 1931–1932, pp. 121–134. Abstr.: *Engineer*, *Lond.*, vol. 153, no. 3973, March 4, 1932, pp. 255–257. *Rly Engr*, vol. 53, no. 628, May 1932, p. 166. *Mech. Engng*, *N.Y.*, vol. 54, no. 5, May 1932, pp. 359–360.

1042. Fatigue des métaux. [The fatigue of metals.] *Rev. gén. Sci. pur. appl.*, vol. 42, 1931, pp. 493–496.

1043. Surfaces. *Metallurgist*, vol. 7, Jan. 30, 1931, pp. 3–4.

1044. Damping of materials under alternating stresses. *Metallurgist*, vol. 7, Jan. 30, 1931, pp. 13–14.

1045. Fatigue cracking of lead sheathed cable. *Distrib. Elect.*, vol. 3, no. 32, Jan. 1931, pp. 554–555.

1046. Fatigue tests on the M.A.N. machine. *Metallurgist*, vol. 7, March 27, 1931, pp. 44–45.

1047. What we know of fatigue—a summary of present knowledge of endurance of metal under repeated stress. *Engng News Rec.*, vol. 106, no. 16, April 16, 1931, pp. 651–653.

1048. Endurance of petrol pipes of various materials when subjected to vibration. *Roy. Aircr. Estab., Rep.*, no. M.T. 5528, Aug. 1931, pp. 32.

1049. Torsional fatigue of welded steel. *Metallurgist*, vol. 7, Aug. 28, 1931, pp. 126–127.

1050. La résistance aux efforts alternés et sa détermination a l'aide des machines d'essais de durée. [The resistance to alternating loads and its determination by means of endurance testing machines.] *Aciers spéc.*, vol. 6, Sept. 1931, pp. 465–473.

1051. Tests simple shapes for both static bend and repeated loading. *Automot. Industr., N.Y.*, vol. 65, no. 26, Dec. 26, 1931, p. 992.

1932

1052. ARMBRUSTER, E. Über den Einfluss von Kerben auf die Festigkeit von Konstruktionsteilen. [On the effect of notches on the strength of structural parts.] *Z. bayer. (Dampfk) RevisVer.*, vol. 36, Sept. 30, 1932, pp. 199–205.

1053. BACON, F. Cracking and fracture in rotary bending tests. *Engineering, Lond.*, vol. 134, no. 3480, Sept. 23, 1932, pp. 372–376. [see also pp. 354–355] Abstr.: *Engineer, Lond.*, vol. 154, Sept. 23, 1932, p. 299.

1054. BANKWITZ, E. Die Abhängigkeit der Werkstoffdämpfung von der Grösse und Geschwindigkeit der Formänderung. [The dependence of material damping on the magnitude and velocity of the deformation.] *Mitt. Wöhler-Inst.*, no. 11, 1932, pp. 53.

1055. BEILHACK, M. Der Dauerschlagbiegeversuch. Abhängigkeit der Schlagzahl von Fallgewicht und Fallhöhe. [The impact-bending fatigue test. Dependence of the impact strength on the falling weight and drop height.] *ForschArb. IngWes.*, no. 354, 1932, pp. 22.

1056. BOSSE, P. Resonanz-drehschwingungsdämpfer mit Werkstoffdämpfung für Triebwerke von Automobil- und Flugzeugmotoren. [Resonant torsional vibration damper with material damping for motors of automobile and aircraft.] *Mitt. Wöhler-Inst.*, no. 13, 1932, pp. 95.

1057. BRENNER, P. Problems involved in the choice and use of materials in airplane construction. *Tech. Memor. nat. Adv. Comm. Aero., Wash.*, no. 658, Feb. 1932, pp. 25.

1058. BROPHY, G. R. Damping capacity of steels and correlation with other properties. *Iron Age*, vol. 130, no. 21, Nov. 24, 1932, pp. 800–802 and no. 26, Dec. 29, 1932, pp. 989, 10 and 12.

1059. BUCHMANN, W. Werkstoffeigenschaften bei Wechselbeanspruchung. [Material properties under fatigue loads.] *Schr. tech. Hochsch., Darmstadt*, no. 4, 1932, pp. 37–44.

1060. BULLEID, C. H. Fatigue tests on cast iron. *Bull. Brit. cast Iron Ass.*, vol. 3, no. 6, Oct. 1932, pp. 150–152.

1061. BURNS, G. The properties of some silicomanganese steels. *J. Iron St. Inst.*, vol. 125, no. 1, 1932, pp. 363–391.

1062. CAZAUD, R. Recherches sur la fatigue des métaux. [Investigations on the fatigue of metals.] *Chim. et Industr.*, vol. 27, no. 3, (special number), March 1932, pp. 390–392.

1063. CAZAUD, R. Résistance à la fatigue de quelques alliages d'aluminium de fonderie. [Fatigue resistance of some aluminium casting alloys.] *Bull. Ass. tech. Fond., Paris*, vol. 6, no. 8, Aug. 1932, pp. 485–486.

1064. COLINET, R. Essais de corrosion par l'eau de Mer. [Corrosion tests by sea water.] *Arcos*, vol. 9, no. 48, 1932, pp. 652–654. German abstr.: *Stahl u. Eisen*, vol. 52, no. 30, July 28, 1932, p. 742.

1065. CUTHBERTSON, J. W. The endurance limit of a 0.33 per cent carbon steel at elevated temperatures. *J. Iron St. Inst.*, vol. 126, no. 2, 1932, pp. 237–265. Abstr.: *Engineering, Lond.*, vol. 134, no. 3481, Sept. 30, 1932, p. 396.

1066. DAVISON, A. E.; INGLES, J. A. *and* MARTINOFF, V. M. Vibration and fatigue in electrical conductors. *Trans. Amer. Inst. elect. Engrs*, vol. 51, Dec. 1932, pp. 1047–1051. Abstr.: *Elect. Engng, N.Y.*, vol. 51, no. 6, June 1932, pp. 412–413.

1067. DAVISON, A. E.; INGLES, J. A. *and* MARTINOFF, V. M. Vibration in electrical conductors. *Elect. Engng, N.Y.*, vol. 51, no. 11, Nov. 1932, pp. 795–798.

1068. DEBUS, F. Erhöhung der Dauerfestigkeit von Schrauben durch zweckmässige Formgebung und Fertigung. [Increase of fatigue strength of bolts by suitable dimensioning and machining.] *Schr. tech. Hochsch., Darmstadt*, no. 4, 1932, pp. 68–75.

1069. DIETRICH, O. *and* LEHR, E. Das Dehnungslinienverfahren ein Mittel zur Bestimmung der für die Bruchsicherheit bei Wechselbeanspruchung massgebenden Spannungsverteilung. [The brittle lacquer process as a method for determining the stress distribution to ensure safety under alternating stress.]

Z. Ver. dtsch. Ing., vol. 76, no. 41, Oct. 8, 1932, pp. 973–982.

1070. DÖRNEN, . Zug-Druck-Dauerversuche mit niedriger Frequenz der Kraftrichtungswechsel. [Tension-compression fatigue tests using low frequency of change of load direction.] *Stahlbau*, vol. 5, no. 21, Oct. 14, 1932, pp. 161–163.

1071. DOREY, S. F. Elastic hysteresis in crankshaft steels. *Proc. Instn mech. Engrs, Lond.*, vol. 123, Dec. 1932, pp. 479–535.

1072. FADE, A. Le calcul a la résistance mécanique des engrenages hélicoidaux à axes parallèles. [The calculation of the mechanical resistance of helical gears on parallel axes.] *Sci. et Industr.*, vol. 16, no. 218, March 1932, pp. 91–96.

1073. FAHRENHORST, W.; MATTHAES, K. *and* SCHMID, E. Über die Abhängigkeit der Dauerfestigkeit von der Kristallorientierung. [On the dependence of the fatigue strength on the crystal orientation.] *Z. Ver. dtsch. Ing.*, vol. 76, no. 33, Aug. 13, 1932, pp. 797–799.

1074. FISCHER, F. P. Vorschlag zur Festlegung der zulässigen Beanspruchungen im Maschinenbau. [Proposal to define the permissable stress in machine construction.] *Z. Ver. dtsch. Ing.*, vol. 76, no. 19, May 7, 1932, pp. 449–455.

1075. FISCHER, G. *Kerbwirkung and Biegestäben [Notch effects in bars under bending.]* Berlin, VDI-Verlag, 1932, pp. 64.

1076. FORCELLA, P. Le type de rupture des rail en service dans ses rapports avec les essais d résilience et ceux aux chocs répétés par flexio alternéé. [The type of rupture of rails in service with a report on tests of resilience an those of repeated impact and alternatin bending.] *Journée int. du Rail*, 2nd Congress, Zurich, 1932, pp. 173–183.

1077. DE FOREST, A. V. *and* HOPKINS, L. W. The testing of rope wire and wire rope. *Proc. Amer. Soc. Test. Mater.*, vol. 32, pt. 2, 1932, pp. 398–412. *Wire and Wire Prod.*, vol. 7, Sept. 1932, pp. 286–288, 305 and Dec. 1932, pp. 421–423, 426, 440.

1078. FOSTER, P. F. Fatigue testing of materials. *Machinery, Lond.*, vol. 40, no. 1029, June 30, 1932, pp. 404–407.

1079. FOSTER, P. F. The Haigh alternating stress testing machine. *Machinery, Lond.*, vol. 40, no. 1036, Aug. 18, 1932, pp. 621–624.

1080. FRANKE, E. Ein Beitrag zur Förderseil-prüfung. [A contribution to hoisting cable testing.] *Int. Bergw.*, vol. 25, Aug. 15, 1932, pp. 106–109.

1081. FRANKE, H. W. Biege- und Dauerbiegever-suche an dünnwandigen, nach dem Arcatom-verfahren geschweissten Stahlrohren. [Bending and bending fatigue tests on thin walled steel tubes welded by the arc-atom process.] *Elektroschweissung*, vol. 3, no. 11, Nov. 1932, pp. 206–207.

1082. FULLER, T. S.; MUMMA, P. F. *and* MOORE, H. F. Corrosion fatigue of metals. *Proc. Amer. Soc. Test. Mater.*, vol. 32, pt. 1, 1932, pp. 139–142. Abstr.: *Iron Steel Ind.*, vol. 5, 12, Sept. 1932, pp. 435–436. *Mech. World*, vol. 92, no. 2388, Oct. 7, 1932, pp. 341–342.

1083. GILLETT, H. W. The resistance of copper and its alloys to repeated stress. *Metals and Alloys*, vol. 3, Sept. 1932, pp. 200–204; Oct. 1932, pp. 236–238; Nov. 1932, pp. 257–262 and Dec. 1932, pp. 275–280.

1084. GOUGH, H. J. Corrosion fatigue of metals. *J. Inst. Met.*, vol. 49, no. 2, 1932, pp. 17–92. *Metal Ind., Lond.*, vol. 41, no. 12, Sept. 16, 1932, pp. 275–277; no. 13, Sept. 23, 1932, pp. 295–298; no.14, Sept. 30, 1932, pp. 319–322 and no. 15, Oct. 7, 1932, pp. 349–350. Abstr. *Engineer, Lond.*, vol. 154, no. 4001, Sept. 16, 1932, pp. 284–286. *Engineering, Lond.*, vol. 134, no. 3479, Sept. 16, 1932, pp. 323–324 [see also no. 3482, Oct. 7, 1932, pp. 409–410.]

1085. GOUGH, H. J. *and* COX, H. L. The behaviour of single crystals of bismuth subjected to alternating torsional stresses. *J. Inst. Met.*, vol. 48, no. 1, 1932, pp. 227–254.

1086. GOUGH, H. J. *and* COX, H. L. Note on a new form of combined stress fatigue testing machine and on a proposed programme of research. *Rep. aero. Res. Comm. Lond.*, no. E. F. 325, Nov. 7, 1932, pp. 17.

1087. GOUGH, H. J. *and* COX, H. L. Some fatigue experiments on specimens of aluminium consisting of two crystals. *Rep. aero. Res. Comm., Lond.*, no. E.F. 328, Dec. 15, 1932, pp. 55.

1088. GOUGH, H. J. *and* FORREST, G. Stressless corrosion followed by fatigue test to destruction on aluminium crystals. *Rep. Memor. aero. Res. Comm., Lond.*, no. 1476, June 1932, pp. 11.

1089. GOUGH, H. J. *and* SOPWITH, D. G. The behaviour of a single crystal of aluminium under alternating torsional stresses while immersed in a slow stream of tap water. *Proc. roy. Soc., (A)*, vol. 135, March 1932, pp. 392–411.

1090. GOUGH, H. J. *and* SOPWITH, D. G. Relative temperatures of brass when subjected to reversed direct stresses in vacuo and in air. *Rep. Memor. aero. Res. Comm., Lond.*, no. 1482, June 1932, pp. 4.

1091. GOUGH, H. J. *and* SOPWITH, D. G. Atmospheric action as a factor in fatigue of metals. *J. Inst. Met.*, vol. 49, no. 2, 1932, pp.93–122. Abstr.: *Engineering, Lond.*, vol.134, no. 3491, Dec. 9, 1932, pp. 694–696 [see also no. 3482, Oct. 7, 1932, pp. 409–410.]

1092. GOUGH, H. J. *and* SOPWITH, D. G. The characteristics of corrosion fatigue of an aluminium specimen consisting of two crystals. *Rep. aero. Res. Comm., Lond.*, no. E.F. 329, Dec. 24, 1932, pp. 36.

1093. GRABER, E. Zusammenwirken von Nietung und Schweissung bei Zug und Druck. [Combination of riveting and welding under tension and compression.] *Bauingenieur*, vol. 13, no. 21/22, May 20, 1932, pp. 290–294.

1094. GRAF, O. Versuche mit Nietverbindungen bei oftmals wiederholter Belastung. [Tests of riveted joints under often repeated loads.] *Z. Ver. dtsch. Ing.*, vol. 76, no. 18, April 30, 1932, pp. 438–442.

1095. GRAF, O. Dauerversuche mit Schweissbindungen. [Fatigue tests with welded joints.] *Bautechnik*, vol. 10, no. 30, July 8, 1932, pp. 395–398 and no. 32, July 22, 1932, pp. 414–417.

1096. GRAF, O. Dauerversuche mit Nietverbindungen aus St. 37 zur Erkundung des Einflusses von $\sigma : \tau$. [Fatigue tests on riveted joints of St. 37 to determine the influence of $\sigma : \tau$.] *Bauingenieur*, vol. 13, no. 29/30, July 15, 1932, pp. 389–393.

1097. GRAF, O. Aus Dauerversuchen mit Lichtbogenschweissungen. Ein Beitrag zur Frage der Bemessung und Anordnung der Kehlnähte. Messung von Spannungen in Schweissverbindungen. [Fatigue tests on arc welds. A contribution to the problem of dimension-ing and grouping fillet welds. Measurement of stresses in welded joints.] *Stahlbau*, vol. 5, no. 23, Nov. 11, 1932, pp. 177–182.

1098. GRENET, L. Résistance des aciers aux chocs répétés. [Resistance of steel to repeated impact.] *Aciers spéc.*, vol. 7, no. 79, March 1932, pp. 82–92.

1099. HANKINS, G. A. *and* BECKER, M. L. The fatigue resistance of spring steels. *Engineering, Lond.*, vol. 133, no. 3446, Jan. 29, 1932, pp. 141–145.

1100. HANKINS, G. A. *and* BECKER, M. L. The fatigue resistance of unmachined forged steels. *J. Iron St. Inst.*, vol. 126, no. 2, 1932, pp. 205–236. Abstr.: *Engineering, Lond.* vol. 134, no. 3481, Sept. 30, 1932, pp. 402–404

1101. HARVEY, W. E. Cadmium plating versus corrosion-fatigue; pickling versus corrosion fatigue. *Metals and Alloys*, vol. 3, no. 3 March 1932, pp. 69–72.

1102. HARVEY, W. E. *and* WHITNEY, F. J A corrosion fatigue study of welded lov carbon steel. *J. Amer. Weld. Soc.*, vol. 1 no. 10, Oct. 1932, pp. 12–19. Abstr.: *Engr News Rec.*, vol. 109, no. 15, Oct. 13, 193 pp. 446–447.

1103. HAVEN, H. E. Some effects of corrosio fatigue on streamline wire for aircraft. *Tra Amer. Soc. mech. Engrs*, vol. 54, AER-54-1 1932, pp. 109–115.

1104. HELDT, P. M. Alloys superior to carb steel in endurance tests of valve sprin *Automot. Industr., N.Y.*, vol. 67, no. Nov. 26, 1932, pp. 672–675, 681.

1105. HELLER, P. A. Die Dauerfestigkeit Gusseisens. [The fatigue strength of cast ir *Giesserei*, vol. 19, nos. 31/32, Aug. 5, 19 pp. 301–305 and nos. 33/34, Aug. 19, 1932, pp. 325–332. English abstr.: *Bull. Brit. cast Iron Ass.*, vol. 3, no. 6, Oct. 1932, pp. 158–160.

1106. HENSEL, F. R. *and* HENGSTENBERG, T. F. Effects of inclusion streaks on the tensile and dynamic properties of wrought iron and similar materials. *Tech. Publ. Amer. Inst. Min. (metall.) Engrs*, no. 488, 1932, pp. 29.

1107. HOCHHEIM, R. Dauerfestigkeitsversuche mit geschweissten Trägen. [Fatigue strength tests with welded girders.] *Mitt. ForschAnst. Gutehoffn., Nürnberg*, vol. 1, no. 10, March 1932, pp. 225–227.

1108. HOFFMANN, W. Die Dauerversuche mit Nietverbindungen. [The fatigue strength of riveted joints.] *Elektroschweissung*, vol. 3, no. 4, April 1932, pp. 65–67.

1109. HOTTENROTT, E. Die Korrosionsschwingungsfestigkeit von Stählen und ihre Erhöhung durch Oberflächendrücken und elektrolytischen Schutz. [The corrosion fatigue strength of steels and its increase by surface pressure and electrolytic protection.] *Mitt. Wöhler-Inst.*, no. 10, 1932, pp. 62.

1110. HOUDREMONT, E. *and* BENNEK, H. Federstähle. [Spring steel.] *Stahl u. Eisen*, vol. 52, no. 27, July 7, 1932, pp. 653–662.

1111. IKEDA, S. On the transverse fissure of rails. *Journée int. du Rail*, 2nd Congress, Zurich, 1932, pp. 209–222.

1112. INGLIS, N. P. *and* LAKE, G. F. Corrosion fatigue tests of nitrided steel and nickel plated steel in River Tees water. *Trans. Faraday Soc.*, vol. 28, pt. 9, Sept. 1932, pp. 715–721.

113. JENNINGS, C. H. Fatigue strength and strain relief of welds. *Weld. Engr*, vol. 17,

no. 2, Feb. 1932, pp. 39–41, 58 and no. 3, March 1932, pp. 38–39.

1114. KAHN, L. Note sur les mouvements de tangage et la fatigue du navire à la mer. [Note on the pitching motion and the fatigue of a ship at sea.] *Bull. Ass. tech. marit.*, no. 36, 1932, pp. 125–150.

1115. KEEBLE, H. W. The failure of steel components by fatigue. *Iron Steel Ind.*, vol. 5, no. 4, Jan. 1932, pp. 165–170.

1116. KOMMERS, J. B. Understressing and notch sensitiveness in fatigue. *Engng News Rec.*, vol. 109, no. 12, Sept. 22, 1932, pp. 353–355.

1117. KREISSIG, E. Ermüdung und Korrosion. [Fatigue and corrosion.] *Verkehrstechnik*, vol. 13, no. 3, Jan. 25, 1932, pp. 42–43.

1118. KÜHNEL, R. Achsbrüche von Eisenbahnfahrzeugen und ihre Ursachen. [Axle fractures of railway vehicles and their causes.] *Glasers Ann. Gew.*, vol. 110, Feb. 15, 1932, pp. 29–37 and March 1, 1932, pp. 41–52. *Stahl u. Eisen*, vol. 52, no. 40, Oct. 6, 1932, pp. 965–969.

1119. KUKANOV, L. *Methods for the fatigue testing of metals.* [R] Moscow, Gosmashtechisdat, 1932, pp. 61.

1120. LEA, F. C. The strength of materials as affected by discontinuities and surface conditions. *J. Soc. Glass Tech.*, vol. 16, 1932, pp. 182–209. Abstr.: *Engineering, Lond.*, vol. 134, Aug. 26, 1932, pp. 256–258 and Sept. 2, 1932, pp. 280–281.

1121. LEHR, E. Minderung der Wechselfestigkeit von Fahrzeugfederblättern durch die

Walzhaut. [Reduction of fatigue strength of road vehicle leaf springs by the rolling scale.] *ForschArb. IngWes.*, vol. 3, 1932, pp. 54–55.

1122. LIPP, TH. Dauerfestigkeit geschweisster und gegossener Konstruktionen. [Fatigue properties of welded and cast designs.] *Schr. tech. Hochsch.*, *Darmstadt*, no. 4, 1932, pp. 73–79.

1123. LUDWIK, P. Fatigue strength and resistance to slip. *Metal Ind., Lond.*, vol. 41, Dec. 23, 1932, pp. 613–615 and Dec. 30, 1932, pp. 635–636.

1124. LUDWIK, P. *and* SCHEU, R. Die Veränderlichkeit der Werkstoffdämpfung. [The variability of material damping.] *Z. Ver. dtsch. Ing.*, vol. 76, no. 28, July 9, 1932, pp. 683–685.

1125. McADAM, D. J. Influence of stress on corrosion. *Trans. Amer. Inst. min. (metall.) Engrs*, vol. 99, 1932, pp. 282–322.

1126. McVETTY, P. G. *and* CROSS, H. C. Correlation of tension, creep and fatigue tests of 0.17% carbon steel at elevated temperatures. *Proc. Amer. Soc. Test. Mater.*, vol. 32, pt. 1, 1932, pp. 153–155.

1127. MAILANDER, R. Dauerbrüche und Dauerfestigkeit. [Fatigue fractures and fatigue strength.] *Krupp. Mh.*, vol. 13, no. 3, March 1932, pp. 56–81.

1128. MATTHAES, K. Corrosion fatigue and anodic oxidation of light metals. *Light Metals Res.*, vol. 2, no. 10, May 31, 1932, pp. 2–5.

1129. MATTHAES, K. Dynamische Festigkeitseigenschaften einiger Leichtmetalle. [Dynam-

ic strength properties of some light alloys.] *Z. Metallk.*, vol. 24, no. 8, Aug. 1932, pp. 176–180.

1130. MIES, O. Fehlerhafte Oberflächenbeschaffenheit als Ursache von Dauerbrüchen. [Defective surface finish as a cause of fatigue fracture.] *Maschinenschaden*, vol. 9, no. 4, 1932, pp. 61–64.

1131. MONROE, R. A. *and* TEMPLIN, R. L. Vibration of overhead transmission lines. *Trans. Amer. Inst. elect. Engrs*, vol. 51, 1932, pp. 1059–1073.

1132. MOORE, H. F. Fatigue of metals—a backward glance. *Metals and Alloys*, vol. 3, Sept. 1932, pp. 195 and 207.

1133. MOORE, H. F.; ROY, N. H. *and* BETTY, B. B. A study of stresses in car axles under service conditions. *Uni. Ill. Engng Exp. Sta. Bull.*, no. 244, May 20, 1932, pp. 80.

1134. MUNDT, R. Oberflächenspannungen und Ermüdrungsbruch bei Wälzlagern. [Surface stresses and fatigue fracture of anti-friction bearings.] *ForschArb. IngWes.*, vol. 3, no. 3, May–June 1932, pp. 127–134.

1135. MUSATTI, I. *and* CALBIANI, G. Le molle elicoidali — materiali adoperati, loro composizione e loro caratteristiche meccaniche — metodi di fabbricazione — sollecitazioni ammissibili e cause di rottura. [Coil springs —suitable materials, composition and mechanical characteristics—methods of manufacture—allowable stresses and causes of failure.] *Metallurg. ital.*, vol. 24, June 1932, pp. 465–484 and July 1932, pp. 549–572.

1136. OCHS, H. Einfluss der Korrosion auf die Dauerfestigkeit. [Effect of corrosion on the fatigue strength.] *Schr. tech. Hochsch.*, *Darmstadt*, no. 4, 1932, pp. 55–59.

1137. Oschatz, H. Zur Steigerung der Dauerhaltbarkeit von Konstruktionsteilen. [Increasing the fatigue strength of constructional parts.] *Schr. tech. Hochsch.*, *Darmstadt*, no. 4, 1932, pp. 44–50.

1138. Peineke, W. Allgemeines über Schwingungsbrüche an elektrischen Antrieben. [General information about fatigue fractures in electrical drives]. *Maschinenbau*, vol. 11, July 7, 1932, pp. 272–273.

1139. Peineke, W. Dauerbiegebrüche an elektrischen Antrieben. [Bending fatigue fractures in electrical drives.] *Maschinenbau*, vol. 11, no. 14, July 21, 1932, pp. 293–297.

1140. Peineke, W. Dauerbrüche durch Drehschwingungen an elektrischen Antrieben. [Fatigue fractures resulting from torsional vibration in electrical drives.] *Maschinenbau*, vol. 11, no. 15, Aug. 4, 1932, pp. 313–316.

1141. Persoz, L. La rupture des pièces de machines par fatigue. [The fatigue fracture of machine parts.] *Aciers spéc.*, vol. 7, no. 86, Oct. 1932, pp. 359–373; no. 88, Dec. 1932, pp. 428–438 and vol. 8, no. 90, Feb. 1933, pp. 36–49.

1142. Pester, F. Festigkeitsprüfung an Stangen und Drähten bei tiefen Temperaturen. [Strength tests of bars and wires at low temperatures]. *Z. Metallk.*, vol. 24, no. 3, March 1932, pp. 67–70 and no. 5, May 1932, pp. 115–120.

1143. Peterson, R. E. Dynamic testing of materials. *Elect. (Cl.) J.*, vol. 29, no. 8, Aug. 1932, pp. 377–379.

1144. Peterson, R. E. Fatigue of shafts having keyways. *Proc. Amer. Soc. Test. Mater.*, vol. 32, pt. 2, 1932, pp. 413–420. Abstr.: *Mech. World*, vol. 92, no. 2393, Nov. 11, 1932, pp. 455–456.

1145. Peterson, R. E. *and* Moore, H. F. The significance and limitations of fatigue test results. *Proc. Amer. Soc. Test. Mater.*, vol. 32, pt. 1, 1932, pp. 142–147. *Iron Steel Ind.*, vol. 5, no. 12, Sept. 1932, pp. 437–438.

1146. Pohl, R. Mechanische Problems bei grossen Turbogeneratoren. [Mechanical problems in large turbo-generator rotors.] *Elektrotech. Z.*, vol. 53, no. 46, Nov. 17, 1932, pp. 1099–1101.

1147. Probst, E. *and* Treiber, F. Eisenbetonbalken unter dem Einfluss häufig wiederholter Belastung. [Reinforced concrete beams under frequently repeated loading.] *Bauingenieur*, vol. 13, no. 21/22, 1932, pp. 285–289.

1148. Puritz, P. Schwingungen an Freileitungsseilen und ihre Dämpfung durch Resonanzschwingungsdämpfer. [Vibration of overhead wires and their damping by means of the resonant vibration damper.] *Mitt. Wöhler-Inst.*, no. 12, 1932, pp. 67.

1149. Regler, F. Statische und dynamische Beanspruchung von Stahlkonstruktionen im Lichte der Röntgenstrahlen. [Static and dynamic stressing of structural steel in the light of X-ray investigations.] *Mitt. tech. VersAmt.*, *Wien*, vol. 21, 1932, pp. 31–46.

1150. Rosenthal, D. Augmentation de la résistance aux chocs répétés des pièces assemblées par soudure. [Increasing the resistance to repeated impact of parts assembled by welding.] *Génie civ.*, vol. 100, Jan. 23, 1932, p. 98.

1151. RUTTMANN, W. Ermüdungsrisse in den Einwalzstellen von Biegerohren und Einwalzbarkeit verschiedener Werkstoffe. [Fatigue cracks in the rolled positions of bent tubes and the rollability of different materials.] *Schr. tech. Hochsch., Darmstadt*, no. 4, 1932, pp. 63–68.

1152. SARAN, W. Die Dauerfestigkeit der Leichtmetall-Sandguss-Legierungen. [The fatigue strength of light metal sand cast alloys.] *Mitt. dtsch. MatPrüfAnst.*, no. 12, 1932, pp. 184–185.

1153. SARAN, W. A machine for determining the fatigue limit of metals. *Engineering, Lond.*, vol. 133, no. 3467, June 24, 1932, pp. 731–734.

1154. SARAN, W. Leichtmetall-Sandguss, seine statische und seine Schwingungsfestigkeit. [Light metal sand castings, their static and fatigue strength.] *Z. Metallk.*, vol. 24, no. 8, Aug. 1932, pp. 181–184 and no. 9, Sept. 1932, pp. 207–210.

1155. SARAN, W. Proprietà statiche e dinamiche delle leghe leggere colate in terra. [Static and dynamic properties of light alloys cast in sand moulds.] *Alluminio*, vol. 1, no. 6, Nov.–Dec. 1932, pp. 368–376.

1156. SCHAECHTERLE, K. Dauerversuche mit Nietverbindungen. [Fatigue tests with riveted joints.] *Stahlbau*, vol. 5, no. 9, April 29, 1932, pp. 65–72.

1157. SCHAECHTERLE, K. Die Nietverbindung. Neue Erkenntnisse aus Dauerversuchen. [The riveted joint. New knowledge from fatigue tests.] *Bautechnik*, vol. 10, no. 22, May 20, 1932, pp. 275–278 and no. 23, May 27, 1932, pp. 290–294.

1158. SCHAECHTERLE, K. Die zulässigen Spannungen bei genieteten und geschweissten Stahlbrücken. [The allowable stresses in riveted and welded steel bridges.] *Bautechnik*, vol. 10, no. 44, Oct. 7, 1932, pp. 590–593 and no. 45, Oct. 14, 1932, pp. 603–605.

1159. SCHRAIVOGEL, K. Dauerbiegeversuche mit Schraubenbolzen. [Bending fatigue tests on bolts.] *Stahl u. Eisen*, vol. 52, no. 48, Dec. 1, 1932, pp. 1189–1193.

1160. SCHWINNING, W. and STROBEL, E. Untersuchungen über die Wärmefestigkeit von Leichtmetallen bei statischer und bei wechselnder Beanspruchung. [Investigations of the effect of high temperature on the strength of light metals under static and alternating stresses.] *Z. Metallk.*, vol. 24, no. 6, June 1932, pp. 132–137 and no. 7, July 1932, pp. 151–153.

1161. SOLBERG, T. A. and HAVEN, H. E. Importance of fatigue of metals to engineering design. *Metals and Alloys*, vol. 3, no. 9, Sept. 1932, pp. 196–199.

1162. STAEDEL, W. Dauerfestigkeit von Schrauben. [Fatigue strength of screws.] *ForschArb. IngWes.*, vol. 3, March–April 1932, pp. 106–108.

1163. SWANGER, W. H. and FRANCE, R. D. Effect of zinc coatings on the endurance properties of steel. *Proc. Amer. Soc. Test., Mater.*, vol. 32, pt. 2, 1932, pp. 430–452. *J. Res. nat. Bur. Stand.*, vol. 9, no. 1, July 1932, pp. 9–24. Abstr.: *Automot. Industr. N.Y.*, vol. 67, no. 6, Aug. 6, 1932, p. 175. *Machinery, N.Y.*, vol. 39, no. 1, Sept. 1932, p. 77.

1164. TAPSELL, H. J. and THORPE, P. L. Report on fatigue tests made at air and elevated temperatures on some valve steels and piston alloys. *Rep. aero. Res. Comm., Lond.* no. E. F. 309, March 1932, pp. 14.

1165. THOMAS, H. R. *and* LOWTHER, J. G. Fatigue failure under repeated compression. *Proc. Amer. Soc. Test. Mater.*, vol. 32, pt. 2, 1932, pp. 421–429. Abstr.: *Iron Steel Ind.*, vol. 5, no. 12, Sept. 1932, pp. 429–431.

1166. THORNTON, G. E. Giving the weld a chance. *J. Amer. Weld. Soc.*, vol. 11, no. 2, Feb. 1932, pp. 9–10 and [discussion by C. H. Jennings] no. 6, June 1932, pp. 10–11.

1167. THUM, A. *and* BUCHMANN, W. Über die Eignung von Werkstoffen für Dauerbruchgefährdete Konstruktionsteile. [On the suitability of materials for structural parts liable to dangerous fatigue fractures.] *Maschinenschaden*, vol. 9, no. 12, 1932, pp. 181–185.

1168. THUM, A. *and* BUCHMANN, W. Dauerfestigkeit und Konstruktion. [Fatigue strength and construction.] *Mitt. MatPrüfAnst., Darmstadt*, no. 1, 1932, pp. 82.

1169. THUM, A. *and* OCHS, H. Die Bekämpfung der Korrosionsermüdung durch Druckvorspannung. [The reduction of corrosion-fatigue by means of pre-compression.] *Z. Ver. dtsch. Ing.*, vol. 76, no. 38, Sept. 17, 1932, pp. 915–916.

1170. THUM, A. *and* OSCHATZ, H. Gesetzmässigkeiten des Dauerbruchweges. [Characteristic features of the path of fatigue cracks.] *Z. Ver. dtsch. Ing.*, vol. 76, no. 6, Feb. 6, 1932, pp. 132–134. English abstr.: *Metallurgist*, vol. 8, Oct. 28, 1932, pp. 155–156.

171. THUM, A. *and* OSCHATZ, H. Steigerung der Dauerfestigkeit bei Rundstäben mit Querbohrungen. [Increasing the fatigue strength of round specimens with transverse holes.] *ForschArb. IngWes.*, vol. 3, no. 2, March–April 1932, pp. 87–93.

172. THUM, A. *and* OSCHATZ, H. Dauerbruchformen und ihre Entstehung. [Forms of fatigue fractures and their origin.] *Maschinenschaden*, vol. 9, no. 8/9, 1932, pp. 121–130.

1173. THUM, A. *and* STAEDEL, W. Über die Dauerfestigkeit von Schrauben. [On the fatigue strength of screws.] *Maschinenbau*, vol. 11, June 2, 1932, pp. 230–232.

1174. THUM, A. *and* WUNDERLICH, F. Die Fliessgrenze bei behinderter Formänderung. Ihre Bedeutung für das Dauerfestigkeitsschaubild. [The yield point in the case of restricted deformation. Its significance in relation to the fatigue strength diagram.] *ForschArb. IngWes.*, vol. 3, no. 6, Nov.–Dec. 1932, pp. 261–270.

1175. TOWNSEND, J. R. *and* GREENALL, C. H. Fatigue studies of telephone cable sheath alloys—II. *Bell Telep. Syst.*, Monograph B-669, 1932, pp. 9.

1176. VOIGT, E. *and* CHRISTENSEN, K. H. Über die Dämpfungsfähigkeit und Schwingungsfestigkeit des Stahles. [On the damping capacity and fatigue strength of steel.] *Mitt. K.-Wilh.-Inst. Eisenforsch.*, vol. 14, no. 11, 1932, pp. 151–167. German abstr.: *Stahl u. Eisen*, vol. 52, no. 44, Nov. 3, 1932, pp. 1077–1078.

1177. WORTHINGTON, R. Nickel tubing resists fatigue. *Automot. Industr., N.Y.*, vol. 66, no. 5, Jan. 30, 1932, p. 164.

1178. WUNDERLICH, F. Über Bruchgefahr von elektrischen Freileitungsseilen. [On the fracture of electric transmission cables.] *Schr. tech. Hochsch., Darmstadt*, no. 4, 1932, pp. 59–63.

1179. The annealing of chains. *Metallurgist*, vol. 8, Jan. 29, 1932, pp. 9–11.

1180. Corrosion-fatigue. *Metallurgist*, vol. 8, Jan. 29, 1932, pp. 11–13.

1181. Crystalline metal. *Metallurgist*, vol. 8, March 25, 1932, pp. 33–34.

1182. Das Problem der Fahrzeugfederung. Wirtschaftliche Bedeutung, Prüftechnik, Fortschrittsmöglichkeiten. [The problem of vehicle springs. Industrial significance, testing techniques, possibility of improvements.] *Tech. Bl., Düsseldorf*, vol. 22, April 10, 1932, pp. 196–197.

1183. Dauerversuche über die Korrosion von Kondensatorrohren im Betrieb. [Corrosion-fatigue tests on condenser tubes in service.] *Arch. Wärmew.*, vol. 13, no. 4, April 1932, p. 109.

1184. Recent American researches on "fatigue of metals". *Metallurgist*, vol. 8, Oct. 28, 1932, pp. 157–158 and Nov. 25, 1932, pp. 167–168.

1185. Ausschuss für Aluminiumleitungen — Bericht. Untersuchung beschädigter Leitungsseile — Schwingungsbrüche. [Committee for Aluminium Conductors—report. Investigation of damaged transmission cables—vibration fractures.] *Z. Metallk.*, vol. 24, no. 11, Nov. 1932, pp. 285–288.

1186. A method of designing machine parts for maximum resistance to fracture. *Engng Progr.*, vol. 13, no. 12, Dec. 1932, pp. 264–267.

1187. Les méthodes modernes d'essais de matériaux dans la construction des moteurs. Machine d'essais dynamiques, système Schenck. [Modern methods for testing materials used in the construction of motors. Schenck dynamic testing machines.] *Génie civ.*, vol. 101, no. 23, Dec. 3, 1932, pp. 550–553.

1933

1188. AIZAWA, T. *and* OSANAI, K. On lead alloys for cable sheaths. [J] *Res. electrotech. Lab., Tokyo*, no. 346, 1933, pp. 32.

1189. ASAKAWA, Y.; KUMAGAYA, T. *and* SUGIE, K. Study of fatigue of metals. [J] *J. Soc. mech. Engrs, Japan*, vol. 36, no. 193, May 1933, pp. 321–322.

1190. BACON, F. Fatigue and corrosion fatigue with special reference to service breakage. *Proc. Instn mech. Engrs, Lond.*, vol. 124, June 1933, pp. 685–736.

1191. BAN, S. Der Ermüdungsvorgang von Beton. [The fatigue process in concrete.] *Bauingenieur*, vol. 14, no. 13/14, 1933, pp. 188–192.

1192. BASTIEN, P. Étude des alliages magnésium-aluminium-cuivre, riches en magnésium — contribution à l'étude des propriétés de fonderie des métaux et alliages. [Study of magnesium–aluminium–copper alloys rich in magnesium—contribution to the study of the properties of cast metals and alloys.] *Publ. sci. Minist. Air*, no. 20, 1933, pp. 9.

1193. BEHRENS, O. Der Einfluss der Korrosion auf die Biegungsschwingungsfestigkeit von Stählen und Reinnickel. [The influence of corrosion on the bending fatigue strength of steels and pure nickel.] *Mitt. Wöhler-Inst.*, no. 15, 1933, pp. 73.

1194. BIERETT, G. Zur Klärung der mechanischen Grundlagen des Dauerbruchs geschweisster Konstruktionen. [The clarification of mechanical principles controlling fatigue fracture in welded structures. *Elektroschweissung* vol. 4, no. 2, Feb. 1933, pp. 21–27.

1195. BIERETT, G. Die Schweissverbindung be dynamischer Beanspruchung. [The welde joint under dynamic loading.] *Elektroschweis sung*, vol. 4, no. 4, April 1933, pp. 61–70ua no. 5, May 1933, pp. 94–97.

1196. BIERETT, G. *and* GRÜNING, G. Spannung zustand und Festigkeit von Stirnkehlnah verbindungen. [Stress conditions and strength of fillet welded joints.] *Stahlbau*, vol. 6, no. 22, 1933, pp. 169–175.

1197. BUCHHOLTZ, H. *and* KREKELER, K. Zur Bekämpfung des Korrosiondauerbruchs. [The prevention of corrosion fatigue failures.] *Stahl u. Eisen*, vol. 53, no. 26, June 29, 1933, pp. 671–674.

1198. BÜHLER, H. *and* BUCHHOLTZ, H. Über die Wirkung von Eigenspannungen auf die Schwingungsfestigkeit. [On the effect of internal stresses on the fatigue strength.] *Mitt. ForschInst. ver. Stahlw. A.-G.*, vol. 3, no. 8, Sept. 1933, pp. 235–248.

1199. BÜHLER, H. *and* BUCHHOLTZ, H. Die Wirkung von Eigenspannungen auf die Biegeschwingungsfestigkeit. [The effect of internal stresses on the bending fatigue strength.] *Stahl u. Eisen*, vol. 53, no. 51, Dec. 21, 1933, pp. 1330–1332. English translation: *Henry Brutcher Tech. Trans.*, no. 53.

1200. BÜHLER, H. *and* SCHEIL, E. Zusammenwirkung von Wärme- und Umwandlungsspannungen in abgeschreckten Stählen. [Combined effects of heat and transformation stresses in quenched steels.] *Arch. Eisenhüttenw.*, vol. 6, no. 7, Jan. 1933, pp. 283–288.

1201. CAQOUT, A. L'exploitation, par l'ingénieur de la résistance de la matière. [The utilization, by the engineer of the resistance of materials.] *Bull. Soc. Ing. civ.*, vol. 86, no. 9/10, Sept.–Oct. 1933, pp. 1088–1124.

1202. CASWELL, J. S. The effect of surface finish on the fatigue limit of mild steel. *Engineering, Lond.*, vol. 136, no. 3526, Aug. 11, 1933, pp. 154–155.

1203. CAZAUD, R. Influence du degré d'écrouissage à l'étirage sur la limite de fatigue de l'acier doux. [Influence of degree of cold-working,

on the fatigue limit of mild steel.] *C. R. Acad. Sci., Paris*, vol. 196, no. 10, March 6, 1933, pp. 696–698.

1204. CHMIELOWIEC, A. Le taux de fatigue à admettre dans les ponts métalliques. [Admissible fatigue stresses in steel bridges.] *Publ. int. Ass. Br. struct. Engng*, vol. 2, 1933–1934, pp. 71–79.

1205. COLLELL, R. Schwingungsfestigkeit. [Fatigue strength.] *Tech. Bl., Düsseldorf*, vol. 23, June 4, 1933, pp. 319–320 and June 11, 1933, p. 329.

1206. COOPER, L. V. New machine for laboratory evaluation of fatigue of rubber compounds flexed under compression. *Industr. Engng Chem. (Anal.)*, vol. 5, no. 5, Sept. 15, 1933, pp. 350–351.

1207. CUTHBERTSON, J. W. Fatigue testing. *Engineering, Lond.*, vol. 136, no. 3523, July 21, 1933, pp. 55–57 and no. 3524, July 28, 1933, pp. 80–82.

1208. CUTHBERTSON, J. W. The fatigue resisting properties of light aluminium alloys at elevated temperatures. *J. Inst. Met.*, vol. 51, no. 1, 1933, pp. 163–181. Abstr.: *Engineering, Lond.*, vol. 135, no. 3506, March 24, 1933, pp. 342–343.

1209. DOREY, S. F. The problem of the water-cooled piston rod in two-stroke cycle double-acting oil engines. *Trans. Instn nav. Archit. Lond.*, vol. 75, 1933, pp. 200–236. *Mot. Ship*, vol. 14, no. 159, May 1933, pp. 71–73; no. 163, Sept. 1933, p. 215 and no. 164, Oct. 1933, pp. 236–238. Abstr.: *Shipb. Shipp. Rec.* vol. 41, no. 16, April 20, 1933, pp. 395–397. *Shipbuilder*, vol. 40, no. 277, April 1933, pp. 269–272.

1210. DUSOLD, T. Der Einfluss der Korrosion auf die Drehschwingungsfestigkeit von Stählen und Nichteisenmetallen. [The influence of corrosion on the torsional fatigue strength of steels and non-ferrous metals.] *Mitt. Wöhler-Inst.*, no. 14, 1933, pp. 89.

1211. EDGERTON, C. T. Fatigue tests of helical springs and notes on sundry results of the tests. *Wire and Wire Prod.*, vol. 8, no. 7, July 1933, pp. 210–211 and 220–221.

1212. ESAU, A. *and* KORTUM, H. Die Veränderlichkeit der Werkstoffdämpfung. [The variability in material damping.] *Z. Ver. dtsch. Ing.*, vol. 77, no. 42, Oct. 21, 1933, pp. 1133–1135.

1213. FAULHABER, R.; BUCHHOLTZ, H. *and* SCHULZ, E. H. Über den Einfluss des Probestabdurchmessers auf die Biegeschwingungsfestigkeit von Stahl. [On the influence of test diameter on the bending fatigue strength of steel.] *Mitt. ForschInst. ver. Stahlw. A.-G.*: vol. 3, no. 6, June 1933, pp. 153–172. Abstr.: *Stahl u. Eisen*, vol. 53, no. 43, Oct. 26, 1933, pp. 1106–1108.

1214. FISCHER, F. Vorschlag zur Festlegung der zulässigen Beanspruchungen im Maschinenbau. [Proposal for determining the allowable stresses in machine components.] *Tech. Mitt. Krupp*, vol. 1, Sept. 1933, pp. 67–83.

1215. FÖPPL, O.; BEHRENS, O. *and* DUSOLD, T. Die Erniedrigung der Schwingungsfestigkeit durch Korrosion und ihre Erhöhung durch Oberflächendrücken. [The reduction of the fatigue strength by corrosion and its increase by surface compression.] *Z. Metallk.*, vol. 25, no. 11, Nov. 1933, pp. 279–282. English abstr.: *Metallurgist*, vol. 9, Dec. 29, 1933, p. 94.

1216. FORSMAN, O. Bidrag till kännedomen om ståls hållfasthet mot utmattning och slag vid låg temperatur. [A contribution to the knowledge of fatigue and impact strength of steel at low temperatures.] *Jernkontor. Ann.*, vol. 88, Nov. 1933, pp. 519–530. English abstr.: *J. Iron St. Inst.*, vol. 129, no. 1, 1934, pp. 581–582. *Iron Coal Tr. Rev.*, vol. 127, no. 3435, Dec. 29, 1933, p. 974.

1217. FRANKENBERG, H. Der Einfluss von Drehschwingungsbeanspruchungen auf die Festigkeit und Dämpfungsfähigkeit von Metallen, besonders von Aluminium-Legierungen. [The influence of repeated torsional fatigue stresses on the strength and damping capacity of metals, particularly aluminium alloys.] *Mitt. Wöhler-Inst.*, no. 16, 1933, pp. 55.

1218. FRENCH, H. J. Fatigue and the hardening of steels. *Trans. Amer. Soc. Steel Treat.*, vol. 21, no. 10, Oct. 1933, pp. 899–946. Abstr.: *Iron Coal Tr. Rev.*, vol. 127, no. 3432, Dec. 8, 1933, p. 865.

1219. FULLER, T. S. Endurance of metal in corrosive surroundings. *Metal Progr.*, vol. 23, no. 6, June 1933, pp. 23–26.

1220. GOUGH, H. J. Crystalline structure in relation to failure of metals—especially by fatigue. *Proc. Amer. Soc. Test. Mater.*, vol. 33, pt. 2, 1933, pp. 3–114. Abstr.: *Metals and Alloys*, vol. 4, no. 9, Sept. 1933, pp. 140, 142.

1221. GOUGH, H. J. *and* COX, H. L. Failure of single crystals of aluminium under combined torsional fatigue stresses and static direct stresses. *Rep. aero. Res. Comm., Lond.*, no. E. F. 329, Jan. 13, 1933, pp. 36.

1222. GOUGH, H. J. *and* SOPWITH, D. G. Fatigue of an aluminium specimen consisting of two crystals in twinned relation. *Rep. aero. Res. Comm., Lond.*, no. E. F. 272a, Feb. 23, 1933, pp. 23.

1223. GOUGH, H. J. *and* SOPWITH, D. G. Some comparative corrosion-fatigue tests employing two types of stressing action. *J. Iron St. Inst.*, vol. 127, no. 1, 1933, pp. 301–335. *Engineering, Lond.*, vol. 136, no. 3523, July 21, 1933, pp. 75–78.

1224. GOUGH, H. J. *and* SOPWITH, D. G. Corrosion-fatigue characteristics of an aluminium specimen consisting of two crystals. *J. Inst. Met.*, vol. 52, no. 2, 1933, pp. 57–72.

1225. GOULD, A. J. The influence of solution concentration on the severity of corrosion fatigue. *Engineering, Lond.*, vol. 136, no. 3537, Oct. 27, 1933, pp. 453–454. Abstr.: *Commonw. Engr*, vol. 21, Jan. 1, 1934, p. 157.

1226. GRAF, O. Über die Dauerfestigkeit von Schweissverbindungen. [On the fatigue strength of welded joints.] *Stahlbau*, vol. 6, no. 11, May 26, 1933, pp. 81–85 and nos. 12/13, June 9, 1933, pp. 89–94.

1227. GRAF, O. Tests on the fatigue limit of welded joints subjected to repeated tension stresses. *J. Amer. Weld. Soc.*, vol. 12, no. 8, Aug. 1933, pp. 30–32.

1228. GRAF, O. Versuchsergebnisse als Grundlage für Bemessungsregeln geschweisster Konstruktionen. [Experimental results as the basis for the dimensioning of welded structures.] *Stahl u. Eisen*, vol. 53, no. 47, Nov. 23, 1933, pp. 1215–1220.

229. GREENWOOD, J. N. Modern researches on the properties of engineering materials. *Mod. Engr*, vol. 7, no. 8, Aug. 20, 1933, pp. 243–248 and no. 9, Sept. 20, 1933, pp. 277–281.

230. GREGER, O. Über die Dauererprobung autogen geschweisster Nähte. [Fatigue tests

on fusion welds.] *Z. Schweisstech.*, vol. 23, no. 6, June 1933, pp. 161–165 and no. 7, July 1933, pp. 180–185.

1231. HANEL, R. Schwingungsfestigkeit von Nickelstählen. [Fatigue strength of nickel steels.] *Nickel-Ber.*, no. 3, 1933, pp. 33–38.

1232. HEROLD, W. Wechselfestigkeit der im Automobilbau verwendeten Stähle. [Fatigue strength of steels used in automobile construction.] *Auto.-tech. Z.*, vol. 36, no. 1, Jan. 10, 1933, pp. 4–9 and no. 2, Jan. 25, 1933, pp. 40–42.

1233. HERTEL, H. Dynamic breaking tests of airplane parts. *Tech. Memor. nat. Adv. Comm. Aero., Wash.*, no. 698, Jan. 1933, pp. 42.

1234. HEUMANN, F. *Verhalten Keramischer Werkstoffe bei Zugdruck-Dauerbeanspruchung. [Behaviour of ceramic materials under tension-compression fatigue loading.]* Berlin, VDI-Verlag, 1933, pp. 42. Abstr.: *Z. Ver. dtsch. Ing.*, vol. 77, no. 14, April 8, 1933, p. 365. English abstr.: *Engng Progr.*, vol. 14, no. 4, April 1933, p. 84.

1235. HOFFMANN, W. Beitrag zur Klärung des Dauerbruches geschweisster Verbindungen. [Contribution to the explanation of fatigue failure of welded joints.] *ForschArb. dtsch. Acetylenver.*, vol. 8, 1933, pp. 68–71. *Autogene Metallbearb.*, vol. 26, no. 7, April 1, 1933, pp. 100–102.

1236. HOHENEMSER, K. *and* PRAGER, W. Zur Frage der Ermüdungsfestigkeit bei mehrachsigen Spannungszuständen. [On the question of fatigue strength under complex stress conditions.] *Metallwirtschaft*, vol. 12, no. 24, June 16, 1933, pp. 342–343.

1237. HOLDT, H. Die Beurteilung von Werkstoffeigenschaften aus Gewaltbrüchen. [Esti-

mation of material properties from fractures.]
Schr. tech. Hochsch., Darmstadt, no. 2, 1933,
pp. 26–30.

1238. HOLZHAUER, C. I. Ermüdungsfestigkeit
von Kesselbaustoffen und ihrer Beeinflussung
durch chemische Einwirkungen. [Fatigue
strength of boiler materials and the effect of
chemical action.] *Mitt. MatPrüfAnst., Darmstadt*, no. 3, 1933, pp. 73.

1239. HOPPE, H. Die Messung dynamischer
Wuchtfehler auf einer Auswuchtmaschine mit
elektro-magnetischen Wuchtvorrichtungen für
Gleich- und Wechselstrom. [The measurement
of errors in dynamic forces using balancing
machines with electro-magnetic force apparatus for direct and alternating current.] *Mitt.
Wöhler-Inst.*, no. 17, 1933, pp. 63.

1240. HUNTER, W. The deflection of a cantilever
bar rotated under end load. *Proc. Camb.
phil. Soc.*, vol. 29, pt. 3, 1933, pp. 423–439.

1241. IKEDA, S. Design of railway vehicle axles
from the viewpoint of repeated stresses in
their force fitted parts. [J] *J. Soc. mech.
Engrs, Japan*, vol. 36, no. 190, Feb. 1933,
pp. 101–109.

1242. JENNINGS, C. H. Physical properties of
welded cast steel. *J. Amer. Weld. Soc.*,
vol. 12, no. 10, Oct. 1933, pp. 25–29.

1243. JOHNSON, J. B. *and* OBERG, T. Mechanical
properties at minus 40 degrees of metals used
in aircraft construction. *Metals and Alloys*,
vol. 4, no. 3, March 1933, pp. 25–30.

1244. KENDRICK, J. A. Crystallization of metals
is not caused by vibration. *Pwr Plant (Engng)*,
vol. 37, no. 9, Sept. 1933, p. 402.

1245. KÖRBER, F. *and* HEMPEL, M. Dauerversuche auf einer hochfrequenten Zug-Druck-

Maschine: Die Änderungen der Frequenz
bzw. des Elastizitätsmoduls und deren Bedeutung für die Spannungsermittlung bei verschiedener Schwingungsweite. [Fatigue tests on a
high-frequency tensile-compression machine;
the variations of the frequency and of the
modulus of elasticity and their importance
in the determination of the stresses at various
amplitudes of vibration.] *Mitt. K.-Wilh.-Inst.
Eisenforsch.*, vol. 15, no. 10, 1933, pp. 119–
135. Abstr.: *Stahl u. Eisen*, vol. 53, no. 31,
Aug. 3, 1933, p. 813. *Z. Ver. dtsch. Ing.*,
vol. 77, no. 36, Sept. 9, 1933, p. 982.

1246. KOHN, P. Zusammengesetzte Beanspruchung und Sicherheit bei statischer und wechselnder Belastung. [Complex stresses and safety under static and fatigue loads.] *Schweiz.
Bauztg.*, vol. 102, no. 17, Oct. 21, 1933,
pp. 203–205.

1247. KOMMERELL, O. Verfahren zur Berechnung
von Fachwerkstäben und auf Biegung beanspruchten Trägern bei wechselnder Belastung.
[Methods of proportioning framed girder
members for bending stresses under alternating loads.] *Bautechnik*, vol. 11, 1933,
pp. 114–116.

1248. KOMMERS, J. B. Failure by fatigue. *Engineering, Lond.*, vol. 135, March 3, 1933,
pp. 238–239.

1249. KOMMERS, J. B. The static and fatigue
properties of brass. *Bull. Univ. Wis. Engng
Ser.*, no. 76, 1933, pp. 38.

1250. LAUTE, K. Korrosion und Ermüdung
[Corrosion and fatigue.] *Ber. Korrosionstag
Ver. dtsch. Eisenhüttenl.*, 1933, pp. 1–11
English abstr.: *Bull. Brit. non-ferr. Met. Ass.*
no. 68, Aug.–Sept. 1934, p. 17.

1251. LEHR, E. Die Wechselfestigkeit des Ven
tilfederdrahtes. [The fatigue strength of valv

spring wire.] *Z. Ver. dtsch. Ing.*, vol. 77, no. 24, June 17, 1933, pp. 648–649.

1252. LEHR, E. *and* PRAGER, W. Dauerprüfmaschine für überlagerte Zug-Druck- und Schub-Wechselbeanspruchung. [Fatigue testing machine for combined alternating loads of tension, compression and shear.] *Forsch-Arb. IngWes.*, vol. 4, no. 5, Sept.–Oct. 1933, pp. 209–214.

1253. LEQUIS, W.; BUCHHOLTZ, H. *and* SCHULZ, E. H. Biegeschwingungsfestigkeit und Kerbempfindlichkeit in ihrer Beziehung zu den übrigen Festigkeitseigenschaften bei Stahl. [Bending fatigue strength and notch sensitivity and their relation to other strength properties of steel.] *Mitt. ForschInst. ver. Stahlw. A.-G.*, vol. 3, no. 6, June 1933, pp. 129–152. *Stahl u. Eisen*, vol. 53, no. 44, Nov. 2, 1933, pp. 1133–1137.

1254. LUDWIK, P. Ermüdung metallischer Werkstoffe. [Fatigue of metallic materials.] *Forsch. Fortschr. dtsch. Wiss.*, vol. 9, June 1933, p. 274.

1255. LUDWIK, P. Das Verhalten metallischer Werkstoffe bei ruhender und wechselnder Beanspruchung. [The behaviour of metallic materials under static and alternating stress.] *Z. Metallk.*, vol. 25, no. 10, Oct. 1933, pp. 221–228.

1256. LUDWIK, P. *and* KRYSTOF, J. Korrosionsschutz bei wechselnder Beanspruchung. [Protection against corrosion under alternating stress.] *Mitt. tech. VersAmt.*, Wien, vol. 22, 1933, pp. 42–49.

1257. LUDWIK, P. *and* KRYSTOF, J. Einfluss der Vorspannung auf die Dauerfestigkeit. [The influence of pre-stressing on the fatigue strength.] *Z. Ver. dtsch. Ing.*, vol. 77, no. 24, June 17, 1933, pp. 629–635. Abstr.: *Anz. Akad. Wiss.*, Wien, vol. 70, no. 6, 1933, pp. 51–52.

1258. MAILÄNDER, R. Über die Dauerfestigkeit von nitrierten Proben. [On the fatigue strength of nitrided parts.] *Z. Ver. dtsch. Ing.*, vol. 77, no. 10, March 11, 1933, pp. 271–274. *Tech. Mitt. Krupp*, vol. 1, no. 2, July 1933, pp. 53–58. *Elektrotech. Z.*, vol. 54, no. 10, 1933, pp. 271–274.

1259. MAILÄNDER, R. Gesetzmässigkeiten des Dauerbruches und Wege zur Steigerung der Dauerhaltbarkeit. [Regularity of fatigue failures and ways of increasing fatigue properties.] *ForschArb. IngWes.*, vol. 4, no. 5, 1933, p. 258.

1260. MATTHAES, K. Eine Planbiege-Dauerprüfmaschine der DVL und die damit erhaltenen Versuchsergebnisse. [A plane bending fatigue testing machine of the DVL and test results obtained with it.] *Metallwirtschaft*, vol. 12, no. 34, Aug. 25, 1933, pp. 485–489. Abstr.: *Z. Ver. dtsch. Ing.*, vol. 77, no. 1, Jan. 7, 1933, pp. 27–28.

1261. MATTHAES, K. Die Dauerfestigkeit der Werkstoffe des Flugzeug- und Flugmotorenbaues. [The fatigue strength of materials for aircraft and aircraft engines.] *Z. Flugtech.*, vol. 24, no. 21, Nov. 14, 1933, pp. 593–598 and no. 22, Nov. 28, 1933, pp. 620–626.

1262. MOORE, H. F. The fatigue of metals—its nature and significance. *J. appl. Mech.*, vol. 1, no. 1, 1933, pp. 15–18.

1263. MOORE, H. F. Corrosion fatigue of metals. *Metals and Alloys*, vol. 4, no. 3, March 1933, pp. 39–40.

1264. MOORE, H. F. *and* WISHART, H. B. An "overnight" test for determining endurance limit. *Proc. Amer. Soc. Test. Mater.*, vol. 33, pt. 2, 1933, pp. 334–347. Abstr.: *Automot. Industr.*, N.Y., vol. 69, no. 8, Aug. 19, 1933, p. 211.

1265. MÜLLER, J. Schwingungsfestigkeit von Stahlrohr-Schweissverbindungen. [Fatigue strength of steel tubing with welded joints.] *Z. Ver. dtsch. Ing.*, vol. 77, no. 26, July 1, 1933, pp. 720–721.

1266. NETTMAN, P. Dauerbruch durch Korrosion. [Fatigue failure caused by corrosion.] *Auto.-tech. Z.*, vol. 36, no. 17, Sept. 10, 1933, pp. 438–439 and no. 18, Sept. 25, 1933, pp. 459–460.

1267. NISHIHARA, T.; SAKURAI, T. *and* WATANABE, T. Fatigue testing of steels for tensile, compression and bending stresses. [J] *J. Soc. mech. Engrs, Japan*, vol. 36, no. 198, Oct. 1933, pp. 673–681.

1268. NISHIHARA, T.; SAKURAI, T. *and* WATANABE, T. Contribution to the theory of the Haigh direct stress fatigue testing machine. [J] *J. Soc. mech. Engrs, Japan*, vol. 36, no. 198, Oct. 1933, pp. 682–687.

1269. OGNEVETSKII, A. The dependence of the endurance limit on the nature of the repeated loading. [R] *Avtogennoe delo*, no. 10, 1933, pp. 9–13.

1270. OSCHATZ, H. Über gesetzmässige Dauerbruchformen. [Types of fatigue fractures.] *Schr. tech. Hochsch., Darmstadt*, no. 2, 1933, pp. 38–44.

1271. OSCHATZ, H. Gesetzmässigkeiten des Dauerbruches und Wege zur Steigerung der Dauerhaltbarkeit. [Characteristic features of fatigue fractures and means of increasing fatigue strength.] *Mitt. MatPrüfAnst., Darmstadt*, no. 2, 1933, pp. 64.

1272. PETERSON, R. E. Model testing as applied to strength of materials. *J. appl. Mech.*, vol. 1, no. 2, 1933, pp. 79–85.

1273. PETERSON, R. E. Stress-concentration phenomena in fatigue of metals. *J. appl. Mech.*, vol. 1, no. 4, 1933, pp. 157–171.

1274. POMP, A. *and* HEMPEL, M. Untersuchungen an Stahlstäben bei wechselnder Zugbeanspruchung. [Investigation of steel bars under pulsating tensile stress.] *Mitt. K.-Wilh.-Inst. Eisenforsch.*, vol. 15, no. 18, 1933, pp. 247–254.

1275. REGLER, F. Über das Wesen der Kristallgitterstörungen und ihre Verteilung in Zug- und Dauerbiegestaben. [On the nature of crystal lattice distortion and its distribution in tensile and fatigue test pieces.] *Mitt. tech. VersAmt., Wien*, vol. 22, 1933, pp. 49–60.

1276. RÜFENACHT, F. *and* LEONHARD, R. Prüfmaschinen für Ermüdungsversuche. [Fatigue testing machines.] *Schweiz. tech. Z.*, vol. 8, April 4, 1933, pp. 197–204.

1277. SACHS, G. Fortschritte im Leichtmetallguss für hohe Beanspruchungen. [Progress in light metal castings for high loads.] *Z. Ver. dtsch. Ing.*, vol. 77, no. 5, Feb. 4, 1933, pp. 115–120.

1278. SCHÄCHTERLE, K. Die Dauerfestigkeit von Niet- und Schweissverbindungen und die Bemessung dynamisch beanspruchter Konstruktionsteile auf Grund der aus Dauerversuchen gewonnenen Erkenntnisse. [On the fatigue strength of riveted and welded joints and the design of dynamically stressed structural members based on conclusions drawn from fatigue tests.] *Congr. int. Ass. Br. struct. Engng*, vol. 2, 1933–1934, pp. 312–379.

1279. SCHÄCHTERLE, K. Die Bemessung von dynamisch beanspruchten Konstruktionsteilen. [The measurement of dynamic loads in structural parts.] *Bauingenieur*, vol. 14, no. 17/18, April 28, 1933, pp. 239–242.

1280. SCHAPER, G. Dauerfestigkeit von Schweiss-verbindungen. [Fatigue strength of welded joints.] *Z. Ver. dtsch. Ing.*, vol. 77, no. 21, May 27, 1933, pp. 556–560. Abstr.: *Stahl u. Eisen*, vol. 54, no. 14, April 5, 1934, p. 351.

1281. SCHULZ, E. H. *and* BUCHHOLTZ, H. Über die Dauerfestigkeit von genieteten und ge-schweissten Verbindungen aus Baustahl St. 52. [On the fatigue strength of riveted and welded joints made of steel St. 52.] *Congr. int. Ass. Br. struct. Engng*, vol. 2, 1933–1934, pp. 380–399. *Stahl u. Eisen*, vol. 53, no. 21, May 25, 1933, pp. 545–553.

1282. SCHUSTER, L. W. Investigation of the mechanical breakdown of prime movers and boiler plants. *Proc. Instn mech. Engrs, Lond.*, vol. 124, April 1933, pp. 337–479. *Engineering, Lond.*, vol. 135, no. 3512, May 5, 1933, pp. 488–489; no. 3514, May 19, 1933, pp. 554–556 and no. 3518, June 16, 1933, pp. 661–664.

1283. SCORAH, R. L. Models reveal changes needed to combat fatigue. *Mach. Design*, vol. 5, no. 11, Nov. 1933, pp. 31–33.

1284. SEED, D. *and* EVANS, B. Bend fatigue tests on duralumin, alclad and M.G.7 alloy. *Roy. Aircr. Estab. Rep.*, no. M 2210, Aug. 1933, pp. 8.

1285. SHELTON, S. M. *and* SWANGER, W. H. Fatigue tests of galvanised wire under pulsating tensile stress. *Proc. Amer. Soc. Test. Mater.*, vol. 33, pt. 2, 1933, pp. 348–363.

1286. SODERBERG, C. R. Service conditions control permissible stress. *Mach. Design*, vol. 5, no. 2, Feb. 1933, pp. 27–31.

1287. SODERBERG, C. R. Working stresses. *Trans. Amer. Soc. mech. Engrs*, vol. 55, 1933, APM pp. 131–144.

1288. SPELLER, F. N. *and* McCORKLE, I. B. Effect of organic coatings in preventing damage to metal subjected to stress and corrosion. *Oil Gas J.*, vol. 32, no. 23, Oct. 26, 1933, pp. 73–74.

1289. SPELLER, F. N. *and* McCORKLE, I. B. Corrosion fatigue and protective cotaings. *Proc. Amer. Petrol. Inst.*, sect. IV, vol. 14, Nov. 1933, pp. 24–28.

1290. STAEDEL, W. Dauerfestigkeit von Schrauben — ihre Beeinflussung durch Form, Herstellung und Werkstoff. [Fatigue strength of bolts—influence of form, preparation and material.] *Mitt. MatPrüfAnst.*, Darmstadt, no. 4, 1933, pp. 102.

1291. STEPHEN, R. A. *and* JONES, W. R. D. Recovery of steel after fatigue testing. *Metallurgist*, vol. 9, June 30, 1933, pp. 36–38.

1292. STYRI, H. Fatigue testing with simple test specimens. *Metals and Alloys*, vol. 4, no. 9, Sept. 1933, pp. 141–142.

1293. SUTTON, H. *and* TAYLOR, W. J. The influence of pickling on the fatigue strength of duralumin. *Roy. Aircr. Estab. Rep.*, no. M 2269, Oct. 1933, pp. 11.

1294. TEICHMANN, A. *and* BORKMANN, K. Bericht über statische und dynamische Belastungs-versuche an Rohrniet- und Schraubenbolzen-Verbindungen von Vierkantrohren aus Hydronalium. [Report on static and dynamic load tests on tubular riveted and bolted joints in square tubes of Hydronalium.] *Zent. wiss. Ber., ForschBer.*, no. 25, Nov. 18, 1933, pp. 7.

1295. TEMPLIN, R. L. The fatigue properties of light metals and alloys. *Proc. Amer. Soc. Test. Mater.*, vol. 33, pt. 2, 1933, pp. 364–386.

1296. THUM, A. Einfluss der Korrosion auf die Dauerfestigkeit von Chrom-Nickel-Legierungen. [Influence of corrosion on the fatigue strength of chromium–nickel alloys.] *Heraeus-Vakuum-Schmelze, Festschrift zur 10-Jahres-Feier*, 1933, pp. 424–434.

1297. THUM, A. Der Einfluss der Formgebung auf die Dauerhaltbarkeit von Konstruktionsteilen. [The effect of shape on the fatigue properties of constructional parts.] *Schr. tech. Hochsch., Darmstadt*, no. 2, 1933, pp. 31–37.

1298. THUM, A. *and* BERG, S. Über die Festigkeit von Rippen bei ruhender, wechselnder und stossartiger Belastung. [On the strength of serrations under static, alternating and impact loads.] *Z. Ver. dtsch. Ing.*, vol. 77, no. 11, March 18, 1933, pp. 281–287.

1299. THUM, A. *and* HOLZHAUER, C. Versuche über Kerbdauerfestigkeit und Korrosionsermüdung an Kesselbaustoffen. [Tests on the notched fatigue strength and corrosion fatigue of boiler plate.] *Wärme*, vol. 56, no. 39, Sept. 30, 1933, pp. 640–642.

1300. THUM, A. *and* HOLZHAUER, C. Zur Frage der Korrosionswirkung bei dauerbeanspruchten Kesselstählen. [The question of the influence of corrosion on the fatigue strength of boiler steels.] *Arch. Wärmew.*, vol. 14, no. 12, Dec. 1933, pp. 319–321.

1301. THUM, A. *and* OCHS, H. Die Frage der Korrosionsermüdung der Metalle. [The problem of corrosion-fatigue of metals.] *Forsch. Fortschr. dtsch. Wiss.*, vol. 9, 1933, pp. 478–479.

1302. THUM, A. *and* OSCHATZ, H. Möglichkeiten zur Steigerung der Dauerhaltbarkeit von Konstruktionsteilen. [The possibility of increasing the fatigue properties of constructional parts.] *Maschinenschaden*, vol. 10, no. 2, 1933, pp. 17–23.

1303. THUM, A. *and* OSCHATZ, H. Der Dauerbruchweg in Kurbelwellen. [The path of fatigue cracks in crankshafts.] *Z. Ver. dtsch. Ing.*, vol. 77, no. 30, July 29, 1933, p. 834.

1304. THUM, A. *and* SCHICK, W. Dauerfestigkeit von Schweissverbindungen bei verschiedener Formgebung. [Fatigue strength of welded joints of various forms.] *Z. Ver. dtsch. Ing.*, vol. 77, no. 19, May 13, 1933, pp. 493–496.

1305. THUM, A. *and* WIEGAND, H. Die Dauerhaltbarkeit von Schraubenverbindungen und Mittel zu ihrer Steigerung. [The fatigue properties of bolted joints and methods of increasing the strength.] *Z. Ver. dtsch. Ing.*, vol. 77, no. 39, Sept. 30, 1933, pp. 1061–1063.

1306. THUM, A. *and* WUNDERLICH, F. Der Einfluss von Einspann- und Kraftangriffstellen auf die Dauerhaltbarkeit der Konstruktionen. [The effect of clamping and intensity of force on the fatigue strength of constructional parts.] *Z. Ver. dtsch. Ing.*, vol. 77, no. 31, Aug. 5, 1933, pp. 851–853 and [discussion] no. 50, Dec. 16, 1933, pp. 1335–1337.

1307. VER, T. The development of slip lines and fatigue cracks under repeated stresses in low and high carbon steels welded by the butt welding method. *Carnegie Schol. Mem.* vol. 22, 1933, pp. 135–156.

1308. WESCOTT, B. B. *and* BOWERS, C. N. Explanation of mechanism of corrosion fatigue and its application to sucker rod failure. *Oil Gas J.*, vol. 32, no. 23, Oct. 26, 1933, pp. 65, 68, 70 and 72.

1309. WESCOTT, B. B. *and* BOWERS, C. N. Corrosion fatigue and sucker-rod failures. *Proc. Amer. Petrol. Inst.*, sect. IV, vol. 14, Nov. 1933, pp. 29–42.

1310. WIEGAND, H. Die Dauerfestigkeit der Schraube in Abhängigkeit von der Mutter form. [The fatigue strength of screws and its dependence on their form.] *Schr. tech. Hochsch., Darmstadt*, no. 2, 1933, pp. 67–72.

1311. WUNDERLICH, F. Der Einfluss von Einspann-, Kraftangriffs- und Nabensitzstellen auf die Dauerhaltbarkeit der Konstruktionen. [On the effects of clamping, intensity of force and hub fits on the fatigue properties of constructional parts.] *Schr. tech. Hochsch., Darmstadt*, no. 2, 1933, pp. 45–50.

1312. ZVEGINTSEV, S. K. Fatigue of welded joints welded with electrodes having a chalk or a LIM covering. [R] *Vest. metalloprom.*, vol. 13, no. 10, 1933, pp. 18–22.

1313. Entwurf II des DVM-Prüfverfahrens A 113, Dauerbiegeversuch. [Draft II of the DVM-testing method A 113, bending fatigue test.] *Z. Metallk.*, vol. 25, no. 1, Jan. 1933, pp. 27–28.

1314. Neuere dynamische Prüfmaschinen. [Modern fatigue testing machines.] *Tech. Bl., Düsseldorf*, vol. 23, Feb. 12, 1933, pp. 82–84 and Feb. 19, 1933, pp. 98–99.

1315. Investigation of tensile and fatigue properties of blade portion of fractured magnesium alloy airscrew, "Bulldog" J. 9580. *Roy. Aircr. Estab., Rep.* no. M. 2268A, Aug. 1933, pp. 4.

1316. Damping capacity and fatigue strength. *Metallurgist*, vol. 9, Aug. 25, 1933, pp. 58–60.

1317. The Davos fatigue testing machine. *Machinery, Lond.*, vol. 43, Oct. 5, 1933, pp. 15–16.

1318. Corrosion-fatigue tests on mild steel when coated with nickel by the Fescolising

process. *Engineering, Lond.*, vol. 136, Oct. 6, 1933, p. 398. *Machinery, Lond.*, vol. 43, no. 1099, Nov. 2, 1933, p. 144.

1319. Fatigue properties of forged and rolled duralumin. *Roy. Aircr. Estab. Rep.* no. M. 2485, Oct. 1933, pp. 5.

1320. Short time fatigue testing. *Metallurgist*, vol. 9, Oct. 27, 1933, pp. 75–76.

1321. Dauerfestigkeits-Schaubilder. [Fatigue strength diagrams.] *Z. Ver. dtsch. Ing.*, vol. 77, no. 42, Oct. 21, 1933, pp. 1146–1147 and no. 50, Dec. 16, 1933, p. 1342.

1934

1322. BACON, F. Cracking and fracture of metals with special reference to service breakages. *Iron Steel Ind.*, vol. 7, no. 6, March 1934, pp. 197–202 and no. 7, April 1934, pp. 237–239.

1323. BACON, F. Service fractures in the light of laboratory research. *J. Inst. cons. Mot. Engrs*, Oct. 1934, pp. 256–262; Nov. 1934, pp. 283–290 and Dec. 1934, pp. 325–334.

1324. BAUDER, R. *and* MACKH, H. Der elektrische Antrieb von Dauerbiegeschwingmaschinen. [The electrical drives of bending fatigue testing machines.] *Elektrotech. u. Maschinenb.*, vol. 52, Sept. 9, 1934, pp. 423–425. English abstr.: *Elect. Rev., Lond.*, vol. 115, no. 2968, Oct. 12, 1934, p. 475.

1325. BAUTZ, W. Eigenspannungen als Ursache gesteigerter Dauerhaltbarkeit. [Internal stresses as a reason for increase in fatigue properties.] *Schr. tech. Hochsch., Darmstadt*, no. 3, 1934, pp. 11–21.

1326. BEHRENS, O. Der Einfluss der Korrosion auf die Biegungsschwingungsfestigkeit von Stählen und Reinnickel. [The influence of corrosion on the bending fatigue strength of steels and pure nickel.] *Metallwirtschaft*, vol. 13, no. 3, Jan. 19, 1934, pp. 44–47. English translation: *Henry Brutcher Tech. Trans.*, no. 238.

1327. BEHRENS, O. *and* DUSOLD, TH. Einfluss der Korrosion auf die Biege- und Drehwechsel-festigkeit von Stählen und Nichteisenmetallen. [Influence of corrosion on the bending and torsional fatigue strength of steels and non-ferrous metals.] *Z. Ver. dtsch. Ing.*, vol. 78, no. 3, Jan. 20, 1934, pp. 94–95.

1328. BERNDT, G. Gewindetoleranzen und Fe-stigkeit von Schraubenverbindungen. [Thread tolerances and strength of bolted joints.] *Z. Ver. dtsch. Ing.*, vol. 78, no. 22, June 2, 1934, pp. 661–662.

1329. BIERETT, G. Die Schweissverbindung bei dynamischer Beanspruchung. [Welded joints under dynamic loading.] *Mitt. dtsch. Mat-PrüfAnst.*, no. 25, 1934, pp. 1–10.

1330. BIERETT, G. Dauerfestigkeit von Schweiss-verbindungen. [Fatigue strength of welded joints.] *Mitt. dtsch. MatPrüfAnst.*, no. 25, 1934, pp. 87–89. *Maschinenbau*, vol. 13, no. 15/16, Aug. 1934, pp. 411–413.

1331. BIERETT, G. *and* GRÜNING, G. Spannungs-zustand und Festigkeit von Stirnkehlnaht-verbindungen. [Stress distribution in and strength of fillet welded joints.] *Mitt. dtsch. MatPrüfAnst.*, no. 25, 1934, pp. 11–22. *Elektroschweissung*, vol. 5, Feb. 1934, pp. 33–34.

1332. BONDY, O. Electrically-welded railway bridges. *Rly Engr*, vol. 55, no. 9, Sept. 1934, pp. 277–281.

1333. BORKMANN, K. Elastizitäts- und Dauer-versuche mit Hydronaliumholmstücken. [Elasticity and fatigue tests with Hydronalium spars.] *Zent. wiss. Ber., Prüfber.*, no. 60, 1934, pp. 9 and no. 64, 1934, pp. 15.

1334. BUCHMANN, W. Die Kerbempfindlichkeit der Werkstoffe. [The notch sensitivity of materials.] *ForschArb. IngWes.*, vol. 5, no. 1, Jan.–Feb. 1934, pp. 36–48 and no. 4, July–Aug. 1934, pp. 192–194.

1335. BÜHLER, H. *and* BUCHHOLTZ, H. Einfluss der Querschnittsverminderung beim Kalt-ziehen auf die Spannungen in Rundstangen. [Influence of reduction of cross-sectional area by cold drawing on residual stresses in round bars.] *Arch. Eisenhüttenw.*, vol. 7, no. 7, Jan. 1934, pp. 427–436.

1336. BÜHLER, H. *and* BUCHHOLTZ, H. Bela-stungs-Dehnungs-Messungen an I-Trägen mit und ohne Aussteifung. [Load-extension meas-urements on I-beams with and without reinforcement.] *Mitt. ForschInst. ver. Stahlw. A.-G.*, vol. 4, no. 6, Nov. 1934, pp. 189–196.

1337. BUSCHMANN, E. Das Biege-Zug-Verfahren Ein neues technologisches Prüfverfahren für Werkstoffe. [The bending-tension process A new testing method for materials.] *Z Metallk.*, vol. 26, no. 12, Dec. 1934, pp. 274–279.

1338. BUSSMANN, K. H. Prüfstabdurchmesse und die Dauerfestigkeit von Stahlen. [Test specimen diameter and the fatigue strength of steels.] *ForschArb. IngWes.*, vol. 5, no. 4 1934, pp. 198–199.

1339. CAZAUD, R. Recherches sur la fatigue de aciers. Relations des actions mecanique exercées sur les aciers avec leur aimantation [Researches on the fatigue of steels. Conside ations of mechanical actions on magnetise steels.] *Thèses presentées a la Faculté d Sciences de l'Université de Paris*, Feb. 1 1934, pp. 158. *Publ. sci. Minist. Air*, no. 3 1934, pp. 158.

1340. CAZAUD, R. Metal fatigue and methods for its measurement. *J. aero. Sci.*, vol. 1, no. 3, July 1934, pp. 137–143.

1341. CAZAUD, R. La fatigue des métaux. Son importance dans la construction aéronautique. [The fatigue of metals. Its importance in aeronautical contruction.] *Bull. Ass. tech. marit.*, no. 38, 1934, pp. 515–533.

1342. CHASTON, J. C. Properties of lead and lead alloy cable sheaths. *Elect. Commun.*, vol. 13, July 1934, pp. 31–50.

1343. CROSS, H. C. High temperature tensile, creep and fatigue of cast and wrought high and low carbon 18Cr–8Ni steel from split heats. *Trans. Amer. Soc. mech. Engrs*, vol. 56, no. 7, July 1934, pp. 533–553. Abstr.: *Iron Age*, vol. 132, no. 26, Dec. 28, 1933, pp. 25, 62.

1344. DAEVES, K.; KAMP, E. *and* HOLTHAUS, K. Zur Eststehung der Brüche an wassengekühlten Kolbenstangen von Dieselmotoren. [Causes of failure in water-cooled piston rods of diesel engines.] *Z. Ver. dtsch. Ing.*, vol. 78, no. 36, Sept. 8, 1934, pp. 1065–1067. English abstr.: *J. Iron St. Inst.*, vol. 130, no. 2, 1934, p. 725.

1345. DAVIDENKOV, N. N. Machines for impact and fatigue testing. [R] *Zav. Lab.*, vol. 3, no. 12, 1934, pp. 1115–1121.

1346. DEBUS, F. Dauerbruchsichere Schraubenverbindungen. [The safety of bolted joints in fatigue fracture.] *Schr. tech. Hochsch., Darmstadt*, no. 3, 1934, pp. 55–61.

1347. DOUSSIN, MLLE L. Essais de fatigue sur les soudeurs l'acier doux et l'acier Cr–Mo employés dans la construction aéronautique.

[Fatigue tests on welds of mild steel and Cr–Mo steel employed in aeronautical construction.] *Bull. Soc. Ing. Soud.*, no. 28, 1934, pp. 1188–1204.

1348. DOWLING, J. J.; DIXON, S. M. *and* HOGAN, M. A. Fatigue tests on hard drawn steel wire. *Engineer, Lond.*, vol. 157, no. 4085, April 27, 1934, pp. 424–426.

1349. DRIESSEN, M. G. Beproevingen van constructie-onderdeelen, die aan wisselende belastingen onderworpen zijn, met behulp van de Pulsatormachine. [Testing of structural components submitted to alternating loads by means of the pulsating testing machine.] *Polyt. Weekbl.*, vol. 28, no. 25, June 21, 1934, pp. 385–387.

1350. DUSOLD, T. Der Einfluss der Korrosion auf die Drehschwingungsfestigkeit von Stählen und Nichteisenmetallen. [The effect of corrosion on the torsional endurance limit of steels and non-ferrous metals.] *Metallwirtschaft*, vol. 13, no. 3, Jan. 19, 1934, pp. 41–44.

1351. ESAU, A. *and* KORTUM, H. Die Dämpfungsmessung als Grundlage eines Verfahrens zur Bestimmung der Schwingungsfestigkeit. [The measurement of damping as the basis of a method for determining the fatigue strength.] *Messtechnik*, vol. 10, no. 2, Feb. 1934, pp. 21–23.

1352. FINK, M. *and* HOFMANN, U. Die Abnutzung metallischer Werkstoffe durch Reiboxydation. Zur Frage nach der Entstehung des Dauerbruchs. [The wear of metals by frictional oxidation. On the problem of the cause of fatigue failure.] *Metallwirtschaft*, vol. 13, no. 36, Sept. 7, 1934, pp. 623–625.

1353. FÖPPL, O. Eine neue Keilform mit besserer Dauerhaltbarkeit der Welle. [A new key shape for improving the fatigue properties of shafts.] *Mitt. Wöhler-Inst.*, no. 20, 1934, pp. 61–68.

1354. FORSMAN, O. Utmattningshållfasthet hos kallvalsat stål för flygplan. [Fatigue strength of cold rolled steel for aircraft.] *Ingen Vetensk-Akad. Handl.*, no. 127, 1934, pp. 16.

1355. FRANKENBERG, H. Der Einfluss von Dreh-schwingungsbeanspruchungen auf die Festig-keit und Dämpfungsfähigkeit von Aluminium-Legierungen. [The influence of repeated tor-sional stresses on the strength and damping capacity of aluminium alloys.] *Metallwirt-schaft*, vol. 13, no. 11, March 16, 1934, pp. 187–191.

1356. FRIEDMANN, W. Bestimmung der Biege-wechselfestigkeit von Drähten. Bau einer entsprechenden Materialprüfungsmaschine. [Determination of the bending fatigue strength of wires. Construction of a suitable materials testing machine.] *Mitt. Wöhler-Inst.*, no. 22, 1934, pp. 93.

1357. GEHLER, W. Über einige Grundbegriffe und Ergebnisse bei Versuchen über Dauer-festigkeit. [Some fundamental conceptions and results of tests on fatigue strength.] *Ge-schweisste Träger*, vol. 1, no. 2, 1934, pp. 3–20. Abstr.: *Stahl u. Eisen*, vol. 55, no. 13, March 28, 1935, p. 367.

1358. GERARD, I. J. *and* SUTTON, H. Corrosion fatigue properties of duralumin with and without protective coatings. *Roy. Aircr. Estab.*, *Rep.*, no. M. T. 5563, Jan. 1934, pp. 36.

1359. GILL, E. T. *and* GOODACRE, R. Some as-pects of the fatigue properties of patented steel wire. *J. Iron St. Inst.*, vol. 130, no. 2, 1934, pp. 293–323. Abstr.: *Engineering, Lond.*, vol. 138, no. 3597, Dec. 21, 1934, pp. 692–694.

1360. GOODACRE, R. The mounting of wire for fatigue testing. *Engineering, Lond.*, vol. 137, no. 3564, May 4, 1934, pp. 503–504.

1361. GOUGH, H. J. Fatigue of metals. *Proc. Staffs. Iron St. Inst.*, vol. 50, 1934–1935, pp. 74–97.

1362. GOUGH, H. J.; COX, H. L. *and* SOPWITH, D. G. A study of the influence of the inter-crystalline boundary on fatigue characteris-tics. *J. Inst. Met.*, vol. 54, no. 1, 1934, pp. 193–228.

1363. GOULD, A. J. The effects of superimpos-ing cyclic stressing upon chemical corrosion. *Engineering, Lond.*, vol. 138, no. 3567, Ju-ly 27, 1934, pp. 79–81.

1364. GRAF, O. Über die Festigkeiten der Schweissverbindungen insbesondere über die Abhängigkeit der Festigkeiten von der Gestalt. [On the strength of welded joints, especially on the relation between strength and shape.] *Autogene Metallbearb.*, vol. 27, no. 1, Jan. 1, 1934, pp. 1–12.

1365. GRAF, O. Über die Dauerfestigkeit von Stahlstäben mit Walzhaut und Bohrung bei Druckbelastung. [On the fatigue strength of steel bars with rolling skin and holes under compressive loads.] *Stahlbau*, vol. 7, no. 2, Jan. 19, 1934, pp. 9–10.

1366. GRAF, O. Über Dauerversuche mit I-Trägen aus St.37. [Fatigue tests of I-beams made of steel St.37.] *Stahlbau*, vol. 7, no. 22, Oct. 26, 1934, pp. 169–171.

1367. GRAF, O. Dauerfestigkeit von Schweiss-verbindungen. [The fatigue strength of welded joints.] *Z. Ver. dtsch. Ing.*, vol. 78, no. 49, Dec. 8, 1934, pp. 1423–1427.

1368. GRAF, O. *and* BRENNER, E. Versuche zur Ermittlung der Widerstandsfähigkeit von Beton gegen oftmals wiederholte Druckbelastung. Schwingungsuntersuchungen an einer Eisenbeton-Pilzdecke des Ford-Neubaues in Köln. [Experiments for investigating the resistence of concrete under often repeated compression loads. Fatigue investigation on the reinforced concrete beams for the Ford-Neubaues in Köln.] *Deutscher Ausschuss für Eisenbeton, Berlin*, Berichte no. 76, 1934, pp. 27.

1369. GRAF, O. *and* BRENNER, E. Widerstandsfähigkeit von Holzverbindungen gegen oftmals wiederholte Belastung. [Resistance of wooden joints to often repeated loads.] *Bautechnik*, vol. 12, no. 43, Oct. 5, 1934, pp. 573–577.

1370. GREENALL, C. H. Testing cable sheath for fatigue. *Bell Lab. Rec.*, vol. 13, no. 1, Sept. 1934, pp. 12–16.

1371. GRÖBL, J. Planung und Bau von Freileitungen. [Design and construction of overhead transmission lines.] *Elektrizitätswirtschaft*, vol. 35, Nov. 25, 1934, pp. 491–496.

1372. GRÖBL, J. *and* WAGNER, F. Das drehungsfreie Seil für Hochspannungs-Freileitungen. [Torsion free cable for high tension overhead lines.] *Elektrizitätswirtschaft*, vol. 35, June 15, 1934, pp. 219–222.

1373. HAIGH, B. P. Fatigue in structural steel. *Engineering, Lond.*, vol. 138, no. 3598, Dec. 28, 1934, pp. 698–701.

1374. HANCKE, A. Über die Beanspruchung beim Dauerschlagversuch. [On the stresses in impact fatigue tests.] *Schr. tech. Hochsch., Darmstadt*, no. 3, 1934, pp. 22–26.

1375. HEMPEL, M. Dämpfungsmessung und Werkstoffprüfung. [Damping measurements and material testing.] *Stahl u. Eisen*, vol. 54, no. 47, Nov. 22, 1934, pp. 1217–1220.

1376. HERBST, H. Zur Bewertung von Drahtbrüchen für die Sicherheit von Förderseilen. [Evaluation of broken wires for safety of pit ropes.] *Bergbau*, vol. 47, 1934, pp. 215–220. English translation: *Henry Brutcher Tech. Trans.*, no. 250.

1377. HEROLD, W. *Die Wechselfestigkeit metallischer Werkstoffe, ihre Bestimmung und Anwendung. [The fatigue strength of metallic materials, its determination and application.]* Vienna, Julius Springer, 1934, pp. 276.

1378. HIRST, G. W. C. An explanation of development of cracks by fatigue stresses in spokes of railway carriage wheels. *J. Instn Engrs Aust.*, vol. 6, no. 8, Aug. 1934, pp. 272–275.

1379. HOFFMANN, W. Statische und dynamische Festigkeit der Schweissverbindungen an Baustählen. [Static and dynamic strength of welded joints in structural steels]. *Stahlbau*, vol. 7, no. 11, 1934, pp. 85–87.

1380. HUDITA, S. Internal viscosity of engineering materials. [J] *J. Soc. mech. Engrs, Japan*, vol. 37, no. 211, 1934, pp. 812–814.

1381. JENKINS, C. H. M. *and* WEST, W. J. Accelerated cracking of mild steel (boiler plate) under repeated bending. Part II—further tests. *J. Iron St. Inst.*, vol. 130, no. 2, 1934, pp. 279–291.

1382. JENNINGS, C. H. The effect of welding current on fatigue strength of welds. *J. Amer. Weld. Soc.*, vol. 13, no. 6, June 1934, p. 7.

1383. JOHNSON, F. F. Effects of groove and fillet shapes on endurance of heat treated steel. *Automot. Industr., N.Y.*, vol. 71, no. 12, Sept. 22, 1934, pp. 352–353.

1384. JOHNSON, F. F. *and* LIPPERT, T. W. Axle fatigue inhibited by truss-graining. *Iron Age*, vol. 133, no. 13, March 29, 1934, pp. 19, 72 and 74.

1385. JOHNSON, J. B. Fatigue characteristics of helical springs. *Iron Age*, vol. 133, no. 11, March 15, 1934, pp. 12–15 and no. 12, March 22, 1934, pp. 24–26.

1386. JOHNSON, J. B. *and* OBERG, T. T. Effects of notches on nitrided steel. *Metals and Alloys*, vol. 5, no. 6, June 1934, pp. 129–130.

1387. JÜNGER, A. Korrosionsbiegewechselfestigkeit von Stahl und ihre Steigerung durch Zusätze zur Korrosionslösung. [Bending corrosion fatigue strength of steel and its increase by the addition of reagents to the corrosive liquid.] *Mitt. ForschAnst. Gutehoffn., Nürnberg*, vol. 3, no. 3, July 1934, pp. 55–84 and no. 4, Aug. 1934, pp. 85–101.

1388. JÜNGER, A. Einfluss von Querbohrungen auf die Dauerfestigkeit eines vergüteten Chrom-Molybdänstahles. [Influence of transverse holes on the fatigue strength of a chromium–molybdenum steel.] *Mitt. ForschAnst. Gutehoffn., Nürnberg*, vol. 3, no. 2, June 1934, pp. 29–32.

1389. KOCH, H. Die Biegewechselfestigkeit einer Keilverbindung (Passfederanordnung) und die Erhöhung der Dauerhaltbarkeit durch das Oberflächendrücken. [The bending fatigue strength of keyed connections and increasing their fatigue properties by surface pressure.] *Mitt. Wöhler-Inst.*, no. 20, 1934, pp. 1–60.

1390. KOMMERELL, O. *and* BIERETT, G. Über die statische Festigkeit und die Dauerfestigkeit genieteter, vorbelasteter und unter Vorlast durch Schweissung verstärkter Stabanschlüsse. [On the static and fatigue strength of riveted joints reinforced by welding in the prestressed condition.] *Stahlbau*, vol. 7, no. 11, May 25, 1934, pp. 81–85 and no. 12, June 8, 1934, pp. 91–95. *Mitt. dtsch. MatPrüfAnst.*, no. 25, 1934, pp. 47–64.

1391. KOSTRON, H. Der Bruchweg in Schraubenköpfen. [The path of fracture in screw heads.] *Metallwirtschaft*, vol. 13, no. 6, Feb. 9, 1934, pp. 100–101.

1392. KROUSE, G. N. A high speed fatigue testing machine and some tests of speed effect on endurance limit. *Proc. Amer. Soc. Test. Mater.*, vol. 34, pt. 2, 1934, pp. 156–164.

1393. KÜHL, P. Vergleich der Festigkeit einer genieteten und einer geschweissten Konstruktion bei Dauerbeanspruchung. [Comparison of the strength of riveted and welded structures under fatigue strees.] *Elektroschweissung*, vol. 5, June 1934, pp. 114–116.

1394. KÜHNEL, R. Eisenbahn-Federn und ihre Fertigung. [Railway springs and their manufacture.] *Stahl u. Eisen*, vol. 54, no. 2, Jan. 11, 1934, pp. 25–29.

1395. KÜHNEL, R. Grenzen der Werkstoffleistung — Dauerbrüche und ihre Ursachen. [Material limitations — fatigue fractures and their causes.] *Glasers Ann. Gew.*, vol. 115, no. 5, Sept. 1, 1934, pp. 33–37 and no. 6 Sept. 15, 1934, pp. 41–48.

1396. KUZNETSOV, V. D.; KONVISAROV, D. V *and* STROKOPITOV, V. I. Increase in the plasticity of metals under cyclic plastic torsion [R] *Dokl. Akad. Nauk SSSR.*, vol. 1, no. 7 1934, pp. 399–402.

1397. LACHASSAGNE, J. Quelques résultants d flexion rotative. [Some rotating bendin;

fatigue test results.] *Bull. Ass. tech. marit.*, no. 38, June 1934, pp. 535–551.

1398. LAUTE, K. Korrosion und Ermüdung. [Corrosion and fatigue.] *Mitt. dtsch. MatPrüf-Anst.*, no. 25, 1934, pp. 27–37.

1399. LEHR, E. *Spannungsverteilung in Konstruktionselementen. [Stress distribution in constructional parts.]* Berlin, V. D. I.-Verlag, 1934, pp. 64.

1400. LEHR, E. Die Auswirkung der neueren Festigkeitsforschung in der Praxis. [The practical advances in new strength investigations.] *Z. Ver. dtsch. Ing.*, vol. 78, no. 13, March 31, 1934, pp. 395–401.

1401. LEHR, E. *and* BUSSMANN, K. H. Aufklärung der Ursache für den Bruch der Kurbelwelle eines Einzylinder-Viertakt-Dieselmotors. [Explanation of the cause of fracture of crankshafts in single cylinder four stroke diesel engines.] *Maschinenschaden*, vol. 11, no. 3, 1934, pp. 45–47.

1402. LEISS, K. *and* TEICHMANN, A. Bericht über die Dauerfestigkeit von Stahlrohren in verschiedenen Verarbeitungszuständen. [Report on the fatigue strength of steel tubes in different manufacturing conditions.] *Zent. wiss. Ber., ForschBer.*, no. 121, Sept. 21, 1934, pp. 27.

1403. LINICUS, W. *and* SCHEUER, E. Die Wechselfestigkeit von Leichtmetallguss. [The fatigue strength of cast light metals.] *Metallwirtschaft*, vol. 13, no. 47, Nov. 23, 1934, pp. 829–836 and no. 48, Nov. 30, 1934, pp. 849–855. English abstr.: *Light Metals Res.*, vol. 3, no. 21, April 6, 1935, pp. 365–367.

1404. LOCK, O. H. Resonanz zwischen Mast- und Leitungsseilschwingungen und die Däm-pfung dieser Schwingungen mit Resonanzdämpfern. [Resonance between mast, and tranmission wire vibration and the damping of vibration with resonant dampers.] *Mitt. Wöhler-Inst.*, no. 21, 1934, pp. 61.

1405. LÖBNER, A. Die Ermüdung der Metalle. [The fatigue of metals.] *Giesserei*, vol. 21, nos. 23/24, June 8, 1934, pp. 248–249.

1406. LOHMANN, W. *and* SCHULTZ, E. H. Eigenschaften von Schweissverbindungen aus Hochbaustählen mit verschiedenen Elektroden. [Properties of welded joints of high strength structural steel made using different electrodes.] *Arch. Eisenhüttenw.*, vol. 7, no. 8, Feb. 1934, pp. 465–471.

1407. McADAM, D. J. *and* CLYNE, R. W. Influence of chemically and mechanically formed notches on fatigue of metals. *J. Res. nat. Bur. Stand.*, vol. 13, no. 4, Oct. 1934, pp. 527–572.

1408. MACNAUGHTAN, D. J. The improvements of white bearing metals for severe service: Some general considerations. *J. Inst. Met.*, vol. 55, no. 2, 1934, pp. 33–47. Abstr.: *Engineering, Lond.*, vol. 138, no. 3583, Sept. 14, 1934, pp. 285–286. *Mech. Engng N.Y.*, vol. 56, no. 10, Oct. 1934, pp. 623–624.

1409. MAIER, A. F. Wechselbeanspruchung von Rohren unter Innerdruck. [Alternating stressing of tubes under internal pressure.] *Stahl u. Eisen*, vol. 54, no. 50, Dec. 13, 1934, pp. 1289–1291. English abstr.: *J. Iron St. Inst.*, vol. 131, no. 1, 1935, p. 484. English translation: *Associated Technical Services Inc., Trans.* no. 596–GJ

1410. MAILÄNDER, R. Die neueren Ergebnisse der Werkstoffprüfung und ihre Anwendbarkeit. [New results from materials tests and their application.] *Tech. Bl., Düsseldorf*, vol. 24, Feb. 25, 1934, p. 102.

1411. MAILÄNDER, R. *and* BAUERSFELD, W. Einfluss der Probengrosse und Probenform auf die Dreh- Schwingungsfestigkeit von Stahl. [Influence of the specimen size and shape on the torsional fatigue limit of steel.] *Tech. Mitt. Krupp.*, vol. 2, no. 5, Dec. 1934, pp. 143–152.

1412. MAILÄNDER, R. *and* RUTTMANN, W. Einfluss von Oberflachenbeschaffenheit, Randschichten und Korrosion auf die Dauerfestigkeit. [Influence of surface condition, surface layers and corrosion on fatigue strength.] *Maschinenbau*, vol. 13, no. 21/22, Nov. 1934, pp. 577–581.

1413. MARCOTTE, E. De l'endurance de matériaux. [The endurance of materials.] *Rev. gén. Sci. pur. appl.*, vol. 45, May 31, 1934, pp. 297–301.

1414. MATTHAES, K. Fatigue strength of airplane and engine materials. *Tech. Memor. nat. Adv. Comm. Aero.*, *Wash.*, no. 743, April 1934, pp. 31.

1415. MEYER, W. Die Drehwechselfestigkeit genuteter Stäbe und die Erhöhung der Dauerhaltbarkeit durch Oberflächendrücken. [The torsional fatigue properties of specimens with keyways, and the increase in fatigue resistance from compressing the surface.] *Mitt. Wöhler-Inst.*, no. 18, 1934, pp. 1–73.

1416. MEYERCORDT, F. Dauerhaltbarkeit gusseiserner Konstruktionselemente und ihre Beeinflussung durch die Gusshaut. [Fatigue strength of cast iron constructional components and the influence of casting skin.] *Schr. tech. Hochsch.*, *Darmstadt*, no. 3, 1934, pp. 76–82.

1417. MIDUTA, Z. On the bending stress of rope and rope-strand. *Mem. Ryoj. Coll. Engng*, *Inouye Commemoration Volume*, Dec. 1934, pp. 371–387.

1418. MOORE, H. F. *and* HENWOOD, P. E. The strength of screw threads under repeated tension. *Uni. Ill. Engng Exp. Sta. Bull.*, no. 264, 1934, pp. 17.

1419. MOORE, H. F. *and* KROUSE, G. N. Repeated stress (fatigue) testing machine used in the testing laboratory of the University of Illinois. *Uni. Ill. Engng Exp. Sta. Circ.*, no. 23, March 27, 1934, pp. 36.

1420. MOORE, H. F. *and* PICCO, J. J. Fatigue tests of high strength cast irons. *Trans. Amer. Foundrym. Ass.*, vol. 42, 1934, pp. 525–542.

1421. MORGAN, P. D. *and* DOUBLE, E. W. W. The effects of working tension and lay ratio upon the fatigue failure of cored aluminium conductor. *Rep. Brit. elect. Ind. Res. Ass.*, no. F/T76, 1934, pp. 20.

1422. NISHIHARA, T. *and* KAWAKURA, Y. Repeated torsion tests of steel [J]. *J. Soc. mech. Engrs, Japan*, vol. 37, no. 209, Sept. 1934, pp. 593–598.

1423. OCHS, H. Die Korrosionsdauerfestigkeit von Stählen und der Verlauf des Korrosionsdauerbruches im Gefüge. [The corrosion fatigue strength of steels and the propagation of corrosion fatigue cracks in structures.] *Schr. tech. Hochsch. Hochsch.*, *Darmstadt*, no. 3, 1934, pp. 32–39.

1424. OCHS, H. Einfluss von Wechselbeanspruchungen auf die chemische Korrosion [Influence of alternating loads on chemical corrosion.] *ForschArb. IngWes.*, vol. 5, no. 5, 1934, pp. 254–255.

1425. ODING, I. A. The effect of cyclic over stressing on the endurance of metals. [R vol. 4, no. 2, 1934, *Zh. tekh. Fiz.*, pp. 405 412.

1426. OGÉE, M. Recherches sur la résistance à la fatigue des aciers au carbone. [Research on the fatigue resistance of carbon steels]. *Publ. sci. Minist. Air*, no. 58, 1934, pp. 69.

1427. ONO, A. Effect of oil and water on the fatigue limit of steel. [J] *J. Soc. mech. Engrs, Japan*, vol. 37, no. 201, Jan. 1934, pp. 8–10.

1428. OSCHATZ, H. Eine Dauerprüfmaschine zur Bestimmung der Dauerhaltbarkeit von Proben und Formelementen. [A fatigue testing machine for determining the endurance of specimens and constructional parts.] *Metallwirtschaft*, vol. 13, no. 25, June 22, 1934, pp. 443–448.

1429. OSHIBA, F. The degree of fatigue and recovery therefrom of carbon steels under repeated impact. [J] *J. Soc. mech. Engrs, Japan*, vol. 37, no. 207, July 1934, pp. 436–443. *Sci. Rep. Tohoku Univ.*, vol. 23, no. 4, Nov. 1934, pp. 589–611. *Kinzoku no kenkyu*, vol. 11, no. 7, July 1934, pp. 328–343. English abstr.: *Iron Age*, vol. 135, March 28, 1935, p. 23.

1430. PERSOZ, L. La fatigue sous corrosion. [Corrosion fatigue.] *Aciers spéc.*, vol. 9, Oct. 1934, pp. 311–328.

1431. PETERSON, R. E. *and* WAHL, A. M. Overcoming fatigue of shafts at press-fitted units. *Mach. Design*, vol. 6, no. 12, Dec. 1934, pp. 25–27.

1432. PEYTON, F. A. Flexure fatigue studies of cast dental gold alloys. *J. Amer. dent. Ass.*, vol. 24, no. 3, March 1934, pp. 394–415.

1433. PFANNENSCHMIDT, G. Die Schwingungsfestigkeit von Gusseisen und ihre Beziehungen zum Aufbau und zur Oberflächenbeschaffen-

heit. [The fatigue strength of cast iron with relation to structure and surface condition.] *Giesserei*, vol. 21, no. 21/22, May 25, 1934, pp. 223–228 and no. 23/24, June 8, 1934, pp. 243–245.

1434. PIPER, E. A. Engineering properties of cast iron. *Iron Age*, vol. 134, no. 16, Oct. 18, 1934, pp. 23–29.

1435. PIQUET, G. Essais d'endurance sur deux aciers mangano-siliceux. [Endurance tests on two manganese-silicon steels.] *Aciers spéc.*, vol. 9, April 1934, pp. 107–114; May 1934, pp. 137–144 and June 1934, pp. 162–175.

1436. POHL, E. *and* EHRT, M. Musterbeispiele von Dauerbrüchen und die daraus zu ziehenden Folgerungen für Konstruktion und Betrieb. [Typical examples of fatigue failures from which can be extracted information for construction and service.] *Mitteilungen der Materialprüfstelle der Allianz u. Stuttgarter Verein Versicherungs A-G., Abteilung für Maschinenversicherung, Berlin, Prüfungs-bericht*, no. 2, March 1934, pp. 28.

1437. POLOTOVSKAYA, L. Methods of determining the fatigue limits of metals and energy loss in specimens vibrating in the dynamic range. [R] *Zav. Lab.*, vol. 3, no. 4, 1934, pp. 342–347.

1438. POMP, A. *and* HEMPEL, M. Dauerprüfung von Stahldrähten unter wechselnder Zugbeanspruchung. [Fatigue tests on steel wires under pulsating tensile stress.] *Naturwissenschaften*, vol. 22, no. 22/24, June 1, 1934, pp. 398–400.

1439. PRIMROSE, J. S. G. Some new fatigue testing machines. *Trans. Manchr Ass. Engrs*, 1934–1935, pp. 147–152.

1440. RÖHRIG, H. *and* KREKELER, K. Zur Verhütung des Angriffs von Wasser auf Leichtmetall. [The prevention of corrosion attacks on light metals by water.] *Aluminium, Berl.*, vol. 17, no. 11, Nov. 1934, pp. 140–141.

1441. Roš, M. *and* Eichinger, A. Festigkeitseigenschaften der Stähle bei höhen Temperaturen. [Strength properties of steels at high temperatures.] *Ber. eidgenöss. MatPrüfAnst.*, no. 87, April 1934, pp. 51.

1442. Roš, M. *and* Eichinger, A. Les caractéristiques de résistance des métaux aux températures élevées. [The properties of metals at elevated temperatures.] *Rev. Métall.*, vol. 31, no. 10, Oct. 1934, pp. 460–470.

1443. Rouchet, M. Contribution à l'étude des phénomènes de fatigue dynamique. [Contribution to the study of the phenomenon of fatigue.] *Bull. Ass. tech. marit.*, no. 38, June 1934, pp. 463–510.

1444. Schaper, G. Die Dauerfestigkeit der Baustähle. [Fatigue strength of structural steels.] *Bautechnik*, vol. 12, no. 2, Jan. 12, 1934, pp. 23–24.

1445. Scherrer, P. Kristallstruktur und Festigkeitseigenschaften. [Crystal structure and strength properties.] *Bull. schweiz. elektrotech. Ver.*, vol. 25, 1934, pp. 458–463.

1446. Schick, W. Untersuchungen an Schweissverbindungen — Einfluss der Formgebung auf die Dauerfestigkeit geschweisster Verbindungen. [Investigation of welded joints— effect of shape on the fatigue strength of welded joints.] *Tech. Mitt. Krupp*, vol. 2, no. 2, March 1934, pp. 43–62.

1447. Schlyter, R. Hållfasthetsegenskaper hos barrvirke för flygplanstillverkning och utmattningshållfasthet hos limfogar. [Strength properties of coniferous timber for aircraft

manufacture and fatigue strength of glued joints.] *IngenVetenskAkad. Handl.*, no. 126, 1934, pp. 31.

1448. Schmid, E. *and* Siebel, G. Über Wechseltorsions-Versuche mit Magnesium-kristallen. [Alternating torsion tests on magnesium crystals.] *Metallwirtschaft*, vol. 13, no. 15, May 18, 1934, pp. 353–356.

1449. Schoenmaker, P. Weerstand van lasschen tegen vermoeiingsschokbelastingen. [Strength of welds under impact fatigue loads.] *Polyt. Weekbl.*, vol. 28, no. 41, Oct. 11, 1934, pp. 646–647.

1450. von Schwarz, M. *and* Koch, G. Untersuchungsbericht über einen beachtenswerten Dauerbruch einer Dieselmotorwelle. [Report of investigation on a remarkable fatigue fracture in a diesel motor shaft.] *Maschinenschaden*, vol. 11, no. 5, 1934, pp. 82–83.

1451. Schwinning, W. *and* Dorgerloh, E. Der Einfluss der Wärmebehandlung beim Aushärten und der langdauernden Lagerung auf die Eigenschaften einer Aluminiumlegierung. [The influence of heat-treatment with age-hardening and prolonged storage on the properties of an aluminium alloy.] *Z. Metallk.* vol. 26, no. 4, April 1934, pp. 91–92.

1452. Schwinning, W. *and* Dorgerloh, E Untersuchungen über das Verhalten von an gebrochenen Aluminium- und Kupfer-Frei leitungsseilen bei Schwingungsbeanspruchung [Investigations on the behaviour of aluminiur and copper electrical transmission lines unde vibrational stress.] *Z. Metallk.*, vol. 2€ no. 7, July 1934, pp. 162–164.

1453. Schwinning, W.; Knock, M. *and* Uhl mann, K. Wechselfestigkeit und Kerbempfin lichkeit der Stähle bei hohen Temperatures [Fatigue strength and notch sensitivity (

steels at high temperatures.] *Z. Ver. dtsch. Ing.*, vol. 78, no. 51, Dec. 22, 1934, pp. 1469–1476.

1454. SCHWINNING, W. *and* STROBEL, E. Untersuchungen über die Warmfestigkeit von Kupfer bei statischer und bei wechselnder Beanspruchung. [Investigation on the hot strength of copper under static and alternating stress.] *Z. Metallk.*, vol. 26, no. 1, Jan. 1934, pp. 1–5.

1455. SCOTT, A. T. The fatigue strength of arc welded butt joints. *Carnegie Schol. Mem.*, vol. 23, 1934, pp. 125–138.

1456. SERENSEN, S. V. The problem of the fatigue of metals. [R] *Vest. metalloprom.*, vol. 14, no. 12, Dec. 1934, pp. 42–46.

1457. SHAPOSHNIKOV, S. I. A radiographic method for studying the phenomena of metal fatigue. [R] *Zav. Lab.*, vol. 3, no. 10, 1934, pp. 927–935.

1458. SIEBEL, E. *and* KOPF, E. Einfluss eines eingepassten Bolzens auf die Dauerfestigkeit eines gebohrten Stabes. [Influence of fitted bolts on the fatigue strength of drilled bars.] *Z. Ver. dtsch. Ing.*, vol. 78, no. 30, July 28, 1934, pp. 918–919.

1459. SMITH, H. A. Fatigue and crystal recovery in aluminium. *Physics*, vol. 5, Dec. 1934, pp. 412–414.

1460. SOLAKIAN, A. G. Why threaded parts fail. *Amer. Mach.*, *N.Y.*, vol. 78, no. 9, March 31, 1934, pp. 126E–127E.

1461. SONDEREGGER, A. Die Bedeutung der Grundspannung für die Dauerfestigkeit von Schweissverbindungen — L'importance de la tension initiale pour l'endurance des soudures. [The importance of initial tension stress on the fatigue strength of welded joints.] *Z. Schweisstech.*, vol. 24, no. 9, Sept. 1934, pp. 215–217.

1462. SPÄTH, W. *Theorie und Praxis der Schwingungsprüfmaschinen. [Theory and practice of fatigue testing machines.]* Berlin, Julius Springer, 1934, pp. 98.

1463. SUTTON, H. *and* TAYLOR, W. J. The influence of pickling on the fatigue strength of duralumin. *J. Inst. Met.*, vol. 55, no. 2, Sept. 1934, pp. 149–165.

1464. TAKASE, K. Fatigue fracture of aircraft materials and their durability. [J] *J. Soc. aero. Sci. Japan*, vol. 1, no. 1, Sept. 1934, pp. 79–174.

1465. TEICHMANN, A. *and* BORKMANN, K. Bericht über Dauerversuche mit einem Blechwandträger mit dünnem Stegblech. [Report of fatigue tests on sheet metal beams with thin web plates.] *Zent. wiss. Ber.*, *Prüfber.*, no. 32, 1934, pp. 5.

1466. TEICHMANN, A. *and* BORKMANN, K. Dauerversuche mit Aufhängungsteilen aus Cottonid. [Fatigue tests with suspended sections of Cottonid.] *Zent. wiss. Ber.*, *Prüfber.*, no. 67, Feb. 26, 1934, pp. 5.

1467. TEICHMANN, A. *and* LEISS, K. Dauerfestigkeit von Elektron (AZM) Profilen. [Fatigue strength of Elektron (AZM) sections.] *Zent. wiss. Ber.*, *Prüfber.*, no. 85, Sept. 22, 1934, pp. 10.

1468. THORNTON, G. E. Fatigue of metals. *J. Amer. Weld. Soc.*, vol. 13, no. 3, March

1934, pp. 20–25. Abstr.: *Automot. Industr. N.Y.*, vol. 69, no. 16, Oct. 14, 1933, p. 456.

1469. THUM, A. *and* BAUTZ, W. Steigerung der Dauerhaltbarkeit gekerbter Konstruktionsteile durch Eigenspannungen. [Increase of the fatigue life of notched constructional members through internal stress.] *Z. Ver. dtsch. Ing.*, vol. 78, no. 31, Aug. 4, 1934, pp. 921–925.

1470. THUM, A. *and* BUCHMAN, W. Kerbempfindlichkeit von Stählen. [Notch sensitivity of steels.] *Arch. Eisenhüttenw.*, vol. 7, no. 11, May 1934, pp. 627–635.

1471. THUM, A. *and* LIPP, T. Zur Frage der Dauerhaltbarkeit geschweisster und gegossener Konstruktionsteile. [The problem of the fatigue properties of welded and cast structural parts.] *Giesserei*, vol. 21, no. 5/6, Feb. 2, 1934, pp. 41–49; no. 7/8, Feb. 16, 1934, pp. 64–71; no. 9/10, March 2, 1934, pp. 89–95 and no. 13/14, March 30, 1934, pp. 131–141. English abstr.: *Found. Tr. J.*, vol. 51, no. 938, Aug. 9, 1934, p. 95.

1472. THUM, A. *and* OSCHATZ, H. Die Beurteilung von Dauerbrüchen. [The analysis of fatigue failures.] *Metallwirtschaft*, vol. 13, no. 1, Jan. 5, 1934, pp. 1–8.

1473. THUM, A. *and* WUNDERLICH, F. Dauerbiegefestigkeit von Konstruktionsteilen an Einspannungen, Nabensitzen und ähnlichen Kraftangriffsstellen. [Bending fatigue strength of structural members in the vicinity of grips, wheel seats and similar places.] *Mitt. MatPrüfAnst.*, Darmstadt, no. 5, 1934, pp. 82.

1474. THUM, A. *and* WUNDERLICH, F. Zur Festigkeitsberechnung von Fahrzeugachsen. [The strength calculation of vehicle axles.] *Z. Ver. dtsch. Ing.*, vol. 78, no. 27, July 7, 1934, pp. 823–824.

1475. TREIBER, F. Das Verhalten von Eisenbeton-Balken unter dem Einfluss dauernd ruhender und häufig wiederholter Belastung. [The behaviour of reinforced concrete beams under the influence of rest periods and frequently repeated loads.] *Bauingenieur*, vol. 15, no. 13/14, 1934, pp. 131–133 and no. 17/18, 1934, pp. 178–182.

1476. WALLMANN, K. Beziehungen zwischen Röntgenbild und Festigkeitseigenschaften, ermittelt an Schweissverbindungen. [Relation between X-ray pictures and strength properties, as determined on welded joints.] *Arch. Eisenhüttenw.*, vol. 8, no. 6, Dec. 1934, pp. 243–247.

1477. WEGELIUS, E. Lentokoneissa käytet tyjen teräsputkien väsytyslujuus. [Results of fatigue tests of welded steel tubes used in aircraft construction.] *Tekn. Aikl.*, vol. 24, no. 11, Nov. 1934, pp. 348–352.

1478. WIECKER, H. Die Biegewechselfestigkeit genuteter Stäbe und die Erhöhung der Dauerhaltbarkeit durch das Oberflächendrücken. [The bending fatigue strength of bars with keyways and the increase in fatigue life by surface pressure.] *Mitt. Wöhler-Inst.*, no. 19, 1934, pp. 52.

1479. WIEGAND, H. Über die Dauerfestigkeit von Schraubenwerkstoffen und Schraubenverbindungen. [On the fatigue strength of bolt materials and bolted joints.] *Bauer und Schauerte A.G., Neuss, Germany, Technical Publication*, no. 14, 1934, pp. 90.

1480. WIEMER, A. Biegeschwingungs- und Verdrehschwingungsprüfung von Naben für Metalluftschrauben und von Kurbelwellenflanschen (Europanundflug 1934). [Bending and torsion fatigue tests on hubs for metal aircraft propellers and on crankshaft flanges.] *Zent wiss. Ber., Prüfber.*, no. 76, May 14, 1934 pp. 9.

1481. WOERNLE, R. Drahtseilforschung—von der 10. Tagung des Ausschusses für Drahtseilforschung des VDI. [Wire rope research—on the 10th Convention of the Ausschusses für Drahtseilforschung des VDI.] *Z. Ver. dtsch. Ing.*, vol. 78, no. 52, Dec. 29, 1934, pp. 1492–1498.

1482. WUNDERLICH, F. Ermüdungsvorgang und Dauerbruch an Kraftangriffstellen. [Fatigue process and fatigue fracture and intensity of applied force.] *Schr. tech. Hochsch., Darmstadt.*, no. 3, 1934, pp. 27–31.

1483. YAMADA, R. *and* MATSUOKA, Y. On the change of mechanical properties in metals under repeated stress and on the recovery of fatigue. [J] *J. Soc. mech. Engrs, Japan*, vol. 37, no. 205, May 1934, pp. 273–281.

1484. ZIMMERLI, F. P. Permissible stress range for small helical spring. *Bull. Dep. Engng Res. Univ. Mich.*, no. 26, July 1934, pp. 81.

1485. ZIMMERMAN, J. H. Note on the fatigue testing of welds. *J. Amer. Weld. Soc.*, vol. 13, no. 9, Sept. 1934, pp. 13–15.

1486. Metal strength increased by graduated stressing. *Steel*, vol. 94, no. 10, March 5, 1934, p. 26.

1487. Fatigue failure and crystalline structure. *Engineering, Lond.*, vol. 137, no. 3556, March 9, 1934, pp. 298–299.

1488. Fatigue testing: practical research on a simple and improved machine. *Auto. Engr*, vol. 24, no. 318, April 1934, p. 134.

1489. Aluminium conductor vibrations. *Elect. Rev., Lond.*, vol. 114, no. 2944, April 27, 1934, pp. 595–596.

1490. Haigh-Robertson fatigue testing machine for wire. *Engineering, Lond.*, vol. 138, no. 3578, Aug. 10, 1934, pp. 139–140. *Engineer, Lond.*, vol. 158, Aug. 17, 1934, pp. 167–168.

1491. Fatigue testing machine. *Iron Coal Tr. Rev.*, vol. 129, Aug. 24, 1934, p. 261.

1492. Fatigue testing of wire. *Aircr. Engng*, vol. 6, no. 67, Sept. 1934, pp. 251–253.

1493. Failures of locomotive axles. *Engineer, Lond.*, vol. 158, no. 4106, Sept. 21, 1934, p. 287.

1494. Fatigue properties of chromium-molybdenum steel tubes. *Roy. Aircr. Estab., Rep.*, no. M 3064, Oct. 1934, pp. 4.

1495. Recent developments in the welding of rails. *Rly Engr*, vol. 55, no. 12, Dec. 1934, pp. 388–391.

1935

1496. AFANASEV, N. N. Modern methods for the fatigue testing of metals. [R] *Zav. Lab.*, vol. 4, no. 8, 1935, pp. 951–958.

1497. ALMEN, J. O. Factors influencing the durability of spiral-bevel gears for automobiles. *Automot. Industr., N.Y.*, vol. 73, no. 20, Nov. 16, 1935, pp. 662–668 and no. 21, Nov. 23, 1935, pp. 696–701.

1498. ALMEN, J. O. *and* BOEGEHOLD, A. L. Rear axle gears: factors which influence their life. *Proc. Amer. Soc. Test. Mater.*, vol. 35, pt. 2, 1935, pp. 99–146. Abstr.: *Heat Treat. Forg.*, vol. 21, no. 7, July 1935, pp. 338–340.

1499. APPENRODT, A. Die Dämpfungsfähigkeit von Kurbelwellenstäben im kalten und warmen Zustand bei Anlieferung und im Dauer-betrieb. [The damping capacity of crankshaft specimens at room temperature and high temperatures after continuous operation.] *Mitt. Wöhler-Inst.*, no. 24, 1935, pp. 98. Abstr.: *Stahl u. Eisen*, vol. 55, no. 36, Sept. 5, 1935, pp. 964–965.

1500. BACON, F. The relation of fatigue to modern engine design. *Mech. World*, vol. 97, no. 2514, March 8, 1935, pp. 227–228.

1501. BAUTZ, W. *and* OCHS, H. Kristallstruktur und Dauerbruch. [Crystal structure and fatigue failure.] *Z. Ver. dtsch. Ing.*, vol. 79, no. 48, Nov. 30, 1935, pp. 1450–1451.

1502. BAYLESS, F. B. Sucker rod fatigue changes with the nature of corrosive elements in oil. *Oil Gas J.*, vol. 34, Aug. 15, 1935, pp. 26, 28.

1503. BENOIT, G. *Zum Gedächtnis an W. A. Julius Albert und die Erfindung seines Drahtseiles. [In memory of W. A. Julius Albert and the invention of his wire rope.]* Berlin, VDI-Verlag, 1935, pp. 34.

1504. BERG, P. Die Steigerung der Dauerhaltbarkeit von Keilverbindungen durch Oberflächendrücken. [The increase of the fatigue properties of keyed joints by surface pressing.] *Mitt. Wöhler-Inst.*, no. 26, 1935, pp. 73.

1505. BOBEK, K. Schweisskonstruktionen für Dauerwechselbeanspruchung. [Welded structures for fatigue loadings.] *Elektroschweissung*, vol. 6, no. 4, April 1935, pp. 81–84.

1506. BONDY, O. Recent developments regarding welded joints and the effects of fatigue. *Weld. Ind.*, vol. 3, March 1935, pp. 60–64.

1507. BONDY, O. Relations between the design and fatigue strength of welded structural

joints. *"Symposium on the welding of iron and steel"*, Iron and Steel Institute, London, 1935, vol. 2, pp. 679–687.

1508. BOONE, W. D. *and* WISHART, H. B. High-speed fatigue tests of several ferrous and non-ferrous metals at low temperature. *Proc. Amer. Soc. Test. Mater.*, vol. 35, pt. 2, 1935, pp. 147–155. Abstr.: *Iron Coal Tr. Rev.*, vol. 131, July 19, 1935, pp. 94–95.

1509. BUCHWALD, A. L'examen des ponts au point de vue des manifestations de fatigue, d'après la détermination du "facteur d'elasticité". [The examination of bridges from the point of view of manifestations of fatigue by the determination of the "factor of elasticity". *Génie civ.*, vol. 107, no. 8 (no. 2767), Aug. 24, 1935, pp. 185–186.

1510. BUCKWALTER, T. V. *and* PATERSON, P. C. Locomotive axle failures and wheel press fits. *Rly mech. Engr*, vol. 109, no. 4, April 1935, pp. 127–131.

1511. BURN, W. S. The relation of fatigue to forging design. *Heat Treat. Forg.*, vol. 21, no. 2, Feb. 1935, pp. 71–76.

1512. BURN, W. S. The relation of fatigue to modern engine design. *Engineer, Lond.*, vol. 160, no. 4147, July 5, 1935, pp. 21–22 and no. 4148, July 12, 1935, pp. 48–50.

1513. CAZAUD, R. Quelques nouvelles recherches sur la fatigue des métaux en construction aéronautique. [Some new researches on the fatigue of metals in aeronautical construction. *Bull. Ass. tech. marit.*, no. 39, June 1935, pp. 513–522.

1514. CAZAUD, R. Les essais de fatigue de métaux. [The fatigue testing of metals.] *In*

Congr. Min. etc., Seventh Congress, Paris, Oct. 1935, vol. 1, pp. 385–398.

1515. Cox, H. L. *and* Clenshaw, W. J. Behaviour of three single crystals of aluminium in fatigue under complex stresses. *Proc. roy. Soc., (A)*, vol. 149, no. 867, April 1, 1935, pp. 312–326.

1516. Cuthbertson, J. W. The load deflection fatigue test. A critical investigation of its application to steels. *Carnegie Schol. Mem.*, vol. 24, 1935, pp. 1–50.

1517. Dennison, R. L. Strength and flexibility of corrugated piping. *Engineer, Lond.*, vol. 160, no. 4158, Sept. 20, 1935, pp. 303–304; no. 4159, Sept. 27, 1935, pp. 332–334; no. 4160, Oct. 4, 1935, pp. 358–360; no. 4161, Oct. 11, 1935, pp. 386–388; no. 4162, Oct. 18, 1935, pp. 412–414 and no. 4163, Oct. 25, 1935, pp. 439–440.

1518. Dixon, S. M. *and* Hogan, M. A. The deterioration of haulage ropes in service. *Pap. Saf. Min. Res. Bd, Lond.*, no. 92, 1935, pp. 32.

1519. Dixon, S. M. *and* Hogan, M. A. Corrosion-fatigue in colliery ropes. *Engineer, Lond.*, vol. 160, no. 4163, Oct. 25, 1935, pp. 436–437.

1520. Dorsey, J. R. The fatigue of metals in gasholder crowns. *Gas World*, vol. 102, no. 2642, March 23, 1935, pp. 294–295.

1521. Doussin, Mlle L. Essais de fatigue sur assemblages soudés et brasés. [Fatigue tests on welded and brazed joints.] *Bull. Soc. Ing. Soud.*, vol. 6, no. 37, Nov.–Dec. 1935, pp. 1939–1959.

1522. Dustin, H. Considérations sur l'endurance des assemblages soudés. [The fatigue strength of welded structures.] *Rev. univ. Min.*, vol. 11, no. 14, Dec. 1935, pp. 521–531.

1523. von Ende, E. Festigkeit genuteter Wellen. [Strength of splined shafts.] *ForschArb. Ing-Wes.*, vol. 6, no. 4, 1935, pp. 206–208.

1524. Erber, G. Berechnung der Kerbdauerfestigkeit aus Zugfestigkeit und Einschnürung. [Calculation of the notched fatigue strength from the tensile strength and reduction in area.] *Arch. Eisenhüttenw.*, vol. 9, no. 2, Aug. 1935, pp. 95–97.

1525. Everett, F. L. Stress analysis of failure in machine parts. *Mech. Engng, N. Y.*, vol. 57, no. 3, March 1935, pp. 157–161. Abstr.: *Mech. World*, vol. 97, no. 2525, May 24, 1935, pp. 491–493.

1526. Föppl, O. *and* Dusold, T. Der drosselgesteuerte Schlingertank. Versuche an einem Schlingertankmodell. Der Schiffskreisel. Oberflächendrücken. Kurbelwellen aus dämpfungsfähigem Stahl. [Throttle-controlled anti-roll tank. Tests with an anti-roll tank model. Ship's gyroscopes. Surface pressure. Crankshafts and damping capacity of steel.] *Mitt. Wöhler-Inst.*, no. 25, 1935, pp. 70.

1527. Föppl, O. *and* Meyer, W. Die Dreh- und Biegewechselfestigkeit genuteter Probestäbe und einer Keilverbindung und die Erhöhung der Dauerfestigkeit durch das Oberflächendrücken. [The torsional and bending fatigue strength of notched bar specimens and keyed connections and the increase in fatigue strength by surface pressure.] *Schweiz. Bauztg.*, vol. 105, no. 14, April 6, 1935, pp. 159–163.

1528. de Forest, A. V. Dynamic strength of machine parts as affected by quality of surface. *Iron Age*, vol. 135, no. 8, Feb. 21, 1935, pp. 18–22.

1529. FRIEDMANN, W. Bestimmung der Biege-wechselfestigkeit von Drähten geringen Durch-messers. [Determination of the bending fatigue strength of thin wires.] *Metallwirtschaft*, vol. 14, no. 5, Feb. 1, 1935, pp. 85–87.

1530. FRIEDMANN, W. Biegewechselfestigkeit von Kupfer- und Aluminiumdrähten. [Bending fatigue strength of copper and aluminium wires.] *Z. Ver. dtsch. Ing.*, vol. 79, no. 34, Aug. 24, 1935, pp. 1046–1047.

1531. GERARD, I. J. and SUTTON, H. Corrosion fatigue properties of duralumin with and without protective coatings. *J. Inst. Met.*, vol. 56, no. 1, 1935, pp. 29–53. Abstr.: *Engineering, Lond.*, vol. 139, no. 3609, March 15, 1935, pp. 289–290.

1532. GEROLD, E. Über die Dauerfestigkeit von Metallen. [On the fatigue strength of metals.] *Forsch. Fortschr. dtsch. Wiss.*, vol. 11, Aug. 10 –20, 1935, pp. 313–314.

1533. GILL, E. T. and GOODACRE, R. Some aspects of the fatigue properties of patented steel wires—II. Note on the effect of low-temperature heat treatment. *J. Iron St. Inst.*, vol. 132, no. 2, 1935, pp. 143–177. *Iron Steel Ind.*, vol. 8, no. 13, Sept. 1935, pp. 506–510. Abstr.: *Engineer, Lond.*, vol. 160, no. 4159, Sept. 27, 1935, p. 325.

1534. GOODACRE, R. The effect of heavy oils and greases on the fatigue strength of steel wire. *Engineering, Lond.*, vol. 139, no. 3616, May 3, 1935, pp. 457–458. *Wire and Wire Prod.*, vol. 10, Oct. 1935, pp. 473–475.

1535. GOUGH, H. J. Fatigue of metals. *Bgham metall. Soc. J.*, vol. 15, no. 1, March 1935, pp. 26–52.

1536. GOUGH, H. J. and CLENSHAW, W. J. The testing of engineering materials. *Trans. Inst. Mar. Engrs*, vol. 47, pt. 10, 1935, pp. 241–276.

1537. GOUGH, H. J. and CLENSHAW, W. J. Cracking of boiler plates. *Mech. World*, vol. 98, no. 2553, Dec. 6, 1935, pp. 551, 553.

1538. GOUGH, H. J. and CLENSHAW, W. J. Fatigue testing at the N. P. L. *Mech. World*, vol. 98, no. 2554, Dec. 13, 1935, pp. 575–577.

1539. GOUGH, H. J. and CLENSHAW, W. J. Corrosion fatigue. *Mech. World*, vol. 98, no. 2555, Dec. 20, 1935, pp. 596, 598.

1540. GOUGH, H. J. and CLENSHAW, W. J. Criteria of failure of metals. *Mech. World*, vol. 98, no. 2556, Dec. 27, 1935, pp. 617–619, 631.

1541. GOUGH, H. J. and POLLARD, H. V. The strength of metals under combined alternating stresses. *Proc. Instn mech. Engrs, Lond.* vol. 131, Nov. 1935, pp. 3–103. *Engineering Lond.*, vol. 140, no. 3643, Nov. 8, 1935 pp. 511–513 and no. 3645, Nov. 22, 1935 pp. 565–567. *Mech. World*, vol. 98, no. 2551 Nov. 22, 1935, pp. 501–502 and no. 2552 Nov. 29, 1935, pp. 527–528, 536. *Metallurgia Manchr*, vol. 13, no. 73, Nov. 1935, pp. 17–20 *Iron Steel Ind.*, vol. 9, Jan. 1936, pp. 132–13 and Feb. 1936, pp. 177–178.

1542. GOUGH, H. J. and SOPWITH, D. G. Som further experiments on atmospheric action i fatigue. *J. Inst. Met.*, vol. 56, no. 1, 193 pp. 55–89. Abstr.: *Engineering, Lond* vol. 139, no. 3609, March 15, 1935, p. 29(

1543. GOUGH, H. J. and SOPWITH, G. D. Fatigu tests on a "safety" type hook. *Engineerin Lond.*, vol. 140, no. 3628, July 26, 193 pp. 101–102.

1544. GRAF, O. Dauerversuche mit Nietve bindungen. [Fatigue tests on riveted joint *Bericht des Ausschusses für Versuche im Sta bau, Berlin*, no. 5, 1935, pp. 51.

1545. GRAF, O. On the fatigue strength of constructional members, and particularly of welded joints. "*Symposium on the welding of iron and steel*", Iron and Steel Institute, London, 1935, vol. 2, pp. 791–794.

1546. GRAF, O. Über Dauerzugversuche mit Flachstäben und Nietverbindungen aus Leichtmetall. [On tensile fatigue tests with flat plates and riveted joints of light metal.] *Stahlbau*, vol. 8, no. 17, Aug. 16, 1935, pp. 132–133.

1547. GRAF, O. Dauerversuche mit grossen Schweissverbindungen bei oftmaligem Wechsel zwischen Zug- und Druckbelastung sowie bei oftmaliger Zugbelastung. [Fatigue tests on full size welded joints under alternating tensile and compressive stresses or under repeated tensile loading.] *Stahlbau*, vol. 8, no. 21, Oct. 11, 1935, pp. 164–165.

1548. GREENALL, C. H. Testing cable sheaths for fatigue. *Wire and Wire Prod.*, vol. 10, no. 7, July 1935, pp. 267–269, 293.

549. HAIGH, B. P. Fatigue strength of butt welds. "*Symposium on the welding of iron and steel*", Iron and Steel Institute, London, 1935, vol. 2, pp. 795–802. *Welder, Lond.*, vol. 7, no. 18, May 1935, pp. 548–552.

550. HANKINS, G. A. Fatigue tests and nondestructive tests on welds. "*Symposium on the welding of iron and steel*", Iron and Steel Institute, London, 1935, vol. 2, pp. 811–816. *Welder, Lond.*, vol. 7, no. 18, May 1935, pp. 571–574.

51. HANKINS, G. A. *and* MILLS, H. R. The resistance of spring steels to repeated impact stresses. *J. Iron St. Inst.*, vol. 131, no. 1, 1935, pp. 165–180. Abstr.: *Engineering, Lond.*, vol. 139, no. 3621, June 7, 1935, pp. 611–613. *Heat Treat. Forg.*, vol. 21, no. 7, July 1935, pp. 322–326.

1552. HARVEY, W. E.; CIASTKEWICZ, A. J. *and* WHITNEY, F. J. Corrosion fatigue study of welded 18-8 stainless steel. *J. Amer. Weld. Soc.*, vol. 14, no. 1, Jan. 1935, pp. 18–23.

1553. HAUTTMANN, H. The relation of laboratory research to engine design. *Trans. N.-E. Cst Instn Engrs Shipb.*, vol. 52, 1935–1936, pp. 103–126.

1554. HEMPEL, M. *and* PLOCK, C. H. Schwingungsfestigkeit und Dämpfungsfähigkeit von unlegierten Stählen in Abhängigkeit von der chemischen Zusammensetzung und der Wärmebehandlung. [Fatigue strength and damping capacity of carbon steels in relation to chemical composition and heat treatment.] *Mitt. K.-Wilh.-Inst. Eisenforsch.*, vol. 17, no. 2, 1935, pp. 19–31. Abstr.: *Stahl u. Eisen*, vol. 55, no. 20, May 16, 1935, pp. 550–551. English abstr.: *J. Iron St. Inst.*, vol. 131, no. 1, 1935, p. 481.

1555. HOBROCK, R. H. Some problems of the materials and methods of joining used in strong light structures. *Publ. Guggenheim Airsh. Inst.*, no. 3, 1935, pp. 64–88. Abstr.: *Mech. Engng, N.Y.*, vol. 58, no. 4, April 1936, pp. 250–251.

1556. HOFFMANN, W. Stahlrohre für den Flugzeugbau und ihre Schweissverbindungen. [Steel tubes for aircraft and their welded joints.] *Z. Ver. dtsch. Ing.*, vol. 79, no. 38, Sept. 21, 1935, pp. 1145–1148.

1557. HOLTSCHMIDT, O. Eine Neue Schwingungs- und Dämpfungsprüfmaschine der M. A. N. [A new fatigue and damping testing machine of the M. A. N.] *Mitt. ForschAnst. Gutehoffn., Nürnberg*, vol. 3, no. 10, 1935, pp. 279–284.

1558. HORGER, O. J. Effect of surface rolling on the fatigue strength of steel. *J. appl. Mech.*, vol. 2, no. 4, Dec. 1935, pp. A128–A136.

1559. INGHAM, E. Some unusual causes of shaft failures. *Mech. World*, vol. 98, no. 2551, Nov. 22, 1935, pp. 498, 502.

1560. INGHAM, E. The prevention of air-compressor breakdowns. *Mech. World*, vol. 98, no. 2555, Dec. 20, 1935, pp. 597–598.

1561. IRMANN, R. Limite d'endurance des alliages d'aluminium. [Fatigue limits of aluminium alloys.] *Int. Congr. Min. etc.*, Seventh Congress, Paris, Oct. 1935, vol. 1, pp. 435–440.

1562. IRMANN, R. Die Ermüdungsfestigkeit der Aluminium-Legierungen. [The fatigue strength of aluminium alloys.] *Aluminium, Berl.*, vol. 18, no. 12, Dec. 1935, pp. 638–643.

1563. KENYON, J. N. The rotating-wire arc fatigue machine for testing small-diameter wire. *Proc. Amer. Soc. Test. Mater.*, vol. 35, pt. 2, 1935, pp. 156–166. Abstr.: *Mech. World*, vol. 98, no. 2536, Aug. 9, 1935, pp. 125, 127.

1564. KÖRBER, F. and HEMPEL, M. Einfluss von Recken und Altern auf das Verhalten von Stahl bei der Schwingungsbeanspruchung. [Influence of cold working and ageing upon the behaviour of steel under alternating stresses.] *Mitt. K.-Wilh.-Inst. Eisenforsch.*, vol. 17, no. 22, 1935, pp. 247–257.

1565. KOMMERS, J. B. Notch sensitivity of steels in fatigue. *Engineering, Lond.*, vol. 139, no. 3607, March 1, 1935, p. 225.

1566. KOMMERS, J. B. Understressing and overstressing in iron and steel. *Engng News Rec.*, vol. 114, April 18, 1935, pp. 550–551.

1567. KONTOROVICH, E. J. Endurance of nitrided steel. *Metal Progr.*, vol. 28, Sept. 1935, p. 64.

1568. KRYSTOF, J. Über die Haltbarkeit von Zinn- und Zinküberzügen bei Korrosiondauerbeanspruchungen. [The effectiveness of tin and zinc coatings under corrosion fatigue stresses.] *Metallwirtschaft*, vol. 14, no. 16, April 19, 1935, pp. 305–307.

1569. KRYSTOF, J. Über die Dauerfestigkeit von Leichtmetallen mit besonderer Berücksichtigung von Duralumin. [On the fatigue strength of light metals with particular regard to duralumin.] *Metallwirtschaft*, vol. 14, no. 35, Aug. 30, 1935, pp. 701–704.

1570. LEHR, E. and MAILÄNDER, R. Einfluss von Hohlkehlen an abgesetzten Wellen auf die Biegewechselfestigkeit. [Influence of fillets at changes of section in shafts on the bending fatigue strength.] *Z. Ver. dtsch. Ing.*, vol. 79, no. 33, Aug. 17, 1935, pp. 1005–1011. *Arch. Eisenhüttenw.*, vol. 9, no. 1, July 1935, pp. 31–35.

1571. MACGREGOR, R. A.; BURN, W. S. and BACON, F. The relation of fatigue to modern engine design. *Trans. N.-E. Cst Instn Engr Shipb.*, vol. 51, no. 4, 1935, pp. 161–228 D100–D136. Abstr.: *Metallurgist*, vol. 10 April 26, 1935, pp. 18–19.

1572. MAHIN, E. G. and HAMILTON, J. W. Endurance limit of black-heart malleable iron. *Trans. Amer. Foundrym. Ass.*, vol. 43, no. Dec. 1935, pp. 41–50.

1573. MAILÄNDER, R. Kerbempfindlichkeit bei Wechselbeanspruchung von legierten und unlegierten Stählen. [Notch sensitivity under alternating stressing of unalloyed and alloyed steel.] *Stahl u. Eisen*, vol. 55, no. 2, Jan. 10 1935, pp. 39–41. English abstr.: *Mech. Engng N.Y.*, vol. 57, no. 5, May 1935, pp. 309–310

1574. MAILÄNDER, R. Über die Kerbempfindlichkeit von Stahl bei wechselnder Bea spruchung. [On the notch sensitivity of steel under alternating stressing.] *Tech. Mitt. Krupp*, vol. 3, no. 3, June 1935, pp. 108–111.

1575. MAILÄNDER, R. *and* RUTTMANN, W. Einfluss von Oberflächenbeschaffenheit, Randschichten und Korrosion auf die Dauerfestigkeit. [Influence of surface condition, surface layers and corrosion on fatigue strength.] *Maschinenbau*, vol. 14, nos. 3/4, Feb. 1935, pp. 73–77.

1576. MALEEV, V. L. Avoid waste of material in parts design. *Mach. Design*, vol. 7, no. 8, Aug. 1935, pp. 23–25, 40.

1577. MATTHAES, K. Eine 5-t-Zug-Druck-Dauerprüfmaschine der DVL. [A 5 ton tension-compression fatigue testing machine of the DVL.] *Luftfahrtforsch.*, vol. 12, no. 2, May 16, 1935, pp. 87–88.

1578. MATTHAES, K. Herabsetzung der Dauerfestigkeit von Nichteisenmetallen an Kraftangriffsstellen. [The reduction of the fatigue strength of non-ferrous metals at the loading points.] *Luftfahrtforsch.*, vol. 12, no. 5, Aug. 31, 1935, pp. 176–179.

1579. MEMMLER, K.; BIERETT, G.; GEHLER, W.; GRAF, O. *and* KOMMERELL, O. *Dauerfestigkeitsversuche mit Schweissverbindungen. (Fatigue strength tests on welded joints.)* Berlin, VDI-Verlag, 1935, pp. 46. Abstr.: *Z. Ver. dtsch. Ing.*, vol. 79, no. 4, Nov. 2, 1935, p. 1348.

580. MOORE, H. F. Progress report on the joint investigation of fissures in railroad rails. *Bull. Amer. Rly Engng Ass.*, vol. 37, no. 376, June 1935, pp. 1–26. *Repr. Ill. Engng Exp. Sta.*, no. 4, 1935, pp. 26.

581. MOORE, H. F. Effect of occasional overload on the strength of metals. *Metals and Alloys*, vol. 6, June 1935, p. 144.

1582. MUNZINGER, F. Dauerversuche für den Stahlbau. [Fatigue testing of steel structures.] *Maschinenschaden*, vol. 12, no. 3, 1935, pp. 41–46.

1583. MUSATTI, I. *and* DAINELLI, L. Influenza del trattamento termico sulla resistenza alla fatica ed alla corrosione del bronzo d'alluminio. [Influence of heat treatment on the fatigue and corrosion resistance of aluminium bronze.] *Alluminio*, vol. 4, no. 1, 1935, pp. 51–63.

1584. NEESE, H. E. Form of welds and fatigue strength of welded joints. *Weld. Ind.*, vol. 3, no. 9, Oct. 1935, pp. 303–308.

1585. NEILD, J. F. What about wear and fatigue in trolley wire? *Mass Transportation*, vol. 31, April 1935, pp. 103–104.

1586. NISIHARA, T.; SAKURAI, T. *and* WATANABE, T. The influence of specimen vibration in the evaluation of the bending fatigue strength. [J] *Trans. Soc. mech. Engrs, Japan*, vol. 1, no. 3, Aug. 1935, pp. 189–196. German summary: *ibid.*, pp. S-59–S-62.

1587. OCHS, H. Steigerung der Korrosions-Biegewechselfestigkeit von Stahl durch Zusätze zur angreifenden Flüssigkeit. [Increase of the corrosion bending fatigue strength of steel by additions of inhibitors to the corroding liquids.] *Z. Ver. dtsch. Ing.*, vol. 79, no. 11, March 16, 1935, pp. 358–359.

1588. OGIEVETSKII, A. S. *The strengths of welded joints under repeated loads (fatigue of metals).* [R] Moscow, Ob'edinennoe Nauchno-Tekhnicheskoe Izdatel'stvo Narodnogo Komissariata Tiazheloi Promyshlennosti SSSR, 1935.

1589. ORR, J. Fatigue tests of butt joints in mild steel and high tensile steels. *Mech. World*, vol. 97, no. 2520, April 19, 1935, pp. 373–374.

1590. PERSOZ, L. Les ruptures des pièces de machines et les moyens de les éviter. La fatigue des métaux et les machines d'essais à la fatigue. [Failures of machine parts and means for their elimination. Fatigue of metals and fatigue testing machines.] *Génie civ.*, vol. 106, no. 11, March 16, 1935, pp. 245–251; no. 12, March 23, 1935, pp. 274–278; no. 13, March 30, 1935, pp. 304–308 and no. 14, April 6, 1935, pp. 328–330.

1591. PETERSON, R. E. *and* WAHL, A. M. Fatigue of shafts at fitted members, with a related photoelastic analysis. *J. appl. Mech.*, vol. 2, no. 1, June 1935, pp. A1–A11.

1592. PIKE, E. J. Special apparatus used for testing aluminium alloys. *Metallurgia, Manchr*, vol. 13, no. 74, Dec. 1935, pp. 35–37.

1593. POMEY, J. *and* ANCELLE, A. Introduction á l'etude de la fatigue-corrosion. [Introduction to the study of corrosion fatigue.] *Mem. Comm. Étud. Corros.*, 1935–1936, pp. 17–42.

1594. PREVER, V. S. *and* LOCATI, L. I fenomeni di fatica nelle molle ad elica in acciaio 'super-armonico'. [The phenomena of fatigue in helical springs made of 'superharmonic' steel.] *Metallurg. ital.*, vol. 27, no. 3, March 1935, pp. 188–204 and no. 4, April 1935, pp. 255–274.

1595. RAINIER, E. T. *and* GERKE, R. H. A fatigue cracking test for tire thread compounds: some of the laws of fatigue. *Industr. Engng Chem. (Anal.)*, vol. 7, no. 6, Nov. 15, 1935, pp. 368–373.

1596. RAMBERG, W.; BALLIF, P. S. *and* WEST, M. J. A method for determining stresses in a non-rotating propeller blade vibrating with a natural frequency. *J. Res. nat. Bur. Stand.*, vol. 14, no. 2, Feb. 1935, pp. 189–215.

1597. REIMANN, W. Konstruktive Fehler als Ursache für Dauerbrüche an Triebwerksteilen und ihre Behebung. [Constructional errors as cause of fatigue fracture of mechanical transmission parts and how to prevent them.] *Maschinenschaden*, vol. 12, no. 4, 1935, pp. 57–60.

1598. REIMANN, W. Konstruktive Fehler als Ursache für Dauerbrüche an Zahnrädern und ihre Behebung. [Constructional errors as cause of fatigue failure in gears and how to prevent them.] *Maschinenschaden*, vol. 12, no. 8, 1935, pp. 121–125.

1599. RIST, R. Vergleichende Untersuchungen an Dauerbrüchen. [Comparative investigations of fatigue fractures.] *Z. bayer. (Dampfk. RevisVer.*, vol. 39, no. 23, Dec. 15, 1935, pp. 203–211.

1600. ROBERTS, A. M. Fatigue of welds "Symposium on the welding of iron and steel" Iron and Steel Institute, London, 1935 vol. 2, pp. 831–841.

1601. ROŠ, M. *and* EICHINGER, A. The strength of welded connections. "*Symposium on the welding of iron and steel*", Iron and Steel Institute, London, 1935, vol. 2, pp. 843–866.

1602. ROŠ, M. *and* EICHINGER, A. Festigke geschweisster Verbindungen. [Strength welded joints.] *Schweiz. Arch. angew. Wiss* vol. 1, no. 3, 1935, pp. 33–43 and no. 5, 193 pp. 77–86.

1603. RUTTMANN, W. Verformungslose Brüch an Kesselteilen. [Fractures without deformation in boiler plate.] *Z. Ver. dtsch. Ing* vol. 79, no. 52, Dec. 28, 1935, pp. 1561–156

1604. SALIGER, R. Dauerversuche an Eise betonbalken mit verschiedenen Stahlbewe rungen. [Fatigue tests on reinforced concrete beams with different types of steel reinforcement.] *Mitt. EisenbetAuss. öst. Ing.- u. Archit-Ver.*, no. 15, 1935, pp. 95. Abstr.: *Z. öst. Ing.- u. ArchitVer.*, vol. 87, no. 7/8, 1935, pp. 45–46.

1605. VON SCHWARZ, M. Dauerbrüche an Pleuelstangenschrauben. [Fatigue failures of connecting-rod bolts.] *Maschinenschaden*, vol. 12, no. 5, 1935, p. 77.

1606. VON SCHWARZ, M. Dauerbrüche von Ventilen infolge unsauberer Bearbeitung des Schaftes. [Fatigue failures of valves resulting from defective treatment of stem.] *Maschinenschaden*, vol. 12, no. 5, 1935, p. 78.

1607. SCHWINNING, W. Die Festigkeitseigenschaften der Werkstoffe bei tiefen Temperaturen. [The strength properties of materials at low temperatures.] *Z. Ver. dtsch. Ing.*, vol. 79, no. 2, Jan. 12, 1935, pp. 35–40.

1608. SCHWINNING, W. *and* DORGERLOH, E. Untersuchungen über die Schwingungsfestigkeit von kalt gezogenen Drähten aus Kupfer und Reinaluminium. [Investigation of the fatigue strength of cold-drawn copper and pure aluminium wire.] *Z. Metallk.*, vol. 27, no. 2, Feb. 1935, pp. 33–37.

1609. SEEGER, G. Wirkung von Druckvorspannungen auf die Dauerfestigkeit metallischer Werkstoffe. [Effect of compressive stress on the fatigue strength of metallic materials.] *Mitteilungen der Materialprüfunganstalt der Technischen Hochschule, Stuttgart*, 1935, pp. 56. Abstr.: *Z. Ver. dtsch. Ing.*, vol. 80, no. 22, May 30, 1936, pp. 698–699.

1610. SERENSEN, S. V. The effect of stress concentrations of fatigue strength. [R] *Tekh. vozdush. Flota*, vol. 9, no. 5, May 1935, pp. 57–72.

1611. SHELTON, S. M. *and* SWANGER, W. H. Fatigue properties of steel wire. *J. Res. nat. Bur. Stand.*, vol. 14, no. 1, Jan. 1935, pp. 17–32.

1612. SODERBERG, C. R. Working stresses. *J. appl. Mech.*, vol. 2, no. 3, Sept. 1935, pp. A106–A108.

1613. SPELLER, F. N. Corrosion fatigue of drill pipe is cut by the chemical treatment of mud. *Oil Gas J.*, vol. 34, no. 26, Nov. 14, 1935, pp. 71–72, 75.

1614. STAUDINGER, H. Dauerversuche mit $\frac{5}{8}''$ Schrauben. [Fatigue tests on $\frac{5}{8}''$ bolts.] *Zent. wiss. Ber., ForschBer.*, no. 216, 1935, pp. 13.

1615. SUTTON, H. *and* PEAKE, T. J. New pickling or etching baths for aluminium alloys. *Roy. Aircr. Estab. Rep.*, no. M.2269B, Jan. 1935, pp. 10.

1616. SUTTON, H. *and* TAYLOR, W. J. The influence of pickling on the fatigue strength of duralumin. *Rep. Memor. aero. Res. Comm., Lond.*, no. 1647, Feb. 1935, pp. 4.

1617. SUTTON, H. *and* TAYLOR, W. J. Corrosion fatigue properties of MG.7 alloy (D.T.D. 194). *Roy. Aircr. Estab. Rep.*, no. M.T.5563a, Oct. 1935, pp. 4.

1618. TANABASHI, R. Tests to determine behaviour of riveted joints of steel structures under alternate bending moments. *Mem. Coll. Engng Kyoto*, vol. 8, no. 4, March 1935, pp. 164–190. Abstr.: *Mech. Engng N.Y.*, vol. 57, no. 8, Aug. 1935, pp. 516–517.

1619. TAYLOR, C. F. An analysis of critical stresses in aircraft engine parts. *J. Soc. automot. Engrs, N.Y.*, vol. 37, no. 5, Nov. 1935, pp. 412–418, 421.

1620. THUM, A. Das heutige Gusseisen und seine Verwendungsmöglichkeiten in der Konstruktion. [Modern cast iron and its possible applications in construction.] *Giesserei*, vol. 22, no. 10, May 10, 1935, pp. 214–218.

1621. THUM, A. Die Sicherheit der Konstruktionen. [The safety of structures.] *Maschinenschaden*, vol. 12, no. 10, 1935, pp. 155–164.

1622. THUM, A. Zusammenwirkung der technischen Physik mit Werkstoff und Festigkeitsforschung zu neuen Konstruktionslehren im Maschinenbau. [Relationship between technical physics of materials and strength research and new constructional information in machine structures.] *Z. tech. Phys.*, vol. 16, no. 12, 1935, pp. 554–561.

1623. THUM, A. *and* BAUTZ, W. Ursachen der Steigerung der Dauerhaltbarkeit gedrückter Stäbe. [Causes of improved fatigue resistance of specimens with compressed surface.] *ForschArb. Ingwes.*, vol. 6, no. 3, May–June 1935, pp. 121–128.

1624. THUM, A. *and* BAUTZ, W. Die "Gestaltsfestigkeit": der Einfluss der Form auf die Festigkeitseigenschaften. [The design strength: the effect of shape on strength.] *Schweiz. Bauztg.*, vol. 106, no. 3, July 20, 1935, pp. 25–30.

1625. THUM, A. *and* BAUTZ, W. Die Gestaltfestigkeit. [The design strength.] *Stahl u. Eisen*, vol. 55, no. 39, Sept. 26, 1935, pp. 1025–1029.

1626. THUM, A. *and* BAUTZ, W. Der Entlastungsübergang: Günstigste Ausbildung des Überganges an abgesetzten Wellen u. dgl. [The transition fillet: the development of a suitable transition for shafts.] *ForschArb. Ingwes.*, vol. 6, no. 6, Nov.–Dec. 1935, pp. 269–273.

1627. THUM, A. *and* DEBUS, F. Die Vorzüge der Dehnschraube. [The advantages of prestressed bolts.] *Z. Ver. dtsch. Ing.*, vol. 79, no. 30, July 27, 1935, pp. 917–919.

1628. THUM, A. *and* MEYERCORDT, F. Einfluss von Form, Oberflächenbeschaffenheit und Werkstoff auf die Dauerhaltbarkeit gegossener und geschweisster Konstruktionen. [Influence of shape, surface condition and material on the fatigue properties of cast and welded constructional parts.] *Giesserei*, vol. 22, no. 5, March 1, 1935, pp. 90–94. English abstr.: *J. Iron St. Inst.*, vol. 131, no. 1, 1935, p. 479.

1629. THUM, A. *and* WUNDERLICH, F. Die Reiboxydation an festen Paarverbindungsstellen und ihre Bedeutung für den Dauerbruch. [Fretting corrosion in fixed joints and its influence on fatigue fracture.] *Z. Metallk.*, vol. 27, no. 12, Dec. 1935, pp. 277–280.

1630. UDE, H. Steigerung der Dauerhaltbarkeit der Konstruktionen. [Increasing the fatigue properties of structures.] *Z. Ver. dtsch. Ing.*, vol. 79, no. 2, Jan. 12, 1935, pp. 47–53.

1631. UDE, H. Zur Geschichte der Eisenbahnwerkstoffe. [The history of railway materials.] *Technikgeschichte*, vol. 24, 1935, pp. 38–61.

1632. ULRICH, M. Verdrehungsfestigkeit und Verschleiss von Keilwellen. 1 Teil. Werden Hinterachsenwellenbrüche in Fahrbetrieb durch Drehschwingung oder durch Gewaltbeanspruchung hervorgerufen? [Torsional strength and wear of spline shafts. Part 1. Are rear axle fractures in service due to torsional fatigue or a small number of large overloads?] *VersBer. ForschInst. KraftFahrw.* no. 11, April 1935, pp. 1–24.

1633. WEIBEL, E. E. Correlation of spring-wire bending and torsion fatigue tests. *Trans. Amer*

Soc. mech. Engrs, vol. 57, no. 8, Nov. 1935, pp. 501–516. *Wire and Wire Prod.*, vol. 10, no. 12, Dec. 1935, pp. 560–577, 588.

1634. WESCOTT, B. B. *and* BOWERS, C. N. Economical selection of sucker rods. *Trans. Amer. Inst. min. (metall.) Engrs*, vol. 114, 1935, pp. 177–192.

1635. WESCOTT, B. B. *and* BOWERS, C. N. Selection of materials for sucker rods important in ultimate economy. *Oil Gas J.*, vol. 34, no. 4, June 13, 1935, pp. 90, 92 and 95.

1636. WEVER, F. *and* MÖLLER, H. Die Arbeiten von F. Regler zur Werkstoffprüfung mit Röntgenstrahlen. [The work of F. Regler on materials testing by means of X-rays.] *Arch. Eisenhüttenw.*, vol. 9, no. 1, July 1935, pp. 47–55.

1637. WIEGAND, H. Dauerbrüche an Nietwerkzeugen. Brucharten, Wesen des Dauerbruches und seine Verhinderung, Verbesserung der Werkzeuge. [Fatigue fractures in riveting tools. Types of fracture, nature of fatigue fractures and their prevention, improvement of tools.] *Maschinenbau*, vol. 14, no. 23/24, 1935, pp. 675–676.

1638. WUNDERLICH, F. Dauerhaltbarkeit von Einspannungen und Sitzstellen bei Maschinenteilen. [Fatigue properties of clamped and fitted positions in machine parts.] *Anz. Bergw.*, vol. 57, no. 78, 1935, pp. 5–6.

1639. VON ZEERLEDER, A. Sullo sviluppo delle prove di fatica su l'alluminio e le sue leghe leggere. [On the development of fatigue tests on aluminium and its light alloys.] *Alluminio*, vol. 4, no. 2, March – April 1935, pp. 95–105.

1640. Notes on fatigue tests on rotating-beam testing machines. *Proc. Amer. Soc. Test. Mater.*, vol. 35, pt. 1, 1935, pp. 113–120.

1641. *Dauerfestigkeitsversuche mit Schweissverbindungen*. Bericht des Kuratoriums für Dauerfestigkeitsversuche im Fachausschuss für Schweisstechnik beim Verein deutscher Ingenieure, durchgeführt 1930 bis 1934. [*Fatigue strength tests on welded joints.*] Berlin, VDI-Verlag, 1935, pp. 46. Abstr.: *Z. Ver. dtsch. Ing.*, vol. 79, no. 4, Nov. 2, 1935, p. 1348.

1642. Fatigue failure of metals—considered with regard to engine design. *Iron Coal Tr. Rev.*, vol. 130, no. 3492, Feb. 1, 1935, pp. 207–208.

1643. Fatigue in marine engines. *Engineering, Lond.*, vol. 139, no. 3603, Feb. 1, 1935, pp. 121–122.

1644. Fatigue tests of butt joints applicable to mild and high tensile steel. *Iron Coal Tr. Rev.*, vol. 130, no. 3495, Feb. 22, 1935, p. 335.

1645. Relation of fatigue to modern engine design. *Metallurgist*, vol. 10, April 26, 1935, pp. 18–19.

1646. The influence of precipitated carbides on the fatigue strength of austenitic steel sheet. *Roy. Aircr. Estab. Rep.*, no. M.1734, May 1935, pp. 3.

1647. Easy with the tongs; corrosion-fatigue failures. *Min. and Metall.*, *N.Y.*, vol. 16, no. 342, June 1935, p. 269.

1648. Fatigue research on steel wire. *Engineering, Lond.*, vol. 139, no. 3621, June 7, 1935, pp. 603–604.

1649. Fatigue testing machine for wire. *Wire and Wire Prod.*, vol. 10, no. 7, July 1935, pp. 272–274, 284–285.

1650. National Physical Laboratory, Engineering Department—Fatigue and corrosion-fatigue research. *Engineering, Lond.*, vol. 140, no. 3626, July 12, 1935, pp. 44–46.

1651. New fatigue tester. *Iron Coal Tr. Rev.*, vol. 131, Aug. 23, 1935, p. 275.

1652. An electromagnetic fatigue tester. *Engineer, Lond.*, vol. 160, no. 4154, Aug. 23, 1935, p. 201. *Machinist*, vol. 79, no. 31, Aug. 31, 1935, pp. 463–466. *Mech. World*, vol. 98, no. 2540, Sept. 6, 1935, p. 225. *Auto. Engr*, vol. 25, no. 337, Oct. 1935, p. 380. *Engineering, Lond.*, vol. 140, no. 3639, Oct. 11, 1935, p. 406.

1653. Marine engineering failures. *Engineering, Lond.*, vol. 140, no. 3649, Dec. 20, 1935, pp. 667–668. *Mech. Engng, N.Y.*, vol. 58, no. 5, May 1936, p. 314.

1654. Relations entre la fatigue et le tracé des organes des moteurs Diesel. [Relations between fatigue and the arrangement of the components of Diesel engines.] *Génie civ.*, vol. 107, no. 26 (no. 2785), Dec. 28, 1935, pp. 610–614.

1936

1655. ADLOFF, K. Korrosion und Konstruktion. [Corrosion and construction.] *Wärme*, vol. 59, no. 3, Jan. 18, 1936, pp. 39–42.

1656. AFANASEV, N. N. Methods of increasing the endurance of components subjected to fatigue loading. [R] *Trud. kievsk. aviats. in-ta.*, vol. 4, 1936, pp. 114–124.

1657. AFANASEV, N. N. *and* BAKHAREV, V. M. On the influence of internal stresses in metal on the endurance under repeated bending. [R] *Trud. kievsk. aviats. in-ta.*, vol. 6, 1936, pp. 102–113.

1658. ALTPETER, H. Korrosion unter Ermüdung bei Drahtseilen. [Corrosion fatigue of wire rope.] *Glückauf*, vol. 72, no. 40, Oct. 3, 1936, pp. 1009–1010. English translation: *Henry Brutcher Tech. Trans.*, no. 346.

1659. ASAKAWA, Y. *and* FUDITA, S. On some questions in the study of fatigue of metals. *Sci. Rep. Tohoku Univ.*, *Honda Anniversary Volume*, Oct. 1936, pp. 1056–1059.

1660. BECKER, M. L. *and* PHILLIPS, C. E. Internal stresses and their effect on the fatigue resistance of spring steels. *J. Iron St. Inst.* vol. 133, no. 1, 1936, pp. 427–453. *Heat Treat. Forg.*, vol. 22, no. 5, May 1936, pp. 227–232.

1661. BOBEK, K. Über die Berechnung von dauernd wechselnd beanspruchte Schweissverbindungen. [On the design of welded joints for fatigue loads.] *Elektroschweissung*, vol. 7, no. 3, 1936, pp. 41–45.

1662. BOETCHER, H. N. Failure of metals due to cavitation under experimental conditions. *Trans. Amer. Soc. mech. Engrs*, vol. 58, 1936, pp. 355–360.

1663. BOHNER, H. Einfluss der Formungsart und der thermischen Vergütung auf die Ermüdungsfestigkeit von vergüteten Aluminium-Legierungen. [Influence of the method of deformation and thermal ageing on the fatigue strength of aged aluminium alloys.] *Metallwirtschaft*, vol. 15, no. 35, Aug. 28, 1936, pp. 813–814.

1664. BOHNER, H. Die Dauerschlagfestigkeit vergüteter Aluminium-Legierungen. [The impact fatigue strength of aged aluminium alloys.] *Metallwirtschaft*, vol. 15, no. 42 Oct. 16, 1936, pp. 983–984.

1665. BROPHY, G. R. Damping capacity, a facto in fatigue. *Trans. Amer. Soc. Metals*, vol. 24 March 1936, pp. 154–185. Abstr.: *Automot*

Industr. N.Y., vol. 73, no. 14, Oct. 5, 1935, p. 452. *Steel*, vol. 97, no. 20, Nov. 11, 1935, pp. 69–70.

1666. BUCHWALD, A. Détermination simplifiée de l'endurance de l'acier et d'autres métaux. [Simplified methods for determining the endurance of steels and other metals.] *Génie civ.*, vol. 109, no. 2813, July 11, 1936, pp. 41–43.

1667. BUCKWALTER, T. V. *and* HORGER, O. J. Stress analysis of locomotive and other large axles. *Iron Steel Engr*, vol. 13, no. 11, Nov. 1936, pp. 23–26.

1668. CASSIE, A. B. D.; JONES, M. *and* NAUNTON, W. J. S., Fatigue in rubber. Part I. *I.R.I. Trans.*, vol. 12, no. 1, June, 1936, pp. 49–80.

1669. CAZAUD, R. Les essais de fatigue des métaux. [The fatigue testing of metals.] *Rev. Métall.*, vol. 33, no. 3, March 1936, pp. 164–177.

1670. CLARK, P. R. Essais mécaniques pour établir la résistance des fils d'acier à la fatigue. [Mechanical tests for establishing the resistance of steel wires to fatigue.] *Monit. Petr. roman*, vol. 37, no. 21, Nov. 1, 1936, pp. 1587–1590 and no. 22, Nov. 15, 1936, pp. 1653–1659.

1671. COLPAERT, H. Ruptura por fadiga de eixos de tenders e carros de estradas de ferro. [Fatigue failure of axles of tenders and railroad cars.] *Bol. Inst. Engen. S. Paulo*, vol. 24, no. 125, July–Aug.–Sept. 1936, pp. 10–15.

1672. CORNELIUS, H. *and* BOLLENRATH, F. Gegossene Nocken- und Kurbelwellen. [Cast camshafts and crankshafts.] *Giesserei*, vol. 23, no. 10, May 8, 1936, pp. 229–236. English translation: *Found. Tr. J.*, vol. 55, Nov. 26, 1936, pp. 411–413.

1673. CUTHBERTSON, J. W. Rapid fatigue test. *Metal Progr.*, vol. 29, no. 6, June 1936, pp. 67, 86.

1674. DAASCH, H. L. Notes on fatigue properties of cast iron. *Trans. Amer. Foundrym. Ass.*, vol. 44, 1936, pp. 528–544.

1675. DAVIDENKOV, N. N. *Dynamic testing of metals.* [R] M. – L. ONTI, Glavn. red. liter. po chern. met., 1936, pp. 394.

1676. DIEPSCHLAG, E.; MATTING, A. *and* OLDENBURG, G. Elastizitätsverhältnisse in Schweissverbindungen und deren Zugschwingungsfestigkeit. [Elasticity conditions in welded joints and their tensile fatigue strength.] *Arch. Eisenhüttenw.*, vol. 9, no. 7, Jan. 1936, pp. 341–345.

1677. DINNER, H. *and* CHRISTEN, H. Dauerfestigkeit. [Fatigue strength.] *Schweiz. tech. Z.*, vol. 11, 1936, pp. 77–81.

1678. DIXON, S. M. *and* HOGAN, M. A. L'Influence de la fatigue sous corrosion sur les câbles de mines. [The influence of corrosion-fatigue on mine cables.] *Rev. Industr. min.*, no. 362, Jan. 15, 1936, pp. 21–25. English abstr.: *Cordage*, vol. 17, 1936, p. 11.

1679. DIXON, S. M.; HOGAN, M. A. *and* ROBERTSON, S. L. The deterioration of colliery winding ropes in service. *Pap. Saf. Min. Res. Bd, Lond.*, no. 94, 1936, pp. 108.

1680. DONALDSON, J. W. The fatigue of cast iron. *Found. Tr. J.*, vol. 55, no. 1037, July 2, 1936, pp. 9–11.

1681. DONALDSON, J. W. Further investigations on the fatigue strength of cast iron. *Found. Tr. J.*, vol. 55, no. 1059, Dec. 3, 1936, pp. 432, 442.

1682. DOREY, S. F. Marine machinery defects—their causes and prevention. *Trans. Inst. Mar. Engrs*, vol. 47, pt. 12, Jan. 1936, pp. 305–383.

1683. DORGELO, H. B. *and* NEETESON, P. A. L'exameu aux rayons X du phénomène de fatigue dans les métaux. [X-ray examination of the fatigue phenomena in metals.] *Génie civ.*, vol. 108, June 13, 1936, pp. 562–563. Dutch version: *Ingenieur, 's Grav.*, vol. 51E, no. 7, Feb. 14, 1936, pp. E.21–E.25.

1684. DUTILLEUL, H. Etude de la résistance à la fatigue des soudures de francbord. [Study of the fatigue resistance of butt welded plates.] *Génie civ.*, vol. 108, no. 2788, Jan. 18, 1936, pp. 62–64.

1685. EHRT, M. Über die Dauerfestigkeit alter Kesselbleche. [On the fatigue strength of boiler plate.] *Maschinenschaden*, vol. 13, no. 11, 1936, pp. 169–173.

1686. EHRT, M. *and* KÜHNELT, G. Der Einfluss einer Auftragschweissung auf die Dauerhaltbarkeit von Stahlwellen. [The influence of built-up welding on the fatigue strength of steel shafts.] *Maschinenschaden*, vol. 13, no. 4, 1936, pp. 57–64.

1687. ENGELHARDT, W. Untersuchung über den Einfluss eines Kadmiumzusatzes auf die Schwingungsfestigkeit von Kupferdraht. [Investigation of the influence of an addition of cadmium on the flexural fatigue strength of copper wire.] *Mitt. ForschAnst. Gutehoffn., Nürnberg*, vol. 4, no. 6, May 1936, pp. 144–146. Abstr.: *Z. Ver. dtsch. Ing.*, vol. 81, no. 9, Feb. 27, 1937, p. 281. English abstr.: *Bull. Brit. non-ferr. Met. Ass.*, no. 91, Oct. 1936, p. 7.

1688. ERLINGER, E. Die Genauigkeit von dynamischen Materialprüfmaschinen. [The accuracy of dynamic testing machines.] *Messtechnik*, vol. 12, no. 6, June 1936, pp. 109–111.

1689. EURINGER, G. Die Wechselbiegefestigkeit von Duralumin bei 350°C. [The fatigue strength of duralumin at 350°C.] *Metallwirtschaft*, vol. 15, no. 24, June 12, 1936, pp. 540–541.

1690. EVERETT, F. L. *and* MAULBETSCH, J. L. Review of research in strength of materials. *J. appl. Mech.*, vol. 3, no. 2, June 1936, pp. A67–A70.

1691. FLAISSIER, M. Réflexions sur la possibilité d'étudier la fatigue à la dilatation d'un tuyautage de vapeur sur modèle réduit. [Considerations of the possibility of investigating expansion fatigue in steam pipes by means of models.] *Bull. Ass. tech. marit.*, no. 40, 1936, pp. 303–321.

1692. FÖPPL, O. Der Unterschied zwischen Oberflächendrücken und Drücken mit allseitigem Druck in bezug auf Dauerhaltbarkeit eines Werkteils. [The difference between surface pressing and pressing with three-dimensional pressure on the fatigue properties of components.] *Mitt. Wöhler-Inst.*, no. 29, 1936, pp. 57–60.

1693. FÖPPL, O. The practical importance of the damping capacity of metals, especially steels. *J. Iron St. Inst.*, vol. 134, no. 2, 1936, pp. 393–455. *Iron Steel Ind.*, vol. 10, Oct. 1, 1936, pp. 45–51.

1694. FÖPPL, O. *and* WAGENBLAST, W. Rüttelprüfung von Schraubverbindungen. [Vibration testing of bolted joints.] *Mitt. Wöhler-Inst.*, no. 27, 1936, pp. 52–64.

1695. FÖPPL, O. *and* WAGENBLAST, W. Bisherige Entwicklung des Oberflächendrückens. [Recent developments in surface pressing.] *Werk-*

zeugmaschine, vol. 40, 1936, pp. 501–505. English abstr.: *J. Iron St. Inst.*, vol. 135, no. 1, 1937, pp. 138A–139A.

1696. DE FOREST, A. V. The rate of growth of fatigue cracks. *J. appl. Mech.*, vol. 3, no. 1, March 1936, pp. A23–A25. [discussion] *ibid.*, no. 3, Sept. 1936, pp. A114–A117. *Mach. Design*, vol. 8, no. 7, July 1936, pp. 32–34.

1697. GILL, E. T. *and* GOODACRE, R. Some aspects of the fatigue properties of patented steel wires. Part III—Note on the effect of low-temperature heat treatment on decarburized wires. Part IV—A study of the endurance properties at high stresses. *Carnegie Schol. Mem.*, vol. 25, 1936, pp. 93–110 and pp. 111–139.

1698. GOUGH, H. J. The strength of metals under combined alternating stresses. *Iron Steel Ind.*, vol. 9, Jan. 1936, pp. 132–137 and Feb. 1936, pp. 177–178.

1699. GOUGH, H. J. Inner characteristics of the deformation and fracture of metals under static and fatigue stresses, as revealed by X-rays. *Ingenieur, 's Grav.*, vol. 51, no. 34, Aug. 21, 1936, pp. Mk 21–22 and no. 38, Sept. 18, 1936, pp. Mk 23–25.

1700. GOUGH, H. J. *and* FORREST, G. A study of the fatigue characteristics of three aluminium specimens, each containing from four to six large crystals. *J. Inst. Met.*, vol. 58, no. 1, Jan. 1936, pp. 97–112.

1701. GOUGH, H. J. *and* POLLARD, H. V. The effect of specimen form on the resistance of metals to combined alternating stresses. *Proc. Instn mech. Engrs, Lond.*, vol. 132, Dec. 1936, pp. 549–573.

1702. GOUGH, H. J. *and* WOOD, W. A. A new attack upon the problem of fatigue of metals,
using X-ray methods of precision. *Proc. roy. Soc.*, *(A)*, vol. 154, 1936, pp. 510–539.

1703. GOUGH, H. J. *and* WOOD, W. A. Strength of metals in the light of modern physics. *J. R. aero. Soc.*, vol. 40, no. 308, Aug. 1936, pp. 586–621.

1704. GOULD, A. J. The influence of temperature on the severity of corrosion-fatigue. *Engineering, Lond.*, vol. 141, no. 3669, May 8, 1936, pp. 495–496.

1705. GRAF, O. Über Dauerzugversuche und Dauerbiegeversuche an Stahlstäben mit brenngeschnittenen Flächen. [Tensile and bending fatigue tests on steel specimens cut by oxyacetylene torch.] *Autogene Metallbearb.*, vol. 29, no. 4, Feb. 15, 1936, pp. 49–57.

1706. GRAF, O. Versuche über den Einfluss der Zahl der minutlich auftretenden Lastwechsel auf die Ursprungszugfestigkeit von Nietverbindungen. [Tests on the effect of regularly repeated reversals of stress on the residual tensile strength of riveted joints.] *Stahlbau*, vol. 9, no. 6, March 13, 1936, p. 48.

1707. GRAF, O. Dauerbiegeversuche mit geschweissten Trägern I30 aus St37. [Bending fatigue tests of welded beams I30 made from steel St37.] *Stahlbau*, vol. 9, no. 9, April 24, 1936, pp. 71–72.

1708. GRAF, O. Versuche über die Längenänderungen und über die Tragfähigkeit von Nietverbindungen aus St52 unter oft wiederkehrenden Wechseln zwischen Zug- und Druckbelastung. [Test on the elongation and strength of riveted joints made of steel St52 subjected to repeated tensile and compressive loads.] *Stahlbau*, vol. 9, no. 24, Nov. 20, 1936, pp. 185–188.

1709. GRAF, O. Dauerfestigkeit von Nietverbindungen. [Fatigue strength of riveted joints.] *Congr. int. Ass. Br. struct. Engng*, Second Congress, Berlin, Munich, Oct. 1936, paper V7, pp. 1001–1011.

1710. GRAF, O. *and* BRENNER, E. Versuche zur Ermittlung der Widerstandsfähigkeit von Beton gegen oftmals wiederholte Druckbelastung. 2. Teil — Versuche über den Einfluss langdauernder Belastung auf die Formänderungen und auf die Druckfestigkeit von Beton- und Eisenbetonsäulen. [Tests for investigating the resistance of concrete under often repeated compression loads. Part 2—Tests on the influence of continued loads on the deformation and on the compressive strength of concrete and reinforced concrete.] *Deutscher Ausschuss für Eisenbeton, Berichte* no. 83, 1936, pp. 87.

1711. GREIS, F. *and* RUPPIK, H. Einfluss der Feuerverzinkens auf die Biegewechselfestigkeit und die Gleichmassigkeit der Festigkeitseigenschaften gezogener Stahldrähte. [Influence of hot-dip galvanizing on the bending fatigue strength and the uniformity of the tensile properties of drawn steel wire.] *Arch. Eisenhüttenw.*, vol. 10, no. 2, Aug. 1936, pp. 69–71.

1712. GÜRTLER, G. Zur Frage der Erhöhung der Dauerfestigkeit von Blechen, Rohren und Profilen aus Elektron. [The problem of increasing the fatigue strength of Elektron sheet, tube and sections.] *Zent. wiss. Ber., Forsch-Ber.*, no. 590, 1936, pp. 24.

1713. HANKINS, G. A.; BECKER, M. L. *and* MILLS, H. R. Further experiments on the effect of surface conditions on the fatigue resistance of steels. *J. Iron St. Inst.*, vol. 133, no. 1, 1936, pp. 399–453. *Heat Treat. Forg.*, vol. 22, no. 7, July 1936, pp. 340–345 and no. 8, Aug. 1936, pp. 394–399. Abstr.: *Engineer, Lond.*, vol. 161, no. 4193, May 22, 1936, pp. 553–554.

1714. HAUTTMANN, H. Shrink fit and fatigue strength. *Mech. World*, vol. 99, no. 2560, Jan. 24, 1936, pp. 81–82.

1715. HAUTTMANN, H. The relation of laboratory research to engine design. *Mar. News*, vol. 22, Feb. 1936, pp. 29–32.

1716. HEMPEL, M. *and* TILLMANNS, H. E. Verhalten des Stahles bei höheren Temperaturen unter wechselnder Zugbeanspruchung. [Behaviour of steels at high temperature under pulsating tensile stress.] *Mitt. K-Wilh.-Inst. Eisenforsch.*, vol. 18, no. 12, 1936, pp. 163–181. English translation: *Metal Treatm.*, vol. 3, Summer 1937, pp. 89–94.

1717. HILPERT, A. *and* BONDY, O. Die Dauerfestigkeit geschweisster Verbindungen des Stahlbaues. [The fatigue strength of welded joints of structural steels.] *T Z prakt. Metallbearb.*, vol. 46, no. 21/22, 1936, pp. 796, 799–802 and no. 23/24, 1936, pp. 862–868.

1718. HOGAN, M. A. Wire ropes research in relation to colliery practice. *Trans. Instn Min Engrs, Lond.*, vol. 91, 1936, pp. 196–219.

1719. HOLLARD, A. Les maladies des métaux. Épidémies et contagions. [The disorders c metals, epidemics and contagions.] *Bull. Soc Enc. Industr. nat., Paris*, vol. 135, Oct.–Nov 1936, pp. 593–608.

1720. HOLTSCHMIDT, O. Eine neue Schwingungs und Dämpfungsprüfmaschine. [A new fatigu and damping testing machine.] *Monta Rdsch.*, vol. 28, no. 16. Aug. 16, 193 pp. 1–3.

1721. HORGER, O. J. *and* MAULBETSCH, J. Increasing the fatigue strength of press-fitt axle assemblies by surface rolling. *J. ap Mech.*, vol. 3, no. 3, Sept. 1936, pp. A9 A98.

1722. IRMANN, R. Limite d'endurance des alliages d'aluminium. [The endurance limits of aluminium alloys.] *Rev. Métall.*, vol. 33, no. 4, April 1936, pp. 231–236.

1723. ITIHARA, M. Impact and static-torsion and bending diagrams of fatigued metals. *Sci. Rep. Tohoku Univ., Honda Anniversary Volume*, Oct. 1936, pp. 1041–1049.

1724. ITIHARA, M. *and* SUGAWARA, T. Hysterograph and the torsion endurance limit. [J] *Trans. Soc. mech. Engrs, Japan*, vol. 2, no. 8, Aug. 1936, pp. 336–339. English summary: *ibid.*, pp. S91–S93.

1725. JOHNSON, H. D. Aluminium stringer failures due to fatigue loading. *Engng News Rec.*, vol. 116, Feb. 27, 1936, pp. 318–320.

726. KARPOV, A. V. Modern stress theories. *Proc. Amer. Soc. civ. Engrs*, vol. 62, no. 8, Oct. 1936, pp. 1128–1153.

727. KENYON, J. N. The effect of the addition of lead on the endurance limit of a certain tin-base bearing alloy. *Proc. Amer. Soc. Test. Mater.*, vol. 36, pt. 2, 1936, pp. 194–200. Abstr.: *Metal Ind., Lond.*, vol. 49, no. 2, July 10, 1936, pp. 33–34.

728. KENYON, J. N. Endurance tests on electrolytic tough pitch and oxygen free copper wire. *Wire and Wire Prod.*, vol. 11, no. 10, Oct. 1936, pp. 576, 593. *Metal Ind., Lond.*, vol. 49, ı o. 21, Nov. 20, 1936, p. 510.

729. KIDANI, Y. On the mechanical properties and crystalline structure of metals. [J] *Trans. Soc. mech. Engrs, Japan*, vol. 2, no. 6, Feb. 1936, pp. 61–67. English summary: *ibid.*, pp. S17–S18.

1730. KIDANI, Y. On the fatigue of metals and the internal friction. *Sci. Rep. Tohoku Univ., Honda Anniversary Volume*, Oct. 1936, pp. 1050–1055.

1731. KLÖPPEL, K. Gemeinschaftsversuche zur Bestimmung der Schwellzugfestigkeit voller, gelochter und genieteter Stäbe aus St37 und St52. [Comprehensive tests to determine the increase in tensile strength of solid, punched and riveted specimens of St37 and St52.] *Stahlbau*, vol. 9, no. 13/14, June 19, 1936, pp. 97–112.

1732. KÖRBER, F. *and* HEMPEL, M. Abhängigkeit der Wechselfestigkeit des Stahles von der Lastwechselfrequenz. [Dependence of the fatigue strength of steel on the frequency of stress reversal.] *Mitt. K.-Wilh.-Inst. Eisenforsch.*, vol. 18, no. 1, 1936, pp. 15–19. Abstr.: *Z. Ver. dtsch. Ing.*, vol. 80, no. 49, Dec. 5, 1936, pp. 1489–1490.

1733. KOMMERELL, O. Einfluss häufig Wechselnder Belastungen auf geschweisste Bauwerke. [The influence of frequently alternating load on welded structures.] *Congr. int. Ass. Br. struct. Engng*, Second Congress, Berlin, Munich, Oct. 1936, paper IIIa1, pp. 349–401.

1734. KRESS,. Zur Ermittlung der Dauerfestigkeit von Baustählen. [Methods of determining the fatigue strength of structural steels.] *Bautechnik*, vol. 14, no. 47, Oct. 30, 1936, pp. 692–694.

1735. KÜHNELT, M. Der Einfluss einer Auftragschweissung auf die Dauerhaltbarkeit von Stahlwellen. [The influence of coated weld rod upon the fatigue properties of steel shafts.] *Mitteilungen der Materialprüfstelle der Allianz und Stuttgarter Verein Versicherungs A.-G., Abteilung für Maschinenversicherung*, Report no. 3, March 1936, pp. 11.

1736. LAUTE, K. Dauerschlagversuche an Leichtmetallen. [Impact fatigue tests on light metals.]

Z. Metallk., vol. 28, no. 8, Aug. 1936, pp. 233–236.

1737. LEA, F. C. Repeated stresses on structural elements. *J. Instn civ. Engrs*, no. 1, Nov. 1936, pp. 93–120.

1738. LEA, F. C. *and* PARKER, C. F. Welding Research Committee second report—report of physical tests. *Proc. Instn mech. Engrs, Lond.*, vol. 133, 1936, pp. 15–63.

1739. LEHR, E. Dauerversuche mit Nietverbindungen. [Fatigue tests on riveted joints.] *Z. Ver. dtsch. Ing.*, vol. 80, no. 30, July 25, 1936, pp. 920–922.

1740. LÖHR, A. Die Veränderung der Schwingungsfestigkeit und der Dämpfungsfähigkeit infolge hydraulischen Druckens. [The changes in fatigue strength and damping capacity as a result of hydraulic pressure.] *Mitt. Wöhler-Inst.*, no. 29, 1936, pp. 3–56.

1741. LÜRENBAUM, K. Stand und Ziel der Forschung zur Frage der Gestaltfestigkeit der Kurbelwelle. [Status and aim of the research into the question of the design strength of crankshafts.] *Jb. Lilienthal-Ges. Luftfahrtf.*, 1936, pp. 348–355.

1742. MACGREGOR, R. A. Fatigue in relation to failures in forgings. *Metal Treatm.*, vol. 2, Winter 1936, pp. 173–180.

1743. MCMULLAN, O. W. Endurance of case hardened gears. *Trans. Amer. Soc. Metals*, vol. 24, June 1936, pp. 262–280.

1744. MAILÄNDER, R. Über die Dauerfestigkeit von Gusseisen, Temperguss und Stahlguss. [On the fatigue strength of cast iron, malleable iron and cast steel.] *Tech. Mitt. Krupp*, vol. 4, no. 3, June 1936, pp. 59–66.

1745. MAILÄNDER, R. Eigenspannungen und Biegewechselfestigkeit verstickter Stahlproben. [Internal stresses and bending fatigue strength of nitrided steel specimens.] *Arch. Eisenhüttenw.*, vol. 10, no. 6, Dec. 1936, pp. 257–261. Abstr.: *Stahl u. Eisen*, vol. 56, no. 51, Dec. 17, 1936, p. 1538. English translation: *SLA Trans. Centre, John Crerar Lib.*, no. 59–20570.

1746. MATTHEW, T. U. Effect of continuous corrosion and abrasion on the fatigue of steel. *J. Roy. tech. Coll. Glasg.*, vol. 3, pt. 4, Jan. 1936, pp. 636–660.

1747. MATTING, A. *and* OLDENBURG, G. Untersuchungen über die Dauerfestigkeit von Schweissverbindungen an St.52. [Investigations on the fatigue strength of welded joints in St.52.] *Elektroschweissung*, vol. 7, no. 6, 1936, pp. 108–111.

1748. MAURER, E. *and* HEINE, H. Festigkeitseigenschaften und Korrosionsverhalten von Hochbaustählen. [Strength properties and corrosion behaviour of high strength structural steels.] *Arch. Eisenhüttenw.*, vol. 9, no. 7, Jan. 1936, pp. 347–357.

1749. MEUTH, H. Zur Ermittlung der Ablegereife von Drahtseilen. [Determination of the correct time for the removal of wire ropes from service.] *Z. Ver. dtsch. Ing.*, vol. 80, no. 21, May 23, 1936, pp. 664–666. English translation *Henry Brutcher Tech. Trans.*, no. 296.

1750. MEYER, W. Drehwechselfestigkeit genuteter Stäbe. [Torsional fatigue strength of keyed shafts.] *Z. Ver. dtsch. Ing.*, vol. 80, no. 32, Aug. 8, 1936, pp. 978–979.

1751. MEYERS, W. G. Stress concentration lead to design failures. *Mach. Design*, vol. 8, no. 1 Jan. 1936, pp. 37–40 and no. 2, Feb. 1936 pp. 34–37.

1752. MIKHAILOV-MIKHEEV, P. B. Resistance to fatigue and heat treatment of turbine shafts. [R] *Vest. metalloprom.*, vol. 16, no. 1, 1936, pp. 11–25.

1753. MIKHAILOVSKII, I. I. Vibration fatigue testing of flat metal specimens simultaneously. [R] *Vest. inzh. Tekh.*, vol. 22, no. 5, May 1936, pp. 309–311.

1754. MOORE, H. F. Correlation between metallography and mechanical testing. *Trans. Amer. Inst. min. (metall.) Engrs*, vol. 120, 1936, pp. 13–35. *Repr. Ill. Engng Exp. Sta.*, no. 9, 1936, pp. 23. Abstr.: *Iron Age*, vol. 137, no. 9, Feb. 27, 1936, pp. 26–29, 106.

1755. MOORE, H. F. How and when does a fatigue crack start? *Metals and Alloys*, vol. 7, no. 11, Nov. 1936, pp. 297–299.

1756. MORTADA, S. A. Beitrag zur Untersuchung der Fachwerke aus geschweisstem Stahl und Eisenbeton unter statischen und Dauerbeanspruchungen. [Contribution to the investigation of lattice girders in welded steel structures and reinforced concrete under static and fatigue loading.] *Ber. eidgenöss. MatPrüfAnst.*, no. 103, Oct. 1936, pp. 86.

1757. NELSON, E. W. Failure of parts in service. *Heat Treat. Forg.*, vol. 22, no. 2, Feb. 1936, pp. 63–65.

1758. NISIHARA, T. *and* SAKURAI, T. Effect of notches on the endurance limit of steel. [J] *Trans. Soc. mech. Engrs, Japan*, vol. 2, no. 9, Nov. 1936, pp. 436–446. English summary: *ibid.*, pp. S115–S116.

1759. ONO, A. Yielding of carbon steel specimens subjected to repeated tensile stresses. [J] *Trans. Soc. mech. Engrs, Japan*, vol. 2, no. 9, Nov. 1936, pp. 457–459. English summary: *ibid.*, pp. S118–S120. *J. Soc. mech. Engrs, Japan*, vol. 39, no. 232, 1936, pp. 435–436.

1760. OSCHATZ, H. Prüfmaschinen zur Ermittlung der Dauerfestigkeit. [Testing machines for the determination of fatigue strength.] *Z. Ver. dtsch. Ing.*, vol. 80, no. 48, Nov. 28, 1936, pp. 1433–1439.

1761. PERSOZ, L. Corrosion et fatigue. Influence de l'atmosphère. [Corrosion and fatigue. Influence of the atmosphere.] *Métaux, Paris*, vol. 11, no. 127, March 1936, pp. 60–65.

1762. PETERSON, R. E. *and* WAHL, A. M. Two- and three-dimensional cases of stress concentration and comparison with fatigue tests. *J. appl. Mech.*, vol. 3, no. 1, March 1936, pp. A15–A22.

1763. PEYCKE, A. H. *and* CLYNE, R. W. Hot formed mechanical springs, manufacture and life. *Metal Progr.*, vol. 29, no. 5, May 1936, pp. 44–49.

1764. PINES, N. V. Fatigue strength of nitrided steel. [R] *Metallurg*, vol. 11, no. 9, Sept. 1936, pp. 51–58.

1765. POMP, A. *and* HEMPEL, M. Dauerfestigkeits-Schaubilder von Stählen bei verschiedenen Zugmittelspannungen unter Berücksichtigung der Prüfstabform. [Fatigue strength diagrams of steels at different mean tensile stresses with consideration of the shape of test specimen.] *Mitt. K.-Wilh.-Inst. Eisenforsch.*, vol. 18, no. 1, 1936, pp. 1–14. Abstr.: *Stahl u. Eisen*, vol. 56, no. 26, June 25, 1936, pp. 735–736. *Z. Ver. dtsch. Ing.*, vol. 80, no. 47, Nov. 21, 1936, pp. 1424–1425.

1766. POMP, A. *and* HEMPEL, M. Dauerfestigkeits-Schaubilder von gekerbten und kalt-verformten Stählen sowie von 1″- und 1⅛″-Schrauben bei verschiedenen Zugmittelspannungen. [Fatigue strength diagrams of notched and cold deformed steels and of 1″ and 1⅛″ bolts at different mean tensile stresses.] *Mitt. K.-Wilh.-Inst. Eisenforsch.*, vol. 18, no. 14, 1936, pp. 205–215. Abstr.: *Stahl u. Eisen*, vol. 57, no. 10, March 11, 1937, pp. 274–275. *Z. Ver. dtsch. Ing.*, vol. 81, no. 29, July 17, 1937, p. 870.

1767. REIMANN, W. Die Dauerbruchgefahr bei der Ausbildung von Wellenabsetzungen als Schultern für Wälzlager und ihre Beseitigung durch Verwendung von Schulterringen. [The danger of fatigue fractures developing from steps in shafts or shoulders of anti-friction bearings and its elimination by the use of shoulder collars.] *Maschinenschaden*, vol. 13, no. 9, 1936, pp. 137–140.

1768. RIST, R. Ursachen von Rissschäden an genieteten Kesseltrommeln. [Causes of fissures in riveted boiler drums.] *Stahl u. Eisen*, vol. 56, no. 23, June 4, 1936, pp. 665–666. *Z. Ver. dtsch. Ing.*, vol. 79, no. 26, June 29, 1935, pp. 812–813.

1769. VON ROESSLER, L. Ersatz von Keilbefestigungen durch Schweissungen (Dauerfestigkeit von Ringnähten bei Scherbeanspruchung.) [Substitution of mechanical keys by welding (fatigue strength of annular joints under shearing stress).] *Elektroschweissung*, vol. 7, no. 11, 1936, pp. 209–215.

1770. ROŠ, M. Ermüdungsfestigkeit und Sicherheit geschweisster Konstruktionen (Brücken- und Hochbauten und Druckrohre). [Fatigue strength and safety of welded structures (bridges, structural steelwork and pressure pipes).] *Congr. int. Ass. Br. struct. Engng*, Second Congress, Berlin, Munich, Oct. 1936, pp. 403–427.

1771. RUSSELL, H. W. Resistance to damage by overstress of precipitation-hardened copper-steel and copper-malleable. *Metals and Alloys*, vol. 7, Dec. 1936, pp. 321–324.

1772. RUSSELL, H. W. *and* WELCKER, W. A. Damage and overstress in the fatigue of ferrous materials. *Proc. Amer. Soc. Test. Mater.*, vol. 36, pt. 2, 1936, pp. 118–138. Abstr.: *Iron Age*, vol. 138, no. 2, July 9, 1936, pp. 84–85.

1773. RUTTMANN, W. Verformungslose Brüche an Kesselteilen. [Fractures without deformation in boiler plates.] *Tech. Mitt. Krupp*, vol. 4, no. 2, March 1936, pp. 23–29.

1774. SCHOENMAKER, P. Dauerfestigkeit vor Schweissungen. Resistance of welds to repeated stresses. *Congr. int. (Carb.) Acét.*, Twelfth Congress, London, 1936, vol. 2, pp. 464–46⁕ [German] and pp. 470–476 [English].

1775. SCHOTTKY, H. *and* HILTENKAMP, H. Mitwirkung des Luftstickstoffs beim Fressen aufeinander gleitender Stahlteile und beim Dauerbruch. [Contribution of nitrogen to the wear and fatigue fracture of sliding steel parts.] *Stahl u. Eisen*, vol. 56, no. 15, April 9, 193⁕ pp. 444–446.

1776. SCHOTTKY, H. *and* HILTENKAMP, H. D Mitwirkung des Luftstickstoffs beim Fresse und beim Dauerbruch. [The contribution ⁕ nitrogen to wear and fatigue fracture.] *Tec Mitt. Krupp*, vol. 4, no. 3, June 193⁕ pp. 74–79.

1777. SCHRAIVOGEL, K. Dauerversuche ⁕ Schraubenbolzen. [Fatigue tests on bolt⁕ *Jb. Lilienthal-Ges. Luftfahrtf.*, 1936, pp. 39 403.

1778. SCHROEDER, W. C. *and* PARTRIDGE, E. Effects of solutions on the endurance of lo⁕ carbon steel under repeated torsion at 482°

Trans. Amer. Soc. mech. Engrs, vol. 58, no. 3, April 1936, pp. 223–231.

1779. SERENSEN, S. V. The influence of surface conditions and heat treatment on steel subjected to fatigue loading. [R] *Tekh. vozdush. Flota*, no. 6, 1936, pp. 46–66.

1780. SIEBEL, E. Statische und dynamische Kerbzähigkeit. [Static and dynamic notch toughness.] *Jb. Lilienthal-Ges. Luftfahrtf.*, 1936, pp. 383–396.

1781. SIEBEL, E. *and* LEYENSLETTER, W. Einfluss der Schnittgeschwindigkeit auf die Dauerfestigkeit von Prüfstäben. [Influence of cutting speed on the fatigue strength of test specimens.] *Z. Ver. dtsch. Ing.*, vol. 80, no. 22, May 30, 1936, pp. 697–698. English abstr.: *J. Iron St. Inst.*, vol. 134, no. 2, 1936, p. 127A.

1782. SMEKAL, A. Dauerbruch und Spröder Bruch. [Fatigue fracture and brittle fracture.] *Sci. Rep. Tohoku Univ.*, *Honda Anniversary Volume*, Oct. 1936, pp. 47–56.

1783. SONNEMANN, H. Die Schwingungsfestigkeit und Dämpfungsfähigkeit von handelsüblichen Stählen und Kupfer und ihre Beeinflussung durch Kaltnietung. [The fatigue strength and damping capacity of commercial steels and copper and the influence of cold riveting.] *Mitt. Wöhler-Inst.*, no. 28, 1936, pp. 3–87.

784. SPÄTH, W. Bemerkungen zur Durchführung von Dauerversuchen. [Observations on the execution of fatigue tests.] *Metallwirtschaft*, vol. 15, no. 4, Jan. 24, 1936, pp. 91–93.

785. SPÄTH, W. Kritik der Kurzzeitverfahren zur Bestimmung der Dauerwechselfestigkeit.

[Discussion of short time methods for determining the fatigue strength.] *Metallwirtschaft*, vol. 15, no. 31, July 31, 1936, pp. 726–729 and no. 32, Aug. 7, 1936, pp. 750–752.

1786. SPELLER, F. N. Report of the special subcommittee on corrosion-fatigue of drill pipe. *Proc. Amer. Petrol. Inst., Sect. 4*, vol. 17, 1936, pp. 67–69.

1787. SPINELLI, F. Sul limite di fatica degli acciai impiegati nella costruzione degli scafi. [The fatigue limit of steels used in the construction of ships.] *Ann. Archit. nav., Roma*, vol. 6, 1936, pp. 109–119.

1788. SUTTON, H. Fatigue properties of nitrided steel. *Metal Treatm.*, vol. 2, no. 6, Summer 1936, pp. 89–92.

1789. SUTTON, H. *and* PEAKE, T. J. Note on pickling or etching baths for duralumin. *J. Inst. Met.*, vol. 59, no. 2, 1936, pp. 59–64.

1790. SWANGER, W. H. *and* WOHLGEMUTH, G. F. Failure of heat-treated steel wire in cables of the Mt. Hope, R. I. suspension bridge. *Proc. Amer. Soc. Test. Mater.*, vol. 36, pt. 2, 1936, pp. 21–84. Abstr.: *Heat Treat. Forg.*, vol. 22, no. 8, Aug. 1936, pp. 391–393.

1791. TERAZAWA, K. Fatigue resistance and allowable stress of ductile materials under combined alternating stresses. *J. Soc. nav. Archit., Japan*, vol. 59, Dec. 1936, pp. 15–44.

1792. THUM, A. Gestaltfestigkeit und Konstruktion. [Form strength and construction.] *Jb. Lilienthal-Ges. Luftfahrtf.*, 1936, pp. 75–93.

1793. THUM, A. Die Dauerbruchsichere Schraubenverbindung. [The fatigue fracture safety of bolted joints.] *Anz. Bergw.*, vol. 58, no. 42, May 26, 1936, pp. 17–19.

1794. THUM, A. *and* BANDOW, K. Die Gusskurbelwelle. [The cast crankshaft.] *Z. Ver. dtsch. Ing.*, vol. 80, no. 1, Jan. 4, 1936, pp. 23–27. English translation: *Metal Treatm.*, vol. 2, Spring 1936, pp. 42–44.

1795. THUM, A. *and* BAUTZ, W. Steigerung der Dauerhaltbarkeit von Formelementen durch Kaltverformung. [Increasing the fatigue properties of machine parts by cold working.] *Mitt. MatPrüfAnst.*, Darmstadt, no. 8, 1936, pp. 92.

1796. THUM, A. *and* DEBUS, F. Vorspannung und Dauerhaltbarkeit von Schraubenverbindungen. [Pre-stressing and the fatigue properties of screwed joints.] *Mitt. MatPrüfAnst.*, Darmstadt, no. 7, 1936, pp. 72.

1797. ULRICH, M. Sind Brüche von Kraftwagen-Hinterachswellen Dauerbrüche? [Are fractures in automobile rear axles due to fatigue?] *Z. Ver. dtsch. Ing.*, vol. 80, no. 7, Feb. 15, 1936, pp. 181–182.

1798. VOIGT, O. Die Prüfung von Förderseilen. [The testing of haulage ropes.] *Arch. tech. Messen*, vol. 5, no. 56, Feb., 1936, pp. T17–T18. English translation: *Henry Brutcher Tech. Trans.*, no. 611.

1799. WALKER, E. V. The fatigue testing of wire. *P.O. elect. Engrs' J.*, vol. 29, no. 3, Oct. 1936, pp. 237–239.

1800. WELTER, G. Bending and tension-compression fatigue tests. [P] *Wiad. Inst. Metal.*, vol. 3, no. 3, 1936, pp. 149–156 and no. 4, 1936, pp. 189–198.

1801. WIEGAND, H. Die Schraubenverbindungen bei Wechselbeanspruchungen. [The bolted joint under fatigue stresses.] *Gleistechnik*, vol. 12, 1936, pp. 2–4.

1802. WIEGAND, H. Möglichkeiten zur Erhöhung der Dauerhaltbarkeit von Schrauben im Eisenbahnoberbau. [Possibility of increasing the fatigue properties of screws in railway overhead structures.] *Gleistechnik*, vol. 12, 1936, pp. 37–39.

1803. WILLIAMS, F. H. Failures of locomotive parts. *Rly mech. Engr*, vol. 110, May 1936, pp. 190–192; June 1936, pp. 249–251; July 1936, pp. 314–317; Oct. 1936, pp. 429–433; Nov. 1936, pp. 481–484, 495; Dec. 1936, pp. 534–536; vol. 111, Jan. 1937, pp. 17–20; Feb. 1937, pp. 59–62 and March 1937, pp. 107–110.

1804. WINTERBOTTOM, A. B. An electrical fatigue-testing machine for conduction of tests in controlled atmospheres. *K. norske vidensk Selsk. Forh.*, vol. 9, 1936, pp. 172–175.

1805. ZINOVEV, V. S. *and* LEVIN, I. A. A new type of machine for axial load fatigue testing at high temperatures. [R] *Zav. Lab.*, vol. 5, no. 4, 1936, p. 494.

1806. The fatigue of hard-drawn aluminium wires. *Rep. Brit. elect. Ind. Res. Ass.*, no. F/T7, 1936, pp. 33.

1807. Failure of crosshead due to corrosion fatigue. *Tech. Rep. Brit. Eng. Boil. Insur. Co*, 1936, pp. 45–46.

1808. X-ray diffraction as a means of detecting impending fatigue failure. *Proc. Amer. So. Test. Mater.*, vol. 36, pt. 1, 1936, pp. 125–12

1809. Second report of the Welding Research Committee. *Proc. Instn mech. Engrs, Lond.*, vol. 133, 1936, pp. 5–126.

1810. Fatigue stress and its diminution. *Mech. World*, vol. 99, no. 2563, Feb. 14, 1936, pp. 161, 168.

1811. Fatigue and fatigue testing machines. *Wire Ind.*, vol. 3, no. 27, March 1936, pp. 115–117.

1812. Deutscher Normenausschuss, DVM 4001. Werkstoffprüfung, Dauerfestigkeitsprüfung, Begriffe und Zeichen. [German Committee on Standards, DVM 4001. Material testing, fatigue strength testing, conceptions and symbols.] *Z. Metallk.*, vol. 28, no. 4, April 1936, pp. 102–104.

1813. Haigh-Robertson fatigue wire testing machine. *Bull. Ont. hydro-elect. Pwr Comm.*, vol. 23, no. 6, June 1936, pp. 192–195.

1937

814. APPELT, W. Steigerung der Dauerhaltbarkeit von Autokurbelwellen durch Oberflächendrücken des Bohrrandes. [Improvement of fatigue strength of automobile crankshafts by application of pressure at the surface of the bore.] *Auto.-tech. Z.*, vol. 40, no. 19, Oct. 10, 1937, pp. 473–475.

815. ASSMANN, H. Wärmebehandlung, Dauerbruchempfindlichkeit und Schneidleistung handelsüblicher Stähle für Pressluftmeissel. [Heat-treatment, susceptibility to fatigue fracture and cutting efficiency of the usual commercial steels for compressed-air chisels.] *Werkst. u. Betr.*, vol. 70, no. 13/14, July 1937, pp. 171–174.

816. BARRETT, C. S. Distortion of grains by fatigue and static stressing. *Metals and Alloys*, vol. 8, no. 1, Jan. 1937, pp. 13–21.

1817. BARRETT, C. S. The application of X-ray diffraction to the study of fatigue in metals. *Trans. Amer. Soc. Metals*, vol. 25, Dec. 1937, pp. 1115–1148.

1818. BASTIEN, P. Etat des surfaces métalliques: qualification et influence sur la résistance à la fatigue. [Condition of metallic surfaces: classification and influence on fatigue resistance.] *Mécanique*, vol. 21, no. 274, Sept.–Oct. 1937, pp. 211–214.

1819. BAUTZ, W. Mittel zur Steigerung der Dauerhaltbarkeit gekerbter Konstruktionen. [Methods of increasing the fatigue strength of notched structures.] *Anz. Bergw.*, vol. 59, no. 95, Nov. 26, 1937, pp. T1–T6.

1820. BEHRENS, P. *and* HUTTER, H. Schwingungsversuche mit Stahlaluminium- und Kupferhohlseilen für Hoch- und Höchstspannungsleitungen. [Fatigue tests with steel–aluminium and hollow copper cables for medium and high tension transmission lines.] *Elektrizitätswirtschaft*, vol. 36, May 15, 1937, pp. 331–336.

1821. BERG, S. Gestaltfestigkeitsversuche der Industrie. [Industrial shape-strength investigations.] *Z. Ver. dtsch. Ing.*, vol. 81, no. 17, April 24, 1937, pp. 483–487.

1822. BERNHARD, R. K. Dynamic tests by means of induced vibrations. *Proc. Amer. Soc. Test. Mater.*, vol. 37, pt. 2, 1937, pp. 634–649.

1823. BERNHARD, R. K. Why dynamic testing? *Machinist*, vol. 81, no. 39, Nov. 6, 1937, pp. 911–914.

1824. BIERETT, G. Über das Verhalten geschweisster Träger bei Dauerbeanspruchung unter besonderer Berücksichtigung der Schweissspannungen. [On the behaviour of welded beams under fatigue stress with particular regard to the welding stresses.] *Ber. Aussch. Stahl., Berl.*, no. 7, 1937, pp. 21. German abstr.: *Z. Ver. dtsch. Ing.*, vol. 81, no. 38, Sept. 18, 1937, pp. 1126–1127.

1825. BOEGEHOLD, A. L. Endurance of gear steels at 250°F. *Trans. Amer. Soc. Metals*, vol. 25, March 1937, pp. 245–259.

1826. BOEGEHOLD, A. L. Appraising a steel for a given duty. *Metal Progr.*, vol. 31, no. 4, April 1937, pp. 403–406.

1827. BOLLENRATH, F. Über die Weiterentwicklung warmfester Werkstoffe für Flugtriebwerke. [On the further development of elevated temperature materials for aircraft engine parts.] *Jb. dtsch. VersAnst. Luftf.*, 1937, pp. 340–347.

1828. BORGMANN, C. W. Corrosion fatigue of drill pipe. *Proc. Amer. Petrol. Inst.*, Sect. 4, vol. 18, 1937, pp. 41–44.

1829. BUCKWALTER, T. V. *and* HORGER, O. J. Investigation of fatigue strength of axles, press-fits, surface rolling and effect of size. *Trans. Amer. Soc. Metals*, vol. 25, no. 1, March 1937, pp. 229–245.

1830. BUCKWALTER, T. V.; HORGER, O. J. *and* SANDERS, W. C. Modern locomotive and axle-testing equipment. *Trans. Amer. Soc. mech. Engrs*, vol. 59, no. 3, RR-59-1, 1937, pp. 225–238.

1831. BUCKWALTER, T. V.; HORGER, O. J. *and* SANDERS, W. C. Improved parts for the modern locomotive. *Heat Treat. Forg.*, vol. 23, no. 5, May 1937, pp. 227–230. Abstr.: *Steel*, vol. 100, no. 21, May 24, 1937, pp. 52, 55.

1832. CALDWELL, F. W. The vibration problem in aircraft propeller designing. *S.A.E. Jl.*, vol. 41, no. 2, Aug. 1937, pp. 372–380.

1833. CAPPER, P. L. Fatigue in shafts under combined bending and torsion. *Engineer, Lond.*, vol. 164, no. 4256, Aug. 6, 1937, pp. 150–152.

1834. CASSIE, A. B. D.; JONES, M. *and* NAUNTON, W. J. S. Fatigue in rubber—part I. *Rubb. Chem. Technol.*, vol. 10, no. 1, Jan. 1937, pp. 29–54.

1835. CAZAUD, R. *and* PERSOZ, L. *La fatigue des métaux. [The fatigue of metals.]* Paris, Dunod, 1937, pp. 190.

1836. CHRISTOL, E. Les essais de fatigue. [Fatigue tests.] *Arts et Mét.*, vol. 91, no. 203, Aug. 1937, pp. 173–183.

1837. CLINEDINST, W. O. Fatigue life of tapered roller bearings. *J. appl. Mech.*, vol. 4, no. 4, Dec. 1937, pp. A143–A150.

1838. CORNELIUS, H. Die Ermüdungsfestigkeit dünnwandiger Rohre für den Flugzeugbau im ungeschweissten und geschweissten Zustand. [The fatigue strength of thin walled tubes for aircraft construction in the unwelded and welded condition.] *Zent. wiss. Ber., ForschBer.* no. 801, April 13, 1937, pp. 24.

1839. CORNELIUS, H. Die Dauerfestigkeit von Schweissverbindungen. [The fatigue strength of welded joints.] *Z. Ver. dtsch. Ing.*, vol. 81, no. 30, July 24, 1937, pp. 883–888.

1840. CORNELIUS, H. *and* BOLLENRATH, F. Die Ermüdungsfestigkeit dünnwandiger Rohre für den Flugzeugbau im ungeschweissten und geschweissten Zustand. [The fatigue strength of thin walled tubes for aircraft construction in the unwelded and welded condition.] *Luftfahrtforsch.*, vol. 14, no. 10, Oct. 1937, pp. 520–526. German abstr.: *Stahl u. Eisen*, vol. 57, no. 45, Nov. 11, 1937, pp. 1283–1284.

1841. DEUTLER, H. *and* HAVERS, A. Die günstigste Gestalt der Hohlkehlen bei verdrehbeanspruchten Wellen. [The most favourable shape of keyways in shafts under torsional loading.] *Jb. dtsch. Luftfahrtf.*, 1937, vol. 2, pp. II-132–II-136.

1842. DOLAN, T. J. The combined effect of corrosion and stress concentration at holes and fillets in steel specimens subjected to reversed torsional stresses. *Uni. Ill. Engng Exp. Sta. Bull.*, no. 293, 1937, pp. 39.

1843. DONALDSON, J. W. Research and the Iron and Steel Industry. IV—fatigue, corrosion fatigue and corrosion. *Iron Steel Ind.*, vol. 10, Feb. 1937, pp. 266–269.

1844. EDGERTON, C. T. Progress report no. 3 on heavy helical springs. *Trans. Amer. Soc. mech. Engrs*, vol. 59, no. 7, Oct. 1937, pp. 609–616.

1845. ERLINGER, E. Eine Prüfmaschine zur Erzeugung wechselnder Zug-Druck-Kräfte. [A testing machine for the production of alternating tension and compression forces.] *Arch. Eisenhüttenw.*, vol. 10, no. 7, Jan. 1937, pp. 317–320. English translation: *R. T. P. Trans.*, no. 661.

1846. FARNHAM, G. S. The fatigue of metals. *Engng J. Montreal*, vol. 20, no. 11, Nov. 1937, pp. 799–802.

1847. FÖPPL, O. Dämpfungsfähigkeit der Werkstoffe. Oberflächendrücken. Resonanz-Schwingungsdämpfer für Kurbelwellen. [Damping properties of materials. Surface compression. Resonant vibration damper for crankshafts.] *Mitt. Wöhler-Inst.*, no. 30, 1937, pp. 61.

848. FÖPPL, O. Damping capacity—its practical importance. *Metal Progr.*, vol. 32, no. 3, Sept. 1937, pp. 262, 300, 302.

1849. FÖPPL, O. Erniedrigung der Dauerhaltbarkeit durch Korrosion und die Mittel zur Verringerung des Korrosionseinflusses. [The reduction of fatigue strength by corrosion and a method of decreasing the corrosive influence.] *Werkzeugmaschine*, vol. 41, no. 19, Oct. 15, 1937, pp. 405–408. English abstr.: *J. Iron St. Inst.*, vol. 137, no. 1, 1938, pp. 159A–161A.

1850. GALIBOURG, J. *and* LAURENT, P. Contribution à l'etude des aciers au nickel et au nickel-chrome. [Contribution to the study of nickel and nickel–chromium steels.] *Rev. Nickel*, vol. 8, July 1937, pp. 105–110.

1851. GARF, S. E. A machine for fatigue testing in rotating bending. [R] *Žav. Lab.*, vol. 6, no. 4, 1937, pp. 480–486.

1852. GLIKMAN, L. A. The effect of repeated uni-directional impact loads on the brittle strength of steel. An approach to the question of the process of fatigue. [R] *Zh. tekh. Fiz.*, vol. 7, no. 14, 1937, pp. 1434–1451.

1853. GOUGH, H. J. Cast crankshafts. *Engineer, Lond.*, vol. 163, no. 4237, March 26, 1937, pp. 360–361.

1854. GOUGH, H. J. *and* POLLARD, H. V. Properties of some materials for cast crankshafts, with special reference to combined stresses. *J. Instn Auto. Engrs*, vol. 5, no. 6, March 1937, pp. 96–166. *Proc. Instn Auto. Engrs.* vol. 31, March 1937, pp. 821–893., Abstr: *Engineering, Lond.*, vol. 143, no. 3723, May 21, 1937, p. 584. *Metal Progr.*, vol. 31, no. 5, May 1937, pp. 516–517.

1855. GOUGH, H. J. *and* SOPWITH, D. G. The resistance of some special bronzes to fatigue and corrosion-fatigue. *J. Inst. Met.*, vol. 60, no. 1, 1937, pp. 143–158.

1856. GOUGH, H. J. *and* SOPWITH, D. G. The influence of the mean stress of the cycle on the resistance of metals to corrosion fatigue. *J. Iron St. Inst.*, vol. 135, no. 1, 1937, pp. 293p–313p. Abstr.: *Engineering, Lond.*, vol. 143, no. 3726, June 11, 1937, pp. 673–674.

1857. GOUGH, H. J. *and* WOOD, W. A. Deformation and fracture of mild steel under cyclic stresses in relation to crystalline structure. *Rep. aero. Res. Comm., Lond.*, no. E.F. 411, (unpublished report no. 2859), Feb. 17, 1937, pp. 19.

1858. GRAF, O. Versuche über das Verhalten von genieteten und geschweissten Stössen in Trägern I 30 aus St.37 bei oftmals wiederholter Belastung. [Tests on the behaviour of riveted and welded joints in I 30 beams of St.37 under often repeated loads.] *Stahlbau*, vol. 10, no. 2, Jan. 15, 1937, pp. 9–16.

1859. GRAF, O. Weitere Versuche über die Dauerbiegefestigkeit von Stahlstäben mit brenngeschnittenen Flächen. [Further experiments on the bending fatigue strength of steel bars with oxyacetylene cut surfaces.] *Autogene Metallbearb.*, vol. 30, no. 19, Oct. 1, 1937, pp. 321–323.

1860. GRAF, O. Über die Veränderlichkeit der Dauerzugfestigkeit von Betonstahl durch Recken und Altern. [On the effect of stretching and ageing on the tensile fatigue strength of steel reinforcing bars.] *Beton u. Eisen*, vol. 36, no. 19, Oct. 5, 1937, pp. 309–310.

1861. GREENALL, C. H. *and* GOHN, G. R. Fatigue properties of non-ferrous sheet metals. *Proc. Amer. Soc. Test. Mater.*, vol. 37, pt. 2, 1937, pp. 160–194.

1862. HAIGH, B. P. Wire rope problems. *Proc. S. Wales Inst. Engrs*, vol. 52, no. 5, 1937, pp. 327–358.

1863. HÄNCHEN, R. Berechnung der geschweissten Maschinenteile auf Dauerhaltbarkeit. [Calculation of the fatigue limit of welded parts.] *Elektroschweissung*, vol. 8, no. 11, Nov. 1937, pp. 201–207 and no. 12, Dec. 1937, pp. 226–230.

1864. HEMPEL, M. Wechselfestigkeits-Schaubilder von Rund- und Flachstäben, T-Trägern und Drähten aus Stahl. [Fatigue strength diagrams of round and flat steel bars, T-beams and wires.] *Arch. Eisenhüttenw.*, vol. 11, no. 5, Nov. 1937, pp. 231–240.

1865. HEMPEL, M. *and* TILLMANNS, H. E. Zugwechselversuche mit Stahl bei Temperaturen bis 600°. [Tension fatigue tests on steels at temperatures up to 600°.] *Arch. Eisenhüttenw.*, vol. 10, no. 9, March 1937, pp. 395–403.

1866. HEROLD, W. Versuche über Drehschwingungsfestigkeit abgesetzter, genuteter und durchbohrter Wellen. [Tests on the effects of upsetting, splines and bored holes on the torsional fatigue strength of shafts.] *Z. Ver. dtsch. Ing.*, vol. 81, no. 18, May 1, 1937, pp. 505–509. English summary: *Auto. Engr*, vol. 34, no. 446, Feb. 1944, p. 68.

1867. HEYER, H. O. Prüfung der Laufeigenschaften von Lagermetallen unter dynamischer Belastung. [Testing the running qualities of bearing metals under dynamic loads.] *Auto.-tech. Z.*, vol. 40, no. 22, Nov. 25, 1937, pp. 551–559 and no. 23, Dec. 10, 1937, pp. 589–595.

1868. HIRST, G. W. C. The development of cracks in the wheel-seats of axles within the hubs of wheels. *J. Instn Engrs Aust.*, vol. 9, no. 6, June 1937, pp. 215–229.

1869. HOWELL, F. M. *and* HOWARTH, E. S. A fatigue machine for testing metals at elevated temperatures. *Proc. Amer. Soc. Test. Mater.*, vol. 37, pt. 2, 1937, pp. 206–217.

1870. HUDSON, R. A. *and* CHICK, J. E. Acceleration of fatigue tests. *Aero. Dig.*, vol. 31, no. 3, Sept. 1937, p. 78.

1871. HULL, E. H. Alternating stress measurement by the resistance strip method. *Gen. Elect. Rev.*, vol. 40, no. 8, Aug. 1937, pp. 379–380.

1872. IGARASHI, I. *and* FUKAI, S. On the decrease of fatigue limit of duralumin and super-duralumin by sea-water corrosion. [J] *Sumit. kinz. kog. kenk.*, vol. 2, no. 10, 1937, pp. 1041–1055.

1873. IGARASHI, I. *and* TAKETOMI, R. Fatigue tests of some light alloys. [J] *Sumit. kinz. kog. kenk.*, vol. 2, no. 9, 1937, pp. 900–910.

1874. JOHNSON, J. B. Aircraft engine materials. *S.A.E. Jl.*, vol. 40, no. 4, April 1937, pp. 153–162.

1875. JOUKOFF, A. S. Les essais d'endurance sur les aciers de construction. [The endurance testing of constructional steels.] *Ossat. métall.*, vol. 6, no. 12, Dec. 1937, pp. 585–590.

1876. JÜNGER, A. Steigerung der Seewasser-Korrosionsfestigkeit von Stahl durch Oberflächendrücken, Nitrieren, Einsatzhärten und durch elektrolytischen Zinkschutz. [Increasing the sea-water corrosion fatigue strength of steel by surface compression, nitriding, case hardening and zinc electroplating.] *Mitt. ForschAnst. Gutehoffn., Nürnberg*, vol. 5, Jan. 1937, pp. 1–12. Abstr.: *Z. Ver. dtsch. Ing.*, vol. 81, no. 33, Aug. 14, 1937, pp. 974–975. English abstr.: *Henry Brutcher Tech. Trans.*, no. 589.

877. KAHNT, H. Über Kerbempfindlichkeit, Verfestigung und Dämpfung von Stählen bei Drehschwingungen. [On notch sensitivity, strain hardening and damping of steels subjected to oscillating torsional stressing.] *Z. tech. Phys.*, vol. 18, no. 8, Aug. 1937, pp. 230–237.

1878. KARELINA, A. G. *and* MIROLUBOV, I. N. The influence of quenching on a steel surface and its fatigue limit. [R] *Zh. tekh. Fiz.*, vol. 7, no. 5, 1937, pp. 492–497.

1879. KAUFMANN, F. Die Dauerfestigkeit von Stumpfnahtverbindungen von Proben mit aufgelegten Raupen und von Laschenverbindungen. [The fatigue strength of butt welded joints, of specimens with built-up welds and of strap joints.] *Tech. Mitt. Krupp*, vol. 5, no. 4, June 1937, pp. 102–125.

1880. KNOWLTON, H. B. Physical properties of axle shafts. *Trans. Amer. Soc. Metals*, vol. 25, March 1937, pp. 260–296. Abstr.: *Automot. Industr., N.Y.*, vol. 75, no. 24, Dec. 12, 1936, pp. 829–831.

1881. KÖRBER, F. *and* HEMPEL, M. Verhalten von geschweissten und geschraubten Steifknotenverbindungen bei ruhender und wechselnder Biegebeanspruchung. [Behaviour of welded and bolted gusset plate joints under static and alternating bending stresses.] *Mitt. K.-Wilh.-Inst. Eisenforsch.*, vol. 19, no. 19, 1937, pp. 273–287.

1882. KOMMERS, J. B. Overstressing and understressing in fatigue. *Engineering, Lond.*, vol. 143, no. 3724, May 28, 1937, pp. 620–622 and no. 3726, June 11, 1937, pp. 676–678.

1883. KONVISAROV, D. R. *and* PRISHCHEPA, M. P. A new method of testing metals under repeated load. [R] *Vest. metalloprom.*, vol. 17, no. 11, 1937, pp. 30–40.

1884. KUKANOV, L. I. Investigation of axle steels subjected to abrasion and fatigue. [R]

Metallurg, vol. 12, no. 2, Feb. 1937, pp. 51–60. English abstr.: *J. Iron St. Inst.*, vol. 137, no. 1, 1938, p. 340A.

1885. KUNTZE, W. Einfluss des durch die Gestalt erzeugten Spannungszustandes auf die Biege-wechselfestigkeit. [Influence of stress condition due to shape on the bending fatigue strength.] *Arch. Eisenhüttenw.*, vol. 10, no. 8, Feb. 1937, pp. 369–373.

1886. KUNTZE, W. *and* LUBIMOFF, W. Gesetz-massige Abhängigkeit der Biegewechselfe-stigkeit von Probengrösse und Kerbform. [Dependence of bending fatigue strength on size of specimen and shape of notch.] *Arch. Eisenhüttenw.*, vol. 10, no. 7, Jan. 1937, pp. 307–311.

1887. LANGER, B. F. Fatigue failure from stress cycles of varying amplitude. *J. appl. Mech.*, vol. 4, Dec. 1937, pp. A160–A162.

1888. LEA, F. C. The effect of discontinuities and surface conditions on failure under repeated stress. *Engineering, Lond.*, vol. 144, no. 3732, July 23, 1937, pp. 87–90 and no. 3734, Aug. 6, 1937, pp. 140–144.

1889. LEA, F. C. *and* WHITMAN, J. G. The failure of girders under repeated stresses. *J. Instn civ. Engrs*, vol. 7, no. 1, Nov. 1937, pp. 119–152.

1890. LEHR, E. Die betriebsmässige Beanspru-chung der Konstruktionswerkstoffe und die Nutzbarmachung der im Laboratorium gefun-denen Gütezahlen. [The working stresses of structural materials and the utilization of laboratory data.] *Congr. int. Ass. Test. Mat., Ninth Congr., London*, 1937, pp. 575–578.

1891. LEHR, E. Dauerhaltbarkeit von Ritzel-wellen. [Fatigue strength of pinion shafts.]

Z. Ver. dtsch. Ing., vol. 81, no. 5, Jan. 30, 1937, pp. 117–118.

1892. LEHR, E. Beitrag zur Berechnung dauer-biegebeanspruchter Wellen. [Contribution to the calculation of bending fatigue stresses in shafts.] *T Z prakt. Metallbearb.*, vol. 47, no. 17/18, Sept. 1937, pp. 698, 700 and no. 19/20, Oct. 1937, pp. 769–772.

1893. LEHR, E. Die wichtigsten Ergebnisse der neueren Festigkeitsforschung. [The most important results of the latest strength investi-gations.] *Maschinenschaden*, vol. 14, no. 9, 1937, pp. 136–142 and vol. 15, no. 1, 1938, pp. 1–7.

1894. LEON, A. Eine neue Prüfmaschine zur Ermittlung der Dauerwechselfestigkeit bei verschiedenen Vorspannungen. [A new testing machine for determining the fatigue strength under different initial stresses.] *Z. öst. Ing.- u. ArchitVer.*, vol. 89, no. 31–32, Aug. 6, 1937, pp. 224–225.

1895. LESSELLS, J. M. New aspects of fatigue and creep. *Metal Progr.*, vol. 32, no. 3, Sept. 1937, pp. 257–262.

1896. LONSDALE, T. The influence of atmospheric conditions on the fatigue strength of metals. *Mech. World*, vol. 102, no. 2641, Aug. 13, 1937, pp. 145–146.

1897. LOSSIER, H. La limite de fatigue des aciers spéciaux dans les constructions en béton armé. [The fatigue limit of special steels used in reinforced concrete.] *Génie civ.*, vol. 110, no. 2845, Feb. 20, 1937, pp. 187–188.

1898. LUCAS, F. F. How flaws occur in meta structures. *Nat. Safety News*, vol. 35, Feb 1937, pp. 12–13, 60.

1899. LÜRENBAUM, K. Einfluss von Formgebung und Werkstoff auf die Gestaltfestigkeit geschmiedeter und gegossener Flugmotoren-Kurbelwellen. [Influence of shape and material on the form strength of forged and cast aero-engine crankshafts.] *Jb. dtsch. VersAnst. Luftf.*, 1937, pp. 490–493. *Jb. dtsch. Luftfahrtf.*, vol. II, 1937, pp. 128–131.

1900. LÜRENBAUM, K. Belastung und Tragfähigkeit von Flugmotoren-Kurbelwellen. [Loads and load carrying capacity of aero-engine crankshafts.] *Jb. Lilienthal-Ges. Luftfahrtf.*, 1937, pp. 296–304.

1901. MACGREGOR, R. A. Deterioration of steel under service stresses. *Trans. Min. geol. Inst. India*, vol. 33, Dec. 1937, pp. 207–219.

1902. MASON, W. Note on certain combined alternating stress systems and a stress criterion of the "fatigue limit". *Phil. Mag.*, ser. 7, vol. 24, no. 162, Oct. 1937, pp. 695–703.

1903. MINAMIÔDI-KEN'ITI. A new testing machine for roller chains. [J] *Trans. Soc. mech. Engrs, Japan*, vol. 3, no. 10, Feb. 1937, pp. 88–93. English summary: *ibid.* pp. S-22–S-23.

1904. MOORE, H. F. Fatigue testing of wire. *Wire and Wire Prod.*, vol. 12, no. 5, May 1937, pp. 235–240.

1905. MÜLLER, W. Neue Versuchseinrichtung zur Prüfung der Ermüdungsfestigkeit von genieteten bzw. geschweissten Knotenpunktverbindungen. [New methods of investigating the fatigue strength of riveted and welded joints.] *Schweiz. Arch. angew. Wiss.*, vol. 3, no. 10, Oct. 1937, pp. 276–279. English translation: *Tech. Memor. nat. Adv. Comm. Aero., Wash.*, no. 947, July 1940, pp. 5.

1906. NEIGAUS, L. D. Investigations into the fatigue of steels. [R] *Vest. inzh., Tekh.*, vol. 23, no. 1, Jan. 1937, pp. 40–43.

1907. NEUBER, H. *Kerbspannungslehre*. [*Theory of notch stresses*.] Berlin, Julius Springer, 1937, pp. 160.

1908. NISIHARA, T. A new alternating tension-compression machine. [J] *Trans. Soc. mech. Engrs, Japan*, vol. 3, no. 11, May 1937, pp. 153–156.

1909. NISIHARA, T. and KOBAYASHI, T. Pitting of steel under lubricated rolling contact and allowable pressure on tooth profiles. [J] *Trans. Soc. mech. Engrs, Japan*, vol. 3, no. 13, Nov. 1937, pp. 292–298. English summary: *ibid.*, p. S-73.

1910. NISIHARA, T. and SAKURAI, T. Endurance test of belts in power transmission. [J] *Trans. Soc. mech. Engrs, Japan*, vol. 3, no. 11, May 1937, pp. 156–162. English summary: *ibid.*, pp. S-33–S-34. .

1911. OBERG, T. T. and JOHNSON, J. B. Fatigue properties of metals used in aircraft construction at 3450 and 10600 cycles. *Proc. Amer. Soc. Test. Mater.*, vol. 37, pt. 2, 1937, pp. 195–205.

1912. OCHS, H. Korrosionsermüdung als Ursache für Schadensfälle an Maschinenteilen. [Corrosion-fatigue as the cause of damage and failure in machine parts.] *Maschinenschaden*, vol. 14, no. 12, 1937, pp. 202–208.

1913. ODING, I. A. Theory of the fatigue limit in metals with asymmetric cycles under complex stressing. [R] *Zav. Lab.*, vol. 6, no. 4, 1937, pp. 471–479.

1914. OSHIBA, F. The degree of fatigue of carbon steels under repeated bending. [J] *Kinzoku no kenkyu*, vol. 14, March 1937, pp. 96–106. *Sci. Rep. Tohoku Univ.*, vol. 26, no. 3, Dec. 1937, pp. 323–340. *Trans. Soc. mech. Engrs,*

Japan, vol. 3, no. 11, May 1937, pp. 145–150. English summary: *ibid.*, pp. S-30–S-31.

1915. PARIS, M. Influence des eaux de chaudières sur la formation des cassures de fatigue. [Influence of boiler feedwater on formation of fatigue cracks.] *Mécanique*, vol. 21, no. 270, Jan.–Feb. 1937, pp. 12–18.

1916. PIGEAUD, G. Lois d'endurance et coefficients de sécurité dans les constructions métalliques soumises à des efforts variables. [Incorporation of fatigue factor in safety coefficients for metallic structures subjected to variable stress.] *Génie civ.*, vol. 110, no. 1, (No. 2838), Jan. 2, 1937, pp. 7–11 and no. 2, (no. 2839), Jan. 9, 1937, pp. 35–38.

1917. PILARSKI, S. *and* JAŹWIŃSKI, S. Influence of non-metallic inclusions on the mechanical properties of chromium-molybdenum structural steels. [P] *Wiad. Inst. Metal.*, vol. 4, no. 2, 1937, pp. 65–72 and nos. 3/4, Sept.–Dec. 1937, pp. 139–144. English abstr.: *J. Iron St. Inst.*, vol. 136, no. 2, 1937, p. 281A and vol. 137, no. 1, 1938, p. 281A.

1918. PIZZUTO, C. Duralumin studied in relation to the rotating-beam endurance limit. *Congr. int. Ass. Test. Mat., Ninth Congr., London*, 1937, pp. 119–121.

1919. POMP, A. *and* HEMPEL, M. Das Verhalten von Gusseisen unter Zug-Druck-Wechselbeanspruchung. [The behaviour of cast iron under tension-compression alternating stress.] *Stahl u. Eisen*, vol. 57, no. 40, Oct. 7, 1937, pp. 1125–1127. English translation: *Brit. cast. Iron Res. Ass. Trans.*, no. 158.

1920. POMP, A. *and* HEMPEL, M. Vergleichende Untersuchung von nickelhaltigen und nickelfreien Stählen auf ihre mechanischen Eigenschaften, insbesondere auf ihr Verhalten bei der Schwingungsprüfung. [Comparative ex-amination of steels having some or no nickel content with respect to their mechanical properties, in particular their behaviour during fatigue tests.] *Mitt. K.-Wilh.-Inst. Eisenforsch.*, vol. 19, 1937, pp. 221–236. *Luftfahrtforschung*, vol. 14, no. 10, Oct. 1937, pp. 511–519. Abstr.: *Z. Ver. dtsch. Ing.*, vol. 82, no. 14, April 2, 1938, pp. 417–419.

1921. POMP, A. *and* HEMPEL, M. Dauerprüfung von Stahldrähten unter wechselnder Zugbeanspruchung. I — Einfluss des Drahtherstellungsverfahrens auf die Zugschwellfestigkeit. [Fatigue testing of steel wires under pulsating tensile stress. I—Influence of method of wire manufacture on the limiting tensile strength.] *Mitt. K.-Wilh.-Inst. Eisenforsch.*, vol. 19, no. 17, 1937, pp. 237–246. Abstr.: *Stahl u. Eisen*, vol. 57, no. 52, Dec. 30, 1937, pp. 1454–1455. English abstr.: *Henry Brutcher Tech. Trans.*, no. 729.

1922. PRICE, W. B. *and* BAILEY, R. W. Fatigue properties of five cold rolled copper alloys. *Trans. Amer. Inst. min. (metall.) Engrs*, vol. 124, 1937, pp. 271–286.

1923. PROT, M. Un nouveau type de machine d'essai des métaux à la fatigue par flexion rotative. [A new type of machine for testing fatigue of metals under rotating bending stress.] *Rev. Métall. Mem.*, vol. 34, no. 7, July 1937, pp. 440–442.

1924. RATHBONE, T. C. Detection of fatigue cracks by Magnaflux method. *Weld. J., Easton, Pa.*, vol. 16, no. 4, April 1937, pp. 22–27. *Mech. Engng, N.Y.*, vol. 59, no. 3, March 1937, pp. 147–152.

1925. REGLER, F. Röntgenographische Feingefügeuntersuchungen an Brückentragwerken. [X-ray fine structure investigation of bridge members.] *Z. Elektrochem.*, vol. 43, no. 8, Aug. 1937, pp. 546–557.

1926. RICHTER, G. Zur Frage des Einflusses der Oberflächenbeschaffenheit auf die Dauerfestigkeit von Aluminiumdrähten. [The influence of surface conditions on the endurance strength of aluminium wires.] *Z. Metallk.*, vol. 29, no. 7, July 1937, pp. 214–217.

1927. RINAGL, F. Bruchgefahr bei Statischer und Wechselnder Beanspruchung. [Danger of failure under static and repeated stresses.] *Congr. int. Ass. Test. Mat., Ninth Congr., London,* 1937, pp. 639–643.

1928. ROBERTSON, S. L. Metallic damping. *Metal Treatm.*, vol. 3, Autumn 1937, pp. 138–142, 150.

1929. ROLFE, R. T. Steels for user—fatigue testing. *Iron Steel Ind.*, vol. 10, no. 6, Feb. 1937, pp. 259–262 and no. 8, April 1937, pp. 345–349. *Heat Treat. Forg.*, vol. 23, no. 7, July 1937, pp. 336–341 and no. 8, Aug. 1937, pp. 394–396.

930. ROŠ, M. Relations entre les Resultats de l'Essai des Materiaux au Laboratoire, d'un part, et la Solicitation Statique Réelle ou la Fatigue et la Stabilité, d'autre part, et leur Importance quant à la Securité des Constructions. [Relations between the results of laboratory tests of materials and the influence of actual static conditions, fatigue and endurance and their importance in regard to the safety of structures.] *Congr. int. Ass. Test. Mat., Ninth Congr., London,* 1937, pp. 583–587.

)31. ROUCHET, M. Une nouvelle machine d'essais des matériaux à la fatigue dynamique. [A new machine for testing materials under dynamic fatigue.] *Bull. Ass. tech. marit.*, no. 41, 1937, pp. 305–316.

32. SCHRAIVOGEL, K.; STAUDINGER, H. *and* HAAS, B. Dauerversuche mit Schraubenbolzen. [Fatigue tests with bolts.] *Zent. wiss. Ber., Untersuch. Mitt.*, no. 482, 1937, pp. 22.

1933. SERENSEN, S. V. Machines for testing endurance under cyclic loading. [R] *Zav. Lab.*, vol. 6, no. 8, 1937, pp. 991–998.

1934. SERENSEN, S. V. On estimating safety under cyclic loading. [R] *Vest. metalloprom.*, vol. 17, nos. 7–8, May 1937, pp. 3–13.

1935. SERENSEN, S. V. *The endurance of metal and the design of machine components.* [R] M.-L. Gos. izd-vo tekhnich. liter., 1937, pp. 252.

1936. SHCHAPOV, N. P. *and* NIKOLAEV, R. S. On the effect of small surface damage on the brittle failure of machine components and structures in service. Fatigue of metals. [R] *Vest. metalloprom.*, vol. 17, no. 12, Aug. 1937, pp. 86–95.

1937. SIGOLAEV, S. YA. The evaluation of fatigue tests using a radiographic method. [R] *Zav. Lab.*, vol. 6, no. 10, 1937, pp. 1243–1246.

1938. SMEKAL, A. Dauerbruch und spröder Bruch. [Fatigue fracture and brittle fracture.] *Metallwirtschaft*, vol. 16, no. 8, Feb. 19, 1937, pp. 189–193. English abstr.: *Light Metals Res.*, vol. 5, no. 15, 1937, pp. 340–341.

1939. SONDEREGGER, A. Materialprüfung und Schweissung. [Material testing and welding.] *Schweiz. Arch. angew. Wiss.*, vol. 3, no. 11, Nov. 1937, pp. 301–304.

1940. SOPWITH, D. G. *and* GOUGH, H. J. The effect of protective coatings on the corrosion-fatigue resistance of steel. *J. Iron St. Inst.*, vol. 135, no. 1, 1937, pp. 315p–351p. *Engineering, Lond.*, vol. 143, no. 3721, May 7, 1937, pp. 533–535.

1941. SPÄTH, W. Betrieb von Wechselprüfmaschinen im Eigenschwingungsbereich der

Proben. [Operation of fatigue testing machines at the natural frequency of vibration of the specimen.] *Arch. Eisenhüttenw.*, vol. 10, no. 7, Jan. 1937, pp. 313–315. English abstr.: *J. Iron St. Inst.*, vol. 135, no. 1, 1937, p. 234A.

1942. Späth, W. Umlauf-Dauerbiegemaschine mit harter Belastungsfeder. [Rotating bending fatigue machine with hard loading springs.] *Z. Ver. dtsch. Ing.*, vol. 81, no. 25, June 19, 1937, pp. 710–712.

1943. Speller, F. N. Report of joint meeting of the Eastern District Committee and Special Committee on corrosion fatigue of drill pipe. *Proc. Amer. Petrol. Inst.*, sect. 4, vol. 18, 1937, pp. 38–40.

1944. Spraragen, W. *and* Claussen, G. E. Fatigue strength of welded joints—a review of the literature to October 1, 1936. *Weld. J., Easton, Pa.*, vol. 16, no. 1, Jan. 1937, pp. 1–44.

1945. Tagawa, A. Study of the corrosion-fatigue of materials for the piston rod of a blast furnace blower. [J] *Tetsu to Hagane*, vol. 23, no. 11, Nov. 25, 1937, pp. 1063–1084. English abstr.: *Metals and Alloys*, vol. 9, no. 6, June 1938, p. MÀ372.

1946. Takahashi, E. *and* Shioya, K. On the fatigue failure of valve spring. [J] *Nippon kink. Gakk.*, vol. 1, no. 8, Dec. 1937, pp. 320–334.

1947. Tatnall, R. R. The development of a fatigue testing method for springs and spring wire. *Wire and Wire Prod.*, vol. 12, no. 6, June 1937, pp. 297–301.

1948. Tatnall, R. R. Fatigue properties of helical springs. *Wire and Wire Prod.*, vol. 12, no. 10, Oct. 1937, pp. 577–579, 582–586, 588–591.

1949. Thum, A. Gusseisen im Automobilbau. [Cast iron in automobile construction.] *T Z prakt. Metallbearb.*, vol. 47, no. 1/2, Jan. 1937, pp. 66, 69–70, 72.

1950. Thum, A. Die Dauerbruchgefahr und ihre Bekämpfung. [The danger of fatigue fracture and its prevention.] *Forsch. Fortschr. dtsch. Wiss.*, vol. 13, 1937, pp. 153–155.

1951. Thum, A. Versuche zur besseren Ausnutzung der erreichten hohen Güteeigenschaften des Gusseisens in unseren Konstruktionen. [Tests to improve the utilization of the high material properties of cast iron in construction.] *Giesserei*, vol. 24, no. 22, Oct. 22, 1937, pp. 533–537.

1952. Thum, A. *and* Bandow, K. Dauerhaltbarkeit geschmiedeter Stahlkurbelwellen und Milter zu ihrer Steigerung. [Fatigue strength of forged steel crankshafts and methods of increasing the strength.] *Auto.-tech. Z.* vol. 40, no. 2, Jan. 25, 1937, pp. 29–33.

1953. Thum, A. *and* Bautz, W. Zeitfestigkeit [Endurance strength.] *Z. Ver. dtsch. Ing* vol. 81, no. 49, Dec. 4, 1937, pp. 1407–1412.

1954. Thum, A. *and* Bergmann, G. Dauerprüfung von Formelementen und Bauteilen in natürlicher Grösse. [Fatigue tests on full size structures and components.] *Z. Ver. dtsch. Ing.*, vol. 81, no. 35, Aug. 28, 1937, pp. 1012–1018.

1955. Thum, A. *and* Erker, A. Einfluss von Wärmeeigenspannungen auf die Dauerfestigkeit. [Influence of internal thermal stresses on fatigue strength.] *Z. Ver. dtsch. Ing.*, vol. 81, no. 9, Feb. 27, 1937, pp. 276–278.

1956. Thum, A.; Greth, A. *and* Jacobi, H. Dauerbiegeversuche mit Kunstharz-Pressst fen. [Bending fatigue tests on moulded synthetic resin plastics.] *Kunst- u. Pressst.*, no. 2, 1937, pp. 16–24. Abstr.: *Z. Ver. dtsch. Ing.*, vol. 81, no. 29, July 17, 1937, pp. 868–870.

1957. THUM, A. *and* JACOBI, H. R. Dauerbiegungsversuche mit Kunstharz-pressstoffen. [Bending fatigue tests on moulded synthetic resin plastics.] *Umschau*, vol. 41, no. 25, 1937, pp. 564–567.

1958. THUM, A.; KAUFMANN, F. *and* SCHÖNROCK, K. Zugschwellfestigkeits Untersuchungen an Proben mit aufgelegten Schweissraupen und an geschweissten Laschenverbindungen. [Tensile fluctuating stress investigations on specimens with inclined welding beads and on welded strap joints.] *Arch. Eisenhüttenw.*, vol. 10, no. 10, April 1937, pp. 469–476. English abstr.: *J. Iron St. Inst.*, vol. 136, no. 2, 1937, pp. 78A–79A.

959. THUM, A. *and* OCHS, H. Korrosion und Dauerfestigkeit. [Corrosion and fatigue strength.] *Mitt. MatPrüfAnst.*, *Darmstadt*, no. 9, 1937, pp. 109.

960. THUM, A. *and* OCHS, H. Die Korrosionsdauerfestigkeit. [The corrosion fatigue strength.] *Korros. Metallsch.*, vol. 13, no. 10/11, Oct.–Nov. 1937, pp. 380–383. English translation: *Henry Brutcher Tech. Trans.*, no. 551.

961. THUM, A. *and* STROHAUER, R. Prüfung von Lagermetallen und Lagern bei dynamischer Beanspruchung. [The testing of bearing metals and bearings under dynamic loading.] *Z. Ver. dtsch. Ing.*, vol. 81, no. 43, Oct. 23, 1937, pp. 1245–1248.

962. TRIPODI, A. R. L' "Indice di durata" dei cavi di comanto flessibili. [The fatigue life of flexible control cables.] *Aerotecnica, Roma*, vol. 17, no. 3, March 1937, pp. 237–245.

1963. WALLS, F. J. Cast camshafts and crankshafts. *S.A.E. Jl.*, vol. 41, no. 1, July 1937, pp. 284–290.

1964. WATERHOUSE, H. The fatigue resistance of lead and lead alloys. *Res. Rep. Brit. non-ferr. Met. Res. Ass.*, no. 440, June 1937, pp. 7.

1965. WEISS, E. *and* HÖVEL, T. Ursprungsfestigkeiten von Schweissungen verschieden legierter St.52. [Basic strength of different welded joints in alloyed steel St.52.] *Bautechnik*, vol. 15, no. 43, Oct. 1, 1937, pp. 549–552.

1966. WELTER, G. Einfluss von mechanischen Schwingungen auf die Festigkeitseigenschaften von Konstruktionswerkstoffen. [Influence of mechanical oscillations on the strength properties of constructional materials.] *Z. Metallk.*, vol. 29, no. 2, Feb. 1937, pp. 60–62.

1967. WELTER, G. Dauer-Biege- und Dauer-Zug-Druck-Versuche. [Bending fatigue and tension-compression fatigue testing.] *Rev. tech. luxemb.*, vol. 29, no. 5, Sept.–Oct. 1937, pp. 101–109.

1968. WELTER, G. Neue Erkenntnisse über Ermüdungseinflüsse bei Zug-Druck-Beanspruchung. [New knowledge on fatigue effects under tensile-compressive stresses.] *Rev. tech. luxemb.*, vol. 29, no. 6, Nov.–Dec. 1937, pp. 294–305.

1969. WELTER, G. Bending and tension-compression fatigue testing—III. [P] *Wiad. Inst. Metal.*, vol. 4, no. 1, 1937, pp. 30–39. French abstr.: *Rev. Métall., Ext.*, vol. 35, no. 3, March 1938, pp. 91–93.

1970. WELTER, G. Biege- und Zug-Druck-Dauerversuche mit Beobachtungen über Temperatureinflüsse. [Bending and tension-compression fatigue testing and observations on the influence of temperature.] *Congr. int. Ass. Test. Mat., Ninth Congr., London*, 1937, pp. 29–32.

1971. WEVER, F. *and* MÖLLER, H. Über den Einfluss der Wechselbeanspruchung auf den Kristallzustand metallischer Werkstoffe. [On the influence of fatigue stressing on the crystal state of metallic materials.] *Naturwissenschaften*, vol. 25, no. 28, July 9, 1937, pp. 449–453.

1972. WILSON, W. M. Fatigue tests of butt welds in structural plates. *Weld. J., Easton, Pa.*, vol. 16, no. 10, Oct. 1937, pp. 23s–27s.

1973. WILSON, W. M. Recent developments in structural welding. *J. West. Soc. Engrs*, vol. 42, no. 6, Dec. 1937, pp. 306–320.

1974. WISHART, H. B. *and* LYON, S. W. Effect of overload on the fatigue properties of several steels at various low temperatures. *Trans. Amer. Soc. Metals*, vol. 25, no. 3, Sept. 1937, pp. 690–701.

1975. WRAY, C. F. An automatic electrical fatigue testing machine. *Engineer, Lond.*, vol. 164, no. 4260, Sept. 3, 1937, pp. 251–254.

1976. ZIMMERLI, F. P. The effect of longitudinal scratches upon the endurance limit in torsion of spring wire. *Wire and Wire Prod.*, vol. 12, no. 3, March, 1937, pp. 133–138 and no. 4, April 1937, pp. 185–191.

1977. Nomenclature for various ranges in stress in fatigue. *Proc. Amer. Soc. Test. Mater.*, vol. 37, pt. 1, 1937, pp. 159–163.

1978. On the carrying capacity and life of ball bearings. *Ball Bearing J.*, no. 3, 1937, pp. 34–44.

1979. Interim report on the fatigue of round cadmium–copper trolley wires. *Rep. Brit. elect. ind. Res. Ass.*, no. F/T 111, 1937, pp. 29.

1980. Pulsator fatigue testing machine. *Machinery, Lond.*, vol. 49, no. 1266, Jan. 14, 1937, pp. 482–483.

1981. Metals do not "crystallize" under vibration. *Letter Circular, nat. Bur. Stand.*, no. 486, Jan. 6, 1937, pp. 2.

1982. Les machines à essayer les métaux pour déterminer la résistance d'endurance. [Machines for determining the fatigue resistance of metals.] *Génie civ.*, vol. 110, no. 11, (no. 2848), March 13, 1937, pp. 252–254.

1983. Cleaning of duralumin in Zonax and trichlorethylene. *Roy. Aircr. Estab. Rep.* no. M.4257, May 1937, pp. 4.

1984. National Physical Laboratory: Fatigue and corrosion fatigue. *Engineering, Lond.* vol. 144, no. 3730, July 9, 1937, pp. 49–50.

1985. The testing of aero-engine materials. *Machinery, Lond.*, vol. 50, no. 1300, Sept. 9 1937, pp. 731–732.

1986. "Rayflex" fatigue testing machine. *Machinist*, vol. 81, Sept. 22, 1937, p. 868.

1987. Amerikanische Untersuchungen über die Oberflächendrücken. [American investigations on surface rolling.] *Werkzeugmaschine* vol. 41, no. 19, Oct. 15, 1937, pp. 408–413.

1988. Fatigue testing machine for reversed plane bending. *J. sci. Instrum.*, vol. 14, no. 12, Dec 1937, pp. 419–421.

1938

1989. AMATULLI, A. *and* HENRY, O. H. Fatigue tests of welds at elevated temperatures. *Weld. J., Easton, Pa.*, vol. 17, no. 6, June 1938, pp. 14s–20s.

1990. BALDWIN, T. The fatigue strength of machined tyre steels—including some general notes on fatigue and related matters. *J. Instn Loco. Engrs, Lewes*, vol. 28, no. 146, Nov.–Dec. 1938, pp. 649–721.

1991. BANDOW, K. Dauerhaltbarkeit von Stahl- und Guss-Kurbelwellen. [Fatigue properties of steel and cast crankshafts.] *Dtsch. Kraftfahrtforsch.*, no. 14, 1938, pp. 35.

1992. BAUTZ, W. Zweckmässige Ausbildung von Querschnittsübergängen. [Suitable forms for transition fillets.] *T Z prakt. Metallbearb.*, vol. 48, no. 23/24, Dec. 1938, pp. 881–884.

1993. BIERETT, G. Die Dauerhaltbarkeit geschweisster Maschinenteile. [The fatigue properties of welded machine parts.] *Maschinenschaden*, vol. 15, no. 9, 1938, pp. 133–137 and no. 10, 1938, pp. 160–163.

1994. BLEAKNEY, W. N. Fatigue testing of wing beams by the resonance method. *Tech. Note nat. Adv. Comm. Aero., Wash.*, no. 660, Aug. 1938, pp. 22.

1995. BÖHM, E. Gussgefüge und Dauerbiegefestigkeit einiger Aluminium-Legierungen. [Cast structure and bending fatigue strength of some aluminium alloys.] *Aluminium, Berl.*, vol. 20, no. 3, March 1938, pp. 168–174.

1996. BOLLENRATH, F. Zeit- und Dauerfestigkeit der Werkstoffe. [Endurance and fatigue strength of materials.] *Gesamm. Vortr. Hauptversamml. Lilienthal—Ges. Luft Forsch.*, 1938, pp. 147–157.

1997. BOLLENRATH, F. *and* BUNGARDT, K. Untersuchungen über die Korrosionsermüdung von Aluminium- und Magnesium-Knetlegierungen. [Investigations on the corrosion-fatigue of wrought aluminium and magnesium alloys.] *Z. Metallk.*, vol. 30, no. 10, Oct. 1938, pp. 357–359.

1998. BOLLENRATH, F. *and* BUNGARDT, K. Einfluss von Randentkohlung und Wärmevorbehandlung auf Dauer- und Zeitfestigkeit von Stahlspanndrähten. [Influence of surface decarburization and prior heat treatment upon the fatigue and endurance strength of steel guy wires.] *Arch. Eisenhüttenw.*, vol. 12, no. 4, Oct. 1938, pp. 213–218.

1999. BOLLENRATH, F. *and* CORNELIUS, H. Zeit- und Dauerfestigkeit ungeschweisster und Stumpfgeschweisster Chrom-Molybdän-Stahlrohre bei verschiedenen Zugmittelspannungen. [Endurance and fatigue strength of unwelded and butt welded chromium–molybdenum steel tubes at different mean tensile stresses.] *Jb. dtsch. Luftfahrtf.*, 1938, pp. I549–I553. *Jb. dtsch. VersAnst. Luftf.*, 1938, pp. 304–308. *Stahl u. Eisen*, vol. 58, no. 9, March 3, 1938, pp. 241–245.

2000. BOLLENRATH, F.; CORNELIUS, H. *and* SIEDENBURG, W. Einfluss der Querschnittsform auf die Dauerfestigkeit von weichem Flussstahl. [Influence of the cross-sectional shape on the fatigue strength of soft steel.] *Luftfahrtforsch.*, vol. 15, no. 4, April 6, 1938, pp. 214–217. *Jb. dtsch. VersAnst. Luftf.*, 1938, pp. 309–312.

2001. BRENNER, P. *and* KOSTRON, H. Der Einfluss der Faserrichtung auf die Dauerfestigkeit einer Al-Cu-Mg Legierung. [The influence of fibre direction on the fatigue properties of an Al–Cu–Mg alloy.] *Luftwissen*, vol. 5, no. 1, Jan. 1938, pp. 15–16.

2002. BUCHMANN, W. Dauerfestigkeitseigenschaften von Elektronlegierungen, insbesondere Kerbempfindlichkeit der Knetlegierungen. [Fatigue strength of Elektron alloys, in particular notch sensitivity of the wrought alloys.] *Jb. dtsch. Luftfahrtf.*, 1938, pp. I524–I528. English translation: *R.T.P. Trans.*, no. 960, July 29, 1939, pp. 15.

2003. BUCHNER, H. Die Elastizitätsgrenze von Stahlen bei Dauerbeanspruchung und ihr Zusammenhang mit der Dauerfestigkeit, Werkstoffdämpfung und Kerbempfindlichkeit. [The elastic limit of steels under fatigue stressing and its relation to fatigue resistance, material damping and notch effect.] *ForschArb. IngWes.*, vol. 9, no. 1, Jan.–Feb. 1938, pp. 14–27. Abstr.: *Z. Ver. dtsch. Ing.*, vol. 82, no. 26, June 25, 1938, pp. 781–782.

2004. BUCKWALTER, T. V.; HORGER, O. J. *and* SANDERS, W. C. Locomotive axle testing. *Trans. Amer. Soc. mech. Engrs*, vol. 60, no. 4, May 1938, pp. 335–345. *Rly mech. Engr*, vol. 112, Oct. 1938, pp. 365–371.

2005. BÜHLER, H. Zur Dauerfestigkeit von Walzträgern. [The fatigue strength of welded steel girders.] *Stahlbau*, vol. 11, no. 2, Jan. 21, 1938, pp. 9–12. English translation: *Weld. J.*, *Easton, Pa.*, vol. 17, no. 3, March 1938, pp. 11–13.

2006. BÜHLER, H. Rundstahl für Eisenbetonbauten. Gütewerte, Längen und Stösse. [Round steel for concrete reinforcing bars. Quality, elongation and joints.] *Beton u. Eisen*, vol. 37, no. 16, Aug. 20, 1938, pp. 258–263. English summary: *Weld. J., Easton, Pa.*, vol. 18, no. 2, Feb. 1939, pp. 49s–51s.

2007. BUNGARDT, K. Dynamische Festigkeitseigenschaften von Leichtmetall-Legierungen bei tiefen Temperaturen. [Dynamic strength properties of light alloys at low temperatures.] *Jb. dtsch. Luftfahrtf.*, 1938, pp. I529–I531. *Jb. dtsch. VersAnst. Luftf.*, 1938, pp. 325–327. *Z. Metallk.*, vol. 30, no. 7, July 1938, pp. 235–237.

2008. BUSSMANN, K. H. Versuche zur Ermittlung der Dauerbiegefestigkeit von Ledertreibriemen. [Tests for determining the bending fatigue strength of leather driving belts.] *Z. Ver. dtsch. Ing.*, vol. 82, no. 43, Oct. 22, 1938, pp. 1249–1250.

2009. CAMPUS, F. Essais de fatigue de joints soudés de rails. [Fatigue tests on welded steel rail joints.] *Rev. univ. Min.*, vol. 14, no. 6, June 1938, pp. 493–499. English summary: *Weld. J.*, *Easton, Pa.*, vol. 17, no. 8, Aug. 1938, pp. 31s–32s.

2010. CAZAUD, R. La fatigue des métaux. Importance et moyens éviter des ruptures qui en rèsultent. [The fatigue of metals. Importance and methods of avoiding the resulting fractures.] *Prat. Industr. méc.*, vol. 20, no. 12, March 1938, pp. 490–494; vol. 21, no. 1, April 1938, pp. 9–12 and no. 2, May 1938, pp. 67–73.

2011. CHASTON, J. C. Mechanical tests and thei engineering significance. *Metal Treatm.* vol. 3, no. 12, 1938, pp. 159–164.

2012. COLLINS, W. L. *and* DOLAN, T. J. Physica properties of four low-alloy high-strengt steels. *Proc. Amer. Soc. Test. Mater.*, vol. 38 pt. 2, 1938, pp. 157–175.

2013. CORNELIUS, H. *and* BOLLENRATH, F. Dau erhaltbarkeit von hohlen Kurbelwellenzapfe mit Innenverstärkung an der Ölbohrun [Fatigue strength of hollow crankshaft jou nals with reinforcement at the oil hole.] *Z. Ve dtsch. Ing.*, vol. 82, no. 30, July 23, 193 pp. 885–889. English abstr.: *Automot. Industr N.Y.*, vol. 79, no. 9, Aug. 27, 1938, p. 252.

2014. CORNELIUS, H. *and* BOLLENRATH, F. D Ermüdungsfestigkeit dünnwandiger Rohre f den Flugzeugbau im ungeschweissten u geschweissten Zustand. [The fatigue streng of thin walled tubes for aircraft construction in the unwelded and welded condition.] *Jb. dtsch. VersAnst. Luftf.*, 1938, pp. 297–303.

2015. DENARO, L. F. Fatigue resistance of welded joints—a summary of published German investigations. *Trans. Inst. Weld.*, vol. 1, no. 1, Jan. 1938, pp. 52–58.

2016. DENARO, L. F. The influence of fatigue on the design of welded connections. *Welding of steel structures*, London, H.M.S.O., 1938, pp. 241–263.

2017. DOLAN, T. J. Simultaneous effects of corrosion and abrupt changes in section on the fatigue strength of steel. *J. appl. Mech.*, vol. 5, no. 4, Dec. 1938, pp. A141–A148.

2018. DOUGLAS, W. D. *and* TAYLOR, W. J. Effect of initial plastic strain on fatigue strength of three light alloys. *Roy. Aircr. Estab. Rep.* no. M. T. 5649, March 1938, pp. 3.

2019. DRAKE, H. C. Flash welding of rails. *Weld. J., Easton, Pa.*, vol. 17, no. 10, Oct. 1938, pp. 17–21.

2020. VAN DER EB, W. J. Een hypothese ter verklaring van de in taaie staalsoorten optredende vermoeidheidsverschijnselen. [An hypothesis to explain the fatigue phenomena that occur in tough steels.] *Ingen. Ned. Ind.*, vol. 5, no. 2, 1938, pp. 35–41 and no. 3, 1938, pp. 50–59.

2021. EDGERTON, C. T. Research on fatigue properties of heavy helical springs. *Wire and Wire Prod.*, vol. 13, no. 1, Jan. 1938, pp. 17–26, 41; no. 2, Feb. 1938, pp. 69–73, 97; no. 3, March 1938, pp. 125–131, 155 and no. 4, April 1938, pp. 183–185, 204.

2022. EHRT, M. *and* KÜHNELT, G. Das Gesicht des Dauerbruches. [The appearance of fatigue fractures.] *Mitteilungen der Materialprüfungsstelle der Allianz und Stuttgarter Verein Versicherungs A.-G., Abteilung für Maschinenversicherung*, no. 4, Nov. 1938, pp. 51. Abstr.: *Maschinenschaden*, vol. 15, no. 11, 1938, pp. 176–179.

2023. ERLINGER, E. Wechselbiegemaschine. [An alternating bending fatigue machine.] *Arch. Eisenhüttenw.*, vol. 11, no. 9, March 1938, pp. 455–456.

2024. FISCHER, G. Über die Kerbwirkung bei Dauerwechselbeanspruchung und den Einfluss der Kaltverformung auf die Dauerhaltbarkeit. [On the notch effect under fatigue stress and the influence of cold working on the endurance.] *Jb. dtsch. VersAnst. Luftf.*, 1938, pp. 328–334. *Jb. dtsch. Luftfahrtf.*, 1938, pp. I517–I523. English translation: *R.T.P. Trans.*, no. 969, pp. 26.

2025. FÖPPL, O. Oberflächenrisse. [Surface cracks.] *Mitt. Wöhler-Inst.*, no. 32, 1938, pp. 75–76.

2026. FÖPPL, O. Oberflächendrucken und Druckeigenspannungen. [Surface compression and compressive internal stresses.] *Mitt. Wöhler-Inst.*, no. 33, 1938, pp. 55–65.

2027. FÖPPL, O. Drücken des Kerbgrundes von gerollten, geschnittenen und geschliffenen Schrauben zum Zwecke der Steigerung der Dauerhaltbarkeit. [Compressing the bottom of threads of bolts by rolling, cutting and grinding for the purpose of increasing their fatigue properties.] *Werkzeugmaschine*, vol. 42, no. 1, 1938, pp. 459–462.

2028. FRYE, J. H. *and* KEHL, G. L. The fatigue resistance of steel as affected by some cleaning methods. *Trans. Amer. Soc. Metals*, vol. 26, March 1938, pp. 192–218.

2029. GARDNER, E. P. S. Regulations and specifications for welded steelwork. *Trans. Inst. Weld.*, vol. 1, no. 2, April 1938, pp. 75–107.

2030. GERARD, I. J. *and* SUTTON, H. Corrosion-fatigue properties of duralumin with and without protective coatings. *Rep. Memor. aero. Res. Comm., Lond.*, no. 1828, April 29, 1938, pp. 2–3.

2031. GERTSRIKEN, S. D. *and* DIKHTYAR, I. YA. X-ray investigation on the fatigue of metals. [R] *Zh. tekh. Fiz.*, vol. 8, no. 20, 1938, pp. 1793–1798.

2032. GISEN, F. *and* GLOCKER, R. Röntgenographische Bestimmungen der zeitlichen Änderung des Eigenspannungszustandes bei Biegewechselbeanspruchung. [X-ray determination of the variations of the internal stresses with time under alternating bending.] *Z. Metallk.*, vol. 30, no. 9, Sept. 1938, pp. 297–298.

2033. GLOCKER, R. *and* KEMMNITZ, G. Spannungsmessungen am Dauerbruchvorgang. [Stress measurements during the fatigue fracture process.] *Z. Metallk.*, vol. 30, no. 1, Jan. 1938, pp. 1–3.

2034. GOUGH, H. J. Mechanism of fatigue failure. *Metal Progr.*, vol. 33, no. 6, June 1938, pp. 620, 622.

2035. GOUGH, H. J. *and* WOOD, W. A. The deformation and fracture of metals. *J. Instn civ. Engrs*, no. 5, March 1938, pp. 249–284.

2036. GOUGH, H. J. *and* WOOD, W. A. The crystalline structure of steel at fracture. *Proc. roy. Soc. (A)*, vol. 165, 1938, pp. 358–371.

2037. GRAF, O. Dauerfestigkeit von Holzverbindungen, Versuche mit Holzverbindungen bei stufenweise gesteigerter Belastung und bei oftmals wiederholter Belastung. [Fatigue strength of wood joints, tests on wood joints with gradually increasing loads and under often repeated loads.] *Mitt. Fachaussch. Holzfr.*, no. 22, 1938, pp. 58.

2038. GRAF, O. Dauerversuche mit Nietverbindungen, welche an den Gleitflächen statt mit einem Anstrich aus Leinöl und Mennige mit einem aufgespritzten Belag aus Leichtmetall versehen waren. [Fatigue tests with riveted joints the slip surfaces of which have been coated with a sprayed-on deposit of light metal instead of a coat of linseed oil and red lead.] *Stahlbau*, vol. 11, no. 3, Feb. 4, 1938, pp. 17–19.

2039. GRAF, O. Über Dauerversuche mit Gurtverstärkerungen an Zugstäben und an Trägern. [On fatigue tests of tension members and girders having reinforced flanges.] *Z. Ver. dtsch. Ing.*, vol. 82, no. 7, Feb. 12, 1938, pp. 158–160.

2040. GRAF, O. German fatigue tests on welds *Weld. J., Easton, Pa.*, vol. 17, no. 3, March 1938, pp. 2s–9s.

2041. GRAF, O. Dauerversuche mit Holzverbindungen. [Fatigue tests on wood joints.] *Hol a. Roh- u. Werkst.*, vol. 1, no. 7, April 1938 pp. 266–269.

2042. GRAF, O. Über die Dauerbiegefestigkei von geschweissten Schienen. [On the bendin fatigue strength of welded rails.] *Autogen Metallbearb.*, vol. 31, Aug. 15, 1938, pp. 255 266 and Sept. 1, 1938, pp. 271–279.

2043. GÜRTLER, G. *and* SCHMID, E. Untersu chungen über die Dauerfestigkeit von Legierun gen der Silumingruppe. [Investigations on th fatigue strength of alloys of the Silumi group.] *Aluminium, Berl.*, vol. 20, no. March 1938, pp. 174–181.

2044. VON HANFFSTENGEL, K. *and* HANEMANN, H. Zur Schwingungsfestigkeit von Blei und Bleilegierungen. [The bending fatigue strength of lead and lead alloys.] *Z. Metallk.*, vol. 30, no. 2, Feb. 1938, p. 51.

2045. HANKINS, G. A. *and* THORPE, P. L. Fatigue experiments on weld metal and welded joints. *Welding of steel structures*, London, H.M.S.O., 1938, pp. 187–195.

2046. HELLWIG, W. Verdrehdauerfestigkeit von Federdrähten aus Beryllium-Nickel und Beryllium-Contracid bei Temperaturen bis 300°C. [Torsional fatigue strength of beryllium–nickel and beryllium-Contracid spring wire at temperatures up to 300°C.] *ForschArb. IngWes.*, vol. 9, no. 4, July–Aug., 1938, pp. 165–176. Abstr.: *Z. Ver. dtsch. Ing.*, vol. 83, no. 20, May 20, 1939, pp. 639–640.

2047. HEMPEL, M. Die Beziehungen zwischen dem Röntgen-Grobgefügebild und der Zugschwellfestigkeit von geschweissten Proben aus St.37. [The relation between the appearance of X-ray radiographs and the tensile strength of welded specimens of St.37.] *Stahl u. Eisen*, vol. 58, no. 28, July 14, 1938, pp. 756–760.

2048. HEMPEL, M. Prüfung der Dauerfestigkeit. [Determining the fatigue strength.] *Rdsch. dtsch. Tech.*, vol. 18, no. 44, 1938, pp. 3–4.

2049. HERBST, H. Bedeutung und Ursachen innerer Drahtbrüche bei Draht-, im besondern Förderseilen. [Significance and causes of internal fractures in wire rope, in particular hoisting cable.] *Glückauf*, vol. 74, no. 40, Oct. 8, 1938, pp. 849–856 and no. 41, Oct. 15, 1938, pp. 878–884.

2050. HÖVEL, T. Einfluss der äusseren Bearbeitung und innerer Poren auf die Dauerfestigkeit elektrisch geschweisster Stumpfnaht-verbindungen. [Influence of external operation and internal porosity on the fatigue strength of electrically welded butt joints.] *Elektroschweissung*, vol. 9, no. 8, Aug. 1938, pp. 144–146.

2051. HOMERBERG, V. O. Nitriding. *Industr. Heat.*, vol. 5, Oct. 1938, pp. 917–920.

2052. HOMÈS, G. A. Relations entre l'endurance et la compacite des soudures a l'arc de l'acier doux — mesure radiographique de la compacite. [Relation between the endurance limit and porosity of arc welds in mild steel—radiographic measurement of porosity.] *Arcos*, vol. 15, no. 89, Nov. 1938, pp. 1951–1967.

2053. HOMÈS, G. A. *and* DUWEZ, P. Distinction entre le mécanisme cristallin de la rupture statique et celui de la rupture dynamique. [Distinction between the crystalline mechanism of static and dynamic fracture.] *Bull. Acad. Belg. cl. Sci.*, ser. 5, vol. 24, no. 3, 1938, pp. 159–162.

2054. HORGER, O. J. Fatigue failure of railroad axles and stresses in hollow cylinders. *Stephen Timoshenko 60th Anniversary Volume*, New York, The Macmillan Co., 1938, pp. 73–80.

2055. HORGER, O. J. Photoelastic analysis practically applied to design problems. *J. appl. Phys.*, vol. 9, no. 7, July 1938, pp. 457–465.

2056. HÜNLICH, R. *and* PÜNGEL, W. Untersuchungen über die Bruchursachen von Ventilfedern. [Investigations on the fracture of valve springs.] *Jb. dtsch. Luftfahrtf.*, 1938, pp. II134–II140.

2057. HUNTER, R. Some aspects of failures in steel. *J. W. Scot. Iron St. Inst.*, vol. 45, pt. IV, Jan. 1938, pp. 41–53.

2058. IGARASHI, I. and FUKAI, S. On the decrease of the fatigue limits of duralumin and super-duralumin caused by sea-water corrosion. [J] *Tetsu to Hagane*, vol. 24, no. 5, 1938, pp. 451–455.

2059. IWAI, S. and OSUGA, T. Endurance tests with forged aluminium alloy under repeated tension and compression. [J] *J. aero. Res. Inst. Tokyo*, no. 166, 1938, pp. 308–311.

2060. JENSEN, A. E. Corrosion fatigue. *J. S. Afr. Inst. Engrs*, vol. 36, no. 12, July 1938, pp. 273–281.

2061. JOHNSON, J. B. and OBERG, T. T. Airplane propeller blade life. *Metals and Alloys*, vol. 9, no. 10, Oct. 1938, pp. 259–262.

2062. KAUL, H. W. Die erforderliche Zeit- und Dauerfestigkeit von Flugzeugtragwerken. [The required endurance and fatigue resistance of wing beam structures.] *Jb. dtsch. Luftfahrtf.*, 1938, pp. I274–I288. *Jb. dtsch. Vers Anst. Luftf.*, 1938, pp. 195–209. English translation: *Tech. Memor. nat. Adv. Comm. Aero., Wash.*, no. 992, Oct. 1941, pp. 39.

2063. KIDANI, Y. On the fatigue of metals and the internal friction. *Mem. Ryoj. Coll. Engng*, vol. 11, no. 8, 1938, pp. 253–269.

2064. KIDANI, Y. The fatigue of metals. Part 1 —statistical theory of fatigue. *Sci. Pap. Inst. phys. chem. Res. Tokyo*, vol. 34, no. 825, Oct. 1938, pp. 1042–1052.

2065. KINNEY, J. S. An investigation of the effects of elevated temperatures on the fatigue properties of two alloy steels. *Proc. Amer. Soc. Test. Mater.*, vol. 38, pt. 2, 1938, pp. 197–201.

2066. KLIMOV, K. N. Microscopic investigation of the deformation under static and dynamic loading and in fatigue. [R] *Metallurg*, vol. 13, no. 10, Oct. 1938, pp. 80–85. English abstr.: *J. Iron St. Inst.*, vol. 140, no. 2, 1939, p. 98A.

2067. KÖRBER, F. Das Verhalten metallischer Werkstoffe im Bereich kleiner Verformungen. [The behaviour of metallic materials subjected to small deformation.] *Proc. Int. Congr. appl. Mech.*, Fifth Congr., Cambridge, Mass., 1938, pp. 20–33.

2068. KÖTZSCHKE, P. Neuere Erkenntnisse über Herstellung und Prüfung von hochwertiger Drähten unter besonderer Berücksichtigung von Ventilfederdraht. [Recent information on the production and testing of high grade wires with special reference to valve spring wire. *Jb. dtsch. Luftfahrtf.*, 1938, pp. II319–II325

2069. KOMMERS, J. B. Design stress diagrams for alternating plus steady loads. *Prod. Engng* vol. 9, no 10, Oct. 1938, pp. 395–397.

2070. KOMMERS, J. B. The effect of overstressing and understressing in fatigue. *Proc. Amer. Soc. Test. Mater.*, vol. 38, pt. 2, 1938 pp. 249–268.

2071. LEA, F. C. and WHITMAN, J. G. Failure of girders under repeated stresses. *J. Instn civ Engrs*, no. 7, June 1938, pp. 301–328.

2072. LEHR, E. and MAILÄNDER, R. Einfluss von Hohlkehlen an abgesetzten Wellen und vc Querbohrungen auf die Biegewechselfestig keit. [Influence of fillets and transverse drille holes on the bending fatigue strength of ba with reduced sections.] *Arch. Eisenhüttenw* vol. 11, no. 11, May 1938, pp. 563–56 Abstr.: *Stahl u. Eisen*, vol. 58, no. 21, May 2 1938, p. 577.

2073. DE LEIRIS, H. L'essai de mises en pressic répétées. [The test by repeatedly applied pre sure.] *Bull. Ass. tech. marit.*, vol. 42, Mémoi no. 766, 1938, pp. 615–628.

2074. LOCATI, L. La resistenza alla fatica per te sione di alcuni acciai da construzion

[The fatigue resistance in torsion of some structural steels.] *Industr. mecc.*, vol. 20, no. 5, May 1938, pp. 341–347; no. 6, June 1938, pp. 471–478 and no. 7, July 1938, pp. 573–578.

2075. LÜRENBAUM, K. Schwingungsbeanspruchung und Dauerhaltbarkeit von Elektron-Luftschrauben. [Alternating stresses in, and fatigue strength of Elektron airscrews.] *Jb. dtsch. VersAnst. Luftf.*, 1938, pp. 457–460. *Jb. dtsch. Luftfahrtf.*, 1938, pp. 1416–1419.

2076. LUGAS'KOV, A. S. The resistance of cast magnesium alloys to dynamic loads. [R] *Aviapromyshlennosti*, vol. 7, no. 5, 1938, pp. 20–24.

2077. MALISIUS, R. *and* MICKEL, E. Untersuchungen der Zugschwellfestigkeit an Abbrenn-Stumpfschweissverbindungen. [Investigation of the tensile fatigue strength of resistance welded butt joints.] *Mitt. ForschAnst. Gutehoffn., Nürnberg*, vol. 6, no. 10, Dec. 1938, pp. 266–268.

2078. ZOEGE VON MANTEUFFEL, R. Versuche über die Dauerhaltbarkeit von Federn. Der Einfluss der Werkstoffauswahl auf die Dauerhaltbarkeit von Federn mit schwarzer Oberfläche. [Tests on the fatigue behaviour of springs. The influence of material selection on the fatigue behaviour of springs with black surfaces.] *Dtsch. Kraftfahrtforsch., Zwischenber.*, no. 49, 1938, pp. 30.

2079. MARIN, J. Working stresses in members under fluctuating loads. *Prod. Engng*, vol. 9, no. 7, July 1938, pp. 251–253.

2080. MARTY, R. Etude des déformations et de la fatigue des ressorts à boudin de forme quelconque soumis à des efforts de directions quelconques. [Study of the deformation and fatigue of coil springs of different shape and under various applied loads.] *Mécanique*, vol. 22, no. 276, Jan.–Feb. 1938, pp. 33–39 and no. 279, July–Aug. 1938, pp. 137–144.

2081. MICKEL, E. Neuere Erkenntnisse über die Gestaltfestigkeit gusseiserner Bauteile. [Recent knowledge of the form strength of cast iron components.] *Giesserei*, vol. 25, no. 16, Aug. 12, 1938, pp. 401–405.

2082. MÖLLER, H. Röntgen-Grobgefügebild und Wechselfestigkeit. [X-ray photographs and fatigue strength.] *Berg- u. hüttenm. Mh.*, vol. 86, no. 6, June 1938, pp. 148–152.

2083. MÖLLER, H. *and* HEMPEL, M. Wechselbeanspruchung und Kristallzustand. [Alternating stress and the crystal structure.] *Mitt. K.-Wilh.-Inst. Eisenforsch.*, Vol. 20, no. 2, 1938, pp. 15–33. Abstr.: *Z. Ver. dtsch. Ing.*, vol. 83, no. 44, Nov. 4, 1939, pp. 1183–1184. English abstr.: *Metallurgia, Manchr*, vol. 19, 10. III, Jan. 1939, p. 82.

2084. MÖLLER, H. *and* HEMPEL, M. Wechselbeanspruchung und Kristallzustand. II—Zur Frage der Kristallverformung beim Dauerbruch. [Alternating stress and the crystal structure. II—the question of crystal deformation for fatigue fracture.] *Mitt. K.-Wilh.-Inst. Eisenforsch.*, vol. 20, no. 17, 1938, pp. 229–238.

2085. MOHR, E. Über die Bestimmung einer der Schwingungsfestigkeit nahestehenden Kennziffer mittels des Biege-Zug-Versuches. [On the determination of a property related to the fatigue strength by means of the bending-tensile test.] *Z. Metallk.*, vol. 30, no. 1, Jan. 1938, pp. 30–35.

2086. MOHR, E. Über den Zusammenhang zwischen Biege-Zug-Festigkeit und Biege-wechselfestigkeit. [On the relation between the bending-tensile strength and the alternating bending fatigue strength.] *Z. Metallk.*, vol. 30, no. 2, Feb. 1938, pp. 71–73.

2087. MOHR, E. Werkstoffprüfung mittels der Biege-Zug-Versuches. [Material testing by means of the bending-tensile test.] *Metallwirtschaft*, vol. 17, no. 20, May 20, 1938, pp. 535–537.

2088. MOORE, H. F. Fourth progress report on the joint investigation of fissures in railroad rails. *Bull. Amer. Rly Engng Ass.*, vol. 40, no. 404, June–July 1938, pp. 1–52. *Repr. Ill. Engng Exp. Sta.*, no. 12, 1938, pp. 52.

2089. MOORE, H. F. and JORDAN, R. L. Stress concentration in steel shafts with semicircular notches. *Proc. Int. Congr. appl. Mech.*, Fifth Congr., Cambridge, Mass., 1938, pp. 188–192. Abstr.: *J. appl. Mech.*, vol. 5, no. 3, Sept. 1938, p. A-117.

2090. MOORE, H. F.; SPRARAGEN, W. and CLAUSSEN, G. E. Fatigue tests. *Welding handbook*, New York, American Welding Society, 1938, pp. 710–731.

2091. MÜLLER-STOCK, H. Der Einfluss dauernd und unterbrochen wirkender schwingender Überbeanspruchung auf die Entwicklung des Dauerbruchs. [The effect of continuous and interrupted vibratory overstressing on the development of fatigue failure.] *Mitt. Kohle- u. Eisenforsch.*, vol. 2, no. 2, March 1938, pp. 83–107.

2092. MÜLLER-STOCK. H.; GEROLD, E. and SCHULZ, E. H. Der Einfluss einer Wechselvorbeanspruchung auf Biegezeit- und Biegewechselfestigkeit von Stahl St.37. [The influence of alternating pre-stressing on the bending life and bending fatigue strength of steel St.37.] *Arch. Eisenhüttenw.*, vol. 12, no. 3, Sept. 1938, pp. 141–148. English translation: *Motor Industry Research Association*, Lindley, England.

2093. NEWPORT, A. J. The possibility of fatigue failure in welded connections of steel floor beams as a result of vibration and shock.

Welding of steel structures, London, H.M.S.O., 1938, pp. 196–213.

2094. NISHIHARA, T. and SAKURAI, T. Endurance limits of cast iron for repeated tension and compression. [J] *Trans. Soc. mech. Engrs, Japan*, vol. 4, no. 15, May 1938, pp. 105–110.

2095. NISSEN, O. Festigkeitsfragen bei der Gestaltung neuzeitlicher Flugzeuge. [Questions of strength in aircraft construction.] *Jb. dtsch. Luftfahrtf.*, 1938, pp. 158–163. English translation: *Aircr. Engng*, vol. 12, no. 140, Oct. 12, 1940, pp. 293–295, 306.

2096. OCHS, H. Der Einfluss der Vorspannung auf Korrosionsermüdung. [The influence of initial stress on corrosion-fatigue.] *ForschArb. IngWes.*, vol. 9, no. 2, 1938, pp. 106–107.

2097. ODING, I. A. Effect of uneven distribution of stress through the section on the limits of plastic flow and fatigue. [R] *Zav. Lab.*, vol. 7, no. 4, 1938, pp. 445–458.

2098. ONO, K. Fatigue of lead and lead alloys. [J] *Nippon kink. Gakk.*, vol. 2, no. 6, June, 1938, pp. 290–295. English abstr.: *Metals and Alloys*, vol. 10, no. 1, Jan. 1939, p. MA48.

2099. OTTITZKY, K. Dauerbiegeversuche an gewinkelten Flachstäben. [Bending fatigue tests on tapered plate specimens.] *Z. Ver. dtsch. Ing.*, vol. 82, no. 17, April 23, 1938, pp. 501–502.

2100. PETERSON, R. E. Methods of correlating data from fatigue tests of stress concentration specimens. *Stephen Timoshenko 60th Anniversary Volume*, New York, The Macmillan Co., 1938, pp. 179–183.

2101. PIRKL, J. and von LAIZNER, H. Wechselfestigkeits-Prüfmaschine mit Antrieb durch bandgeführte Differentialrollen. [Fatigue testing machine with drive through rack with differential sectors.] *Arch. Eisenhüttenw.*, vol. 12, no. 6, Dec. 1938, pp. 305–308. English abstr.: *J. Iron St. Inst.*, vol. 139, no. 1, 1939, pp. 175A–176A.

2102. PIWOWARSKY, E. Einfluss von Korrosion auf die Festigkeitseigenschaften von Gusseisen. [Influence of corrosion upon the strength properties of cast iron.] *Z. Ver. dtsch. Ing.*, vol. 82, no. 13, March 26, 1938, pp. 370–372.

2103. POMP, A. *and* HEMPEL, M. Dauerprüfung von Stahldrähten unter wechselnder Zugbeanspruchung. II—Einfluss der Ziehbedingungen auf die Zugschwellfestigkeit von Stahldraht. [Fatigue testing of steel wires under pulsating tensile stress. II—Effect of drawing conditions on the tensile strength of steel wire.] *Mitt. K.-Wilh.-Inst. Eisenforsch.*, vol. 20, no. 1, 1938, pp. 1–14.

2104. PORTEVIN, A. Les états de surface et la corrosion. [The condition of the surface and corrosion.] *Metaux et Corros.*, vol. 13, no. 151, March 1938, pp. 43–64.

2105. VON RAJAKOVICS, E. Dauerfestigkeit von Aluminium-Knetlegierungen. [Fatigue strength of wrought aluminium alloys.] *Vortr. dtsch. Ges. Metallk.*, 1938, pp. 74–77.

2106. RAVILLY, E. Contribution a l'etude de la rupture des fils métalliques soumis a des torsions alternées. [Contribution to the study of the fracture of metallic wires subjected to alternating torsion.] *Publ. sci. Minist. Air*, no. 120, 1938, pp. 187.

2107. REGLER, F. Grundlagen und neue Ergebnisse der röntgenographischen Feingefügeuntersuchung in verformten und ermüdeten Werkstoffen. [Basis of new results of X-ray analysis of deformed and fatigued materials.]

Berg- u. hüttenm. Mh., vol. 86, 1938, pp. 133–145.

2108. ROARK, R. J. Factors of stress concentration for elastic stress (k), repeated stress (k_f) and rupture (k_r). *Prod. Engng*, vol. 9, no. 4, April 1938, pp. 153–156 and no. 5, May 1938, pp. 198–200.

2109. SCHMIDT, M. Einfluss der Vergütung und des Verschmiedungsgrades auf die Biegewechselfestigkeit legierter Baustähle. [Effect of degree of forging and heat treating on the bending fatigue strength of various alloy steels.] *Arch. Eisenhüttenw.*, vol. 11, no. 8, Feb. 1938, pp. 393–400. English translation: *SLA Trans. Centre, John Crerar Lib.*, no. 59-20574.

2110. SEAGER, G. C. *and* TAIT, W. H. Apparatus for high-speed repeated impact testing of metals under separately applied stress. *Engineering, Lond.*, vol. 146, no. 3802, Nov. 25, 1938, p. 611.

2111. SEELIG, R. P. Recent developments in European research on fatigue of metals. *Bull. Amer. Soc. Test. Mat.*, no. 94, Oct. 1938, pp. 23–30 and no. 95, Dec. 1938, pp. 15–22.

2112. SEHRING, J. Ausbildung der dauerfesten Autogennaht. [Improvement of fatigue strength of autogenous joints.] *Autogene Metallbearb.*, vol. 31, no. 4, Feb. 15, 1938, pp. 49–56.

2113. SERENSEN, S. V. Hypotheses of endurance under cyclic loading. [R] *Izv. Akad. Nauk SSSR Otd. tekh. nauk.*, no. 8–9, 1938, pp. 3–16.

2114. SERENSEN, S. V. Theory of strength under variable loading. [U] *Akademii Nauk Ukrain-*

skoi SSR, Naukovi pratsi Instytuta Budivel'noi Mekhaniky, no. 33, 1938, pp. 5–30. German abstr.: *Stahl u. Eisen*, vol. 60, no. 13, March 28, 1940, p. 285.

2115. SERENSEN, S. V. On the endurance of machine components under alternating loads. [R] *M.–L. Izd-vo Akad. Nauk SSSR*, 1938, pp. 37.

2116. SIEBEL, E. Statische und dynamische Werkstoffprüfung. [Static and dynamic testing of materials.] *Vortr. dtsch. Ges. Metallk.*, 1938, pp. 11–15.

2117. STOY, W. Über die Tragfähigkeit der Bauhölzer und Holzverbindungen und über die Dauerfestigkeit der letzteren. [On the bearing strength of structural timbers and timber joints and on the fatigue strength of the latter.] *Bautechnik*, vol. 16, no. 51, Dec. 2, 1938, pp. 689–692.

2118. STREIFF, F. Passungrost. [Fitting rust.] *Schweiz. Arch. angew. Wiss.*, vol. 4, no. 1, 1938, pp. 17–19.

2119. SUTTON, H. *and* TAYLOR, W. J. Corrosion-fatigue properties of MG.7 alloy (DTD 194) with and without protective coatings. *Roy. Aircr. Estab. Rep.*, no. M.T. 5563b, July 1938, pp. 9.

2120. SUTTON, H. *and* TAYLOR, W. J. Corrosion-fatigue properties of magnesium alloys D.T.D. 127 and D.T.D. 129 with and without protective coatings. *Roy. Aircr. Estab. Rep.*, no. M.T. 5563c, July 1938, pp. 13.

2121. THOMPSON, F. C.; KENNEFORD, A. S. *and* SEAGER, G. C. Tensile stresses in bearing metal cast on to a strip and the "fatigue" failure of bearings. *Engineering, Lond.*, vol. 146, no. 3789, Aug. 26, 1938, pp. 235–236 and no. 3791, Sept. 9, 1938, p. 295.

2122. THUM, A. Gewaltbruch, Zeitbruch und Dauerbruch. Bruchaussehen und Bruckverlauf bei Zug-, Biege- und Verdrehbeanspruchung. [Impact, creep and fatigue fracture. The appearance and course of the fracture due to tensile, bending and torsional stresses.] *Forsch-Arb. IngWes.*, vol. 9, no. 2, March–April 1938, pp. 57–67.

2123. THUM, A. Research on materials and modern design. *Engineering, Lond.*, vol. 146, no. 3785, July 29, 1938, pp. 143–146. *Mech. World*, vol. 104, no. 2699, Sept. 23, 1938, pp. 295–296.

2124. THUM, A. "Time resistance" of machine parts. *Mech. World*, vol. 104, no. 2703, Oct. 21, 1938, pp. 393–394, 402.

2125. THUM, A. Querbrüche an Kardanwellen und Behebung ihrer Ursachen. [Transverse fractures of Cardan shafts and the elimination of the causes.] *Dtsch. Kraftfahrtforsch.*, no. 6, 1938, pp. 8.

2126. THUM, A. *and* BRUDER, E. Dauerbruchgefahr an Hohlkehlen von Wellen und Achsen und ihre Verminderung. [Danger of fatigue failure in grooves of shafts and axles and its reduction.] *Dtsch. Kraftfahrtforsch.*, no. 11, 1938, pp. 10.

2127. THUM, A. *and* ERKER, A. Dauerbiegefestigkeit von Kehl- und Stumpfnahtverbindungen. [Bending fatigue strength of fillet and butt welds.] *Z. Ver. dtsch. Ing.*, vol. 82, no. 38, Sept. 17, 1938, pp. 1101–1106. English abstr.: *Weld. J., Easton, Pa.*, vol. 17, no. 11, Nov. 1938, pp. 27s–28s.

2128. THUM, A. *and* JACOBI, H. R. Die Dauerfestigkeit von Kunstharzpressstoffen. [The fatigue strength of moulded synthetic resin plastics.] *Maschinenschaden*, vol. 15, no. 6, 1938, pp. 85–91 and no. 7, 1938, pp. 101–105.

2129. THUM, A. *and* JACOBI, H. R. Die Biegefestigkeit von stahlbewehrtem Panzerholz. [The bending strength of steel covered wood.] *Holz a. Roh- u. Werkst.*, vol. 1, no. 9, June 1938, pp. 335–339.

2130. THUM, A. *and* WEISS, H. Versuche zur Steigerung der Verdrehdauerhaltbarkeit quergebohrter Wellen durch Kaltverformung. [Experiments on increasing the torsional fatigue resistance of shafts with transverse holes by cold working.] *Auto.-tech. Z.*, vol. 41, no. 24, Dec. 25, 1938, pp. 629–633.

2131. THUM, A. *and* WEISS, H. Dauerverdrehversuche an Kurbelwellenstählen. [Torsional fatigue tests on steel crankshafts.] *Zent. wiss. Ber.*, *ForschBer.*, no. 966, 1938, pp. 45.

2132. TIMOSHUK, L. T. Apparatus for testing metals subjected to repeated plastic deformation under the corroding action of the surrounding medium. [R] *Zav. Lab.*, vol. 7, no. 7, July 1938, pp. 819–822.

2133. UNDERWOOD, A. F. Automotive bearing materials and their application. *S.A.E. Jl.*, vol. 43, no. 3, Sept. 1938, pp. 385–392.

2134. VASILEVSKII, P. F. The evaluation of metal fatigue tests using a radiographic method. [R] *Zav. Lab.*, vol. 7, no. 10, 1938, pp. 1214–1216.

2135. VATER, M. Wasserschlag-Dauerversuche an reinem Eisen. [Cavitation fatigue tests on pure iron.] *Z. Ver. dtsch. Ing.*, vol. 82, no. 22, May 28, 1938, pp. 672–674.

2136. VOLK, C. Das erweiterte Wöhlerbild, das neue Spannungsbild. [The amplified Wöhler diagram, the new stress diagram.] *Metallwirtschaft*, vol. 17, no. 44, Nov. 4, 1938, pp. 1167–1169.

2137. WEDEMEYER, E. Die Steigerung der Dauerhaltbarkeit von Schrauben durch Gewindedrücken. [Increasing the fatigue strength of screw by cold rolling.] *Mitt. Wöhler-Inst.*, no. 33, 1938, pp. 3–54. Abstr.: *Metallwirtschaft*, vol. 17, no. 49, Dec. 9, 1938, pp. 1313–1314.

2138. WESCOTT, B. B. Fatigue and corrosion fatigue of steels. *Mech. Engng, N.Y.*, vol. 60, no. 11, Nov. 1938, pp. 813–822, 828. Abstr.: *Metallurgia, Manchr.*, vol. 19, no. 110, Dec. 1938, p. 46.

2139. WESTHOFF, H. Kritische Zusammenstellung der neuesten und wichtigsten Dauerfestigkeitsuntersuchungen von Aluminium-Knetlegierungen. [Critical summary of the latest and most important investigations on the fatigue of wrought aluminium alloys.] *Z. Metallk.*, vol. 30, no. 8, Aug. 1938, pp. 258–265. English translation: *Metal Treatm.*, vol. 4, no. 15, Autumn 1938, pp. 129–134.

2140. WEVER, F.; HEMPEL, M. *and* MÖLLER, H. Die Änderungen des Kristallzustandes wechselbeanspruchter Metalle im Röntgenbild. [The changes in the crystalline state of metals under alternating stressing as revealed by X-rays.] *Arch. Eisenhüttenw.*, vol. 11, no. 7, Jan. 1938, pp. 315–318.

2141. WILLIAMS, F. H. Failures of locomotive parts. *Rly mech. Engr*, vol. 112, Jan. 1938, pp. 15–18.

2142. WILLIAMS, F. H. Railway equipment service failures. *Rly mech. Engr*, vol. 112, May 1938, pp. 174–175, 181 and Aug. 1938, pp. 293–294, 296.

2143. WILLIAMS, F. H. Combination lever service failures. *Rly mech. Engr*, vol. 112, Oct. 1938, pp. 377–379.

2144. WILSON, W. M. Fatigue tests of riveted joints. *Civ. Engng, Easton*, vol. 8, no. 8, Aug. 1938, pp. 513–516.

2145. WILSON, W. M. *and* THOMAS, F. P. Fatigue tests of riveted joints. *Uni. Ill. Engng Exp. Sta. Bull.*, no. 302, 1938, pp. 114. Abstr.: *Engineering, Lond.*, vol. 146, no. 3807, Dec. 30, 1938, p. 773.

2146. ZHUKOV, S. L. The influence of prior stressing on the strength of steel under bending fatigue. [R] *Tekh. vozdush. Flota*, no. 7, 1938, pp. 38–46.

2147. ZIMMERLI, F. P. Relation of Wahl correction factor to fatigue tests on helical compression springs. *Trans. Amer. Soc. mech. Engrs*, vol. 60, no. 1, Jan. 1938, pp. 43–44.

2148. ZIMMERLI, F. P.; WOOD, W. P. *and* WILSON, G. D. The effects of longitudinal scratches on valve spring wire. *Trans. Amer. Soc. Metals*, vol. 26, Dec. 1938, pp. 997–1018.

2149. Passenger car axle tests $5\frac{1}{2}'' \times 10''$ journals. *First progress report, Association of American Railroads, Operation and Maintenance Department, Mechanical Division*, May 1, 1938, pp. 123.

2150. Revolving cantilever beam testing machine. *J. sci. Instrum.*, vol. 15, no. 1, Jan. 1938, pp. 28–29.

2151. Preliminary fatigue studies on aluminium alloy aircraft girders. *Tech. Note nat. Adv. Comm. Aero., Wash.*, no. 637, Feb. 1938, pp. 36.

2152. Rayflex fatigue testing machine. *Automot. Industr., N.Y.*, vol. 78, no. 7, Feb. 12, 1938, pp. 211–212.

2153. Essais de fatigue der métaux par flexion rotative. [Rotating bending fatigue tests on metals.] *Ministere de l'Air, France, Standard*, no. MC AIR 0830, March 23, 1938, pp. 8.

2154. Effect of mat finish on the fatigue strength of duralumin. *Roy. Aircr. Estab. Rep.*, no. M.4657, May 1938, pp. 6.

2155. Dynamic fatigue-testing machines. *Machinery, Lond.*, vol. 52, no. 1344, July 14, 1938. pp. 469–471.

1939

2156. AFANASEV, N. N. Fatigue of boiler-plate iron. [R] *Vest. metalloprom.*, vol. 19, no. 3 1939, pp. 28–34. English abstr.: *J. Iron St Inst.*, vol. 140, no. 2, 1939, pp. 217A–218A

2157. ARNSTEIN, K. *and* SHAW, E. L. Fatigu problems in the aircraft industry. *Metals an Alloys*, vol. 10, no. 7, July 1939, pp. 203–20%

2158. BLANK, H. E. Valve spring material *Automot. Industr. N.Y.*, vol. 80, no. 3, Jan. 2 1939, pp. 72–77.

2159. BLEAKNEY, H. H. Fatigue in theory an practice. *Iron Age*, vol. 143, no. 12, March 2 1939, pp. 29–32, 100.

2160. BLUMBERG, H. Über die Dauerfestigke von Schweissverbindungen. [On the fatig strength of welded joints.] *Autogene Meta bearb.*, vol. 32, no. 16, Aug. 15, 193 pp. 249–253.

2161. BOLLENRATH, F. *and* BUNGARDT, K. Unt suchung über die Korrosions-Ermüdung v Aluminium- und Magnesium-Knetlegieru gen. [Investigation on the corrosion-fatig of aluminium and magnesium wrought alloys.] *Jb. 1939 dtsch. Luftfahrtf.*, pp. I586–I588.

2162. BOLLENRATH, F. *and* BUNGARDT, K. Dauerfestigkeit einiger Leichtmetallknetlegierungen bei verschiedenen Arten der Beanspruchung; Einfluss der Kaltverformung. [Fatigue strength of light metal forging alloys under different types of stressing; influence of cold working.] *Jb. 1939 dtsch. Luftfahrtf.*, pp. I595–I599. *Metallwirtschaft*, vol. 18, no. 1, Jan. 6, 1939, pp. 2–6.

2163. BOLLENRATH, F.; CORNELIUS, H. *and* SIEDENBURG, W. Festigkeitseigenschaften von Leichtmetallschrauben. [Strength properties of light metal screws.] *Z. Ver. dtsch. Ing.*, vol. 83, no. 44, Nov. 4, 1939, pp. 1169–1173.

2164. BRETNEY, C. E. Failure of press tools under fatigue and impact. *Metal Progr.*, vol. 36, no. 6, Dec. 1939, pp. 751–754.

2165. BUCKINGHAM, F. Universal fatigue tester. *Steel*, vol. 104, no. 14, April 3, 1939, pp. 38–40.

2166. BÜHLER, H. Static and fatigue strength of welded reinforcing bars. *Weld. J., Easton, Pa.*, vol. 18, no. 2, Feb. 1939, pp. 49s–51s.

2167. BUSSMANN, K. H. Der Einfluss verschiedenartiger Nachbehandlung auf die Dauerzugfestigkeit gas-schmelzgeschweisster Kesselbleche. [The influence of different kinds of treatment on the tensile-fatigue strength of gas welded joints in boiler plate.] *Wiss. Abh. dtsch. MatPrüfAnst.*, vol. 1, no. 2, 1939, pp. 59–64.

2168. CASSIE, W. F. The fatigue of concrete. *J. Instn civ. Engrs*, vol. 11, no. 4, Feb. 1939, pp. 165–167.

2169. CAZAUD, R. Influence des revêtements métalliques projecteurs sur la résistance à la fatigue, à l'air et sous corrosion de l'acier doux. [The influence of metallic protective coatings on the resistance of mild steel to fatigue in air and in a corrosive medium.] *Chim. et Industr.*, vol. 41, no. 4, special number April 1939, pp. 381–384. English translation: *Henry Brutcher Tech. Trans.*, no. 756.

2170. CLEFF, T. *and* ERKER, A. Dauerhaltbarkeit geschweisster und genieteter Eckverbindungen. [Fatigue resistance of welded and riveted gusset joints.] *Dtsch. Kraftfahrtforsch.*, no. 35, 1939, pp. 13–26. English summary: *Weld. J., Easton, Pa.*, vol. 19, no. 8, Aug. 1940, pp. 311s–312s.

2171. CLYNE, R. W. Some fatigue problems of the railroad industry. *Metals and Alloys*, vol. 10, no. 10, Oct. 1939, pp. 316–323.

2172. CORNELIUS, H. Versuchsmethoden und Versuchseinrichtungen für Kurbelwellen schnellaufender Verbrennungsmaschinen. [Methods and equipment for testing crankshafts of high speed internal combustion engines.] *Auto.-tech. Z.*, vol. 42, April 10, 1939, pp. 190–197. English abstr.: *Metals and Alloys*, vol. 10, Sept. 1939, p. MA 550.

2173. CORNELIUS, H. Berechnung und Gestaltung schnellaufender Kurbelwellen. [Calculation and design of high-speed crankshafts.] *Auto.-tech. Z.*, vol. 42, no. 14, July 25, 1939, pp. 385–393.

2174. CORNELIUS, H. *and* BOLLENRATH, F. Verdreh-Dauerhaltbarkeit von einsatzgehärteten, hohlen Kurbelwellenzapfen mit Innenvestärkung an der Ölbohrung. [The torsional fatigue strength of case-hardened hollow crankpins having a strengthened oil hole.] *Z. Ver. dtsch. Ing.*, vol. 83, no. 48, Dec. 2, 1939, pp. 1257–1258.

2175. DOREY, S. F. Strength of marine engine shafting. *Trans. N.-E. Cst Instn Engrs Shipb.*, vol. 55, 1939, pp. 203–294. Abstr.: *Engineering, Lond.*, vol. 147, no. 3826, May 12, 1939, pp. 563–564. *Metallurgia, Manchr*, vol. 19, no. 114, April 1939, pp. 221–222.

2176. DUCKWITZ, C. A. Zusammenhänge und Auswirkungen der Werkstoff-Forschung für die Konstruktion. [Dependence and results of material research on construction.] *Metallwirtschaft*, vol. 18, no. 6, Feb. 10, 1939, pp. 125–127.

2177. VON ENDE, E. Bemerkungen zur Dauerfestigkeit geschweisster Stabanschlüsse an Fachwerkträgern im Kranbau. [Observations on the fatigue strength of welded joints in fabricated lattice girders in crane construction.] *Wiss. Abh. dtsch. MatPrüfAnst.*, vol. 1, no. 2, 1939, pp. 41–44.

2178. ERKER, A. Beeinflussung der Dauerhaltbarkeit von Rahmenträgen durch Bohrungen, Nietungen und Schweissungen. [The effect of holes, rivets and welds on the fatigue strength of structures.] *Dtsch. Kraftfahrtforsch.*, no. 35, 1939, pp. 1–12.

2179. ERLINGER, E. Une machine d'essais de fatigue par résonance pour efforts de torsion alternée. [An alternating torsion fatigue machine.] *Métaux et Corros.*, vol. 14, no. 161, Jan. 1939, pp. 11–14.

2180. ERLINGER, E. Prüfanlagen zur Ermittlung der Wechselfestigkeit von Maschinenteilen. [Testing machines for determining the fatigue strength of machine parts.] *Arch. Eisenhüttenw.*, vol. 12, no. 12, June 1939, pp. 613–621.

2181. FERCHAUD, A. La rupture des métaux par fatigue vibratoire. Les ressorts de soupape. [The fracture of metals by fatigue. Valve springs.] *Tech. aéro.*, vol. 30, no. 152, 1939, pp. 79–85.

2182. FLACK-TÖNNESSEN, R. Beitrag zur Beurteilung der Wärmespannungen und ihr Einfluss auf die Dauerfestigkeit von Schweissverbindungen. [Contribution to the estimation of heating stresses and their influence on the fatigue strength of welded joints.] *Stahlbau*, vol. 12, no. 23/24, 1939, pp. 166–168.

2183. FÖPPL, O. Von was hängen Fliessbeginn und Bruchfestigkeit eines Werkstoffes ab? [What does the yield strength and ultimate strength of materials depend upon?] *Mitt. Wöhler-Inst.*, no. 35, 1939, pp. 56–70.

2184. FÖPPL, O. Oberflächendrücken zum Zweke der Steigerung der Dauerhaltbarkeit mit Hilfe des Stahlkugelgebläses. [Increasing the fatigue resistance of pins by surface compression using steel shot.] *Mitt. Wöhler-Inst.*, no. 36, 1939, pp. 44–59.

2185. FÖPPL, O. Oberflächendrücken zum Zweke der Steigerung der Dauerhaltbarkeit durch das Stahlkugelgebläse. [Increasing the fatigue resistance of pins by surface compression using blown steel shot.] *Werkzeugmaschine*, vol. 43, no. 8, April 30, 1939, pp. 167–169.

2186. FOX, F. A. Fatigue—a study from the engineer's point of view. *Machinery, Lond.*, vol. 54, no. 1390, June 1, 1939, pp. 265–268.

2187. GALIBOURG, J. Fatigue. Caractéristiques mécaniques et corrosion. [Fatigue. Mechanical characteristics and corrosion.] *Rev. Nickel*, vol. 10, no. 5, Sept.–Oct. 1939, pp. 130–145.

2188. GARF, S. E. A machine for determining the bending fatigue strength of flat test pieces. [R] *Zav. Lab.*, vol. 8, no. 10/11, Oct.–Nov. 1939, pp. 1163–1167. English abstr.: *J. Iron St. Inst.*, vol. 142, no. 2, 1940, p. 66A.

2189. GASSNER, E. Über bisherige Ergebnisse aus Festigkeits-Versuchen im Sinne der Betriebsstatistik. [On earlier results of strength tests treated in terms of operational statistics.] *Berichte no. 106 (part I) Lilienthal Gesellschaft für Luftfahrtforschung*, Jan. 1939, pp. 9–14.

190. GASSNER, E. Festigkeitsversuche mit wiederholter Beansprunchung im Flugzeugbau. [Strength tests with repeated loads in the construction of aircraft.] *Luftwissen*, vol. 6, no. 2, Feb. 1939, pp. 61–64.

191. GEHLER, W. Grundbeziehungen für sie Dauerfestigkeit geschweisster Stabverbindungen und spröder Stoffe im allgemeinen. [The fundamental principles of the fatigue strength of welded test specimens and brittle materials in general.] *Wiss. Abh. dtsch. MatPrüfAnst.*, vol. 1, no. 2, 1939, pp. 1–11.

192. GEIGER, J. Über die Dämpfung bei Gusseisen mit besonderer Berücksichtigung gegossener Kurbelwellen. [The damping capacity of cast iron with special reference to cast crankshafts.] *Mitt. ForschAnst. Gutehoffn., Nürnberg*, vol. 7, Dec. 1939, pp. 215–231. *Giesserei*, vol. 27, Jan. 12, 1940, pp. 1–9 and Jan. 26, 1940, pp. 30–32.

193. GLOCKER, R.; KEMMNITZ, G. and SCHAAL, A. Röntgenographische Spannungsmessung bei dynamischer Beanspruchung. [X-ray methods of stress measurement under dynamic loads.] *Arch. Eisenhüttenw.*, vol. 13, no. 2, Aug. 1939, pp. 89–92.

194. GÖLER, V. and JUNG–KÖNIG, W. Die Erhöhung der Wechselfestigkeit von Leichtmetallen durch Oberflächendrücken. [Increasing the fatigue strength of light metals by surface compression.] *Jb. 1939 dtsch. Luftfahrtf.*, pp. I631–I635.

195. GOETZEL, C. G. Some properties of oxygen-free high conductivity copper (OFHC).

Trans. Amer. Soc. Metals, vol. 27, June 1939, pp. 458–478.

2196. GOODGER, A. H. Some points in the design and inspection of pressure plant. *Chem. and Ind.*, vol. 58, no. 15, April 15, 1939, pp. 352–358.

2197. GOUGH, H. J. and WOOD, W. A. Deformation and fracture of mild steel under cyclic stresses in relation to crystalline structure. *Proc. Instn mech. Engrs, Lond.*, vol. 141, 1939, pp. 175–185.

2198. GOULD, A. J. and EVANS, U. R. A scientific study of corrosion fatigue. Preliminary report of experiments at Cambridge University. *Spec. Rep. Iron St. Inst., Lond.*, no. 24, sect. XI, 1939, pp. 325–342.

2199. GRAF, O. Versuche mit geschweissten Eisenbahnschienen. [Tests on welded steel railway rails.] *Z. Ver. dtsch. Ing.*, vol. 83, no. 48, Dec. 2, 1939, pp. 1250–1253.

2200. GÜRTLER, G.; JUNG-KÖNIG, W. and SCHMID, E. Über die Dauerbewährung der Leichtmetalle bei verschiedenen Temperaturen. [On the fatigue behaviour of light metals at different temperatures.] *Aluminium, Berl.*, vol. 21, no. 3, March 1939, pp. 202–208.

2201. GÜRTLER, G. and SCHMID, E. Temperaturabhängigkeit der Dauerbewährung metallischer Werkstoffe bei ruhender und wechselnder Beanspruchung. [The influence of time and temperature on the static and alternating strength of metallic materials.] *Z. Ver. dtsch. Ing.*, vol. 83, no. 25, June 24, 1939, pp. 749–752. English translation: *Central Electricity Generating Board, London, Translation* no. 2808.

2202. HABART, H. and CAUGHEY, R. H. 18–8: Effect of grain size on fatigue strength. *Metal Progr.*, vol. 35, no. 5, May 1939, pp. 469–470.

2203. HAIGH, B. P. Electric welding as an integral part of structural design. *Trans. N.-E. Cst Instn Engrs Shipb.*, vol. 55, pt. 2, 1939, pp. 43–82. *Engineering, Lond.*, vol. 149, no. 3860, Jan. 5, 1940, pp. 21–25 and no. 3861, Jan. 12, 1940, pp. 49–51.

2204. HAIGH, B. P. *and* ROBERTSON, T. S. Fatigue in structural steel plates with riveted or welded joints. *Trans. Instn nav. Archit., Lond.*, vol. 81, 1939, pp. 84–131. *Engineering, Lond.*, vol. 147, April 14, 1939, pp. 451–453 and April 28, 1939, pp. 513–517. Abstr.: *Engineer, Lond.*, vol. 167, no. 4355, June 30, 1939, p. 793.

2205. HARISS, R. H. Mysteries of fatigue stresses in steel. *Mech. World.* vol. 106, no. 2744, Aug. 4, 1939, pp. 102–103.

2206. HEILMANN, W. Dauerversuche an Drähten und Seilen. [Fatigue tests on wires and wire ropes.] *Wiss. Abh. dtsch. MatPrüfAnst.*, vol. 1, no. 3, 1939, pp. 27–30.

2207. HEMPEL, M. Einfluss der Beanspruchungsart auf die Wechselfestigkeit von Stahlstäben mit Querbohrungen und Kerben. [Influence of manner of loading on the fatigue strength of steel bars with transverse holes and notches.] *Arch. Eisenhüttenw.*, vol. 12, no. 9, March 1939, pp. 433–444.

2208. HEMPEL, M. Einfluss der Probenform, Prüfmaschine und Versuchsführung auf die Wechselfestigkeit. [Effect of shape of specimen, testing machine and test procedure on the fatigue strength.] *Mitt. K.-Wilh.-Inst. Eisenforsch.*, vol. 21, no. 1, 1939, pp. 21–26.

2209. HEMPEL, M. Zur Frage des Dauerbruches: Magnetpulverbild und Dauerbruchanrisse. [On the question of fatigue fracture: magnetic powder diagrams and fatigue cracks.] *Mitt. K.-Wilh.-Inst. Eisenforsch.*, vol. 21, no. 9,

1939, pp. 147–162. German abstr.: *Stahl u Eisen*, vol. 59, no. 23, June 8, 1939, pp. 692–693.

2210. HEMPEL, M. *and* ARDELT, F. Verhalten de Stahles in der Wärme unter Zugdruck-Wechselbeanspruchung. [Behaviour of steel at high temperature under tension-compression fatigue stresses.] *Mitt. K.-Wilh.-Inst. Eisen forsch.*, vol. 21, no. 7, 1939, pp. 115–132 German abstr.: *Z. Ver. dtsch. Ing.*, vol. 83 no. 40, Oct. 7, 1939, pp. 1109–1111.

2211. HEMPEL, M. *and* ARDELT, F. Verhalten vo Stahl gegen Zug-Druck-Wechselbeanspru chung bei 500°. [Behaviour of steel under ten sion-compression fatigue stresses at 500° *Arch. Eisenhüttenw.*, vol. 12, no. 11, May 1939 pp. 553–564.

2212. HENDERSON, O. *and* SEENS, W. B. Prepa ration of rotating beam fatigue specimen *Metals and Alloys*, vol. 10, no. 3, March 1939 pp. 82–84.

2213. HORGER, O. J. *and* NEIFERT, H. R. Fatigu strength of machined forgings 6 to 7 in. di meter. *Proc. Amer. Soc. Test. Mater.*, vol. 39 1939, pp. 723–740. *Heat Treat. Forg.*, vol. 2 no. 7, July 1939, pp. 329–334.

2214. HOUDREMONT, E. Einige Gesichtspunk zur Entwicklung der Forschung auf de Gebiet der Dauerfestigkeit. [Some notes c the development of fatigue strength inves gations.] *Tech. Mitt. Krupp A*, vol. 2, 193 appendix pp. 1–8. English abstr.: *J. Iron S Inst.*, vol. 141, no. 1, 1940, p. 251A.

2215. HOWAT, D. D. *and* MITCHELL, E. A fatig failure with some unusual characteristi *Iron and Steel*, vol. 12, no. 8, April 193 pp. 376–380.

2216. IGARASHI, I. *and* FUKAI, S. Influence size, surface finish and sand blasting of spe men on results of fatigue tests on duralumin and super-duralumin. [J] *Trans. Soc. mech. Engrs, Japan*, vol. 5, no. 18, Feb. 1939, pp. 30–35.

217. IGARASHI, I. *and* FUKAI, S. On the relation between fatigue strength and elastic limit. I—for cold drawn materials of various reduction percentages. [J] *Trans. Soc. mech. Engrs, Japan*, vol. 5, no. 20, Aug. 1939, pp. 3–16. *Sumit. Kinz. Kog. Kenk.*, vol. 3, no. 6, 1939, pp. 553–576.

218. ITIHARA, M. Fixed-stress type hysterograph and torsion fatigue tests of duralumin and nickel–chromium–tungsten steel. [J] *Nippon kink. Gakk.*, vol. 3, no. 1, 1939, pp. 14–21.

219. KARPOV, A. V. Fatigue problems in structural designs. *Metals and Alloys*, vol. 10, no. 11, Nov. 1939, pp. 346–352 and no. 12, Dec. 1939, pp. 381–388.

220. KAUTZ, K. Über die Dauerfestigkeit von Stumpf- und Kehlnahtverbindungen. [On the fatigue strength of butt and fillet welded joints.] *Elektroschweissung*, vol. 10, no. 4, April 1939, pp. 74–76.

221. KEMNITZ, G. Röntgenographische Spannungsmessung am Dauerbruchvorgang. [Stress measurements by X-rays in the process of fatigue failure.] *Z. tech. Phys.*, vol. 20, no. 5, May 1939, pp. 129–140.

222. KIDANI, Y. On the fatigue of metals and internal friction. [J] *Trans. Soc. mech. Engrs, Japan*, vol. 5, no. 18, Feb. 1939, pp. 36–41.

223. KIES, J. A. *and* QUICK, G. W. Effect of service stresses on impact resistance, X-ray diffraction patterns, and microstructure of 5S aluminium alloy. *N.A.C.A. Rep.* no. 659, 1939, pp. 22.

2224. KÖRBER, F. Über den Dauerbruch metallischer Werkstoffe. [On the fatigue fracture of metallic materials.] *Schr. dtsch. Akad. Luftfahrtf.*, no. 15, 1939, pp. 1–53.

2225. KÖRBER, F. Das Verhalten metallischer Werkstoffe im Bereich kleiner Verformungen. [The behaviour of metallic materials in the range of small deformations.] *Stahl u. Eisen*, vol. 59, no. 21, May 25, 1939, pp. 618–626.

2226. KÖRBER, F. *and* HEMPEL, M. Zugdruck-, Biege- und Verdrehwechselbeanspruchung an Stahlstäben mit Querbohrungen und Kerben. [Tension-compression, bending and torsion fatigue stressing of steel bars with holes and notches.] *Mitt. K.-Wilh.-Inst. Eisenforsch.*, vol. 21, no. 1, 1939, pp. 1–19. *Arch. Eisenhüttenw.*, vol. 12, no. 9, March 1939, pp. 433–444. English summary: *Metallurgia, Manchr*, vol. 22, no. 131, Sept. 1940, pp. 139–140.

2227. KUDRYAVTSEV, I. V. *and* CHERNYAK, V. S. The fatigue strength of steels at low temperatures. [R] *Vest. metalloprom.*, vol. 19, no. 12, 1939, pp. 40–44. English abstr.: *J. Iron St. Inst.*, vol. 143, no. 1, 1941, p. 183A.

2228. LAUDERDALE, R. H.; DOWELL, R. L. *and* CASSELMAN, K. Endurance of annealed gold and a quenched dental alloy. *Metals and Alloys*, vol. 10, no. 1, Jan. 1939, pp. 24–25.

2229. LEHR, E. Nutzbarmachung der Ergebnisse der neueren Festigkeitsforschung für den Konstrukteur. [The application of new strength investigations to practice.] *Metallwirtschaft*, vol. 18, July 7, 1939, pp. 595–599 and July 14, 1939, pp. 617–623.

2230. LEHR, E. *and* BUSSMANN, K. H. Dauer-Zugfestigkeit von Stabköpfen. [Tension fatigue strength of specimen ends.] *Z. Ver. dtsch. Ing.*, vol. 83, no. 18, May 6, 1939, pp. 513–514.

2231. LLOYD, W. S. *and* SIMPSON, A. W. Exhibit "H1"—corrosion fatigue of drill pipe. *Proc. Amer. Petrol. Inst.*, sect. IV, vol. 20, no. 224, 1939, pp. 114–119.

2232. LOEPELMANN, F. Einfluss einer Oberflächenvorbehandlung und Kaltverformung auf die Dauerbiegefestigkeit einer Legierung der Gattung Al DIN 1713. [Influence of surface treatment and cold forming on the bending fatigue strength of alloys of the type Al DIN 1713.] *Zent. wiss. Ber., Untersuch. Mitt.*, no. 568, 1939, pp. 7.

2233. MAILÄNDER, R. Ergebnisse von Dauerversuchen an Stählen. [Results of fatigue tests on steels.] *Tech. Mitt. Krupp A.*, vol. 2, 1939, appendix pp. 8–15.

2234. MAILLARD, P. Les contraintes thermiques dans les parois de la chambre de combustion des moteurs à explosion. [Thermal strains in the walls of the combustion chambers of explosion motors.] *Chal. et Industr.*, vol. 20, 1939, pp. 161–167.

2235. ZOEGE VON MANTEUFFEL, R. Versuche über die Dauerhaltbarkeit von Federn. Der Einfluss der Formgebung auf die Dauerhaltbarkeit verdrehbeanspruchter Federn. [Tests on the fatigue life of springs. The influence of forming on the fatigue strength of torsionally strained springs.] *Dtsch. Kraftfahrtforsch. Zwischenbericht.*, no. 69, 1939, pp. 35.

2236. MECHLING, W. B. *and* JACK, S. S. Some mechanical tests on aluminium alloys 14S-T and 24S-T. *Proc. Amer. Soc. Test. Mater.*, vol. 39, 1939, pp. 769–779.

2237. MIZUTA, Z. On the safety factor of winding rope. [J] *Mem. Ryoj. Coll. Engng*, vol. 12, 1939, pp. 115–203. [English summary pp. 199–203.]

2238. MOORE, H. F. Fatigue of metals—developments in the United States. *Metals and Alloys* vol. 10, no. 5, May 1939, pp. 158–162 an no. 6, June 1939, pp. 180–183.

2239. MOORE, H. F. Ten years of metallurgic progress, 1929 to 1939; endurance limit unde repeated stress. *Metals and Alloys*, vol. 1 no. 10, Oct. 1939, p. A.70.

2240. MOORE, R. R. Strength of screw threa under tensile fatigue and impact loads. *Pro Engng*, vol. 10, no. 11, Nov. 1939, pp. 475–47

2241. MOORE, R. R. High speed fatigue testin *Steel*, vol. 105, no. 22, Nov. 27, 193 pp. 40–41.

2242. MÜLLER, W. Verschleissversuch an Schrau verbindungen mit Schrauben aus ein Aluminiumlegierung und mit Parker-Kalo Stahl-Schrauben. [Wear tests on screw joints using aluminium alloy screws a Parker-Kalon steel screws.] *Aluminium, Ber* vol. 21, no. 1, 1939, pp. 37–43. English tra lation: *Roy. Aircr. Estab. Lib. Trans.*, no. 4

2243. MÜLLER, W. Influenza dei fori e de chiodatura sulla resistenza a fatica di lami e profilati in lega di alluminio. [The influen of holes and of riveting on the fatigue stren of aluminium alloy rolled sheets and section *Alluminio*, vol. 8, no. 2, March–April 19 pp. 76–81.

2244. MÜLLER, W. Massnahmen zur Verbes rung der Ermüdungsfestigkeit genieteter K tenpunk-Verbindungen aus Aluminiumle rungen für den Flugzeug-, Karosserie- Kranenbau. [Investigations on improving fatigue strength of riveted joints in alumini alloys for aircraft, coachwork and crane str tures.] *Schweiz. Arch. angew. Wiss.*, vol no. 10, Oct. 1939, pp. 294–307.

2245. MUZZOLI, M. Saggi di resistenza alla fatica con particolare riguardo agli acciai per assali ferroviari e tramviari. [Studies on fatigue resistance with particular regard to the steel axles of trains and trams.] *Ric. di Ing.*, vol. 7, no. 1, Jan.–Feb. 1939, pp. 1–37.

2246. NAGASHIMA, K. A study of fatigue limit of shafts at force-fitted parts. [J] *Trans. Soc. mech. Engrs, Japan*, vol. 5, no. 18, Feb. 1939, pp. 80–86.

2247. NAUNTON, W. J. S. *and* WARING, J. R. S. Fatigue in rubber. Part II. *Rubb. Chem. Technol.*, vol. 12, no. 2, April 1939, pp. 332–343.

2248. NAUNTON, W. J. S. *and* WARING, J. R. S. Fatigue in rubber. Part III—fatigue and reinforcement. *I.R.I. Trans.*, vol. 14, no. 6, April 1939, pp. 340–364. *Rubb. Chem. Technol.*, vol. 12, no. 4, Oct. 1939, pp. 845–860.

2249. NEPOTI, A. Considerazioni sul limite di fatica delle strutture in acciaio con particolare riguardo alle strutture saldate in aeronautica —confronti con le costruzioni in duraluminio. [Considerations of the fatigue limit of steel structures with special reference to welded aircraft structures—comparison of steel and duralumin structures.] *Metallurg. ital.*, vol. 31, no. 5, May 1939, pp. 295–307.

2250. NISHIHARA, T. *and* KAWAMOTO, M. The detection of the fatigued zone by a corrosion method. [J] *Trans. Soc. mech. Engrs, Japan*, vol. 5, no. 20, Aug. 1939, pp. 43–50.

2251. NISHIHARA, T. *and* KOJIMA, K. Diagram of endurance limit of duralumin for repeated tension and compression. [J] *Trans. Soc. mech. Engrs, Japan*, vol. 5, no. 20, Aug. 1939, pp. 1–2.

2252. NISHIHARA, T. *and* KORI, T. Further experiments on the pitting of metals. [J] *Trans. Soc. mech. Engrs, Japan*, vol. 5, no. 21, Nov. 1939, pp. 89–97.

2253. NISHIHARA, T. *and* SAKURAI, T. Fatigue strength of steel for repeated tension and compression. [J] *Trans. Soc. mech. Engrs, Japan*, vol. 5, no. 18, Feb. 1939, pp. 25–29.

2254. NISHIHARA, T.; SAWARAGI, Y. *and* TAGA, Y. Relation between the speed of strain and the deflection due to internal friction of metals. [J] *Trans. Soc. mech. Engrs, Japan*, vol. 5, no. 20, Aug. 1939, pp. 34–43.

2255. ONO, M. *and* ONO, A. Experimental investigation of the effect of notches on the fatigue strength of steel. [J] *Trans. Soc. mech. Engrs, Japan*, vol. 5, no. 19, May 1939, pp. 167–174.

2256. O'NEILL, H. *and* JOHANSEN, F. C. Endurance tests on special joints with heat treated or machined welds. *Trans. Inst. Weld.*, vol. 2, no. 4, Oct. 1939, pp. 222–225.

2257. OROWAN, E. Theory of the fatigue of metals. *Proc. roy. Soc. (A)*, vol. 171, no. 944, 1939, pp. 79–106.

2258. PESKIN, L. C. Studies on conductor vibration. *Bull. Edison elect. Inst.*, vol. 7, no. 8, Aug. 1939, pp. 404–414.

2259. PETERSON, R. E. Fatigue problems in the electrical industry. *Metals and Alloys*, vol. 10, no. 9, Sept. 1939, pp. 276–279.

2260. PREVER, V. S. *and* LOCATI, L. Comportamento alla flessione rotante di provette con anelli variamente calettati. [Behaviour in rotating bending of specimens with various coupling rings.] *Industr. mecc.*, vol. 21, no. 3, 1939, pp. 199–203 and no. 4, 1939, pp. 294–300.

2261. REGLER, F. Verformung und Ermüdung metallischer werkstoffe im Röntgenbild. [Deformation and fatigue of metallic materials as detected by X-rays.] *ForschArb. Metallk.*, no. 26, 1939, pp. 98. English translation: *R.T.P. Trans.*, no. 1179, 1939.

2262. REICHEL, K. Spot welding light metals in a German aircraft works. *Weld. J., Easton, Pa.*, vol. 18, no. 5, May 1939, pp. 182–184.

2263. ROELIG, H. Dynamische Bewertung der Dämpfung und Dauerfestigkeit von Vulkanisation. [Dynamic estimation of the damping and fatigue strength of vulcanisation.] *Kautschuk*, vol. 15, no. 1, 1939, pp. 7–10 and no. 2, 1939, pp. 32–34.

2264. Roš, M. *and* BRANDENBERGER, E. Erfahrungen mit röntgendurchstrahlten, geschweissten Druckleitungen und deren festigkeitstechnische Sicherheit. [Experiences with radiographically examined welded pressure piping and their strength and safety.] *Ber. eidgenöss. MatPrüfAnst.*, no. 122, May 1939, pp. 26.

2265. SACHS, G. Improving aircraft propellers by surface rolling. *Metals and Alloys*, vol. 10, no. 1, Jan. 1939, pp. 19–23. Summary: *Metallurgia, Manchr*, vol. 19, no. 114, April 1939, p. 228.

2266. SACHS, G. Internal stresses in piston-rods of a large diesel engine ocean liner. *Trans. Amer. Soc. Metals*, vol. 27, Sept. 1939, pp. 821–836.

2267. SAXTON, R. Wire rope fatigue. *Mech. Handl.*, vol. 26, no. 8, Aug. 1939, pp. 241–242.

2268. SCHERER, R. Geschmiedete Arbeitswalzen für Kaltwalzwerke und ihre Herstellung. [Ductility of rolls for cold working and their manufacture.] *Stahl u. Eisen*, vol. 59, no. 40, Oct. 5, 1939, pp. 1105–1111.

2269. SCHRADER, H. Über praktische Erfahrungen in der Anwendung des Magnetpulververfahrens zur Rissprüfung. [Practical experience in the application of the magnetic powder method for the discovery of cracks.] *Tech. Mitt. Krupp A.*, vol. 2, 1939, appendix pp. 18–28.

2270. SCHUSTER, L. W. The relationship between the mechanical properties of materials and the liability for failure in service. *Trans. Lpool Engng Soc.*, vol. 61, 1939–1940, pp. 36–68. *Bull. Lpool Engng Soc.*, vol. 13, no. 4, Nov. 1939, pp. 8–47. *Metallurgia, Manchr*, vol. 19, no. 109, Nov. 1938, pp. 25–28; no. 110, Dec. 1938, pp. 51–54 and no. 111, Jan. 1939, pp. 91–96.

2271. SERENSEN, S. V. The development and design of modern fatigue testing machines. [R] *Zav. Lab.*, vol. 8, no. 8, Aug. 1939, pp. 845–856.

2272. SERENSEN, S. V. New machines for fatigue testing. [R] *Zav. Lab.*, vol. 8, no. 12, Dec. 1939, pp. 1291–1297.

2273. SIMPSON, A. W. Exhibit "H2"—corrosion fatigue of drill pipe. *Proc. Amer. Petrol. Inst.* sect. IV, vol. 20, no. 224, 1939, pp. 119–123.

2274. SMITH, J. O. The effect of range of stress on the torsional fatigue strength of steel. *Un. Ill. Engng Exp. Sta. Bull.*, no. 316, 1939, pp. 35.

2275. SPENCER, R. G. An X-ray study of the changes that occur in malleable iron during the process of fatiguing. *Phys. Rev.*, vol. 55, no. 11, June 1, 1939, pp. 991–994.

2276. TEICHMANN, A. Grundsätzliche Betrachtungen über Festigkeitsversuche im Sinne der Betriebsstatistik. [Systematic considerations of strength tests as related to service statistics.] *Berichte no. 106 (part I) Lilienthal Gesellschaft für Luftfahrtforschung*, Jan. 1939, pp. 6–8.

2277. TEMPLIN, R. L. Fatigue of light metal alloys. *Metals and Alloys*, vol. 10, no. 8, Aug. 1939, pp. 243–245. *Metal Ind., Lond.*, vol. 55, no. 14, Oct. 6, 1939, pp. 315–317.

2278. TEMPLIN, R. L. Fatigue machines for testing structural units. *Proc. Amer. Soc. Test. Mater.*, vol. 39, 1939, pp. 711–722. Summary: *Mech. World*, vol. 106, no. 2745, Aug. 11, 1939, pp. 128–129.

2279. THUM, A. Der Werkstoff in der konstruktiven Berechnung. [Materials for structural design.] *Stahl u. Eisen*, vol. 59, no. 9, March 2, 1939, pp. 252–263.

2280. THUM, A. *and* BRUDER, E. Gestaltung und Dauerhaltbarkeit von geschlossenen Stabköpfen und ähnlichen Bauteilen. [Design and fatigue strength of bar ends and similar parts.] *Dtsch. Kraftfahrtforsch.*, no. 20, 1939, pp. 10. German abstr.: *Z. Ver. dtsch. Ing.*, vol. 83, no. 40, Oct. 7, 1939, p. 1111.

2281. THUM, A. *and* ERKER, A. Wechselverdrehfestigkeit von Kehlnahtverbindungen. [The torsional fatigue strength of fillet welds.] *Elektroschweissung*, vol. 10, no. 11, Nov. 1939, pp. 205–209.

2282. THUM, A. *and* ERKER, A. Bemessung von Kehlnahtverbindungen bei Wechselbiegebeanspruchung. [The dimensioning of fillet welds by bending fatigue tests.] *Z. Ver. dtsch. Ing.*, vol. 83, no. 51, Dec. 23, 1939, pp. 1293–1297.

2283. THUM, A. *and* FEDERN, K. *Spannungszustand und Bruchausbildung. [Stress conditions and crack development.]* Berlin, Julius Springer, 1939, pp. 78.

2284. THUM, A. *and* JACOBI, H. R. Mechanische Festigkeit von Phenol-Formaldehyd-Kunststoffen. [Mechanical strength of phenol-formaldehyde plastics.] *Forschungsh. Ver dtsch. Ing.*, no. 396, 1939, pp. 39.

2285. THUM, A. *and* JACOBI, H. R. Festigkeitseigenschaften von hochfesten Kunstharz-Pressstoffen. [Strength properties of highly rigid moulded plastics.] *Z. Ver. dtsch. Ing.*, vol. 83, no. 37, Sept. 16, 1939, pp. 1044–1048.

2286. TOMLINSON, G. A.; THORPE, P. L. *and* GOUGH, H. J. An investigation of the fretting corrosion of closely-fitting surfaces. *Proc. Instn mech. Engrs, Lond.*, vol. 141, 1939, pp. 223–249. *Engineer, Lond.*, vol. 167, no. 4339, March 10, 1939, pp. 315, 324–326 and no. 4340, March 17, 1939, pp. 355–356. Abstr.: *Engineering, Lond.*, vol. 147, no. 3817, March 10, 1939, pp. 293–295. *Metal Progr.*, vol. 35, no. 5, May 1939, pp. 468, 510. *Mech. Engng, N.Y.*, vol. 61, no. 5, May 1939, pp. 386–387.

2287. VATER, M. *and* SORDERGER, W. Wersktoff-Zerstörung durch Wasserschlag bei Dauer- und Einzelschlag-Beanspruchung. [Destruction of material by water impact under fatigue and single impact stressing.] *Z. Ver. dtsch. Ing.*, vol. 83, no. 24, June 17, 1939, pp. 725–728.

2288. VOLK, C. Dauerfestigkeit und Belastungsgrenze geschweisster Proben. [Fatigue strength and load limits of welded specimens.] *Elektroschweissung*, vol. 10, no. 3, March 1939, pp. 54–57.

2289. Volk, C. Lebensdauer der Bauteile — Ausnutzung der Werkstoffes. [Durability of structural members — efficient use of materials.] *Metallwirtschaft*, vol. 18, July 21, 1939, pp. 636–639.

2290. Wampler, C. P. *and* Alleman, N. J. Fatigue tests of wire. *Bull. Amer. Soc. Test. Mat.*, no. 101, Dec. 1939, pp. 13–18. *Wire and Wire Prod.*, vol. 14, Nov. 1939, pp. 649–653.

2291. Way, S. How to reduce surface fatigue. *Mach. Design*, vol. 11, no. 3, March 1939, pp. 42–45.

2292. Wever, F.; Hempel, M. *and* Möller, H. Die Veränderungen des Kristallzustandes von Stahl bei Wechselbeanspruchung bis zum Dauerbruch. [Changes in the crystal structure of steel during alternating loading and at fatigue fracture.] *Stahl u. Eisen*, vol. 59, no. 2, Jan. 12, 1939, pp. 29–33.

2293. Wever, F. *and* Martin, G. Verhalten Spannungsbehafteter Werkstücke bei Wechselbeanspruchung. [Behaviour of residual stresses in components under alternating loads.] *Mitt. K.-Wilh.-Inst. Eisenforsch.*, vol. 21, no. 14, 1939, pp. 213–218.

2294. White, A. E. Changes in a high pressure drum to eliminate recurrence of cracks due to corrosion fatigue. *Trans. Amer. Soc. mech. Engrs*, vol. 61, no. 6, Aug. 1939, pp. 507–516.

2295. Wiegand, H. Hartverchromung von Stahl und Leichtmetall unter Berücksichtigung ihres Einflusses auf die Dauerfestigkeit. [Hard chroming of steel and light metal with consideration of the influence on fatigue strength.] *Jb. 1939 dtsch. Luftfahrtf.*, pp. II322–II326.

2296. Wiegand, H. Mutterwerkstoff und Dauerhaltbarkeit der Schraubenverbindungen. [Nut material and the fatigue life of bolted joints.] *Z. Ver. dtsch. Ing.*, vol. 83, no. 2, Jan. 14, 1939, pp. 64–65.

2297. Wiegand, H. Innere Kerbwirkung und Dauerfestigkeit. [Internal notch effect and fatigue strength.] *Metallwirtschaft*, vol. 18, no. 4, Jan. 27, 1939, pp. 83–85.

2298. Wiegand, H. *and* Göbel, F. Temperatureinflüsse in Schwingungsbeanspruchten Gummifedern. [Influence of temperature on the fatigue strength of rubber springs.] *Dtsch. MotorZ.*, vol. 16, no. 9, 1939, pp. 278–282.

2299. Wiegand, H. *and* Scheinost, R. Einfluss der Einsatzhärtung auf die Biege- und Verdrehwechselfestigkeit von glatten und quergebohrten Probestäben. [Influence of case hardening on the bending and torsional fatigue strength of plain and transversely drilled test specimens.] *Arch. Eisenhüttenw.*, vol. 12, no. 9, March 1939, pp. 445–448. Abstr.: *Z. Ver. dtsch. Ing.*, vol. 83, no. 39, Sept. 30, 1939, p. 1089.

2300. Wiegand, H. *and* Scheinost, R. Dauerfestigkeit hartverchromter Teile. [The fatigue strength of hard chromium plated components.] *Z. Ver. dtsch. Ing.*, vol. 83, no. 21, May 27, 1939, pp. 655–659.

2301. Wilson, W. M. *and* Coombe, J. V. Fatigue tests of connection angles. *Uni. Ill. Engng Exp. Sta. Bull.*, no. 317, 1939, pp. 22.

2302. Wilson, W. M. *and* Wilder, A. B. Fatigue tests of butt welds in structural steel plates. *Uni. Ill. Engng Exp. Sta. Bull.*, no. 31 (vol. 36, no. 42), 1939, pp. 58.

2303. Zimmermann, J. H. Flame treating. *Heat Treat. Forg.*, vol. 25, no. 12, Dec. 1939 pp. 609–615.

2304. Strength of spline shafts; a survey of published information. *Rep. Instn Auto. Engrs Res. Comm.*, no. 9161, July 1939, pp. 12.

2305. National Physical Laboratory, Engineering Department. Fatigue properties of materials. *Engineering, Lond.*, vol. 148, no. 3838, Aug. 4, 1939, p. 144.

2306. Stress concentration effects in fatigue. *Metal Treatm.*, vol. 5, no. 19, Autumn 1939, pp. 108–114.

2307. Avery-Schenck Push-Pull fatigue testing machine. *Engineering, Lond.*, vol. 148, no. 3847, Oct. 6, 1939, p. 379.

1940

2308. AFANASEV, N. N. A statistical theory of the fatigue strength of metals. [R] *Zh. tekh. Fiz.*, vol. 10, no. 19, 1940, pp. 1553–1568.

2309. BAILEY, R. W. Plastic yielding and fatigue of ductile metals—a criterion and its application. *Proc. Instn mech. Engrs. Lond.*, vol. 143, 1940, pp. 101–107.

310. BARDGETT, W. E. Modern methods of mechanical testing. *J. W. Scot. Iron St. Inst.*, vol. 47, Jan. 1940, pp. 47–66.

311. BAUTZ, W. Untersuchungen über die Schadenslinie bei Leichtmetallen. [Investigation of the damage line in light metals.] *Z. Ver. dtsch. Ing.*, vol. 84, no. 47, Nov. 23, 1940, pp. 918–919.

312. BECK, E. H. *The technology of magnesium and its alloys.* London, F. A. Hughes and Co. Ltd., 1940, pp. 512. [pages 200–238 are devo-
ted to "Fatigue Strength" of magnesium alloys.]

2313. BERTRAM, W. Die Dauerhaltbarkeit von Gewinden bei verschiedenen Temperaturen und ihre Beeinflussung durch Oberflächendrücken. [The fatigue life of screw threads at different temperatures and the influence of surface pressing.] *Mitt. Wöhler-Inst.*, no. 37, 1940, pp. 3–52. *Werkzeugmaschine*, vol. 44, no. 23, Dec. 1, 1940, pp. 512–515.

2314. BOLLENRATH, F. Einflüsse auf die Zeit- und Dauerfestigkeit der Werkstoffe. [The effects of the endurance and fatigue strength of materials.] *Luftfahrtforsch.*, vol. 17, no. 10, Oct. 26, 1940, pp. 320–328. English translation: *Tech. Memor. nat. Adv. Comm. Aero., Wash.*, no. 987. Sept. 1941, pp. 20.

2315. BOLLENRATH, F. *and* CORNELIUS, H. Der Einfluss von Betriebspausen auf die Zeit- und Dauerfestigkeit metallischer Werkstoffe. [The effect of rest periods on the endurance and fatigue strength of metallic materials.] *Z. Ver. dtsch. Ing.*, vol. 84, no. 18, May 4, 1940, pp. 295–299. English translation: *Saf. Min. Res. Estab., Lond., Trans. no.* 4115, 1958.

2316. BOLLENRATH, F. *and* CORNELIUS, H. Zeit- und Dauerfestigkeit einfach gestalteter metallischer Bauteile. [Endurance and fatigue strength of simple structural parts.] *Z. Ver. dtsch. Ing.*, vol. 84, no. 24, June 15, 1940, pp. 407–412. English abstr.: *Weld. J., Easton, Pa.*, vol. 20, no. 3, March 1941, pp. 159s–160s.

2317. BOLLENRATH, F. *and* CORNELIUS, H. Einfluss des Oberflächendrückens auf die Verdrehzeitfestigkeit quergebohrter Leichtmetall-Wellen. [Effect of surface compression on the torsional fatigue strength of light metal shafts with transverse holes.] *Z. Metallk.*, vol. 32, no. 7, July 1940, pp. 249–252.

2318. BOLLENRATH, F. *and* CORNELIUS, H. Verdrehwechselfestigkeit von Wellen aus unlegiertem und legiertem Stahl. [Torsional fatigue strength of shafts of carbon and alloy steels.] *Arch. Eisenhüttenw.*, vol. 14, no. 6, Dec. 1940, pp. 283–287.

2319. BRICK, R. M. *and* PHILLIPS, A. Fatigue studies of aircraft sheet metals. *Heat Treat. Forg.*, vol. 26, no. 11, Nov. 1940, pp. 551–554 and no. 12, Dec. 1940, pp. 598–600.

2320. CLARK, F. H. Developments in fatigue, creep, age-hardening, diffusion, microscopy, borocarbides, powders, electrodeposition and die castings. *Min. and Metall. N.Y.*, vol. 21, no. 397, Jan. 1940, pp. 18–23.

2321. CORNELIUS, H. Verhalten einiger Kurbel-wellen-Gusswerkstoffe bei Dauerbeanspru-chung durch Biegung und Verdrehung. [Behaviour of some cast crankshaft materials under fatigue stressing in bending and torsion.] *Giesserei*, vol. 27, no. 25, Dec. 13, 1940, pp. 491–499. English abstr.: *Engrs' Dig.*, vol. 2, no. 5, May 1941, p. 206.

2322. CORNELIUS, H. *and* BOLLENRATH, F. Kerbwirkungszahl kaltgereckter Stähle bei Biegewechselbeanspruchung. [Notch sensitiv-ity coefficient of cold worked steels under bending fatigue stress.] *Arch. Eisenhüttenw.*, vol. 14, no. 6, Dec. 1940, pp. 289–292.

2323. COTTON, A. F.; MATHEWS, F. M. *and* FRASER, N. C. Introductory study of fatigue in steels. *Heat Treat. Forg.*, vol. 26, no. 3, March 1940, pp. 130–135.

2324. DAEVES, K.; GEROLD, E. *and* SCHULZ, E. H. Beeinflussung der Lebensdauer wechsel-beanspruchter Teile durch Ruhepausen.[Effect of rest periods on fatigue life of components.] *Stahl u. Eisen*, vol. 60, no. 5, Feb. 1, 1940, pp. 100–103. *Jb. 1940 dtsch. Luftfahrtf.*, pp. II343–II348.

2325. DALE, D. R. *and* JOHNSON, D. O. Endur-ance of sucker rods. *Metal Progr.*, vol. 37, no. 6, June 1940, pp. 680, 686.

2326. DANSE, L. A. Surface decarburization. *Steel*, vol. 106, no. 6, Feb. 5, 1940, pp. 60–61.

2327. DAYTON, R. W. High surface finishes; preparation, specification and effect on wear and fatigue. *Metal Progr.*, vol. 38, no. 1, July 1940, pp. 54–56.

2328. DEHLINGER, U. Dauerfestigkeit, Wechsel-festigkeit und ihr Zusammenhang mit der wahren Kriechgrenze. [Fatigue strength, endurance limit, and their relation to the true creep limit.] *Z. Metallk.*, vol. 32, no. 6, June 1940, pp. 199–200. English translation: *R.T.P. Trans.*, no. 1943. English abstr.: *Auto. Engr*, vol. 33, no. 436, May 1943, p. 202.

2329. DEHLINGER, U. Zur Theorie der Wech-selfestigkeit. [The theory of the endurance limit.] *Z. Phys.*, vol. 115, no. 11/12, 1940, pp. 625–638.

2330. DEICHMÜLLER, F. Über die Bestimmung der Wechselfestigkeit der Werkstoffe im Biege-Zug-Versuch. [On the determination of the fatigue strength of materials in bending tension tests.] *Die Abnahme — Sonderteil de Anzeigers für Maschinenwesen für die Abnahm von Werkstoffen und Betriebsbedarf*, vol. 3 no. 6, June 11, 1940, pp. 41–44.

2331. DITTRICH, W. Statische und dynamisch Untersuchungen von Schraubensicherunger [Static and dynamic investigations of screwe devices.] *Zbl. Mech.*, vol. 9, 1940, pp. 25–2

2332. DOLAN, T. J. *and* BENNINGER, H. H. Th effect of protective coatings on the corrosio fatigue strength of steel. *Proc. Amer. Soc. Tes Mater.*, vol. 40, 1940, pp. 658–669. *He*

Treat. Forg., vol. 26, no. 7, July 1940, pp. 326–330. *Mach. Design*, vol. 12, no. 9, Sept. 1940, pp. 52–54, 97.

2333. DOLAN, T. J. *and* PRICE, B. R. Properties and machinability of a leaded steel. *Metals and Alloys*, vol. 11, no. 1, Jan. 1940, pp. 20–27.

2334. DONALDSON, J. W. The fatigue and impact strengths of cast iron. *Found. Tr. J.*, vol. 63, no. 1251, Aug. 8, 1940, pp. 91–93, 96.

2335. DRAKE, H. C. Is the transverse fissure problem growing? *Rly Age, Chicago*, vol. 108, no. 6, Feb. 10, 1940, pp. 277–281.

2336. DURANT, L. B. *and* ENNIS, J. F. Investigation of the fatigue strength of weld metal and welded butt joints in the as-welded and stress-relieved conditions. *Weld. J., Easton, Pa.*, vol. 19, no. 2, Feb. 1940, pp. 61s–65s.

2337. FIELD, G. H. *and* SUTTON, H. Investigations on the spot-welding of light alloys. *Trans. Inst. Weld.*, vol. 3, no. 3, July 1940, pp. 157–174.

2338. FÖPPL, O. Die grundsätzliche Verschiedenheit zwischen Zerreissfestigkeit und Wechselfestigkeit eines Werkstoffes mit Beziehung auf das Oberflächendrücken zur Steigerung der Dauerhaltbarkeit. [The fundamental distinction between the resistance of a material to tensile and to alternating stresses, in relation to the application of surface pressure as a means of increasing fatigue resistance.] *Werkstattstechnik*, vol. 34, no. 1, Jan. 1, 1940, pp. 4–7.

339. FÖPPL, O. Das Oberflächendrücken. [Surface compression.] *Metallwirtschaft*, vol. 19, no. 9, 1940, pp. 162–164 and no. 10, 1940, pp. 182–185.

2340. FÖPPL, O. Oberflächendrücken zum Zwecke der Steigerung der Dauerhaltbarkeit mit Hilfe des Stahlkugelgebläses. [Surface compression of pins, and increasing their fatigue resistance by means of shot blasting.] *Werkzeugmaschine*, vol. 44, 1940, pp. 123–128.

2341. FORREST, G. Fatigue with reference to wrought aluminium alloys. *Metallurgia, Manchr*, vol. 22, no. 128, June 1940, pp. 51–53.

2342. FRANK, N. B.; OBERG, T. T. *and* FULLER, F. B. Mechanical properties of various aluminium alloy forgings. *Tech. Rep. U.S. Army Air Force*, no. 4560, Aug. 13, 1940, pp. 31.

2343. FRANKLAND, F. H. European developments in the study of impact and fatigue in structures. *Civ. Engng, Easton, Pa.*, vol. 10, no. 4, Apri 11940, pp. 208–210. *J. Ass. Chin. Amer. Engrs*, vol. 21, no. 3, May–June 1940, pp. 105–112.

2344. GALLIK, I. Dauerfestigkeit der eingekerbten Stabe und der Nietverbindungen. [Fatigue strength of notched bars and riveted joints.] *Anyagvizsgálók Közlönye*, vol. 18, Jan.–Feb. 1940, pp. 1–28 and March–April 1940, pp. 33–62. Abstr.: *Stahl u. Eisen*, vol. 60, no. 26, June 27, 1940, p. 581 and no. 48, Nov. 28, 1940, p. 1096.

2345. GEROLD, A. Über den Fortgang der Werkstoffzerstörung bei hohen Wechselbeanspruchungen und der Einfluss von Entlastungspausen auf die Lebensdauer von Stahl. [On the progress in materials destruction under high alternating loads, and the influence of rest periods on the life of steel.] *Jb. 1940 dtsch. Luftfahrtf.*, pp. 1938–1943.

2346. GILLETT, H. W. Toughness (impact strength) as affected by alternating stresses (fatigue). *Metal Progr.*, vol. 38, no. 6, Dec. 1940, pp. 806–807.

2347. GÖBEL, F. Das Verhalten von Gummi-federn bei zügiger und wechselnder Bean-spruchung unter besonderer Berücksichtigung der Verhältnisse bei der federnden Flugmoto-renlagerung. [The behaviour of rubber springs under tensile and alternating loads with par-ticular reference to their application as the springs for aircraft engine mountings.] *Jb. 1940 dtsch. Luftfahrtf.*, pp. II110–II129.

2348. GOETZEL, C. G. *and* SEELIG, R. P. Fatigue of porous metals. *Proc. Amer. Soc. Test. Mater.*, vol. 40, 1940, pp. 746–761. *Mech. World*, vol. 108, no. 2802, Sept. 13, 1940, pp. 196–198.

2349. GRDINA, YU. *and* GOVOROV, A. Machine for fatigue testing of rails. [R] *Stal'*, vol. 10, no. 5–6, May–June 1940, pp. 67–69. English abstr.: *J. Iron St. Inst.*, vol. 144, no. 2, 1941, p. 182A.

2350. GÜRTLER, G. Untersuchungen über die Schadenslinie bei Leichtmetallen. [Investiga-tions on the "damage lines" of light metals.] *Z. Metallk.*, vol. 32, no. 2, Feb. 1940, pp. 21–30. Abstr.: *Z. Ver. dtsch. Ing.*, vol. 84, no. 47, Nov. 23, 1940, pp. 918–919.

2351. GUEST, J. J. Recent research on combined stress. *J. Instn Auto. Engrs*, vol. 9, no. 3, Dec. 1940, pp. 33–72.

2352. GUTFREUND, K. Schwingungsprüfung von Motorenbauteilen. [Vibration testing of en-gine components.] *Die Abnahme — Sonderteil des Anzeigers für Maschinenwesen für die Abnahme von Werkstoffen und Betriebsbedarf*, vol. 3, no. 9, 1940, pp. 65–68.

2353. HAASE, C. Punkt- und Nahtschweissung von Leichtmetallen. [Spot and seam welding of light metals.] *Z. Ver. dtsch. Ing.*, vol. 84,

no. 6, Feb. 10, 1940, pp. 89–96. English sum-mary: *Weld. J.*, *Easton, Pa.*, vol. 19, no. 8, Aug. 1940, pp. 307s–310s.

2354. HÄNCHEN, R. Berechnung der Bolzen, Achsen und Wellen auf Dauerhaltbarkeit. [Calculation of the fatigue durability of bolts, axles and shafts.] *Fördertechnik*, vol. 33, no. 19/20, 1940, pp. 145–153; no. 21/22, 1940, pp. 166–173 and no. 23/24, 1940, pp. 185–190.

2355. HAGUE, F. T. Vibratory stresses: tempera-ture, impact and speed effects by optical methods. *Prod. Engng*, vol. 11, no. 4, April 1940, pp. 174–175.

2356. DE HALLER, P. La corrosion par cavitation et par choc de gouttes liquides. [Corrosion caused by cavitation and by impact from liquid drops.] *Schweiz. Arch. angew. Wiss.*, vol. 6, no. 3, March 1940, pp. 61–68.

2357. HELD, H. Das plastiche Verhalten wech-selbeanspruchter Zinn-Einkristalle bei reiner Schubverformung. [The plastic behaviour of a tin single crystal alternately stressed in pure shear.] *Z. Metallk.*, vol. 32, no. 6, 1940, pp. 201–209. English abstr.: *Metallurg. Abstr.*, vol. 8, 1941, pp. 319–320.

2358. HEMPEL, M. Magnetpulverbild und Dauer-haltbarkeit von Schraubenfedern. [Magnetic powder pictures and the fatigue durability of helical springs.] *Arch. Eisenhüttenw.*, vol. 13, no. 11, May 1940, pp. 479–487.

2359. HOLLEY, E. G. The static and fatigue tor-sion strengths of various steels with circular square, and rectangular sections. *Proc. Instr mech. Engrs, Lond.*, vol. 143, 1940, pp. 237–246 and [discussion] vol. 145, 1941, pp. 135–137.

2360. HORGER, O. J. Fatigue strength of mem-bers influenced by surface conditions. *Proc*

Engng, vol. 11, no. 11, Nov. 1940, pp. 490–493; no. 12, Dec. 1940, pp. 562–565 and vol. 12, no. 1, Jan. 1941, pp. 22–24.

2361. HORGER, O. J. *and* BUCKWALTER, T. V. Fatigue strength of 2 inch diameter axles with surfaces metal coated and flame hardened. *Proc. Amer. Soc. Test. Mater.*, vol. 40, 1940, pp. 733–745.

2362. HORGER, O. J. *and* BUCKWALTER, T. V. Photoelasticity as applied to design problems. *Iron Age*, vol. 145, no. 21, May 23, 1940, pp. 42–50.

2363. HORGER, O. J. *and* BUCKWALTER, T. V. Fatigue strength of axles with surfaces metal coated and flame hardened. *Heat Treat. Forg.*, vol. 26, no. 7, July 1940, pp. 321–325, 336.

2364. HOTTENROTT, E. Zur Dauerfestigkeit von Duralblech-Nietungen. [The fatigue strength of riveted joints in duralumin sheets.] *Luftfahrtforsch.*, vol. 17, no. 8, Aug. 20, 1940, pp. 247–250. English summary: *Engrs' Dig.*, vol. 2, no. 2, Feb. 1941, pp. 77–79.

2365. HRUSKA, J. H. Welds and the testing of their endurance. *Iron Age*, vol. 145, no. 20, May 16, 1940, pp. 33–37.

2366. IGARASHI, I. *and* FUKAI, S. On the influence of Almite treatment (anodising) on fatigue strength of some aluminium alloys. [J] *Trans. Soc. mech. Engrs, Japan*, vol. 6, no. 22, Feb. 1940, pp. I-12–I-20. *Sumit. Kinz. Kog. Kenk.*, vol. 4, no. 2, 1940, pp. 134–151.

2367. IGARASHI, I. *and* FUKAI, S. Fatigue tests of light alloy sheets. [J] *Trans. Soc. mech. Engrs, Japan*, vol. 6, no. 22, Feb. 1940, pp. I-20–I-30. English abstr.: *Metallurg. Abstr.*, vol. 7, 1940, p. 394.

2368. INGLIS, N. P. Nitriding of steels. *Wild-Barfield Heat Treatment Journal*, vol. 4, no. 27, Dec. 1940, pp. 26–27.

2369. IVERONOVA, V. I. *and* KOSTETSKAYA, T. X-ray study of fatigue in samples subjected to a known alternating bending force. [R] *Zh. tekh. Fiz.*, vol. 10, no. 4, 1940, pp. 304–308.

2370. JÄGER, K. Die Festigkeit leichtgekrümmter Druckstäbe aus Stahl bei schwingender Belastung. [The strength of lightly curved steel columns under oscillating loading.] *Stahlbau*, vol. 13, no. 23/24, Nov. 1, 1940, pp. 128–131 and vol. 14, no. 6/7, March 7, 1941, p. 32.

2371. JONES, J. Fatigue provisions in riveted joints. *Civ. Engng, Easton, Pa.*, vol. 10, no. 6, June 1940, pp. 344–345.

2372. KAGAN, G. *and* TERMINASOV, YU. X-ray study of the mechanism of fracture of metals due to fatigue. [R] *Zh. tekh. Fiz.*, vol. 10, 1940, pp. 781–785.

2373. KAUFMANN, F. *and* JÄNICHE, W. Beitrag zur Dauerhaltbarkeit der Schraubenverbindung. [Contribution to the fatigue strength of screwed joints.] *Tech. Mitt. Krupp A.*, vol. 3, no. 11, Oct. 1940, pp. 147–159.

2374. KEHL, G. L. *and* OFFENHAUER, C. M. The fatigue resistance of steel as affected by acid pickling. *Trans. Amer. Soc. Metals*, vol. 28, no. 1, March 1940, pp. 238–256.

2375. KENYON, J. N. A corrosion-fatigue test to determine the protective qualities of metallic platings. *Proc. Amer. Soc. Test. Mater.*, vol. 40, 1940, pp. 705–716.

2376. KENYON, J. N. A pulsating tension-fatigue machine for small diameter wires. *Proc. Amer.*

Soc. Test. Mater., vol. 40, 1940, pp. 762–770. Abstr.: *Mech. World*, vol. 108, no. 2805, Oct. 4, 1940, pp. 253–254.

2377. KENYON, J. N. The endurance properties of hard drawn wires of various kinds of copper. *Wire and Wire Prod.*, vol. 15, no. 10, Oct. 1940, pp. 587, 632.

2378. KIDANI, Y. A new fatigue tester and the testing of tempered steel. [J] *Trans. Soc. mech. Engrs, Japan*, vol. 6, no. 22, Feb. 1940, pp. I-36–I-40.

2379. KIRMSER, W. Versuche zur Dauerbeanspruchung von Gummi-Metallverbindungen bei Zug-, Druck- und Wechselbeanspruchung. [Repeated load tests on rubber-metal joints under tension, compression and alternating loads.] *Dtsch. Kraftfahrtforsch., Zwischenberichte* no. 81, 1940, pp. 20.

2380. KOCH, J. J. *and* BIEZENO, C. B. Over een verkorte methode ter bepaling van der kerffactor β bij wisselnde buig- en wringbelasting. [On a rapid method for determining the notch factor β by alternating bending and torsion loads.] *Ingenieur, 's Grav.*, vol. 55, 1940, pp. W91–95.

2381. KÖPKE, G. Der Einfluss gedrehter Oberflächen auf die Wechsel- und Zeitfestigkeit von Stahl. [The influence of distorted surfaces on the alternating and endurance strength of steel.] *Metallwirtschaft*, vol. 19, no. 47, 1940, pp. 1049–1055; no. 49, 1940, pp. 1107–1114 and no. 50, 1940, pp. 1129–1139.

2382. KÖRBER, F. Über den Dauerbruch metallischer Werkstoffe. [The fatigue fracture of metallic materials.] *Schr. dtsch. Akad. Luftfahrtf.*, no. 15, 1940, pp. 53.

2383. KNOWLTON, H. B. *and* SNYDER, E. H. Selection of steel and heat treatment for spur gears. *Trans. Amer. Soc. Metals*, vol. 28, Sept. 1940, pp. 687–713.

2384. KROON, R. P. Turbine blade fatigue testing. *Mech. Engng, N.Y.*, vol. 62, no. 7, July 1940, pp. 531–535 and no. 12, Dec. 1940, pp. 919–921.

2385. KÜCH, W. Zeit- und Dauerfestigkeit von Lagenhölzern. [Endurance and fatigue strength of laminated wood.] *Jb. 1940 dtsch. Luftfahrtf.*, pp. I1114–I1118.

2386. KÜCH, W. Über den Einfluss des Harzgehaltes auf die statischen und dynamischen Festigkeitseigenschaften von geschichteten Kunstharz-Pressstoffen. [On the influence of resin content on the static and dynamic strength properties of laminated moulded plastics.] *Jb. 1940 dtsch. Luftfahrtf.*, pp. I1126–I1132.

2387. LEA, F. C. Repeated stresses on structures. Part I—repeated stresses on steel structures. Part II—repeated stresses on reinforced concrete. *Struct. Engr*, vol. 18, no. 1, Jan. 1940, pp. 483–495 and no. 2, Feb., 1940, pp. 511–520. Summary [Part II only]: *Concr. constr Engng*, vol. 35, no. 3, March 1940, pp. 164–166.

2388. LEA, F. C. *and* WHITMAN, J. G. Fatigue tests of structural steel flats and bars cut from butt-welded boiler plate. *Proc. Instn mech Engrs, Lond.*, vol. 144, 1940, pp. 132–139.

2389. LIPPACHER, K. Der Steigerung der Verdreh dauerhaltbarkeit von Kerbverzahnten Drehstabfedern durch Oberflächendrücken. [The increase in the torsional fatigue resistance of splined torsion bars by means of surface compression.] *Werkstatttechnik*, vol. 34, no. 22, Nov. 15, 1940, pp. 369–371.

2390. LOCATI, L. La fatica dei materiali metallic [The fatigue of metallic materials.] *Tecn. e Org.*, vol. 4, April–Aug. 1940, pp. 58–6

2391. LORIG, C. H. and SCHNEE, V. H. Damping capacity, endurance, electrical and thermal conductivities of some grey cast irons. *Trans. Amer. Foundrym. Ass.*, vol. 48, Dec. 1940, pp. 425–448. Abstr.: *Found. Tr. J.*, vol. 63, no. 1264, Nov. 7, 1940, pp. 297–299.

2392. MCADAM, D. J. and GEIL, G. W. Influence of cyclic stress on corrosion pitting of steels in fresh water, and influence of stress corrosion on fatigue limit. *J. Res. nat. Bur. Stand.*, vol. 24, no. 6, June 1940, pp. 685–722.

2393. MCDOWELL, J. F. The fatigue endurance of killed, capped and rimmed steels. *Metals and Alloys*, vol. 11, no. 1, Jan. 1940, pp. 27–32. Abstr.: *Metallurgia, Manchr*, vol. 21, no. 125, March 1940, p. 153.

2394. MAILÄNDER, R.; SZUBINSKI, W. and WIESTER, H. J. Biegewechselversuche und metallographische Untersuchungen an geschweissten Dünnblechen aus Stählen höherer Festigkeit. [Bending fatigue tests and metallographic investigations with welded thin sheets of high tensile steel.] *Tech. Mitt. Krupp A*, vol. 3, 1940, pp. 199–221. *Jb. 1940 dtsch. Luftfahrtf.*, pp. I944–I954. English summary: *Weld. J., Easton, Pa.*, vol. 21, no. 1, Jan. 1942, pp. 30s–33s.

2395. MARTIN, P. R. Fatigue tests on bent, straightened and heat-treated duralumin bars. *Roy. Aircr. Estab. Rep.* no. M.T. 5649a (revised), Aug. 1940, pp. 2.

2396. MEYER, H. Entwicklung eines dynamischen Pleuelprüfstandes. [Development of dynamic connecting rod testing equipment.] *Auto.-tech. Z.*, vol. 43, no. 15, 1940, pp. 367–376.

2397. MÖNCH, E. Dauerbeanspruchung und magnetoelastische Eigenschaften von Stählen. [Fatigue stressing and magnetoelastic properties of steel.] *ForschArb. IngWes.*, vol. 11, no. 6, Nov.–Dec. 1940, pp. 324–334.

2398. MOORE, H. F.; THOMAS, H. R. and CRAMER, R. E. Second progress report—joint investigation of continuous welded rails. *Weld. J., Easton, Pa.*, vol. 19, no. 8, Aug. 1940, pp. 293s–302s. *Repr. Ill. Engng Exp. Sta.*, no. 17, 1940, pp. 20.

2399. MÜLLER, W. New equipment for testing the fatigue strength of riveted and welded joints. *Tech. Memor. nat. Adv. Comm. Aero.*, Wash., no. 947, July 1940, pp. 5.

2400. NEUGEBAUER, G. H. Biegeschwingungsuntersuchungen an den Kurbelwellen eines Fahrzeugmotors. [Bending fatigue investigations on automobile engine crankshafts.] *Auto.-tech. Z.*, vol. 43, July 25, 1940, pp. 339–350.

2401. NEWBERRY, C. W. An investigation into the occurrence and causes of locomotive tyre failures. *Proc. Instn mech. Engrs, Lond.*, vol. 142, 1939, pp. 289–303. Abstr.: *Found. Tr. J.*, vol. 62, no. 1229, March 7, 1940, p. 184. *Mech. Engng, N.Y.*, vol. 62, no. 4, April 1940, pp. 324–325. *Locomotive*, vol. 46, no. 573, May 15, 1940, pp. 135–136.

2402. NISHIHARA, T. and KAWAMOTO, M. The fatigue of steel under combined bending and torsion. [J] *Trans. Soc. mech. Engrs, Japan*, vol. 6, no. 24, Aug. 1940, pp. I-8–I-11. [English summary p. S-2.] English translation: *Caterpillar Tractor Co.*, Translation no. 95.

2403. NISHIHARA, T. and KAWAMOTO, M. Some experiments on the fatigue of steel. [J] *Trans. Soc. mech. Engrs, Japan*, vol. 6, no. 25, Nov. 1940, pp. I-47–I-51.

2404. NISHIHARA, T. and KOJIMA, K. Diagram of endurance limit of super-duralumin for repeated tension and compression. [J] *Trans. Soc. mech. Engrs, Japan*, vol. 6, no. 24, Aug. 1940, pp. 35–36. English abstr.: *Metallurg. Abstr.*, vol. 8, pt. 6, June 1941, p. 135.

2405. NISSEN, O. Fatigue in aeroplane structures. *Aircr. Engng*, vol. 12, no. 140, Oct. 1940, pp. 293–295, 306.

2406. ONO, A. Fatigue tests on iron, steel and light alloy chiefly under bending. [J] *Trans. Soc. mech. Engrs.*, *Japan*, vol. 6, no. 24, Aug. 1940, pp. I-19–I-26. English abstr.: *Metallurg. Abstr.*, vol. 8, pt. 6, June 1941, p. 135.

2407. ONO, A. Strength of materials under repeated combined stress. [J] *Trans. Soc. mech. Engrs*, *Japan*, vol. 6, no. 25, Nov. 1940, pp. I–30–I-39.

2408. OSCHATZ, H. Ventilfeder-Prüfmaschine. [Valve spring testing machine.] *Z. Ver. dtsch. Ing.*, vol. 84, no. 33, Aug. 17, 1940, p. 598.

2409. OSHIBA, F. On the change of hardness caused by repeated stress and the effect of ageing on its recovery. *Sci. Rep. Tohoku Univ.*, vol. 28, no. 2, Feb. 1940, pp. 370–385.

2410. OSHIBA, F. Recovery from fatigue caused by annealing. *Sci. Rep. Tohoku Univ.*, vol. 29, no. 1, June 1940, pp. 69–112. *Nippon kink. Gakk.*, vol. 4, Jan. 1940, pp. 13–30. Abstr.: *Chem. Abstr.*, vol. 34, 1940, col. 3634.

2411. PIWOWARSKY, E. Beitrag zur Schlag- und Ermüdungsfestigkeit von hochwertigem Grauguss. [Contribution to the impact and fatigue strength of high-duty grey cast iron.] *Giesserei*, vol. 27, no. 4, Feb. 23, 1940, pp. 59–61.

2412. POGODIN-ALEKSEEV, G. I. Effect of grain size of austenite upon the fatigue limit and cold chortness of axle steel. [R] *Metallurg*, vol. 15, no. 9, 1940, pp. 30–36. English abstr.: *Chem. Abstr.*, vol. 36, 1942, col. 2826.

2413. POMP, A. *and* HEMPEL, M. Über die Dauerhaltbarkeit von Schraubenfedern mit und ohne Oberflächenverletzungen. [On the fatigue resistance of helical springs with and without surface defects.] *Mitt. K.-Wilh.-Inst. Eisenforsch.*, vol. 22, no. 4, 1940, pp. 35–56. *Jb. 1940 dtsch. Luftfahrtf.*, pp. II204–II224. English abstr.: *J. Iron St. Inst.*, vol. 141, no.1, 1940, pp. 298A–299A.

2414. POMP, A. *and* HEMPEL, M. Biegewechselversuche an Chrom-Molybdän-Vergütungs- und Einsatzstählen im Vergleich zu nickelhaltigen Stählen. [Bending fatigue tests on heat treatable and case hardening chromium-molybdenum steels in comparison with nickel steels.] *Mitt. K.-Wilh.-Inst. Eisenforsch.* vol. 22, no.10, 1940, pp. 149–168. English abstr.: *J. Iron St. Inst.*, vol. 144, no. 2, 1941 p. 216A.

2415. POMP, A. *and* HEMPEL, M. Über da Verhalten von Gusseisen und Tempergus unter wechselnder Beanspruchung. [The behaviour of cast iron and malleable cast iron under alternating stresses.] *Mitt. K.-Wilh. Inst. Eisenforsch.*, vol. 22, no. 11, 1940 pp. 169–201. English abstr.: *J. Iron St. Inst.* vol. 144, no. 2, 1941, pp. 216A–217A.

2416. PORTEVIN, A. M. Toughness (impac strength) unaffected by alternating stresse (fatigue). *Metal Progr.*, vol. 37, no. 5, Ma 1940, pp. 563–564.

2417. PRINGLE, B. High-speed Wöhler test gea B. T.-H. Activ., vol. 16, no. 5, 1940, pp. 161 163.

2418. VON RAJAKOVICS, E. Die Schwingung festigkeit von Aluminium-Legierungen b einer Grenzlastspielzahl von 50 Millione [The fatigue strength of aluminium alloys f a limiting number of 50 million cycles *Metallwirtschaft*, vol. 19, no. 42, Oct. 1 1940, pp. 929–932.

2419. Roš, M. *and* Theodorius, Ph. Statischer Bruch and Ermüdungsfestigkeit genieteter Fachwerke aus Avional 'SK'. [Static fracture and fatigue strength of riveted girders of Avional 'SK'.] *Ber. eidgenöss. MatPrüfAnst.*, no. 126, Feb. 1940, pp. 1–55.

2420. Rossheim, D. B. *and* Markl, A. R. C. The significance of, and suggested limits for, the stress in pipe lines due to the combined effects of pressure and expansion. *Trans. Amer. Soc. mech. Engrs*, vol. 62, 1940, pp. 443–460.

2421. Sachs, G. Stress raisers: review of their effects. *Metal Treatm.*, vol. 6, no. 24, Winter 1940–1941, pp. 159–166, 177.

2422. Sachs, G. Stress raisers. *Iron Age*, vol. 146, no. 5, Aug. 1, 1940, pp. 31–34 and no. 6, Aug. 8, 1940, pp. 34–37.

2423. Sachs, G. Fatigue failure. *Iron Age*, vol. 146, no. 10, Sept. 5, 1940, pp. 31–34 and no. 11, Sept. 12, 1940, pp. 36–40.

2424. Samokhotskii, A. I. *The fatigue of ferrous and non-ferrous metals.* [R] Moscow and Leningrad, Gostekhizdat, 1940, pp. 200.

2425. Savage, E. G. *and* Taylor, W. J. The effect of anodic oxidation on the fatigue properties of RR 56 alloy. *Roy. Aircr. Estab. Rep.* no. M.5541, March 1940, pp. 13.

2426. Savage, E. G. *and* Taylor, W. J. The influence of cleaning and protective treatments on the fatigue properties of magnesium alloy D.T.D. 129. *Roy. Aircr. Estab. Rep.* no. M.2946, June 1940, pp. 24.

2427. Schaal, A. Das Spannungsverhalten von Stahl und Leichtmetall bis zum Bruchanriss bei Wechselverdrehbeanspruchung. [The stress behaviour of steel and light metal at fracture under alternating torsional loads.] *Z. tech. Phys.*, vol. 21, no. 1, 1940, pp. 1–7.

2428. Serensen, S. V. The effect of processing quality on metal strength under variable loading. [R] *Vest. metalloprom.*, vol. 20, no. 1, Jan. 1940, pp. 35–37.

2429. Serensen, S. V. The fatigue testing of structural alloy steels. [R] *Stal'*, vol. 10, no. 3, March 1940, pp. 31–38. English abstr.: *J. Iron St. Inst.*, vol. 144, no. 2, 1941, p. 182A.

2430. Shevandin, E. *and* Kaganovich, L. Influence of surface cold-working by cutting on the fatigue limit of steel. [R] *Zh. tekh. Fiz.*, vol. 10, no. 4, 1940, pp. 295–303.

2431. Shleier, E. V. *and* Oding, I. A. The influence of surface hardening by high frequency currents on the mechanical strength of structural steeels. [R] *Vest. metalloprom.*, vol. 20, no. 7, July 1940, pp. 7–17.

2432. Siebel, E. *and* Wellinger, K. Festigkeitsverhalten von Rohrschweissungen bei Dauerbelastung. [Strength behaviour of welded tubes under fatigue loading.] *Z. Ver. dtsch. Ing.*, vol. 84, no. 4, Jan. 27, 1940, pp. 57–59.

2433. Smith, A. M. Screw threads—the effect of method of manufacture on the fatigue strength. *Iron Age*, vol. 146, no. 8, Aug. 22, 1940, pp. 23–28.

2434. Sopwith, D. G. The resistance of aluminium and beryllium bronzes to fatigue and corrosion fatigue. *Rep. Memor. aero. Res. Comm., Lond.*, no. 2486, April 18, 1940, pp. 10. Abstr.: *Metal Progr.*, vol. 60, no. 4, Oct. 1951, pp. 190, 192, 194.

2435. Swift, H. W. Fatigue and static failure. *Engineering, Lond.*, vol. 150, no. 3896, Sept. 13, 1940, pp. 218–220.

2436. TANAKA, S.; SHINODA, G. and TAKABAYASI, J. X-ray studies on the fatigue of super-dur-alumin. [J] *Trans. Soc. mech. Engrs, Japan*, vol. 6, no. 24, Aug. 1940, pp. I-37–I-39. English abstr.: *Metallurg. Abstr.*, vol. 8, 1941, pp. 134–135.

2437. TATNALL, R. R. Factors in the fatigue of helical springs. *Mech. Engng, N.Y.*, vol. 62, no. 4, April 1940, pp. 289–292. *Engineering, Lond.*, vol. 150, no. 3887, July 12, 1940, pp. 36–37.

2438. TERMINASOV, YU. and KAGAN, G. X-ray study of the mechanism of fracture of rail steel by fatigue. [R] *Zh. tekh. Fiz.*, vol. 10, no. 11, 1940, pp. 873–885.

2439. THUM, A. and BRUDER, E. Flanschwellen-Dauerbrüche und ihre Ursachen. [Fatigue failures of flanged shafts and their causes.] *Dtsch. Kraftfahrtforsch.*, no. 41, 1940, pp. 10. Abstr.: *Z. Ver. dtsch. Ing.*, vol. 84, no. 30, July 27, 1940, pp. 542–543.

2440. THUM, A. and LANGE, S. Dauerversuche an Leichtmetallstäben mit einer aufgespritzten Stahlschicht. [Fatigue tests of light metal rods protected with sprayed steel coatings.] *Z. Ver. dtsch. Ing.*, vol. 84, no. 38, Sept. 21, 1940, pp. 718–720.

2441. THUM, A. and LORENZ, H. Versuche an Schrauben aus Magnesium-Legierungen. [Tests on magnesium alloy screws.] *Z. Ver. dtsch. Ing.*, vol. 84, no. 36, Sept. 7, 1940, pp. 667–673.

2442. THUM, A. and WÜRGES, M. Die zweck-mässige Vorspannung in Schraubenverbin-dungen. [The initial tension of bolted joints.] *Dtsch. Kraftfahrtforsch.*, no. 43, 1940, pp. 20. Abstr.: *Z. Ver. dtsch. Ing.*, vol. 84, no. 46, Nov. 16, 1940, pp. 896–897.

2443. UNCKEL, H. Versuche über den Einfluss der Probenlage zur Walzrichtung auf die Dauerfestigkeit. [Researches on the effect of specimen orientation relative to the direction of rolling on the fatigue strength.] *Metall-wirtschaft*, vol. 19, no. 43, Oct. 25, 1940, pp. 949–951.

2444. URBACH, E. Drehschwingungsuntersu-chungen an Kurbelwellen eines Fahrzeug-motors. [Torsional fatigue investigations on crankshafts of vehicle engines.] *Auto.-tech. Z.*, vol. 43, no. 13, July 10, 1940, pp. 315–326.

2445. VIKKER, I. V. X-ray study of the fatigue of mild steel. [R] *Zh. tekh. Fiz.*, vol. 10, no. 16, 1940, pp. 1353–1357.

2446. WAGENKNECHT, W. E. Über den Einfluss der Walzdrahtvorglühung, der Patentie-rungs-Temperaturen und der Randentkohlung auf die Eigenschaften von Stahldraht. [On the influence of the wire rod annealing, the patenting temperature and the surface car-burization on the properties of steel wire.] *Mitt. Kohle- u. Eisenforsch.*, vol. 2, 1940, pp. 157–184.

2447. WAHL, A. M. How holes and notches affect flat springs. *Mach. Design*, vol. 12, no. 5 May 1940, pp. 40–43, 108.

2448. WAHL, A. M. Spring materials. *Mach. Design*, vol. 12, no. 10, Oct. 1940, pp. 46–49, 94.

2449. WATT, D. G. Fatigue tests on zinc-coated steel wire. *Proc. Amer. Soc. Test. Mater.* vol. 40, 1940, pp. 717–732. *Wire and Wire Prod.*, vol. 16, no. 5, May 1941, pp. 280–28 294–295.

2450. WEVER, F. Abbau von Eigenspannungen durch Wechselbeanspruchung. [Relaxation internal stress under alternating loads.] *Jb. 1940 dtsch. Luftfahrtf.*, pp. I1107–I1113.

2451. WIEGAND, H. *Oberfläche und Dauerfestigkeit. [Surfaces and fatigue strength.]* Berlin-Spandau, BMW-Flugmotorenbau, G.m.b.H., 1940, pp. 83. English translation: *R.T.P. Trans.*, no. 1772, pp. 32.

2452. WIEGAND, H. Oberflächengestaltung und -behandlung dauerbeanspruchter Maschinenteile. [Surface finish and the fatigue strength of machine parts.] *Z. Ver. dtsch. Ing.*, vol. 84, no. 29, July 20, 1940, pp. 505–510.

2453. WILLIAMS, C. G. *and* BROWN, J. S. Fatigue strength of crankshafts. *Rep. Instn Auto. Engrs Res. Comm.* no. 9195B, Oct. 1940, pp. 24. *Engineering, Lond.*, vol. 154, no. 3992, July 17, 1942, pp. 58–59 and no. 3993, July 24, 1942, pp. 78–80.

2454. WILSON, W. M. *and* WILDER, A. B. Fatigue tests of welded joints in structural plates. *Weld. J., Easton, Pa.*, vol. 19, no. 3, March 1940, pp. 100s–108s.

2455. WOOD, W. A. *and* THORPE, P. L. Behaviour of the crystalline structure of brass under slow and rapid cyclic stresses. *Proc. roy. Soc., (A)*, vol. 174, no. 958, Feb. 1940, pp. 310–321.

2456. WÜRGES, M. Vorspannung von Schraubenverbindungen. [Initial tension of bolted joints.] *Z. Ver. dtsch. Ing.*, vol. 84, no. 46, Nov. 16, 1940, pp. 896–897.

2457. ZEYEN, K. L. Untersuchungen über statische Festigkeit, Kerbschlagzähigkeit und Dauerfestigkeit von geschweissten Baustahl St.52 nach verschiedenen Wärmebehandlungen und nach Schweissung unter Vorwärmung. [Investigations of static strength, notch-impact toughness and fatigue strength of welded structural steel St.52 after different heat treatments and after welding with preheating.] *Tech. Mitt. Krupp A.*, vol. 3, no. 6, June 1940, pp. 87–98. *Bautechnik*, vol. 18, no. 24, June 7, 1940, pp. 269–273. *Stahl u. Eisen*, vol. 60, no. 21, May 23, 1940, pp. 456–461. English abstr.: *Weld. J., Easton, Pa.*, vol. 20, no. 2, Feb. 1941, p. 120s.

2458. ZIMMERLI, F. P. How shot blasting increases fatigue life. *Mach. Design*, vol. 12, no. 11, Nov. 1940, pp. 62–63. *Heat Treat. Forg.*, vol. 26, no. 11, Nov. 1940, pp. 534–536.

2459. ZIMMERMANN, J. H. Flame-treating. *Weld. J., Easton, Pa.*, vol. 19, no. 2, Feb. 1940, pp. 104–110.

2460. ZIMMERMAN, J. H. Flame strengthening. *Iron Age*, vol. 145, no. 5, Feb. 1, 1940, pp. 38–39.

2461. Fatigue properties of Electron alloys. *Light Metals*, vol. 3, no. 25, Feb. 1940, pp. 43–45.

2462. The size effect in fatigue. *Metals and Alloys*, vol. 11, March 1940, pp. A19, 94.

2463. Surface rolling to minimise fatigue—German research data. *Aircr. Prod., Lond.*, vol. 2, no. 3, March 1940, p. 88.

2464. Fillets for piston ring grooves. *Automot. Industr. N.Y.*, vol. 82, no. 8, April 15, 1940, p. 382.

2465. Fatigue of metals—a collection of abstracts compiled by the I. A. E. Research Department. *Rep. Instn Auto. Engrs Res. Comm.*, no. 9187B, May 1940, pp. 44.

2466. Fatigue testing machine. *J. sci. Instrum.*, vol. 17, no. 5, May 1940, pp. 135–136.

2467. The problem of fatigue fracture: magnetic powder pictures, and fatigue fracture cracks. *Metallurgia, Manchr*, vol. 22, no. 130, Aug. 1940, pp. 125–126.

2468. Causes and cures—torsional fatigue failure. *Prod. Engng*, vol. 11, no. 10, Oct. 1940, p. 445.

2469. Surface pressing. The use of shot blasting to increase fatigue strength. *Aircr. Prod., Lond.* vol. 2, no. 10, Oct. 1940, p. 341.

2470. 18:8 and notched fatigue. *Metals and Alloys*, vol. 12, Dec. 1940, pp. 733, 768.

2471. Magnetic fatigue test for sucker rods. *Petrol. World*, vol. 37, no. 12, Dec. 1940, pp. 17–18, 49.

1941

2472. AFANASEV, N. N. The diagram of extension of metals and their notch sensitivity under alternating loading. [R] *Zh. tekh. Fiz.*, vol. 11, no. 4, 1941, pp. 349–358.

2473. ANDERSON, A. R. *and* SMITH, C. S. Fatigue tests on some copper alloys. *Proc. Amer. Soc. Test. Mater.*, vol. 41, 1941, pp. 849–858. *Engng Insp.*, vol. 8, no. 2, April–June 1942, pp. 15–21.

2474. BAKHAREV, V. M. Determining the loads for a fatigue test on steels under a symmetric cycle. [R] *Zav. Lab.*, vol. 10, no. 4, April 1941, pp. 406–408.

2475. BATTELLE MEMORIAL INSTITUTE. *Prevention of the failure of metals under repeated stress.*

New York, John Wiley and Sons, Inc., 1941, pp. 277.

2476. BAUTZ, W. Kritik der Dauerfestigkeit als Bemessungsgrundlage. [Critical consideration of fatigue strength as a basis for design.] *ForschArb. IngWes.*, vol. 12, no. 4, 1941, pp. 162–166.

2477. BEERWALD, A. Über die Dauerfestigkeit von hartverchromten Dural. [On the fatigue strength of hard chromium plated duralumin.] *Luftfahrtforsch.*, vol. 18, no. 10, Oct., 1941, pp. 368–371. English translation: *R.T.P Trans.*, no. 1383, 1941, pp. 6. English abstr.: *Engrs' Dig.*, vol. 3, no. 7, July 1942, pp 252–254.

2478. BERG, S. Zum Technik des Schwingversuches. [Technique of fatigue testing.] *Z. Ver dtsch. Ing.*, vol. 85, no. 27, July 5, 1941 pp. 605–608. English translation: *R.T.P Trans.*, no. 1269, 1941, pp. 5. Englisl abstr.: *Engrs' Dig.*, vol. 3, no. 8, Aug. 1942 pp. 275–277.

2479. BERNHARD, R. K. Testing material in th resonance range. *Proc. Amer. Soc. Test Mater.*, vol. 41, 1941, pp. 747–757.

2480. BERTSCHINGER, R. *and* PIWOWARSKY, E Werkstoffe Gusseisen: Ermüdungsfestigke und Kerbsicherheit. I. Einfluss der Frequen; der Temperatur und der metallischen Grund masse. [Cast iron: fatigue strength and notc sensitivity. I. The effect of frequency, ten perature and the metal matrix.] *Giessere* vol. 28, Aug. 22, 1941, pp. 365–372 and Sept. ! 1941, pp. 385–389. English translation: *Bri cast. Iron Res. Ass. Trans.*, no. 201. Englis abstr.: *J. Iron St. Inst.*, vol. 145, no. 1, 194: p. 241A.

2481. BOLLENRATH, F. Factors influencing th fatigue strength of materials. *Tech. Memo* nat. Adv. Comm. Aero., Wash., no. 987, Sept. 1941, pp. 20.

2482. BOLLENRATH, F. and BUNGARDT, W. Korrosionsermüdung einiger Aluminium-Knetlegierungen bei Einwirkung heisser Flüssigkeiten. [Corrosion fatigue of some aluminium alloys under the action of hot liquids.] Luftfahrtforsch., vol. 18, no. 12, Dec. 1941, pp. 417–424.

2483. BOLLENRATH, F. and CORNELIUS, H. Verdrehwechselfestigkeit von Wellen aus unlegiertem und legiertem Stahl. [Torsional fatigue strength of shafts of unalloyed and alloyed steel.] Jb. 1941 dtsch. Luftfahrtf., pp. II333–II337.

2484. BOLLENRATH, F. and CORNELIUS, H. Fatigue tests of welded aircraft tubes. Weld. J., Easton, Pa., vol. 20, no. 3, March 1941, pp. 159s–160s.

2485. BRICK, R. M. "Quick, Watson, the Micro" —what broke the shaft? Metal Progr., vol. 40, no. 1, July 1941, pp. 49–51.

2486. BRICK, R. M. and PHILLIPS, A. Nonferrous physical metallurgy—present status of metallurgical thought on plastic deformation, strain-hardening and recrystallization, copper–gas reactions, fatigue and magnesium alloys. Min. and Metall., N.Y., vol. 22, no. 410, Feb. 1941, pp. 87–92.

2487. BRICK, R. M. and PHILLIPS, A. Fatigue and damping studies of aircraft sheet materials: duralumin alloy 24ST, alclad 24ST, and several 18-8 type stainless steels. Trans. Amer. Soc. Metals, vol. 29, no. 2, June 1941, pp. 435–469. Sheet Metal Ind., vol. 15, no. 168, April 1941, pp. 511–516 and no. 169, May 1941, pp. 635–638, 653.

2488. BUCHMANN, W. Dauerfestigkeitseigenschaften der Magnesiumlegierungen. [The fatigue properties of magnesium alloys.] Z. Ver. dtsch. Ing., vol. 85, no. 1, Jan. 4, 1941, pp. 15–20. English abstr.: Metals and Alloys, vol. 14, Dec. 1941, p. 936.

2489. BUCHMANN, W. Einfluss der Querschnittsgrösse auf die Dauerfestigkeit (besonders Biegedauerfestigkeit) von Leichtmetallen. [Influence of size of cross section on the fatigue strength (especially bending fatigue strength) of light metals.] Metallwirtschaft, vol. 20, no. 38, Sept. 19, 1941, pp. 931–937.

2490. BÜRNHEIM, H. Über den Einfluss von Bohrungen mit Gewinden und Kerbverzahnungen auf die Zeit- und Dauerfestigkeit von Leichtmetall-Flachstäben. [On the effect of screw threaded holes and holes with longitudinal serrations on the endurance and fatigue strength of flat light metal strips.] Luftfahrtforsch., vol. 18, no. 2/3, March 29, 1941, pp. 102–106. English translation: Tech. Memor. nat. Adv. Comm. Aero., Wash., no. 994, Nov. 1941, pp. 10. English abstr.: Aircr. Prod., Lond., vol. 3, no. 34, Aug. 1941, p. 293.

2491. BÜRNHEIM, H. Der Dauerhaltbarkeit von Stabköpfen aus einer hochfesten Al–Cu–Mg-Legierung. [The fatigue strength of bar ends of high strength Al–Cu–Mg alloy.] Aluminium, Berl., vol. 23, no. 4, April 1941, pp. 208–213. English abstr.: Metallurg. Abstr., vol. 8, Nov. 1941, p. 294.

2492. BÜRNHEIM, H. and MECHEL, R. Rissbeobachtungen an der Oberfläche wechselbiegebeanspruchter, plattierter Bleche aus Aluminium-Kupfer-Magnesium-Legierungen. [Cracks on the surface of composite sheets of aluminium–copper–magnesium alloys subjected to alternating bending loads.] Z. Metallk., vol. 33, no. 1, Jan. 1941, pp. 25–27.

2493. BUSSMANN, K. H. Versuchsergebnisse bei der Dauerprüfung von ganzen Bauteilen.

[Test results from the fatigue testing of complete structural parts.] *Metallwirtschaft*, vol. 20, no. 42, Oct. 17, 1941, pp. 1035–1043.

2494. COLLACOTT, R. A. Vibration and fatigue failure of turbine blades. *Shipb. Shipp. Rec.*, vol. 58, no. 5, July 31, 1941, pp. 105–106.

2495. COLLINS, W. L. *and* SMITH, J. O. Fatigue and static load tests of a high-strength cast iron at elevated temperatures. *Proc. Amer. Soc. Test. Mater.*, vol. 41, 1941, pp. 797–810.

2496. CORNELIUS, H. Verhalten einiger Kurbelwellen-Gusswerkstoffe bei Dauerbeanspruchung durch Biegung und Verdrehung. [Behaviour of some cast crankshaft materials under fatigue loads in bending and torsion.] *Jb. 1941 dtsch. Luftfahrtf.*, pp. II325–II332.

2497. CORNELIUS, H. Einfluss von Betriebspausen auf die Zeitfestigkeit von Stählen mit Ferrit. [Influence of rest periods on the fatigue strength of steels containing ferrite.] *Luftfahrtforsch.*, vol. 18, no. 8, Aug. 1941, pp. 284–288. English abstr.: *Engrs' Dig.*, vol. 3, no. 2, Feb. 1942, pp. 60–62.

2498. CORNELIUS, H. *and* BOLLENRATH, F. Einfluss von Einspannungen auf die Wechselfestigkeit von unlegiertem Stahl. [Influence of clamping on the fatigue strength of carbon steel.] *Arch. Eisenhüttenw.*, vol. 14, no. 7, Jan. 1941, pp. 335–340. Abstr.: *Z. Ver. dtsch. Ing.*, vol. 86, no. 5/6, Feb. 7, 1942, pp. 91–92.

2499. CORNELIUS, H. *and* KRAINER, H. Festigkeitseigenschaften von Chrom-Mangan-Molybdän-Vergütungsstählen. [Strength properties of heat-treated chromium–manganese–molybdenum steels.] *Stahl u. Eisen*, vol. 61, no. 38, Sept. 18, 1941, pp. 871–877.

2500. COX, H. N. Dynamic testing in Germany. *Proc. Instn mech. Engrs, Lond.*, vol. 146, no. 1, Nov. 1941, pp. 39–41.

2501. DAEVES, K. Anstrengungen und Ermüdung bei Werkstoffen und Körperleistungen. [Straining and fatigue of materials and behaviour of solids.] *Metallwirtschaft*, vol. 20, no. 30, 1941, pp. 760–761.

2502. DUCKWITZ, C. A. Zur Frage der Schwingungsfestigkeit von Bohrgestängen. [The problem of fatigue strength of drilling rods.] *Öl u. Kohle*, vol. 37, no. 23, 1941, pp. 454–461.

2503. DUCKWITZ, C. A. *and* BUCHHOLTZ, H. Beitrag zur Frage der Sicherheit von Kesselrohren gegenüber Innendruck bei hohen Temperaturen. [Contribution to the problem of the resistance of boiler tubes to internal pressure at high temperatures.] *Arch. Eisenhüttenw.*, vol. 15, no. 5, Nov. 1941, pp. 235–242.

2504. EDGERTON, C. T. Recommended Code of Procedure for fatigue testing of hot-wound helical compression springs. *Trans. Amer. Soc. mech. Engrs*, vol. 63, no. 6, Aug. 1941, pp. 553–560.

2505. ERLINGER, E. Eine Schwingungsprüfmaschine für 200 Kg Zug- oder Druck-Kraft [A fatigue testing machine for 200 Kg tensile or compressive loads.] *Metallwirtschaft*, vol. 20, no. 17, April 25, 1941, pp. 414–416.

2506. ERLINGER, E. Dauerprüfmaschinen für grosse Proben. [Fatigue testing machines for large specimens.] *Luftwissen*, vol. 8, no. 6, June 1941, pp. 177–181. English abstr.: *Metallurg. Abstr.*, vol. 9, 1942, pp. 81–82.

2507. ERLINGER, E. Eine neue Umlaufbiegemaschine. [A new reversed bending fatigue machine.] *Metallwirtschaft*, vol. 20, no. 30, July 25, 1941, pp. 748–749.

2508. FEHR, R. O. *and* SCHABTACH, C. Resonant vibration testing. *Steel*, vol. 109, no. 19, Nov. 10, 1941, pp. 64–65, 96, 102.

2509. FINDLEY, W. N. Mechanical tests of cellulose acetate. *Proc. Amer. Soc. Test. Mater.*, vol. 41, 1941, pp. 1231–1245.

2510. FINK, K. *and* LANGE, H. Wechselbeanspruchung und magnetische Eigenschaften. [Fatigue stress and magnetic properties.] *Phys. Z.*, vol. 42, no. 6, April 1941, pp. 90–95.

2511. FOCKE, A. E. *and* WARRICK, W. A. Fatigue effects in roller chain. *Proc. Amer. Petrol. Inst.*, sect. IV, vol. 22, no. 227, 1941, pp. 11–22.

2512. FÖPPL, O. Die zweckmassigste Art der Durchführung des Oberflächendrückens. [The suitability of the application of surface compression.] *Mitt. Wöhler-Inst.*, no. 38, 1941, pp. 54–68.

2513. FRY, A.; KESSNER, A. *and* OETTEL, R. Die Bedeutung der Streckgrenze für die Wechselfestigkeit bei Stählen höherer Festigkeit. [The significance of the elastic limit for the fatigue strength of steels of high strength.] *Arch. Eisenhüttenw.*, vol. 14, no. 11, May 1941, pp. 571–576.

2514. GASSNER, E. Auswirkung betriebsähnlicher Belastungsfolgen auf die Festigkeit von Flugzeugbauteilen. [Effects of load sequences similar to those occurring in service on the strength of aircraft structures.] *Jb. 1941 dtsch. Luftfahrtf.*, pp. I472–I483. *Ber. dtsch. VersAnst. Luftf., ForschBer.*, no. 1461, Aug. 30, 1941, pp. 111.

2515. GASSNER, E. *and* PRIES, H. Zeit- und Dauerfestigkeitsschaubilder für stabartige Bauteile aus Cr–Mo-Stahl, Duralumin, Hydronalium und Elektron. [Endurance and fatigue diagrams for cylindrical specimens made of Cr–Mo steel, duralumin, hydronalium and Elektron.] *Luftwissen*, vol. 8, no. 3, March 1941, pp. 82–85. English translation: *R.T.P. Trans.*, no. 1395, 1941, pp. 5.

2516. GAUSS, F. Der Dauerbiegehaltbarkeit von Hierth-Verbindungen und einigen Vergleichsverbindungen. [The bending fatigue behaviour of Hierth joints and some comparable joints.] *Zent. wiss. Ber., ForschBer.*, no. 1364, 1941, pp. 81.

2517. GELLER, YA. A. *and* SHREIBER, G. K. The effect of residual stresses on the fatigue limit of hardened steel. [R] *Zh. tekh. Fiz.*, vol. 11, no. 8, 1941, pp. 700–710.

2518. GLOCKER, R.; LUTZ, W. *and* SCHAABER, O. Nachweis der Ermüdung wechselbeanspruchter Metalle durch Bestimmung der Oberflächenspannungen mittels Röntgenstrahlen. [Detection of fatigue of metals under alternating stress by determination of surface stresses by X-ray diffraction.] *Z. Ver. dtsch. Ing.*, vol. 85, no. 39/40, Oct. 4, 1941, pp. 793–800. English translation: *R.T.P. Trans.*, no. 1371, 1941, pp. 13. *Iron St. Inst. Trans. Ser.*, no. 79, 1942.

2519. GODFREY, H. J. The fatigue and bending properties of cold drawn steel wire. *Trans. Amer. Soc. Metals*, vol. 29, no. 1, March 1941, pp. 133–168.

2520. GÖBEL, F. Verhalten von Hülsengummifedern bei zügiger und wechselnder Beanspruchung. [Behaviour of rubber springs under tension and alternating loads.] *Z. Ver. dtsch. Ing.*, vol. 85, no. 29, July 19, 1941, pp. 631–635. English summary: *Engrs' Dig.*, vol. 2, no. 10, Oct. 1941, pp. 388–391.

2521. GOUGH, H. J. *and* POLLARD, H. V. Some experiments on the behaviour of specimens of boiler plate and boiler joints subjected to slow cycles of repeated bending stresses while immersed in a boiling aqueous solution. *J. Iron St. Inst.*, vol. 143, no. 1, 1941, pp.136P–158P.

2522. GOUGH, V. E. *and* PARKINSON, D. Dunlop fatigue test for rubber compounds. *I.R.I. Trans.*, vol. 17, no. 4, Dec. 1941, pp. 168–242.

2523. GOULD, A. J. *and* EVANS, U. R. A scientific study of corrosion-fatigue. Preliminary report of experiments at Cambridge University. *Spec. Rep. Iron. St. Inst., Lond.*, no. 24, 1941, pp. 325–342.

2524. GRAF, O. Versuche mit Nietverbindungen. [Tests on riveted joints.] *Ber. dtsch. Aussch. Stahlb.*, no. 12, 1941, pp. 45.

2525. GROD, P. The modern theory of elasticity and strength—the effect of fatigue on metals. *Pract. Engng*, vol. 4, no. 92, Oct. 23, 1941, pp. 374–376; no. 95, Nov. 13, 1941, pp. 454–456, 460 and vol. 5, no. 111, March 5, 1942, pp. 178–180.

2526. GRODZINSKI, P. Investigations on shaft fillets. *Engineering, Lond.*, vol. 152, no. 3954, Oct. 24, 1941, pp. 321–324.

2527. HAAS, B. Das Verhalten von Werkstoff und Bauteilen bei statischer und dynamischen Beanspruchung. [The behaviour of materials and structures under static and dynamic stresses.] *Luftwissen*, vol. 8, Nov. 1941, pp. 338–343. English translation: *R.T.P. Trans.*, no. 1969, 1941, pp. 10.

2528. HÄNCHEN, R. Berechnung der Schweiss-konstruktionen auf Dauerhaltbarkeit. [Design of welded structures for fatigue durability.] *Glasers Ann. Gew.*, vol. 65, no. 13, 1941, pp. 199–203 and no. 14, 1941, pp. 207–215.

2529. HAMM, H. W. Endurance limit, its impor tance in determining allowable stress. *Proc Engng*, vol. 12, no. 12, Dec. 1941, pp. 650 651.

2530. HARTMANN, E. C. Fatigue test results– their use in design calculations. *Proc Engng*, vol. 12, no. 2, Feb. 1941, pp. 74–78.

2531. HARTMANN, E. C. Fatigue test result *Machinist*, vol. 85, no. 28, Oct. 4, 194 pp. 226E–227E and no. 32, Nov. 1, 194 pp. 250E–251E.

2532. HARTMANN, E. C. *and* TEMPLIN, R. Fatigue data used in design. *Metal Progr* vol. 39, no. 5, May 1941, pp. 604, 606, 61

2533. HAUTTMANN, H. Mit Silizium und Al minium beruhigter härterer Thomas-Ba stahl. [Silicon and aluminium killed harden Thomas constructional steel.] *Stahl u. Eise* vol. 61, no. 6, Feb. 6, 1941, pp. 129–137 ar no. 7, Feb. 13, 1941, pp. 164–170.

2534. HEMPEL, M. Gusseisen und Tempergu unter Wechselbeanspruchung. [Cast iron ar malleable cast iron under alternating stre ing.] *Z. Ver. dtsch. Ing.*, vol. 85, no. 1 March 22, 1941, pp. 290–292. English trar lation: *R.T.P. Trans.*, no. 1214, 194 pp. 6.

2535. HEMPEL, M. Die Dämpfung von Gusseis bei Zug-Druck-Beanspruchung. [The dampi of cast iron under tensile-compressi stresses.] *Arch. Eisenhüttenw.*, vol. 14, no. April 1941, pp. 505–511.

2536. HEMPEL, M. Dauerversuche an Dräht und Seilen und ihre Prüfeinrichtungen. [I tigue tests on wires and ropes and their me ods of testing.] *Stahl u. Eisen*, vol. 61, no. May 22, 1941, pp. 521–522.

2537. HEMPEL, M. *and* LUCE, J. Verhalten von Stahl bei tiefen Temperaturen unter Zugdruck-Wechselbeanspruchung. [Behaviour of steel at low temperatures under tension-compression fatigue stresses.] *Mitt. K.-Wilh.-Inst. Eisenforsch.*, vol. 23, no. 5, 1941, pp. 53–79. Abstr.: *Z. Ver. dtsch. Ing.*, vol. 85, no. 51/52, Dec. 27, 1941, pp. 992–993. English translation: *Roy. Aircr. Estab., Lib. Trans.*, no. 303, March 1949, pp. 42.

2538. HOFMANN, W. *and* HANEMANN, H. Verfestigung und Aushärtung bei Blei-Tellur-Legierungen. [Work-hardening and precipitation hardening of lead-tellurium alloys.] *Z. Metallk.*, vol. 33, no. 2, Feb. 1941, pp. 62–63. English abstr.: *Metals and Alloys*, vol. 14, no. 1, July 1941, pp. 110, 112.

2539. HOFMEIER, H. Über die Bruchursache von Rotary-Gestängerohrverbindungen. [On the causes of failure in joints of rotating tubular bars.] *Mitt. Kohle- u. Eisenforsch.*, vol. 3, no. 2, 1941, pp. 63–104.

2540. HORGER, O. J. Fatigue strength—III. *Prod. Engng*, vol. 12, no. 1, Jan. 1941, pp. 22–24.

2541. HORGER, O. J. Effect of burnishing on fatigue strength. *Machinist*, vol. 85, July 26, 1941, pp. 154E–156E and Aug. 9, 1941, pp. 170E–171E.

2542. HORGER, O. J. *and* BUCKWALTER, T. V. Fatigue comparison of 7-in. diameter solid and tubular axles. *Proc. Amer. Soc. Test. Mater.*, vol. 41, 1941, pp. 682–695.

2543. HORGER, O. J. *and* BUCKWALTER, T. V. Improving engine axles and piston rods. *Metal Prog.*, vol. 39, no. 2, Feb. 1941, pp. 202–206.

2544. HORGER, O. J. *and* BUCKWALTER, T. V. Fatigue strength improved by flame treatment. *Iron Age*, vol. 148, no. 25, Dec. 18, 1941, pp. 47–53.

2545. HORGER, O. J. *and* NEIFERT, H. R. Effect of surface conditions on fatigue properties. *Surface treatment of metals*, Cleveland, Ohio, American Society for Metals, 1941, pp. 279–298. Abstr.: *Oil Gas J.*, vol. 39, no. 34, Jan. 2, 1941, pp. 50. *Automot. Industr. N. Y.*, vol. 85, no. 11, Dec. 1, 1941, p. 78.

2546. HOSKINS, H. G. Fatigue tests on duralumin; notes on the variation in results of tests on wrought aluminium alloys of this type. *Aircr. Engng*, vol. 13, no. 147, May 1941, p. 132.

2547. HUKAI, S. A micrographic study on fatigue fracture of brass. [J] *Nippon kink. Gakk.*, vol. 5, no. 6, 1941, pp. 227–237.

2548. JAMES, H. F. Testing aluminium alloys —need for standardization of fatigue testing methods. *Metallurgia, Manchr*, vol. 24, no. 144, Oct. 1941, pp. 191–192. *Engineering, Lond.*, vol. 152, no. 3955, Oct. 31, 1941, p. 355.

2549. JÜNGER, A. Die Bedeutung der Streckgrenze für die Wechselfestigkeit bei Stählen Höherer Festigkeit. [The significance of the elastic limit for the fatigue strength of steels of high strength.] *Arch. Eisenhüttenw.*, vol. 15, no. 4, Oct. 1941, pp. 201–202.

2550. KARAS, F. Dauerfestigkeit von Laufflächen gegenüber Grübchenbildung. [Fatigue strength of bearing surfaces against pitting.] *Z. Ver. dtsch. Ing.*, vol. 85, no. 14, April 15, 1941, pp. 341–343. English translation: *R.T.P. Trans.*, no. 1206, 1941, pp. 8.

2551. KAUFMANN, F. *and* JÄNICHE, W. Dauerhaltbarkeit von Schraubenverbindungen mit Muttern aus der Magnesium-Legierung GMg –

Al. [Fatigue durability of bolted joints with nuts of magnesium alloy GMg–Al.] *Z. Ver. dtsch. Ing.*, vol. 85, no. 22, May 31, 1941, pp. 504–505. English translation: *R.T.P. Trans.*, no. 1359, 1941, pp. 5. *Iron St. Inst. Trans. Ser.*, no. 78, 1942.

2552. KAUL, H. W. Statistical analysis of the time and fatigue strength of aircraft wing structures. *Tech. Memor. nat. Adv. Comm. Aero., Wash.*, no. 992, Oct. 1941, pp. 39.

2553. KENYON, J. N. A pulsating tension-fatigue machine for small diameter wires. *Engng Insp.*, vol. 6, no. 4, 1941, pp. 8–14.

2554. KIKUKAWA, M. Some experiments on the fatigue of the mild steel. [J] *Trans. Soc. mech. Engrs, Japan*, vol. 7, no. 28, Aug. 1941, pp. 1–9.

2555. KÜHNEL, R. Entwicklung der Stähle für Reichsbahnbauwerke und -fahrzeuge sowie Oberbau in Wechselwirkung von Beanspruchung und Formgebung. [Development of steels for railroad structures and vehicles, together with the design of superstructures under alternating loads.] *Glasers Ann. Gew.*, vol. 65, no. 1, Jan. 1, 1941, pp. 1–8.

2556. LEA, F. C. *and* WHITMAN, J. G. Fatigue tests of structural steel flats and bars cut from butt-welded boiler plate. *Proc. Instn mech. Engrs, Lond.*, vol. 141, no. 3, Jan. 1941, pp. 132–139.

2557. LEHR, E. Dauerbiegefestigkeit von Wellen mit Nabensitzen. [Bending fatigue strength of shafts with hub seats.] *TZ prakt. Metallbearb.*, vol. 51, no. 1/2, 1941, pp. 48–52 and no. 3/4, 1941, pp. 134–137.

2558. LEHR, E. Formgebung und Werkstoffausnutzung. [Design of sections and material utilization.] *Stahl u. Eisen*, vol. 61, no. 43, Oct. 23, 1941, pp. 965–975.

2559. LESSELLS, J. M. *and* MURRAY, W. M. The effect of shot blasting and its bearing on fatigue. *Proc. Amer. Soc. Test. Mater.*, vol. 41, 1941, pp. 659–681.

2560. LESSELLS, J. M. *and* MURRAY, W. M. Effect of shot blasting on strength of metals. *Heat Treat. Forg.*, vol. 27, no. 8, Aug. 1941, pp. 383–384; no. 10, Oct. 1941, pp. 516–517, 533 and no. 11, Nov. 1941, pp. 557–558, 567–568.

2561. LIPSON, C. Photoelasticity in automobile engineering. *Automot. Industr. N.Y.*, vol. 85, no. 2, July 15, 1941, pp. 36–38, 68–69.

2562. LYON, S. W. Mechanical properties of rail steel. *Bull. Amer. Rly Engng Ass.*, vol. 42, no. 423, Feb. 1941, pp. 726–743.

2563. McADAM, D. J. *and* GEIL, G. W. Pitting and its effect on the fatigue limit of steels corroded under various conditions. *Proc. Amer. Soc. Test. Mater.*, vol. 41, 1941, pp. 696–732.

2564. McADAM, D. J. *and* GEIL, G. W. Influence of stress on the corrosion pitting of aluminium bronze and Monel metal in water. *J. Res. nat. Bur. Stand.*, vol. 26, no. 2, Feb. 1941, pp. 135–159.

2565. McBRIAN, R. *and* ARCHIBALD, P. A. Railroad failures in tracks and cars. *Metal Progr.*, vol. 40, Aug. 1941, pp. 189–191.

2566. MÄDING, H. Berechnung von Bauteilen auf Dauerhaltbarkeit. [Design of structural members for fatigue durability.] *Werkst. u*

Betr., vol. 74, no. 8, Aug. 1941, pp. 201–204. English abstr.: *Engrs' Dig.*, vol. 2, no. 10, Oct. 1941, pp. 384–386.

2567. ZOEGE VON MANTEUFFEL, R. Dauerhaltbarkeit von Kraftfahrzeugfedern und Möglichkeiten zur ihre Beeinflussung. [Fatigue durability of automobile springs and the possibilities for influencing it.] *Dtsch. Kraftfahrtforsch.*, no. 49, 1941, pp. 51. Abstr.: *Z. Ver. dtsch. Ing.*, vol. 86, no. 9/10, March 7, 1942, p. 154.

2568. MARIN, J. Designing shafting for static or fatigue loading. *Mach. Design*, vol. 13, no. 8, Aug. 1941, pp. 55–60, 132, 134.

2569. MAUKSCH, W. Die Biegewechselfestigkeit eloxierter Aluminiumlegierungen. [The bending fatigue strength of anodised aluminium alloys.] *Aluminium, Berl.*, vol. 23, no. 6, June 1941, pp. 285–288. English abstr.: *Light Metals*, vol. 5, no. 53, June 1942, pp. 223–224.

2570. METTLER, E. Biegeschwingungen eines Stabes mit Kleiner Vorkrümmung, exzentrisch Angreifender Pulsierender Axiallast und Statischer Querbelastung. [Bending fatigue of specimens with small curvature, eccentric straining, pulsating axial load and static transverse loading.] *Forschungsh. Stahlb.*, no. 4, 1941, pp. 1–23.

2571. MEWES, K. F. Betriebsbeanspruchung und Dauerprüfung. [Service stresses and fatigue testing.] *Die Abnahme — Sonderteil des Anzeigers für Maschinenwesen für die Abnahme von Werkstoffen und Betriebsbedarf*, vol. 4, no. 3, March 1941, pp. 21–22 and no. 4, April 1941, pp. 31–32.

2572. MOHR, E. Über eine neue Drahtprüfmethode. [On a new method for testing wire.] *Metallwirtschaft*, vol. 20, no. 49/50, 1941, pp. 1199–1201.

2573. MOORE, H. F. The effect of type of testing machine on fatigue test results. *Proc. Amer. Soc. Test. Mater.*, vol. 41, 1941, pp. 132–153.

2574. MOORE, H. F. Seventh progress report of the joint investigation of fissures in railroad rails. *Bull. Amer. Rly Engng Ass.*, vol. 42, no. 423, Feb. 1941, pp. 681–751. *Repr. Ill. Engng Exp. Sta.*, no. 21, Dec. 31, 1940, pp. 79.

2575. MOORE, H. F. The development of internal fissures in laboratory tests of rails and in rails in service. *Bull. Amer. Rly Engng Ass.*, vol. 42, no. 423, Feb. 1941, pp. 686–691.

2576. MÜLLER, H. Berechnung der Biegewechselfestigkeit aus der Zugfestigkeit. [Calculation of bending fatigue strength from the tensile strength.] *Stahl u. Eisen*, vol. 61, no. 6, Feb. 6, 1941, pp. 143–144.

2577. MURRAY, W. M. A photoelastic study in vibrations. *J. appl. Phys.*, vol. 12, no. 8, Aug. 1941, pp. 617–622.

2578. MUZZOLI, M. Indagini sulle caratteristiche dell' affaticamento e sulle cause determinanti le rotture per cimenti alterni negli acciai duri e durissimi. [Causes and characteristics of fatigue fractures in hard steels as determined by alternating tests of long duration.] *Metallurg. ital.*, vol. 33, no. 11, 1941, pp. 478–492; vol. 34, no. 1, 1942, pp. 5–29, no. 2, 1942, pp. 50–64 and no. 3, 1942, pp. 90–106. German abstr.: *Stahl u. Eisen*, vol. 63, no. 14, April 8, 1943, pp. 288–289. English translation: *Henry Brutcher Tech. Trans.*, no. 1489.

2579. NISHIHARA, T. and KAWAMOTO, M. The fatigue testing of duralumin under combined bending and torsion. [J] *Nippon kink. Gakk.*, vol. 5, no. 3, March 1941, pp. 110–115. English abstr.: *Metallurg. Abstr.*, vol. 8, pt. 9, Sept. 1941, pp. 238–239.

2580. NISHIHARA, T. and KAWAMOTO, M. The strength of metals under combined alternating bending and torsion. *Mem. Coll. Engng Kyoto*, vol. 10, no. 6, 1941, pp. 117–201. Abstr.: *Metallurg. Abstr.*, vol. 10, pt. 4, April 1943, p. 124.

2581. NISHIHARA, T. and KAWAMOTO, M. The strength of metals under combined alternating bending and torsion. [J] *Trans. Soc. mech. Engrs, Japan*, vol. 7, no. 29, Nov. 1941, pp. I-85–I-95.

2582. O'NEILL, H. White spots in fractures and transverse fissures in rails. *Metallurgist*, vol. 13, April 25, 1941, pp. 12–15.

2583. ONO, M. and ONO, A. Further experiments on the effect of notching on the fatigue of steel. [J] *Trans. Soc. mech. Engrs, Japan*, vol. 7, no. 29, 1941, pp. 120–125.

2584. OSCHATZ, H. Eine Maschine zur Schwingprüfung von Ventilfedern. [A machine for vibration testing of valve springs.] *M T Z*, vol. 3, no. 4, April 1941, pp. 123–124.

2585. PATTERSON, W. and PIWOWARSKY, E. Der Einfluss von Korrosion auf die Festigkeitseigenschaften metallischer Werkstoffe, besonders des Gusseisens. [The influence of corrosion on the strength properties of metallic materials, in particular cast irons.] *Arch. Eisenhüttenw.*, vol. 14, no. 11, May 1941, pp. 561–570. English translation: *Henry Brutcher Tech. Trans.*, no. 1206.

2586. PETERSON, R. E. Some examples of failure due to stress concentration. *J. appl. Phys.*, vol. 12, no. 8, Aug. 1941, pp. 624–625.

2587. PHILLIPS, A. J.; SMITH, A. A. and BECK, P. A. The properties of certain lead-bearing alloys. *Proc. Amer. Soc. Test. Mater.*, vol. 41, 1941, pp. 886–896.

2588. PÖSCHL, TH. Elementare Theorie der Schwingungsfestigkeit. [An elementary theory of fatigue strength.] *Ingen.-Arch.*, vol. 12, no. 2, April 1941, pp. 71–76. English translation: *R.T.P. Trans.*, no. 1275, 1941, pp. 8. *Iron St. Inst. Trans. Ser.*, no. 62, 1941.

2589. POMP, A. and HEMPEL, M. Biegewechselfestigkeit von molybdän- und nickelhaltigen Baustählen. [Bending fatigue strength of molybdenum and nickel containing structural steels.] *Arch. Eisenhüttenw.*, vol. 14, no. 8, Feb. 1941, pp. 403–413.

2590. POMP, A. and HEMPEL, M. Beanspruchungsart und Wechselfestigkeit von Gusseisen und Temperguss. [The strength under alternating stresses of cast iron and malleable cast iron.] *Arch. Eisenhüttenw.*, vol. 14, no. 9, March 1941, pp. 439–449 and no. 10, April 1941, pp. 505–511.

2591. VON RAJAKOVICS, E. Über den Einfluss eines holzfaserartigen Bruchgefüges auf die dynamischen Festigkeitseigenschaften von Blechen aus Al–Cu–Mg-Legierungen. [On the effect of "wood-fibre" fracture on the dynamic strength properties of sheet Al–Cu–Mg alloys.] *Metallwirtschaft*, vol. 20, no. 43, Oct 24, 1941, pp. 1049–1051.

2592. VON RAJAKOVICS, E. Dauerversuche mi Leichtmetall-Pleuelstangen. [Fatigue inves tigations on light metal connecting rods.] Z *Ver. dtsch. Ing.*, vol. 85, no. 43/44, Nov. 1 1941, pp. 867–868. English abstr.: *Light Metals*, vol. 5, no. 56, Sept. 1942, pp. 338 340.

2593. REICHEL, E. Spot welding developmen at the Arado works. *Aircr. Engng*, vol. 1? no. 144, Feb. 1941, pp. 49–54.

2594. ROŠ, M. G. Festigkeit und Berechnun von Schweissverbindungen. [Strength and d sign of welded joints.] *Ber. eidgenöss. Mat-PrüfAnst.*, no. 135, July 1941, pp. 56. *Schweiz. Arch. angew. Wiss.*, vol. 7, no. 9, 1941, pp. 245–271.

595. Roš, M. G. *and* Eichinger, A. Festigkeitseigenschaften der Stähle bei hohen Temperaturen. [Strength properties of steel at high temperatures.] *Ber. eidgenöss. MatPrüfAnst.*, no. 138, Nov. 1941, pp. 55.

596. Russell, H. W. *and* Lowther, J. G. Corrosion fatigue of notched specimens—class 40 cast iron. *Metals and Alloys*, vol. 13, no. 2, Feb. 1941, pp. 169–171. Abstr.: *Metallurgia, Manchr*, vol. 23, no. 138, April 1941, p. 170.

597. Sachs, G. *and* Stefan, P. Chafing fatigue strength of some metals and alloys. *Trans. Amer. Soc. Metals*, vol. 29, no. 2, June 1941, pp. 373–401.

98. Savage, E. G. *and* Taylor, W. J. The effect of anodic oxidation on the fatigue properties of RR.56 alloy (transverse tests). *Roy. Aircr. Estab. Rep.* no. M.5541A, Feb. 1941, pp. 6.

99. Savage, E. G. *and* Taylor, W. J. The influence of cleaning and protective treatments on the fatigue properties of magnesium alloy A.Z. 855. *Roy. Aircr. Estab. Rep.* no. M.2946A, June 1941, pp. 16.

0. Schwerber, P. Sicherheit beim Leichtbau durch Festigkeit und Gestaltung. [Safety in light construction through strength and design.] *Aluminium, Berl.*, vol. 23, no. 12, Dec. 1941, pp. 571–582.

1. Setz, H. L. The influence of corrosion on propeller shaft maintenance. *J. Amer. Soc. Nav. Engrs*, vol. 53, no. 4, Nov. 1941, pp. 735–744.

2602. Siebel, E. Das Verhalten der Werkstoffe bei schwingender Beanspruchung. [The behaviour of materials under alternating stress.] *Metallwirtschaft*, vol. 20, no. 17, April 25, 1941, pp. 409–414.

2603. Siebel, E. *and* Stähli, G. Prüfung von Kurbelwellen. [Testing of crankshafts.] *Giesserei*, vol. 28, no. 7, April 4, 1941, pp. 145–150. English translation: *Engrs' Dig.*, vol. 2, no. 7, July 1941, pp. 280–282 and no. 8, Aug. 1941, pp. 312–314.

2604. Spencer, R. G. *and* Marshall, J. W. An X-ray study of the changes that occur in aluminium during the process of fatiguing. *J. appl. Phys.*, vol. 12, no. 3, March 1941, pp. 191–196.

2605. Stickley, G. W. The effect of alternately high and low repeated stresses on the fatigue strength of 25 ST aluminium alloy. *Tech. Note nat. Adv. Comm. Aero., Wash.*, no. 792, Jan. 1941, pp. 5.

2606. Tagawa, A. The endurance strength of low-carbon manganese structural steels. [J] *Tetsu to Hagane*, vol. 27, no. 1, Jan. 25, 1941, pp. 9–34. English abstr.: *J. Iron St. Inst.*, vol. 144, no. 2, 1941, p. 52A.

2607. Taketomi, R. On the repeated impact test in corrosive media of "Albrac" and some metallic materials. [J] *Sumit. Kinz. Kog. Kenk.* vol. 4, no. 4, 1941, pp. 353–364. English abstr.: *Metallurg. Abstr.*, vol. 8, 1941, p. 223.

2608. Taylor, W. J. *and* Simpkin, G. F. Effect of mean stress on fatigue range of airscrew materials. *Roy. Aircr. Estab. Rep.* no. M.T. 11,216b, June 1941, pp. 4.

2609. Teichmann, A. Grundsätzliches zum Betriebsfestigkeitsversuch. [Fundamentals of the service strength test.] *Jb. 1941 dtsch. Luftfahrtf.*, pp. 1467–1471.

2610. THUM, A. *and* LORENZ, H. Vorspannung und Dauerhaltbarkeit von Schraubenverbindungen mit einer und mehreren Schrauben. [Prestressing and the fatigue strength of single and multiple screwed joints.] *Dtsch. Kraftfahrtforsch.*, no. 56, 1941, pp. 18, Abstr.: *Z. Ver. dtsch. Ing.*, vol. 86, no. 17/18, May 2, 1942, pp. 285–286.

2611. THUM, A. *and* PETERSEN, C. Die Vorgänge im zügig und wechselnd beanspruchten Metallgefüge. I — Zur Mechanik der Festigkeits und Brucherscheinungen. [The structural changes in metals under tensile and alternating stresses. I—on the mechanics of the phenomena of strength and fracture.] *Z. Metallk.*, vol. 33, no. 7, July 1941, pp. 249–259. English abstr.: *Metallurg. Abstr.*, vol. 9, 1942, p. 270.

2612. THUM, A. *and* RICHARD, K. Versprödung und Schädigung warmfeste Stähle bei Dauerbeanspruchung. [Embrittlement and damaging of high temperature steels under fatigue loading.] *Arch. Eisenhüttenw.*, vol. 15, no. 1, 1941, pp. 33–45.

2613. VIETORISZ, J. Design and manufacture with reference to the endurance limit of materials. [H] *Anyagvizsgálók Közlönye*, vol. 19, 1941, pp. 45–84. English abstr.: *Chem. Abstr.*, vol. 37, 1943, p. 3034.

2614. WARLOW-DAVIES, E. J. Fretting corrosion and fatigue strength. Brief results of preliminary experiments. *Proc. Instn mech. Engrs, Lond.*, vol. 146, no. 1, Nov. 1941, pp. 32–38 and vol. 147, no. 2, May 1942, pp. 83–87.

2615. WATT, D. G. Fatigue tests on zinc coated steel wire. *Wire and Wire Prod.*, vol. 16, no. 5, May 1941, pp. 280–285, 294–295.

2616. WELCH, W. P. *and* WILSON, W. A. A new high temperature fatigue machine. *Proc. Amer Soc. Test. Mater.*, vol. 41, 1941, pp. 733–746 Abstr.: *Steel*, vol. 109, no. 21, Nov. 24 1941, pp. 62–63.

2617. WEST, H. G. Fatigue of metals. *Aus Engr*, vol. 41, no. 306, Nov. 7, 1941, pp. 127 130.

2618. WIEGAND, H. Einflüsse der Oberfläche behandlung auf die Festigkeitseigenschafte von Leichtmetallen. [Influence of surface treatment on the strength properties of ligh metals.] *Metallwirtschaft*, vol. 20, no. Feb. 14, 1941, pp. 165–168. English tran lation: *R.T.P. Trans.*, no. 1450, pp. 7.

2619. WIEGAND, H. Einfluss der Oberfläche verletzungen auf die Dauerhaltbarkeit vo Ventilfedern. [Influence of surface dama on the fatigue endurance of valve spring *Z. Ver. dtsch. Ing.*, vol. 85, no. 7, Feb. 1 1941, pp. 174–175.

2620. WIEGAND, H. Oberflächenhärten hoc beanspruchter Maschinenteile. [Surface har ening high strength machine parts.] *Masc nenbau*, vol. 20, 1941, pp. 69–71.

2621. WIEGAND, H. Nitrieren im Motorenb [Nitriding in engine construction.] *Härter tech. Mitt.*, vol. 1, 1941, pp. 166–185.

2622. WIEGAND, H. Oberflächenhärtung Mittel zur Leistungssteigerung, Werkstof sparnis und Werkstoffumstellung. [Surf hardening to produce improved performan economy and interchangeability of materia *ForschArb. IngWes.*, vol. 12, no. 4, July–A 1941, pp. 195–202.

2623. WILDERMANN, M. Einiges aus Theorie Praxis des Schwarzgusses. [Some theories practice concerning black malleable cast iron.] *Giesserei*, vol. 28, no. 11, May 30, 1941, pp. 252–255.

2624. WILLIAMS, C. G. *and* BROWN, J. S. Fatigue failure of oil engine crankshafts. *Gas Oil Pwr*, vol. 36, no. 431, Aug. 1941, pp. 150–153.

2625. WILLIAMS, C. G. *and* BROWN, J. S. Crankshaft failures. *Auto. Engr*, vol. 31, no. 417, Nov. 1941, pp. 401–405.

2626. WILLIAMS, G. T. Selection of steels as affected by endurance limit. *Metal Progr.*, vol. 39, no. 4, April 1941, pp. 464–466, 508.

2627. WILSON, W. M. Influence of the fatigue strength of structural members upon design of steel bridges. *Bull. Amer. Rly Engng Ass.*, vol. 43, no. 426, Sept.–Oct. 1941, pp. 1–19.

2628. WILSON, W. M.; BRUCKNER, W. H.; COOMBE, J. V. *and* WILDE, R. A. Fatigue tests of welded joints in structural steel plates. *Uni. Ill. Engng Exp. Sta. Bull.*, no. 327, Feb. 25, 1941, pp. 86. *Weld. J.*, *Easton, Pa.*, vol. 20, no. 8, Aug. 1941, pp. 352s–357s.

2629. WUNDERLICH, F. Festigkeitsberechnung von Ventilfedern. [Strength calculations for valve springs.] *ForschArb. IngWes.*, vol. 12, no. 4, 1941, pp. 202–204.

2630. ZEYEN, K. L. Fatigue tests of welded low-alloy steel. *Weld. J.*, *Easton, Pa.*, vol. 20, no. 2, Feb. 1941, p. 120s.

2631. ZIMMERLI, F. P. Shotblasting and its effect on fatigue life of springs. *Surface treatment of metals*, Cleveland, Ohio, American Society for Metals, 1941, pp. 261–278. Abstr.: *Heat Treat. Forg.*, vol. 26, no. 11, Nov. 1940, pp. 534–536.

2632. Effect of anodic treatments on the fatigue properties of wrought aluminium alloy bar B.B.S. 5L1 (Duralumin). *Roy. Aircr. Estab. Rep.* no. M.4079B, Jan. 1941, pp. 5.

2633. Belgian fatigue tests of defective welds. *Weld. J.*, *Easton, Pa.*, vol. 20, no. 1, Jan. 1941, p. 18.

2634. Investigate joint bar failures and give consideration to the revision of design and specifications. *Bull. Amer. Rly Engng Ass.*, vol. 42, no. 423, Feb. 1941, pp. 666–672.

2635. Nitriding offsets notch weakness. *Iron Age*, vol. 147, no. 15, April 10, 1941, p. 63.

2636. Diesel crankshaft failures. *Auto. Engr*, vol. 31, no. 410, May 1941, p. 142.

2637. The action of the Haigh fatigue testing machine. *Engineer*, *Lond.*, vol. 171, no. 4455, May 30, 1941, pp. 350–351.

2638. Fractured engine mount from Beaufighter. *Roy. Aircr. Estab. Rep.* no. M.6939, Oct. 1941, pp. 8.

2639. Vorrichtung zur Bestimmung der Biegeschwingungsfestigkeit bei erhöhten Temperaturen und bei Korrosion. [Equipment for determining the bending fatigue strength at elevated temperatures and under corrosive conditions.] *Schweiz. Arch. angew. Wiss.*, vol. 7, no. 11, Nov. 1941, pp. 334–335. English abstr.: *Metallurg. Abstr.*, vol. 11, 1944, pp. 193–194.

2640. The effect of type of testing machine on fatigue test results. *Proc. Amer. Soc. Test. Mater.*, vol. 41, 1941, pp. 133–153.

2641. Fatigue of cadmium–copper trolley wires. Second report. *Rep. Brit. elect. Ind. Res. Ass.*, no. F/T147, 1941, pp. 19.

2642. Effect of various surface finishes on the fatigue resistance of steel strip. *Sci. Tech. Memor.*, no. 1/41, 1941, pp. 12.

1942

2643. ALMEN, J. O. Facts and fallacies of stress determination. *S.A.E. Jl.*, vol. 50, no. 2, Feb. 1942, pp. 52–61.

2644. ANDERSON, A. R. *and* SMITH, C. S. Fatigue tests on some copper alloys. *Engng Insp.*, vol. 8, no. 2, April–June, 1942, pp. 15–21, Abstr.: *Metals and Alloys*, vol. 17, no. 3, March 1943, p. 628.

2645. AXILROD, B. M. Strength and fatigue tests on a laminated paper-base plastic proposed for use in molding propellers. *Advance Restricted Report, nat. Adv. Comm. Aero., Wash.*, no. (unnumbered), Aug. 1942, pp. 7.

2646. BAUD, R. V. Die Berechnung fester Flanschverbindungen von Autoklaven, Rohrleitungen und dergleichen. [The calculation of the strength of flanged joints for autoclaves, piping and the like.] *Schweiz. Arch. angew. Wiss.*, vol. 8, no. 3, 1942, pp. 67–76; no. 4, 1942, pp. 122–129; no. 9, 1942, pp. 274–288 and no. 10, 1942, pp. 315–322.

2647. BAURHENN, A. Vorspannung und Dauerhaltbarkeit an Schraubenverbindungen mit einer und mehreren Schrauben. [Initial tension and fatigue durability of screwed joints with single and multiple screws.] *Auto.-tech. Z.*, vol. 45, no. 22, Nov. 25, 1942, pp. 610–612.

2648. BEERWALD, A. The fatigue strength of heavily chromium plated dural. *Sheet Metal Ind.*, vol. 16, no. 188, Dec. 1942, pp. 1889–1891 and vol. 17, no. 189, Jan. 1943, p. 68.

2649. BEHRENS, P. Schwingungsversuche mit Tragklemmen für Hochspannungs-Freileitun-gen. [Fatigue tests on supporting clamps for high-voltage transmission lines.] *Aluminium, Berl.*, vol. 24, no. 8, Aug. 1942, pp. 269–271.

2650. BERG, S. Fatigue tests by means of vibrators. *Engrs' Dig.*, vol. 3, no. 8, Aug. 1942, pp. 275–277.

2651. BIERETT, G. *and* ALBERS, K. Vergleichende Dauerbiegeversuche an geschweissten Vollwandträgern mit verschiedenen Gurtprofilen und an genieteten Vollwandträgern. [Comparative bending fatigue tests on welded web girders with different flange profiles and riveted web girders.] *Ber. dtsch. Aussch. Stahl., Berl.*, no. 13, 1942, pp. 22.

2652. BOELK, G. Über Streuungen bei Zeitfestigkeitsversuchen an Flugzeugbauteilen und ihre Auswirkung. [On the variation in endurance strength tests on aircraft structures and their effect.] *Ber. Lilienthal-Ges. Luftfahrtf.*, no. 152, 1942, pp. 77–85.

2653. BOLLENRATH, F. *and* CORNELIUS, H. Die Ermittlung der Schadenslinie einer ausgehärteten Aluminium – Kupfer – Magnesium-Knetlegierung. [Determination of the damage line of an age hardened aluminium–copper–magnesium wrought alloy.] *Z. Metallk.*, vol. 34, no. 7, July 1942, pp. 150–156. English abstr.: *Bull. Brit. non-ferr. Met. Ass.*, no. 164, Feb. 1943, p. 59.

2654. BOLLERNATH, F. *and* CORNELIUS, H. Die Ermittlung der Schadenslinie von Stahl. [The determination of the damage line of steel.] *Arch. Eisenhüttenw.*, vol. 16, no. 2, Aug. 1942, pp. 49–56. German abstr.: *Metallwirtschaft*, vol. 22, no. 11/12, 1943, p. 178.

2655. BOULTON, B. C. X-ray of aircraft castings —its control and value. *J. aero. Sci.*, vol. 9, no. 8, June 1942, pp. 271–283.

2656. BRIMELOW, E. I. The influence of various methods of heat treatment on the fatigue properties of wrought aluminium alloy bar B.S.S. 5L1 (duralumin). *Roy. Aircr. Estab., Rep.* no. M.2269A, March 1942, pp. 10.

2657. BRUDER, E. *and* GIMBEL, G. Dauerbruch-schäden an Hauptpleueln durch hypozykloid-artige Bewegungen von geschlossenen Lager-stützschalen. [Fatigue fracture damage in main connectors as a result of hypocycloidal movements of closed bearing support shells.] *Luftwissen,* vol. 9, no. 12, 1942, pp. 353–355.

2658. BUCKWALTER, T. V. *and* HORGER, O. J. Fatigue strength of welded aircraft joints. *Weld. J., Easton, Pa.,* vol. 21, no. 11, Nov., 1942, pp. 525s–539s.

2659. BUNGARDT, W. Gefüge und Dauerfestig-keit von Al–Cu–Mg-Knetlegierungen nach DIN 1713 mit erhöhten Mangan- und Magne-sium gehalten. [The structure and fatigue strength of Al–Cu–Mg wrought alloys (DIN 1713) with relatively large manganese and magnesium content.] *Luftfahrtforsch.,* vol. 19, no. 5, May 30, 1942, pp. 174–177. English abstr.: *J. Instn Prod. Engrs,* vol. 21, no. 8, Aug. 1942, pp. 73–74.

2660. BUSSE, W. F.; LESSIG, E. T.; LOUGH-BOROUGH, D. L. *and* LARRICK, L. Fatigue of fabrics. *J. appl. Phys.,* vol. 13, no. 11, Nov. 1942, pp. 715–724.

2661. BUSSMANN, K. H. Prüfung von Schweiss-verbindungen auf Dauerfestigkeit und der Einfluss einer Nachbehandlung. [Testing of welded joints to determine fatigue resistance and the influence of subsequent treatment.] *Autogene Metallbearb.,* vol. 35, no. 11, June 1, 1942, pp. 164–169.

2662. CHASMAN, B.; MARTELL, G. M. *and* OBERG, T. T. Variations in properties of cold-rolled corrosion-resistant 18-8 steel sheet and effects of low temperature heat treatment. *Tech. Rep. U.S. Army Air Force,* no. 4773, May 12, 1942, pp. 47.

2663. CHEVENARD, P. Nouveaux appareils pour l'étude thermomécanique et micromécanique des métaux. Micromachine pour essais de torsion alternée. [New apparatus for thermo-mechanical and micromechanical study of metals. Micromachine for alternating torsion tests.] *Rev. Métall.,* vol. 30, no. 3, March 1942, pp. 65–72.

2664. CLARK, F. H. Metallurgical problems in the telegraph industry. *Min. and Metall. N.Y.,* vol. 23, no. 421, Jan. 1942, pp. 18–20.

2665. COLLINS, W. L. *and* SMITH, J. O. The notch sensitivity of alloyed cast irons subjected to repeated and static loads. *Proc. Amer. Soc. Test. Mater.,* vol. 42, 1942, pp. 639–658.

2666. CORNELIUS, H. Verdrehwechselfestigkeit und Schadenslinie einer aushärtbaren, ver-schieden vorbehalten Aluminium–Kupfer–Magnesium-Knetlegierung. [Torsional fatigue strength and damage lines of an age-hardened aluminium–copper–magnesium wrought alloy after various treatments.] *Metallwirtschaft,* vol. 21, no. 25/26, June 26, 1942, pp. 363–366.

2667. CORNELIUS, H. *and* BOLLENRATH, F. Ver-drehwechselfestigkeit von Stahlwellen mit hoher Zugfestigkeit. [Torsional fatigue strength of high tensile steel crankshafts.] *Z. Ver. dtsch. Ing.,* vol. 86, no. 7/8, Feb. 21, 1942, pp. 103–108. English abstr.: *Engrs' Dig.,* vol. 3, no. 12, Dec. 1942, pp. 427–429. English translation: *R.T.P. Trans.,* no. 1496, pp. 8.

2668. COX, H. N. Dynamic testing in Germany. *J. Amer. Soc. nav. Engrs,* vol. 54, no. 2, May 1942, pp. 296–315.

2669. DAASCH, H. L. Notch sensitivity of welds under repeated loading. *Weld. J., Easton, Pa.*, vol. 21, no. 2, Jan. 1942, pp. 60s–64s.

2670. DEGENHARDT, H. Ergebnisse von Betriebs-festigkeitsversuchen mit Flugzeugbauteilen. [Results of service strength tests on aircraft structures.] *Ber. Lilienthal-Ges. Luftfahrtf.*, no. 152, Aug. 11, 1942, pp. 23–28.

2671. DICK, W. Werkstoffliche Probleme bei der Weiterentwicklung des Baustoffs Stahl. [Materials problems in the further development of structural steel.] *Mitt. ForschAnst. Gutehoffn., Nürnberg*, vol. 9, 1942, pp. 150–160.

2672. DIRKSEN, B. Über die Entwicklung des Spiels in Bolzenverbindungen unter Dauer-beanspruchung. [On the development of play in bolted joints under fatigue loads.] *Luftfahrtforsch.*, vol. 19, no. 4, May 6, 1942, pp. 153–156. English abstr.: *Aircr. Prod., Lond.*, vol. 4, no. 47, Sept. 1942, p. 578.

2673. DOLAN, T. J. Effects of range of stress and of special notches on fatigue properties of alloys suitable for airplane propellers. *Tech. Note nat. Adv. Comm. Aero., Wash.*, no. 852, June 1942, pp. 26.

2674. ERDMANN-JESNITZER, H. H.; HANEMANN, H. and KOHLMEYER, J. Beitrag zur Dauer-biegefestigkeit von gewalztem Zink und von Zink-legierungen. [Contribution to the bending fatigue behaviour of rolled zinc and zinc alloys.] *Z. Metallk.*, vol. 34, no. 9, Sept. 1942, pp. 222–224. English abstr.: *Metallurg. Abstr.*, vol. 10, pt. 8, Aug. 1943, p. 245.

2675. ERKER, A. Zeit- und Dauerfestigkeit von Schweissverbindungen. [Endurance and fatigue strength of welded joints.] *Ber. Lilienthal-Ges. Luftfahrtf.*, no. 152, 1942, pp. 58–66.

2676. ERLINGER, E. Umlaufbiegemaschine für Drahtproben. [Rotating bending machine for wire specimens.] *Fördertechnik*, vol. 35, no. 5/6, 1942, pp. 43–45. German abstr.: *Metallwirtschaft*, vol. 22, no. 11/12, 1943, p. 178.

2677. ERLINGER, E. Machine d'essai pour la production des forces alternées de traction-compression. [Testing machine for the production of alternating tension-compression loads.] *Metaux, Corros-Usure*, vol. 17, no. 207, Nov. 1942, pp. 209–214.

2678. FERGUSSON, H. B. Strength of welded T-joints for ships' bulkhead plates. *Trans. Instn nav. Archit., Lond.*, vol. 84, 1942, pp. 140–147.

2679. FÖPPL, O. Das Oberflächendrücken als Mittel zur Steigerung der Dauerhaltbarkeit der im Kraftfahrzeugbau verwendeten Federn. [Surface compression as a means of increasing the fatigue strength of springs used in motor vehicles.] *Auto.-tech. Z.*, vol. 45, no. 12, June 25, 1942, pp. 321–325. English abstr.: *Engrs' Dig.*, vol. 4, no. 4, April 1943, pp. 111–112.

2680. FÖPPL, O. and HOLZER, R. Das Oberflächendrücken von Drähten zur Steigerung ihrer Dauerhaltbarkeit. [Surface compression of wires to increase their fatigue life.] *Werkstattstechnik*, vol. 36, no. 3/4, 1942, pp. 62–65

2681. FÜRST, O. and GIMBEL, G. Statisch beansp ruchte Schrauben trotz dynamischer Flansch belastung. [Statically stressed screws in spit of dynamic flange loading.] *Metallwirtschaft* vol. 21, no. 39/40, 1942, pp. 594–596.

2682. GABRIELLI, G. Sulle recenti ricerche dell sollecitazioni in volo dei velivoli e sui nuov criteri per il proporzionamento delle strutture [Some recent investigations on flight loads i aircraft and some new criteria for the pro portioning of structures.] *Aerotecnica, Rom* vol. 22, no. 9/10, Sept.–Oct. 1942, pp. 429– 436.

2683. GASSNER, E. Ergebnisse aus Betriebsfestig-keits Versuchen mit Stahl- und Leichtmetall Bauteilen. [Results of service strength tests on steel and light metal components.] *Ber. Lilienthal-Ges. Luftfahrtf.*, no. 152, 1942, pp. 13–23.

2684. GODFREY, H. J. The physical properties of steel wire as affected by variations in drawing operations. *Proc. Amer. Soc. Test. Mater.*, vol. 42, 1942, pp. 513–531.

2685. GOUGH, V. E. *and* PARKINSON, D. Dunlop fatigue test for rubber compounds. *Rubb. Chem. Technol.*, vol. 15, no. 4, Oct. 1942, pp. 905–964.

2686. GRAF, O. Versuche über das Verhalten von geschweissten Trägen unter oftmals wiederholter Belastung. [Tests on the behaviour of welded beams under often repeated loads.] *Ber. dtsch. Aussch. Stahl.*, Berl., no. 14, 1942, pp. 21.

2687. GRAF, O. Über Versuche mit Baustählen. [On tests with structural steels.] *Bauingenieur*, vol. 23, no. 5/6, 1942, pp. 31–44.

2688. GÜTH, H. Einfluss der Oberflächenbehand-lung auf die Dauerfestigkeit einer Mg–Al-Knetlegierung (Magnewin). [Influence of surface treatment on the fatigue strength of a Mg–Al wrought alloy (Magnewin).] *Metall-wirtschaft*, vol. 21, no. 35/36, Sept. 4, 1942, pp. 523–526.

2689. GUILLERY, R. Machine de fatigue pour les barrettes Caquot. [Fatigue machine for Caquot test pieces.] *Rev. Métall.*, vol. 30, no. 1, Jan. 1942, pp. 27–29.

2690. HARTMANN, E. C. *and* STICKLEY, G. W. The direct-stress fatigue strength of 17S-T aluminium alloy throughout the range from ½ to 500,000,000 cycles of stress. *Tech. Note nat. Adv. Comm. Aero.*, Wash., no. 865, Sept. 1942, pp. 6.

2691. HARTMANN, E. C. *and* STICKLEY, G. W. Summary of results of tests made by Alumi-nium Research Laboratories of spot welded joints and structural elements. *Tech. Note nat. Adv. Comm. Aero.*, Wash., no. 869, Nov. 1942, pp. 15.

2692. HEMPEL, M. Dauerprüfung von Gusseisen und Temperguss. [Fatigue testing of cast iron and malleable iron.] *Giesserei*, vol. 29, Sept. 4, 1942, pp. 302–311.

2693. HEMPEL, M. Dauerversuche an Stählen bei tiefen Temperaturen. [Fatigue tests on steels at low temperatures.] *Zent. wiss. Ber., Forsch-Ber.*, no. 1704/1, Dec. 18. 1942, pp. 64. English translation: *U.S. Air Materiel Command Translation* no. F–TS–1855–RE, July 1948, pp. 73.

2694. HEMPEL, M. *and* KRUG, H. Zug-Druck-Dauerversuche an Stahl bei höheren Tempe-raturen und ihre Auswertung nach verschiede-nen Verfahren. [Tension-compression fatigue tests on steel at high temperatures and their evaluation according to various methods.] *Mitt. K.-Wilh.-Inst. Eisenforsch.*, vol. 24, no. 7, 1942, pp. 71–95. *Arch. Eisenhüttenw.*, vol. 16, no. 7, Jan. 1943, pp. 261–268.

2695. HEMPEL, M. *and* KRUG, H. Einfluss der Streckgrenze auf die Biegewechselfestigkeit von Stahl. [The influence of the yield point on the bending fatigue strength of steel.] *Mitt. K.-Wilh.-Inst. Eisenforsch.*, vol. 24, no. 7, 1942, pp. 97–103. *Arch. Eisenhüttenw.*, vol. 16, no. 1, July 1942, pp. 27–30. English translation: *Iron St. Inst. Trans. Serv.*, no. 386, 1949.

2696. HEMPEL, M. *and* KRUG, H. Dauerfestigkeit und Dehnverhalten von Stählen in der Wärme. [Fatigue strength and elongation behaviour of steel at high temperatures.] *Z. Ver. dtsch. Ing.*, vol. 86, no. 39/40, Oct. 3, 1942, pp. 599–605.

2697. HEMPEL, M. *and* LUCE, J. Verhalten von Stahl bei tiefen Temperaturen unter Zug-druck-Wechselbeanspruchung. [Behaviour of steel at low temperatures under tension-compression alternating stresses.] *Arch. Eisenhüttenw.*, vol. 15, no. 9, March 1942, pp. 423–430.

2698. HENRY, O. H. *and* CALAMARI, P. L. The effect on the endurance limit of submerging resistance-welded fatigue specimens in a cold chamber. *Weld. J., Easton, Pa.*, vol. 21, no. 6, June 1942, pp. 291s–292s.

2699. HENRY, O. H. *and* CANNIZZARO, S. The effect on the endurance limit of submerging welded fatigue specimens in a cold chamber. *Weld. J., Easton, Pa.*, vol. 21, no. 8, Aug. 1942, pp. 387s–388s.

2700. HENRY, O. H. *and* COYNE, T. D. The effect on the endurance limit of submerging fatigue specimens in a cold chamber. *Weld. J., Easton, Pa.*, vol. 21, no. 5, May 1942, pp. 249s–254s.

2701. HEYER, K. Mehrstufenversuche an Kon-struktions-Elementen. [Multi-stage tests on structural elements.] *Ber. Lilienthal-Ges. Luftfahrtf.*, no. 152, Aug. 11, 1942, pp. 29–54.

2702. HORGER, O. *and* NEIFERT, H. R. Correlation of residual stresses in the fatigue strength of axles. *J. appl. Mech.*, vol. 9, no. 2, June 1942, pp. A85–A90.

2703. HUGE, E. C. Fatigue tests of full thickness plates with and without butt welds. *Weld. J., Easton, Pa.*, vol. 21, no. 10, Oct. 1942, pp. 507s–514s, and [discussion] vol. 22, no. 1, Jan. 1943, p. 42s.

2704. IRMANN, R. Dauerwechselfestigkeit von Aluminium und Aluminiumlegierungen. [Fa-tigue strength of aluminium and aluminium alloys.] *Schweiz. Arch. angew. Wiss.*, vol. 8, no. 2, Feb. 1942, pp. 52–64.

2705. JOHANSEN, F. C. Mechanical detection of wheel-seat flaws in railway axles. *Engineering, Lond.*, vol. 154, no. 3995, Aug. 7, 1942, pp. 101–104, and [discussion] no. 4000, Sept. 11, 1942, pp. 215–216.

2706. KELTON, E. H. Fatigue testing of zinc base alloy die castings. *Proc. Amer. Soc. Test. Mater.*, vol. 42, 1942, pp. 692–707.

2707. KIES, J. A. *and* HOLSHOUSER, W. L. Evalu-ation of fatigue damage of steel by supple-mentary tension-impact tests. *Proc. Amer. Soc. Test. Mater.*, vol. 42, 1942, pp. 556–567.

2708. KRAINER, H. Einfluss des Verschmiedung-grades auf die Biegewechselfestigkeit vo legiertem Baustahl längs und quer zur Schmie defaser. [The effects of the degree of reductio by forging on the bending fatigue strengt of low alloy steel, parallel and at right angle to the forging direction.] *Arch. Eisenhüttenw* vol. 15, no. 12, June 1942, pp. 543–54 English abstr.: *Iron Age*, vol. 152, no. 1 Oct. 7, 1943, p. 60.

2709. KÜCH, W. Zeit- und Dauerfestigkeit vc Lagenhölzern. [Endurance and fatigue streng of laminated wood.] *Holz a. Roh- u. Werks.* vol. 5, no. 2/3, Feb.–March 1942, pp. 69–7

2710. LANDAU, D. *Fatigue of metals—sor facts for the designing engineer.* Second e tion. New York, The Nitralloy Corporatio 1942, pp. 88.

2711. LAZAN, B. J. Behaviour of plastics und vibrations. *Mod. Plast.*, vol. 20, no. 3, N 1942, pp. 83–88, 136, 138, 140, 142 and 1

2712. LOCATI, L. Terminologia nella scienza della fatica dei metalli. [Terminology in the science of fatigue of metals.] *Metallurg. ital.*, vol. 34, no. 6, 1942, pp. 237–241.

2713. McADAM, D. J. The influence of the combination of principal stresses in fatigue of metals. *Proc. Amer. Soc. Test. Mater.*, vol. 42, 1942, pp. 576–594.

2714. McFARLAN, R. L. Vibration fatigue tests for seam-sealing compounds. *Aero. Dig.*, vol. 41, no. 3, Sept. 1942, pp. 211–212, 289.

2715. MAILÄNDER, R.; SZUBINSKI, W. *and* WIESTER, H. J. Metallographic investigations and bending fatigue tests on welded thin plates of high tensile steel. *Weld. J., Easton, Pa.*, vol. 21, no. 1, Jan. 1942, pp. 30s–33s.

2716. MARIN, J. Interpretation of experiments on fatigue strength of metals subjected to combined stresses. *Weld. J., Easton, Pa.*, vol. 21, no. 5, May 1942, pp. 245s–248s.

2717. MARIN, J. Designing for bending, twisting and axial loads. *Mach. Design*, vol. 14, no. 9, Sept. 1942, pp. 79–83 and no. 10, Oct. 1942, pp. 78–80, 260 and 262.

2718. MARIN, J. Strength of steel subjected to biaxial fatigue stresses. *Weld. J., Easton, Pa.*, vol. 21, no. 11, Nov. 1942, pp. 554s–559s.

2719. MARIN, J. Code for working stresses facilitates design. *Mach. Design*, vol. 14, no. 11, Nov. 1942, pp. 70–74, 158 and 160, and no. 12, Dec. 1942, pp. 99–103. *Engrs' Dig.*, vol. 4, no. 1, Jan. 1943, pp. 2–5.

2720. MARTIN, P. R. Effect of mean stress on fatigue strength of spot welds in alclad sheet.
Roy. Aircr. Estab. Rep. no. Mat/N/5655c, Sept. 1942, pp. 6.

2721. MARTINAGLIA, L. Schraubenverbindungen — Stand der Technik. [Bolted joints—stage of technique.] *Schweiz. Bauztg.*, vol. 119, no. 10, March 7, 1942, pp. 107–112, and no. 11, March 14, 1942, pp. 122–126. English translation: *Engrs' Dig.*, vol. 3, no. 7, July 1942, pp. 231–238.

2722. MOORE, H. F. *and* MORKOVIN, D. Progress report on the effect of size of specimen on fatigue strength of three types of steel. *Proc. Amer. Soc. Test. Mater.*, vol. 42, 1942, pp. 145–154.

2723. MÜLLER, W. Massnahmen zur Erhöhung der Gestaltfestigkeit von Aluminium-Knotenpunktverbindungen. [Means of increasing the form strength of aluminium spot joints.] *Schweiz. Bauztg.*, vol. 119, no. 5, Jan. 31, 1942, pp. 49–52 and no. 6, Feb. 9, 1942, pp. 65–68.

2724. MÜLLER, W. J. Beitrag zur Bestimmung des Einflusses des Beizens bezw. der anodischen Oxydation auf die Dauerwechselbiegefestigkeit verschiedener Aluminiumlegierungen. [Influence of pickling and of anodic oxidation respectively, upon the bending fatigue strength of different aluminium alloys.] *Korros. Metallsch.*, vol. 18, no. 2, Feb. 1942, pp. 56–62.

2725. MUTCHLER, W. H. *and* KIES, J. A. Fatigue tests as a means of evaluating corrosion damage of sheet metals. *Proc. Amer. Soc. Test. Mater.*, vol. 42, 1942, pp. 568–575.

2726. NISHIHARA, T. *and* KAWAMOTO, M. The effect of static tension on the torsional fatigue strength of metals. [J] *Nippon kink. Gakk.*, vol. 6, no. 6, 1942, pp. 316–330.

2727. NISHIHARA, T. *and* KOBAYASHI, A. An X-ray study on the fatigue of carbon steel. [J] *Trans. Soc. mech. Engrs, Japan*, vol. 8, no. 33, Nov. 1942, pp. 194–202.

2728. PEMBERTON, H. N. The testing of welds. *Welding, Lond.*, vol. 10, June 1942, pp. 99–104 and July 1942, pp. 131–133.

2729. VON PHILIPP, H. A. Einfluss von Querschnittsgrösse und Querschnittsform auf die Dauerfestigkeit bei ungleichmässig verteilten Spannungen. [The influence of the size and shape of cross-section on the fatigue strength under non-uniform stress distribution.]*Forsch-Arb. IngWes.*, vol. 13, no. 3, May–June 1942, pp. 99–111.

2730. VON PHILIPP, H. A. Der dynamische Fehler bei der Lastanzeige einer Universal-Schwingungsprüfmaschine. [The dynamic error in the load indicator of a universal fatigue testing machine.] *ForschArb. IngWes.*, vol. 13, no. 4, 1942, pp. 169–171.

2731. POMP, A. *and* HEMPEL, M. Dauerfestigkeit des Stahles bei erhöhten Temperaturen. [Fatigue strength of steels at high temperatures.] *Mitt. dtsch. Akad. Luftfahrtf.*, no. 5, 1942, pp. 17.

2732. PRIBRAM, E. Failure: a new definition needed. *Aircr. Engng*, vol. 14, no. 163, Sept. 1942, pp. 258–259.

2733. PUSCH, A. Grosszahl-Untersuchungen über die Ursachen von Radreifen- und Schienenbrüchen. [Qualitative investigations on the causes of fracture in tyres and rails.] *Stahl u. Eisen*, vol. 62, no. 49, Dec. 3, 1942, pp. 1022–1033.

2734. VON RAJAKOVICS, E. *and* MAIER, H. O. Untersuchungen über die Warmfestigkeits-eigenschaften von Aluminiumknetlegierungen. [Investigation on the high temperature properties of wrought aluminium alloys.] *Z. Metallk.*, vol. 34, no. 8, Aug. 1942, pp. 173–187. English translation: *R.T.P. Trans.*, no. 2317, 1942, pp. 20.

2735. VON RÖSSING, G. Die Biegewechselfestigkeit von Schmiedestücken aus legiertem Stahl in Quer- und Längsfaser. [The bending fatigue strength of forged pieces of alloy steel in the transverse and longitudinal directions.] *Arch. Eisenhüttenw.*, vol. 15, no. 9, March 1942, pp. 407–412. English translation: *R.T.P. Trans.*, no. 2333, 1942, pp. 7. *Iron St. Inst. Trans. Ser.*, no. 225, 1945.

2736. ROŠ, M. G. Festigkeit und Verformung von auf Biegung beanspruchten Eisenbetonbalken, bewehrt mit "TOR-Stahl", hochwertigem Stahl "St52", Normalstahl "St N". [Strength and deformation in bending of reinforced concrete beams, reinforced with "TOR-steel", good quality steel "St52", and standard steel "St N".] *Ber. eidgenöss. MatPrüfAnst.*, no. 141, Oct. 1942, pp. 84. [Supplement], December 1948, pp. 27.

2737. SANDERS, T. P. Drill-pipe inspection as basis for improving drilling technique. *Oil Gas J.*, vol. 41, no. 5, June 11, 1942, pp. 33–34, 37.

2738. SANDERSON, H. Corrosion fatigue testing of metals. *Pract. Engng*, vol. 5, no. 126, June 18, 1942, pp. 569–570.

2739. SAVAGE, E. G. *and* MARTIN, P. R. The corrosion fatigue properties of RR.77 alloy (D.T.D. 363) with and without protective coatings. *Roy. Aircr. Estab. Rep.*, no. M.5424, Oct. 1942, pp. 13.

2740. SAVAGE, E. G. *and* SIMPKIN, G. F. The influence of cold, hot and electrolytic chromate treatments on the fatigue properties o magnesium alloy A.Z. 855. *Roy. Aircr. Estab. Rep.*, no. M.2946E, March 1942, pp. 17.

741. SCHAAL, A. Schwingungsfestigkeit und statische Streckgrenze. [Fatigue strength and static yield point.] *Arch. Eisenhüttenw.*, vol. 16, no. 1, July 1942, pp. 21–26.

742. SCHIEFER, H. F. *and* BOYLAND, P. M. Note on flexural fatigue of textiles. *J. Res. nat. Bur. Stand.*, vol. 29, no. 1, July 1942, pp. 69–71.

743. SCHMIDT, R. Dauerbiegeversuche, Dehnungs- und Röntgenspannungsmessungen an gerichteten Kurbelwellen. [Bending fatigue strength, strain and X-ray stress measurements of straightened crankshafts.] *Luftwissen*, vol. 9, no. 9, Sept. 1942, pp. 263–267. English translation: *R.T.P. Trans.*, no. 1831, 1942, pp. 8. *Iron St. Inst. Trans. Ser.*, no. 157, 1943.

744. SCHNEIDEWIND, R. *and* HOENICKE, E. C. A study of the chemical, physical and mechanical properties of permanent mold gray iron. *Proc. Amer. Soc. Test. Mater.*, vol. 42, 1942, pp. 622–634.

745. SCHÖLLHAMMER, K. Dauerhaltbarkeit und Beanspruchungsverhältnisse des Befestigungs-gewindes am Flugmotor, insbesondere bei Schrauben aus Vergütungs-stahl. [Fatigue durability and stress behaviour of clamping threads in aircraft engines, in particular tempered steel bolts.] *Zent. wiss. Ber., Tech. Ber.* vol. 9, no. 2, 1942, pp. 51–56.

746. SEELHORST, H. Erfahrungen über des Auftreten von Ermüdungsschäden. [Experiences on the occurrence of fatigue damage.] *Ber. Lilienthal. Ges. Luftfahrtf.*, no. 152, 1942, pp. 3–9.

747. SIEBEL, E. *and* STÄHLI, G. Versuche zum Nachweis von Schädigung und Verfestigung im Gebiet der Zeitfestigkeit. [Experiments to determine the damage and work hardening during endurance tests.] *Arch. Eisenhüttenw.* vol. 15, no. 11, May 1942, pp. 519–527., English abstr.: *Metallurg. Abstr.*, vol. 13, pt. 8, Aug. 1946, pp. 291–292.

2748. SIEBEL, E.; STEURER, W. *and* STÄHLI, G. Prüfung von Leichtmetall-Legierungen bei höheren Temperaturen unter gleichzeitig ruhender und schwingender Beanspruchung. [Testing of light metal alloys at high temperatures under simultaneous static and fatigue stresses.] *Z. Metallk.*, vol. 34, no. 7, July 1942, pp. 145–150. English abstr.: *Metallurg. Abstr.*, vol. 10, pt. 4, April 1943, p. 124.

2749. SHANLEY, F R. Importance of stressing. *West. Flying*, vol. 22, no. 7, July 1942, pp. 47 and 64.

2750. SIMS, C. E. *and* DAHLE, F. B. Comparative quality of converter cast steel. *Proc. Amer. Soc. Test. Mater.*, vol. 42, 1942, pp. 532–555.

2751. SMITH, J. O. The effect of range of stress on the fatigue strength of metals. *Uni. Ill. Engng Exp. Sta. Bull.*, no. 334, 1942, pp. 52. Abstr.: *Iron Coal Tr. Rev.*, vol. 145, no. 3892, Oct. 2, 1942, pp. 885–886.

2752. SPRARAGEN, W. *and* ROSENTHAL, D. Fatigue strength of welded joints. Review of literature from Oct. 1936 to Sept. 1941. *Weld. J., Easton, Pa.*, vol. 21, no. 7, July 1942, pp. 297s–348s.

2753. STERNER-RAINER, E. *and* JUNG-KÖNIG, W. Über die Dauerbiegefestigkeit einiger Al-Legierungen unter Korrosionseinfluss im Vergleich zu Gussbronze und Rotguss. [On the bending fatigue strength of some aluminium alloys under the influence of corrosion, compared with cast bronze and red brass.] *Korros. Metallsch.*, vol. 18, no. 10, Oct. 1942, pp. 337–343.

2754. Stickley, G. W. The fatigue strengths of some wrought aluminium alloys. *N.A.C.A. War. Rep.*, no. W-83, June 1942, pp. 7.

2755. Stickley, G. W. Improvement of fatigue life of an aluminium alloy by overstressing. *Tech. Note nat. Adv. Comm. Aero., Wash.*, no. 857, Aug. 1942, pp. 11.

2756. Taylor, W. J. *and* Munro, W. Effect of mean stress on fatigue range of airscrew materials. Further tests of magnesium alloys. *Roy. Aircr. Estab. Rep.*, no. Mat/N/4/11216, July 1942, pp. 5.

2757. Thum, A. Beanspruchungsmechanismus und Gestaltfestigkeit von Nabensitzen. [Stressing mechanism and design strength of hub seats.] *Dtsch. Kraftfahrtforsch.*, no. 73, 1942, pp. 36.

2758. Thum, A. *and* Erker, A. Einfluss von Kerben und Eigenspannungen auf die Dauerhaltbarkeit von Schweissverbindungen. [Influence of notches and internal stresses on the fatigue durability of welded joints.] *Autogene Metallbearb.*, vol. 35, no. 4, Feb. 15, 1942, pp. 49–56.

2759. Thum, A. *and* Erker, A. Zeit- und Dauerfestigkeit und deren Beeinflussung durch eine einmalige hohe überlastung. [Endurance and fatigue strength and the influence of a single high overload.] *Z. Ver. dtsch. Ing.*, vol. 86, no. 11/12, March 21, 1942, pp. 171–174.

2760. Thum, A. *and* Erker, A. Die Berücksichtigung von Massenkräften bei Resonanzschwingprüfständen. [The consideration of inertia forces in resonance fatigue testing machines.] *Arch. tech. Messen*, vol. 136, V9115–5, Oct. 1942, pp. T104–105.

2761. Thum, A. *and* Erker, A. Gestaltfestigkeit von Schweissverbindungen. [Design strength of welded joints.] *Mitt. MatPrüfAnst. Darmstadt*, no. 10, 1942, pp. 146.

2762. Thum, A. *and* Lorenz, H. Vorspannung und Dauerhaltbarkeit an Schraubenverbindungen mit einer und mehreren Schrauben. [Prestressing and fatigue durability of screwed joints with one and multiple screws.] *Auto. tech. Z.*, vol. 45, no. 22, Nov. 25, 1942, pp. 610–612.

2763. Thum, A. *and* Petersen, C. Die Vorgänge im zügig und wechselnd beanspruchten Metallgefüge. II—Betrachtungen zur Dämpfungfähigkeit. [The structural changes in metal under static and alternating stresses. II—Observations on the damping capacity.] *Z. Metallk.*, vol. 34, no. 2, Feb. 1942, pp. 39–46.

2764. Tsuda, S. X-ray study on fatigue of light alloys. 1st report. *J. Cent. Aero. Res. Inst. Mitaka*, vol. 1, no. 3, June 1942, pp. 160–17

2765. Ulrich, M. Steigerung der Dauerschwingungsfestigkeit von Zahnrädern durch besondere Gestaltung, Härtung und Bearbeitung d Zahngrundes. [Increasing the fatigue strength of gear wheels by special shaping, hardening and machining of the tooth base.] *Luftwisse* vol. 9, no. 11, Nov. 1942, pp. 311–312. English translation: *R.T.P. Trans.*, no. 2159, 194 pp. 4. *Iron St. Inst. Trans. Ser.*, no. 198, 194

2766. Unckel, H. Ermüdungsversuche an Stäben aus verschiedenen Leichtmetallen Längs- und Querrichtung. [Fatigue tests bars of different light metals in the longitudinal and transverse directions.] *Metallwirtschaft*, vol. 21, no. 29/30, July 24, 194 pp. 427–429.

2767. Vernay, J. Cassures obtenues sur mé surchauffé. Essais de régénération du mét Influence de la surchauffé sur la résilien

Influence de la surchauffé sur la limite de fatigue. [Cracks obtained by overheating metal. Tests on the regeneration of metal. Influence of overheating on resilience. Influence of overheating on the fatigue limit.] *Bull. cerc. Étud. Metaux*, vol. 3, no. 5, March 15, 1942, pp. 159–169.

2768. VOLK, C. Zeitfestigkeit und Betriebshaltbarkeit. [Endurance strength and service durability.] *Metallwirtschaft*, vol. 21, no. 21/22, May 29, 1942, pp. 303–305.

2769. VOLK, C. Die Wöhlerlinie und die Normungszahlen. [The Wöhler curve and standard figures.] *Werkstattstechnik*, vol. 36, no. 7/8, 1942, pp. 160–162.

2770. WARBURTON-BROWN, D. Phenol-formaldehyde plastics under conditions of bending fatigue. *Plastics, Lond.*, vol. 6, no. 62, July 1942, pp. 210–218.

2771. WILLIAMS, F. S. Endurance tests of metals —effect of chromium plating. *Report U.S. Navy Naval Aircraft Factory, Aeronautical Materials Laboratory*, no. AML (M)-595, July 28, 1942, pp. 7.

2772. WILSON, W. M. Fatigue strength of commercial butt welds in carbon steel plates. *Weld. J., Easton, Pa.*, vol. 21, no. 10, Oct. 1942, pp. 491s–496s.

2773. WILSON, W. M.; BRUCKNER, W. H. *and* McCRACKIN, T. H. Tests of riveted and welded joints in low alloy structural steels. *Uni. Ill. Engng Exp. Sta. Bull.*, no. 337, Sept. 22, 1942, pp. 76.

2774. Causes and cures—fatigue cracks. *Prod. Engng*, vol. 13, no. 1, Jan. 1942, p. 31.

2775. Calculation and graphical representation of the fatigue strength of structural joints. *Weld. J., Easton, Pa.*, vol. 21, no. 2, Feb. 1942, pp. 87s–93s.

2776. Tank parts shotblasted to improve fatigue life. *Steel*, vol. 110, May 25, 1942, pp. 51–52.

2777. Bending fatigue strength of anodized aluminium. *Light Metals*, vol. 5, no. 53, June 1942, pp. 223–224.

2778. Oberflächendrückmaschine. [Machine for compressing the surface.] *Werkzeugmaschine*, vol. 6, no. 11, June 1942, p. 331. English translation: *Engrs' Dig.*, vol. 4, no. 1, Jan. 1943, pp. 19–20.

2779. Fatigue strength of crankshafts. *Engineering, Lond.*, vol. 154, no. 3992, July 17, 1942, pp. 58–59 and no. 3993, July 24, 1942, pp. 78–80.

2780. Endurance tests on light alloy connecting rods. *Light Metals*, vol. 5, no. 56, Sept. 1942, pp. 338–340.

2781. Dauerbrüche an geschweissten Rahmenlängsträgern. [Fatigue fractures in welded frame girders.] *Autogenschweisser*, vol. 15, no. 9, Sept. 1942, pp. 65–67.

2782. Repeated flexural stress (fatigue) test. *Mod. Plast.*, vol. 20, no. 4, Dec. 1942, pp. 95–96, 132, 134, 136 and 138.

2783. The "Pulsator" fatigue testing machine. *Engineer, Lond.*, vol. 174, no. 4537, Dec. 25, 1942, pp. 517–518. *Machinery, Lond.*, vol. 61, no. 1577, Dec. 31, 1942, pp. 742–743.

2784. Fatigue properties of light alloy sheet materials and the influence of aluminium cladding. *Sci. Tech. Memor.*, no. 3/42, 1942, pp. 9.

2785. High strength aluminium alloys for propeller blades. *Sci. Tech. Memor.*, no. C.5/42, 1942, pp. 6.

1943

2786. ALLEMAN, N. J. First progress report of the investigation of fatigue failures in rail joint bars. Rolling load tests of joint bars. *Uni. Ill. Engng Exp. Sta. Bull.*, vol. 40, no. 35, April 20, 1943, (Reprint series no. 26), pp. 12. *Proc. Amer. Rly Engng Ass.*, vol. 44, no. 437, Feb. 1943, pp. 586–596.

2787. ALMEN, J. O. Peened surfaces improve endurance of machine parts. *Metal Progr.*, vol. 43, no. 2, Feb. 1943, pp. 209–215, 270.

2788. ALMEN, J. O. Fatigue of steels as influenced by design and internal stresses. *Steel*, vol. 112, no. 10, March 8, 1943, pp. 88–89, 132–136 and no. 14, April 5, 1943, pp. 112, 146–149. *Automot. Industr. N.Y.*, vol. 88, no. 3, Feb. 1, 1943, pp. 28–32, 84 and no. 4, Feb. 15, 1943, pp. 28–31, 88.

2789. ALMEN, J. O. Fatigue failures in common machine parts. *Metal Progr.*, vol. 43, no. 5, May 1943, pp. 737–740.

2790. ALMEN, J. O. Effects of residual stress on the fatigue of metals. *Prod. Engng*, vol. 14, no. 6, June 1943, pp. 348–352.

2791. ALMEN, J. O. Improving fatigue strength of machine parts. *Iron Age*, vol. 151, no. 23, June 10, 1943, pp. 65–69, 125, 128, 131. *Mech. Engng, N.Y.*, vol. 65, no. 8, Aug. 1943, pp. 553–563. *Engrs' Dig.*, vol. 4, no. 9, Sept. 1943, pp. 249–255.

2792. ALMEN, J. O. Shot blasting to increase fatigue resistance. *S.A.E. Jl.*, vol. 51, no. 7, July 1943, pp. 248–268.

2793. ALMEN, J. O. The useful data to be derivec from fatigue tests. *Metal Progr.*, vol. 44, no. 2 Aug. 1943, pp. 254–261.

2794. ALMEN, J. O. Bolt failure as affected b tightening. *Mach. Design*, vol. 15, no. 8, Aug 1943, pp. 133–134.

2795. ALMEN, J. O. Endurance of machine under a few heavy loads. *Metal Progr.*, vol. 44 no. 3, Sept. 1943, pp. 435–440.

2796. ALTON, F. The physical testing of powe plant materials. *Engng Boil. Ho. Rev.*, vo 57, no. 5, May 1943, pp. 122–127.

2797. ALTSHULER, L. V.; RESHETKINA, N. A SPEKTOR, A. G. and TSUKERMAN, B. A. Plast deformation and surface fatigue of hardene steel in antifriction bearings. [R] *Zh. tek. Fiz.*, vol. 13, no. 6, 1943, pp. 265–280.

2798. ANDREWS, H. J. and STICKLEY, G. V Effect of scratches on fatigue strength Alclad sheet. *Aviation, N.Y.*, vol. 42, no. June 1943, pp. 154–155, 157.

2799. ARNOLD, S. M. Effect of screw threads fatigue. *Mech. Engng, N.Y.*, vol. 65, no. July, 1943, pp. 497–505. *Mech. Worr* vol. 114, no. 2966, Nov. 5, 1943, pp. 523–52

2800. ARNOLD, S. M. Screw threads. *Au Engr*, vol. 33, no. 444, Dec. 1943, pp. 54 546.

2801. BARDGETT, W. E. Factors affecting t influence of nitriding on fatigue streng *Metal Treatm.*, vol. 10, no. 34, Summer 19. pp. 87–101. Abstr.: *Iron Age*, vol. 153, no. Jan. 27, 1944, p. 55.

2802. BLAND, R. B. and SANDORFF, P. E. The control of life expectancy in airplane structures. *Aeronaut. Engng Rev.*, vol. 2, no. 8, Aug. 1943, pp. 7, 9, 11, 13, 15–16, 21.

2803. BRUDER, E. Maschine zur Erzeugung synchroner, kombinierter Biege-Verdreh-Dauerbeanspruchungen. [Machine for the application of synchronized combined bending and torsion fatigue stresses.] *Z. Ver. dtsch. Ing.*, vol. 87, no. 5/6, Feb. 6, 1943, p. 82, English abstr.: *Engrs' Dig.*, vol. 5, no. 10, Oct. 1944, p. 302. English translation: *R.T.P. Trans.*, no. 2155, p. 1.

2804. BRUDER, E. Hohlkehlen-Rundschleifgerät für Dauerbiege-Probestäbe. [Fillet cylindrical grinding equipment for bending fatigue specimens.] *Z. Ver. dtsch. Ing.*, vol. 87, no. 15/16, April 17, 1943, p. 224.

2805. BRUNNER, P. Die bestimmenden Einflüsse auf die Biegewechselfestigkeit von Schweissverbindungen unter Ausschaltung der Formeinflüsse. [The main factors influencing the bending fatigue strength of welded joints, excluding the influence of form.] *Elektroschweissung*, vol. 14, no. 7, July 1943, pp. 85–97.

2806. BUCHMANN, W. Einfluss der Querschnittsgrösse auf die Dauerfestigkeit. [Influence of section size on the fatigue strength.] *Z. Ver. dtsch. Ing.*, vol. 87, no. 21/22, May 29, 1943, pp. 325–327. English abstr.: *Engrs' Dig.*, vol. 6, no. 1, Jan. 1945, pp. 22–23.

2807. BUNGARDT, W. and OSSWALD, E. Presseffekt und Biegedauerfestigkeit einiger Aluminium-Kupfer-Magnesium-Knetlegierungen mit verschiedenen Mangangehalten. [Press effect and the bending fatigue strength of some aluminium–copper–magnesium wrought alloys with various manganese contents.] *Zent. wiss. Ber., Untersuch. Mitt.*, no. 1139, Dec. 1943, pp. 8.

2808. BÜRNHEIM, H. Über das oberflächendrükken gekerbter Probestäbe aus dem Cr–Mn–V stahl VCV 100 (Einfluss der Drückbedingungen, der Lastspeilfrequenz, der Lastspeilzahl und einer zügigen Vorbeanspruchung.) [The surface compression of notched specimens of the Cr–Mn–V steel VCV 100 (influence of the compression conditions, frequency of loading, number of load fluctuations and prior tensile stressing.)] *Luftfahrtforsch.*, vol. 20, no. 1, Jan. 20, 1943, pp. 16–21.

2809. BURGHOFF, H. L. and BLANK, A. I. Fatigue tests on some copper alloys in wire form. *Proc. Amer. Soc. Test. Mater.*, vol. 43, 1943, pp. 774–784. Abstr.: *Metallurgia, Manchr*, vol. 32, no. 190, Aug. 1945, p. 189.

2810. CAZAUD, R. and PERSOZ, L. *La fatigue des métaux.* [*The fatigue of metals.*] Second edition. Paris, Dunod, 1943, pp. 259.

2811. CLEFF, T. Leichtbaugestaltung im Grossfahrzeugbau. [Light alloys in the design of large transport vehicles.] *Z. Ver. dtsch. Ing.*, vol. 87, no. 25/26, June 26, 1943, pp. 377–384.

2812. COLLACOTT, R. A. Fatigue strength of machinery under alternating stresses. *Steam Engr*, vol. 12, no. 144, Sept. 1943, pp. 345–347.

2813. CORNELIUS, H. Festigkeitseigenschaften, Korrosions- und Witterungsverhalten hochwertiger Stahlbänder. [Strength properties, corrosion and atmospheric durability of high quality steel band.] *Luftfahrtforsch.*, vol. 20, no. 1, Jan. 20, 1943, pp. 1–15.

2814. CORNELIUS, H. and BOLLENRATH, F. Festigkeitseigenschaften hochfester Lichtbogenschweissverbindungen aus Stahl. [Strength properties of high strength arc-welded steel joints.] *Luftfahrtforsch.*, vol. 20, no. 6, June

30, 1943, pp. 175–180. English abstr.: *Engrs' Dig.*, vol. 5, no. 2, Feb. 1944, pp. 51–53.

2815. CRAMER, R. E. First progress report of the investigation of Shelly spots in railroad rails. *Uni. Ill. Engng Exp. Sta. Bull.*, vol. 40, no. 34, April 13, 1943, (Reprint series no. 25), pp. 12. *Proc. Amer. Rly Engng Ass.*, vol. 44, no. 437, Feb. 1943, pp. 601–610.

2816. CRATE, H. A preliminary study of machine-countersunk flush rivets subjected to a combined static and alternating shear load. *Nat. Adv. Comm. Aero.*, *Wash.*, *Restricted Bulletin* no. RB 3LO1, Dec. 1943, pp. 3.

2817. DAEVES, K. Zur untersuchung von Dauerbrüchen. [The investigation of fatigue fractures.] *Wärme*, vol. 66, no. 10, 1943, pp. 81–82.

2818. DANNENMULLER, M. Nouveautés dans les essais sur antifrictions. [New methods for testing bearing metals.] *Métaux, Corros-Usure*, vol. 18, no. 217, Sept. 1943, pp. 155–169.

2819. DEARDEN, J. The influence of welding defects on the resistance to fatigue of welded steel joints. *Trans. Inst. Weld.*, vol. 6, no. 3–4, July–Oct. 1943, pp. 120–122.

2820. DEARDEN, J. Welded and riveted joints compared. *Metal Treatm.*, vol. 10, no. 36, Winter 1943–1944, pp. 207–210, 232.

2821. DIETZ, A. G. H. *and* GRINSFELDER, H. Behaviour of plywood under repeated stresses. *Trans. Amer. Soc. mech. Engrs*, vol. 65, April 1943, pp. 187–191.

2822. DOLAN, T. J. Certain mechanical strength properties of aluminium alloys 25S-T and X76S-T. *Tech. Note nat. Adv. Comm. Aero.*, *Wash.*, no. 914, Oct. 1943, pp. 30.

2823. DONALDSON, J. W. The fatigue and corrosion-fatigue of copper alloys. *Metal Ind.*, *Lond.*, vol. 63, no. 12, Sept. 17, 1943, pp. 178–180 and no. 13, Sept. 24, 1943, pp. 198–200.

2824. DUGGAN, F. W. *and* FLIGOR, K. K. Fatigue resistance of flexible plastic sheetings. *Industr. Engng Chem.*, vol. 35, no. 2, Feb. 1943, pp. 172–176.

2825. ENOMOTO, S. Internal friction and fatigue of metallic materials. 1st report—mechanism of fatigue in annealed mild steel. [J] *J. Cent Aero. Res. Inst.*, *Mitaka*, vol. 2, no. 7, July 1943, pp. 177–190.

2826. ENOMOTO, S. Internal friction and fatigue of metallic materials. 2nd report—mechanism of internal friction. [J] *J. Cent. Aero. Res. Inst.*, *Mitaka*, vol. 2, no. 10, Oct. 1943 pp. 305–322.

2827. ERIKSEN, E. L. *and* HANSEN, H. M. Final report on fatigue strength of selected gun steels under combined stress: results of combined bending and torsion fatigue tests o SAE 4340 steel. *Rep. U.S. Off. Sci. Re Developm.*, no. 1523, Ser. no. M-61, June 1943, pp. 19.

2828. ERLINGER, E. Eine neue Baureihe schwingender Zug-Druck-Maschinen. [A new type of tensile-compression fatigue testing machine.] *Metallwirtschaft*, vol. 22, no. 1/2, Jan 8, 1943, pp. 12–17.

2829. FERGUSSON, H. B. Strength of welded T-joints for ships' bulkhead plates. *Weld. J Easton, Pa.*, vol. 22, no. 2, Feb. 194 pp. 57s–62s.

2830. FIELD, G. H.; SUTTON, H. *and* DIXON H. E. Second interim report of the R.52 Sub committee on the Spot Welding of Ligh

Alloys. *Trans. Inst. Weld.*, vol. 6, no. 2, April 1943, pp. 49–68.

2831. FIELD, P. M. Basic physical properties of laminates. *Mod. Plast.*, vol. 20, no. 12, Aug. 1943, pp. 91–102, 126, 128, 130.

2832. FINDLEY, W. N. Mechanical tests of macerated phenolic molding material. *Nat. Adv. Comm. Aero., Wash. Advance Restricted Report* No. 3F19 (*Wartime Report* No. W-99), June 1943, pp. 36.

2833. FINDLEY, W. N. Mechanical tests of cellulose acetate. *Mod. Plast.*, vol. 20, no. 7, March 1943, pp. 99–105, 138. *Trans. Amer. Soc. mech. Engrs*, vol. 65, July 1943, pp. 479–487.

2834. FINDLEY, W. N. *and* HINTZ, O. E. The relation between results of repeated blow impact tests and of fatigue tests. *Proc. Amer. Soc. Test. Mater.*, vol. 43, 1943, pp. 1226–1239. *Mod. Plast.*, vol. 21, no. 4, Dec. 1943, pp. 119–123.

2835. FINK, C. G.; TURNER, W. D. *and* PAUL, G. T. Zinc yellow in the inhibition of corrosion fatigue of steel in sodium chloride solution. *Trans. electrochem. Soc.*, vol. 83, 1943, pp. 377–401.

2836. FÖPPL, O. Die günstige Wirkung des Oberflächendrückens auf die Dauerhaltbarkeit. [The favourable effect of surface compression on fatigue durability.] *Fertigungstechnik, Münch.*, vol. 1, no. 5, 1943, pp. 107–109.

2837. FÖPPL, O. Einfluss von Querschnittsgrösse und form auf die Dauerfestigkeit bei ungleichmässig verteilten Spannungen. [Influence of size and shape of cross section on the fatigue strength under asymmetrically distributed stress.] *Metallwirtschaft*, vol. 22, no. 39/41, Dec. 20, 1943, pp. 552–553.

2838. FULLER, F. B. *and* OBERG, T. T. Fatigue characteristics of natural and resin-impregnated, compressed, laminated woods. *J. aero. Sci.*, vol. 10, no. 3, March 1943, pp. 81–85.

2839. GABRIEL, A. *and* PIWOWARSKY, E. Dauerschlagfestigkeit von Grauguss und Stahlguss bei Querschnittsabmessungen von Flugmotoren-Kolbenringen. [Impact fatigue strength of grey iron and cast steel with sectional dimensions of aero-engine piston rings.] *Kolbenring*, no. 15, 1943, pp. 96–100. German abstr.: *Stahl u. Eisen*, vol. 64, no. 25, June 22, 1944, pp. 405–406.

2840. GASSNER, E. *and* TEICHMANN, A. Zur Aufstellung von Bemessungsregeln für Bauteile aus Flugwerkstoff 1620.5. Mitteilung Nr. 1. [The establishment of rules for the design of structural members of aircraft materials 1620.5. Report No. 1.] *Ber. dtsch. VersAnst. Luftf.*, no. Cf 407/10, March 1943, pp. 18.

2841. GAUSS, F. Die Biegedauerhaltbarkeit von Kurbelwellenverbindungen. [The bending fatigue life of crankshaft joints.] *Luftfahrtforsch.*, vol. 19, no. 10/12, Jan. 11, 1943, pp. 347–352.

2842. GEIGER, J. Untersuchung von verschiedenen Tempergusssorten bezüglich ihrer Drehwechselfestigkeit im ungekerbter und gekerbten Zustand und ihrer Dämpfungsfähigkeit. [Investigation of different types of malleable cast iron with reference to their torsional fatigue strength in the unnotched and notched condition, and their damping capacity.] *Giesserei*, vol. 30, no. 6, March 19, 1943, pp. 85–92.

2843. GREEN, V. E. The mechanical testing of materials with special reference to the testing of welds. *Engng Insp.*, vol. 8, no. 1, Spring 1943, pp. 24–28, 32.

2844. GROVER, H. J. The use of electric strain gages to measure repeated stress. *Proc. Soc. exp. Stress Anal.*, vol. 1, no. 1, 1943, pp. 110–115.

2845. GUERKE, R. M. *and* KNAPP, G. P. Electronic test method is found for simulating operation of propellers. *S.A.E. Jl.*, vol. 51, no. 5, May 1943, pp. 35–36.

2846. GUEVARA LIZAUR, J. Conferencia sobre fatiga de corrosión. [Conference on corrosion fatigue.] Ministerio de Aire, Publicaciones del Instituto Nacional de Tecnica Aeronautica, Madrid, 1943, pp. 31.

2847. HARTMANN, E. C.; WESTCOAT, C. F. *and* BRENNECKE, M. W. Prescription for head cracks in 24ST rivets. *Aviation, N.Y.*, vol. 42, no. 11, Nov. 1943, pp. 139–143, 295–296.

2848. HEMPEL, M. Wechselbeanspruchung von Stahl in der Kälte und Wärme. [Fatigue stressing of steel at low and elevated temperatures.] *Technische Mitteilungen HDT, Essen*, vol. 36, no. 1/2, 1943, pp. 5–7.

2849. HEMPEL, M. Einfluss von Querschnittsgrösse und -form auf die Dauerfestigkeit. [Influence of cross-sectional size and shape on the fatigue strength.] *Stahl u. Eisen*, vol. 63, no. 20, May 20, 1943, pp. 402–404.

2850. HEMPEL, M. *and* KRUG, H. Wechselfestigkeits-Schaubilder von Stählen bei höheren Temperaturen. [Fatigue strength diagrams for steels at high temperatures.] *Arch. Eisenhüttenw.*, vol. 16, no. 7, Jan., 1943, pp. 261–268. English abstr.: *J. Iron St. Inst.*, vol. 150, no. 2, 1944, p. 88A.

2851. HENRY, O. H. *and* STIRBA, A. The effect on the endurance limit of submerging fatigue specimens in a cold chamber. *Weld. J., Easton, Pa.*, vol. 22, no. 8, Aug. 1943, pp. 372s–373s.

2852. L'HERMITE, R. La résistance des matières plastiques aux efforts bruts et aux efforts répétés. [The resistance of plastics to single and repeated loads.] *Plastiques*, vol. 1, no. 1, July 1943, pp. 4–8.

2853. L'HERMITE, R. *and* SÉFÉRIAN, D. Les essais d'endurance en flexion répétée comme critère de soudabilité des tôles. [Endurance tests in repeated bending as a criterion for the weldability of sheet steel.] *Rapp. Gr. franc. Rech aéro.*, no. 10, 1943, pp. 56.

2854. HIRST, G. W. C. Cracks in wheel seat within the hubs of wheels. *Engineering, Lond.* vol. 156, no. 4067, Dec. 24, 1943, pp. 501–502

2855. HORGER, O. J. *and* BUCKWALTER, T. V Fatigue strength of normalised and tempered versus as-forged full size railroad car axles *Trans. Amer. Soc. Metals*, vol. 31, no. 3 Sept. 1943, pp. 559–581.

2856. HORGER, O. J.; NEIFERT, H. R. *an* REGEN, R. R. Residual stresses and fatigu studies. *Proc. Soc. exp. Stress Anal.*, vol. 1 no. 1, 1943, pp. 10–18.

2857. HUGONY, E. *and* MONTICELLI, M. Com portamento alle sollecitazioni meccaniche statiche e dinamiche, di 2 leghe tipo durallu minio allo stato di getto e di estruso. [Th behaviour under mechanical stress, static an dynamic, of two duralumin type alloys, in th cast and extruded condition.] *Alluminie* vol. 12, no. 1, 1943, pp. 1–7.

2858. IGARASHI, I. *and* HUKAI, S. A microscop study of fatigue fracture. [J] *Trans. So mech. Engrs, Japan*, vol. 9, no. 37, Nov. 194. pp. I-161–I-167.

2859. JACKER, O. Einsatz von Werkstoffprüf-maschinen für Bauteilversuche im Flugzeug-bau. [Construction of material testing machines for tests on components of aircraft structures.] *Metallwirtschaft*, vol. 22, no. 21/23, June 20, 1943, pp. 326–334.

2860. JACKSON, L. R.; GROVER, H. J.; BEAVER, W. W. *and* RUSSELL, H. W. Fatigue properties of magnesium alloys and structures: Fatigue properties of magnesium alloy sheet—part I. *Rep. U.S. Off. Sci. Res. Developm.*, no. 3033, Dec. 23, 1943, pp. 141.

2861. JACQUESSON, R. Contribution à l'étude des dislocations cristallines produites dans les métaux par des torsions simples et alternées. [Contribution to the study of crystalline dislocations produced in metals by simple and alternating torsion.] *Publ. sci. Serv. Rech. Aéro.*, no. 188, 1943, pp. 91. English abstr.: *Bull. Brit. non-ferr. Met. Ass.*, no. 205, July 1946, p. 206.

2862. JOHNSTONE, W. W. Fatigue tests on flash-welded chrome–molybdenum steel tubing. *Note Div. Aeronautics, Aust.*, no. S & M 50, Feb. 1943, pp. 4.

2863. KELLER, H. *and* KLEIN, R. Der Einfluss von Schweissfehlern auf die Statische und Dynamische Festigkeit von Schweissverbindungen aus St.52 und die Grenzen der Röntgenuntersuchung Fehlerhafter Schweissungen. [The influence of weld defects on the static and dynamic strength of welded joints of St.52 and the limitations of X-ray examinations of defective welds.] *Schiff u. Werft*, vol. 44/24, no. 17/18, 1943, pp. 257–261.

2864. KENYON, J. N. The fatigue properties of some cold-drawn nickel alloy wires. *Proc. Amer. Soc. Test. Mater.*, vol. 43, 1943, pp. 765–773. Abstr.: *Metallurgia, Manchr*, vol. 32, no. 187, May 1945, pp. 42–43.

2865. KENYON, J. N. The service behaviour of plated steel wire as indicated by corrosion fatigue test. *Wire and Wire Prod.*, vol. 18, no. 8, Aug. 1943, pp. 449–450, 471.

2866. KHRUSHCHOV, M. M. *Fatigue of babbits.* [R] Moscow, Izdatel'stvo Akademii Nauk SSSR, 1943, pp. 139.

2867. KIES, J. A. *and* HOLSHOUSER, W. L. Effects of prior fatigue-stressing on the impact resistance of chromium–molybdenum aircraft steel. *Tech. Note nat. Adv. Comm. Aero., Wash.*, no. 889, March 1943, pp. 47.

2868. KLÖPPEL, K. Zur Frage der Vorschriften für geschweisste Fachwerkkrane. [On the question of regulations for welded framework cranes.] *Z. Ver. dtsch. Ing.*, vol. 87, no. 31/32, Aug. 7, 1943, pp. 501–503.

2869. KNOWLTON, H. B.; SAILER, F. *and* SNYDER, E. H. Axles and shafts—alternate and ideal steels: medium carbon, heat treated steels. *Metal Progr.*, vol. 44, no. 6, Dec. 1943, pp. 1104–1111.

2870. KOMMERS, J. B. The effect of overstressing and understressing on fatigue. *Proc. Amer. Soc. Test. Mater.*, vol. 43, 1943, pp. 749–764.

2871. KOMMERS, J. B. Progressive changes in fatigue failure. *Mach. Design*, vol. 15, no. 9, Sept. 1943, pp. 108–110, 196.

2872. KOMMERS, W. J. Effect of 5000 cycles of repeated bending stresses on five-ply Sitka spruce plywood. *Rep. For. Prod. Lab., Madison*, no. 1305, Feb. 1943, pp.

2873. KOMMERS, W. J. Effect of ten repetitions of stress on the bending and compressive strengths of Sitka spruce and Douglas fir. *Rep. For. Prod. Lab.*, *Madison*, no. 1320, April 1943, pp. 5.

2874. KOMMERS, W. J. The fatigue behaviour of wood and plywood subjected to repeated and reversed bending stresses. *Rep. For. Prod. Lab.*, *Madison*, no. 1327, Oct. 1943, pp. 9.

2875. LANDWERLIN, H. Les phénomènes de fatigue des métaux. [The phenomenon of metal fatigue.] *Tech. mod.*, vol. 35, no. 23/24, Dec. 1/15, 1943, pp. 189–190. English abstr.: *Engrs' Dig.*, vol. 5, no. 10, Oct. 1944, p. 312.

2876. LAZAN, B. J. Some mechanical properties of plastics and metals under sustained vibrations. *Trans. Amer. Soc. mech. Engrs*, vol. 65, no. 2, 1943, pp. 87–104.

2877. LEHR, E. and RUEF, R. Beitrage zur Frage der Dauerhaltbarkeit der Kurbelwellen von Gross-Dieselmotoren. [Contribution to the problem of the fatigue durability of crankshafts of large diesel engines.] *M T Z*, vol. 5, no. 11/12, Dec. 1943, pp. 349–357. English abstr.: *Engrs' Dig.*, vol. 5, no. 10, Oct. 1944, pp. 285–288.

2878. LEHR, E. and SKIBA, A. Dauerprüfmaschine zur Ermittlung der Drehwechselfestigkeit grosser Konstruktionsteile. [Fatigue testing machine for determining the alternating torsional strength of large constructional components.] *M T Z*, vol. 5, no. 6/7, July 1943, pp. 175–182. German abstr.: *Z. Ver. dtsch. Ing.*, vol. 88, no. 3/4, Jan. 22, 1944, p. 54. English abstr.: *Engrs' Dig.*, vol. 6, no. 4, April 1945, pp. 85–88.

2879. LEWIS, D. Progress report on the corrosion fatigue failure of aircraft control cables—effect of lubrication on fatigue properties. *Rep. U.S. off. Sci. Res. Developm.*, no. 1137, Ser. no. M-31, Jan. 12, 1943, pp. 45.

2880. LEWIS, D. Progress report on the corrosion fatigue of aircraft control cables: the effect of metallic coatings and lubricants on fatigue properties. *Rep. U.S. off. Sci. Res. Developm.*, no. 1525, Ser. no. M-83, June 15, 1943, pp. 101.

2881. LEWIS, D.; FLURY, A. H. and GODRFEY, H. J. Progress report on the corrosion-fatigue failure of aircraft cables: the effect of sheave diameter on the fatigue life of aircraft cables. *Rep. U.S. off. Sci. Res. Developm.*, no. 1610, June 4, 1943, pp. 33.

2882. McCLELLAND, A. E. The maintenance of colliery wire ropes. *Colliery Guard.*, vol. 166, no. 4284, Feb. 5, 1943, pp. 153–159. Abstr.: *J. Iron St. Inst.*, vol. 147, 1943, pp. 184A–185A.

2883. MARTINAGLIA, L. Structural durability of crankshafts. *Sulzer tech. Rev.*, no. 2, 1943, pp. 19–28.

2884. MATTSON, R. L. and ALMEN, J. O. Progress report on effect of shot blasting on mechanical properties of steel (NA-15). *Rep. U.S. off. Sci. Res. Developm.*, no. 1205, Ser. no. M-40, Feb. 10, 1943, pp. 29.

2885. MOORE, H. F. and MORKOVIN, D. Second progress report on the effect of size of specimen on fatigue strength of three types of steel. *Proc. Amer. Soc. Test. Mater.*, vol. 43, 1943, pp. 109–124.

2886. NEERFELD, H. and MÖLLER, H. Zur Frage des Spannungsabbaus durch Schwingungsbeanspruchung. [On the question of stress rupture by means of fatigue stressing.] *Jb. 1943 dtsch. Luftfahrtf.*, pp. 46–54.

2887. NEUGEBAUER, G. H. Stress concentration factors and their effect in design. *Prod.*

Engng, vol. 14, no. 2, Feb., 1943, pp. 82–87 and no. 3, March 1943, pp. 168–172.

2888. NIEMANN, G. Walzenfestigkeit und Grübchenbildung von Zahnrad- und Wälzlager-Werkstoffen. [Rolling strength and the formation of pitting in materials used for gears and rolling bearings.] *Z. Ver. dtsch. Ing.*, vol. 87, no. 33/34, Aug. 21, 1943, pp. 521–523.

2889. NISHIHARA, T. *and* KAWAMOTO, M. Studies on fatigue of mild steel by a corrosion method. *Mem. Coll. Engng Kyoto*, vol. 11, no. 3, Nov. 1943, pp. 31–60.

2890. NISHIHARA, T. *and* KAWAMOTO, M. The effect of Almite on the fatigue resistance of aluminium under combined alternating bending and torsion. [J] *Nippon kink. Gakk.*, vol. 7, no. 12, 1943, pp. 539–544.

2891. NORTON, J. T. *and* ROSENTHAL, D. An investigation of the behaviour of residual stresses under external load and their effect on safety. *Weld. J., Easton, Pa.*, vol. 22, no. 2, Feb. 1943, pp. 63s–78s.

2892. OBERG, T. P.; SCHWARTZ, R. T. *and* SHINN, D. A. Mechanical properties of plastics at normal and subnormal temperatures. *Mod. Plast.*, vol. 20, no. 8, April 1943, pp. 87–100, 122, 124, 126 and 128.

2893. OETTEL, R. Zügig beanspruchte Schraubenverbindungen mit überlagerter Wechselbeanspruchung. [Screwed connections under tensile stress with superimposed alternating stress.] *Zent. wiss. Ber., Tech. Ber.*, no. 10, 1943, pp. 353–360.

2894. OSCHATZ, H. Kleine französische Schwingprüfmaschinen. [Small French fatigue testing machine.] *Metallwirtschaft*, vol. 22, no. 39/41, Dec. 20, 1943, pp. 558–559.

2895. OSTERSPEY, J. P. Untersuchung über die Spannungsverteilung in Stangenköpfen und deren Dauerhaltbarkeit. [Investigation of the stress distribution in bar ends and of their fatigue durability.] *ForschArb. IngWes.*, vol. 14, no. 3, 1943, pp. 65–77. German abstr.: *Z. Ver. dtsch. Ing.*, vol. 88, no. 35/36, Sept. 2, 1944, pp. 492–493.

2896. PETERSON, R. E. Application of stress concentration factors in design. *Proc. Soc. exp. Stress Anal.*, vol. 1, no. 1, 1943, pp. 118–127.

2897. PHILLIPS, C. E. Rotating bending fatigue tests on four airscrew materials extended to endurances of 500 million cycles. *Rep. aero. Res. Comm., Lond.*, no. 7128, July 28, 1943, pp. 6.

2898. PIWOWARSKY, E. Gusseisen als Werkstoff. [Cast irons as materials.] *Giesserei*, vol. 30, no. 12/13, June 1943, pp. 141–152. English abstr.: *Engrs' Dig.*, vol. 5, no. 3, March 1944, pp. 61–64.

2899. POMP, A. *and* HEMPEL, M. Dauerversuche an Stählen bei tiefen Temperaturen. [Fatigue tests on steels at low temperatures.] *Jb. 1943 dtsch. Luftfahrtf.*, pp. 1–19.

2900. POTTER, E. V. An automatic frequency-controlled oscillator and amplifier for driving mechanical vibrators. *Rev. sci. Instrum.*, vol. 14, no. 7, July 1943, pp. 207–215.

2901. PÜNGEL, W.; GEROLD, E. *and* BEIDERMÜHLE, A. Einfluss der Dicke auf die Eigenschaften von Stahlseilen. [Influence of thickness on the properties of steel cables.] *Z. Ver. dtsch. Ing.*, vol. 87, no. 31/32, Aug. 7, 1943, pp. 493–497.

2902. VON RAJAKOVICS, E. Über Einflüsse auf die Schwingungsfestigkeit von Aluminium-

Legierungen. [Factors influencing the fatigue strength of aluminium alloys.] *Metallwirtschaft*, vol. 22, no. 15/17, April 20, 1943, pp. 225–239. Abstr.: *Z. Ver. dtsch. Ing.*, vol. 87, no. 27/28, July 10, 1943, p. 418. English abstr.: *Metals and Alloys*, vol. 19, no. 6, June 1944, pp. 1508, 1510.

2903. von Rajakovics, E. Über die Prüfung der Laufeigenschaften von Lagermetallen. [On the testing of the properties of bearing metals.] *Metallwirtschaft*, vol. 22, no. 24/26, 1943, pp. 361–368.

2904. von Rajakovics, E. *and* Teubler, A. Untersuchungen über die Scherfestigkeit von Leichtmetallnieten bei dynamischer Beanspruchung. [Investigation of the shear strength of light metal rivets under dynamic stress.] *Metallwirtschaft*, vol. 22, no. 9/10, March 1943, pp. 129–134.

2905. Roberts, A. M. *and* Grayshon, R. J. W. The effect of surface finish on the fatigue strength of two high tensile steels. *Sci. Tech. Memor.*, no. 6/43, 1943, pp. 8.

2906. Roberts, A. M. *and* Mather, I. S. The fatigue properties of welded joints in D.T.D. 347 steel tubing. *Sci. Tech. Memor.*, no. 4/43, 1943, pp. 11.

2907. Rudorff, D. W. Effects and prevention of surface corrosion of aluminium alloys. *Metallurgia, Manchr*, vol. 28, no. 166, 1943, pp. 157–161.

2908. Russell, H. W. Progress report on fatigue properties of magnesium alloys and structures: fatigue properties of magnesium alloy sheet. *Rep. U.S. off. Sci. Res. Developm.*, no. 1146, Ser. no. M-36, Jan. 19, 1943, pp. 36.

2909. Russell, H. W. Progress report on fatigue properties of magnesium alloys and structures:

fatigue properties of magnesium alloy sheet. *Rep. U.S. off. Sci. Res. Developm.*, no. 1457, Ser. no. M-76, May 17, 1943, pp. 78.

2910. Russell, H. W. Quarterly progress report on fatigue properties of magnesium alloys and structures: fatigue properties of magnesium alloy sheet (NA-145). *Rep. U.S. off. Sci. Res. Developm.*, no. 1858, June 1943, pp. 83.

2911. Russell, H. W. Fatigue strength and related characteristics of spot-welded joints in 24S-T alclad sheet. *N.A.C.A. Advance Restricted Report* no. 3L01 (*Wartime Report* no. W-61), Dec. 1943, pp. 29.

2912. Russell, H. W.; Gillett, H. W.; Jackson, L. R. *and* Foley, G. M. The effect of surface finish on the fatigue performance of certain propeller materials. *Tech. Note nat. Adv. Comm. Aero., Wash.*, no. 917, Dec. 1943, pp. 11.

2913. Russell, H. W. *and* Jackson, L. R. Progress report on fatigue of spot-welded aluminium. *N.A.C.A. War. Rep.*, no. W-38, Feb. 1943, pp. 26.

2914. Russell, H. W.; Jackson, L. R.; Grover, H. J. *and* Beaver, W. W. Fatigue characteristics of spot-welded 24S-T aluminium alloy. *N.A.C.A. Advance Restricted Report* no. 3F16 (*Wartime Report* no. W-64), June 1943, pp. 58.

2915. Sawert, W. Verhalten der Baustähle bei wechselnder mehrachsiger Beanspruchung [Behaviour of structural steels under reversed multi-axial stress.] *Z. Ver. dtsch. Ing.*, vol. 87 no. 39/40, Oct. 2, 1943, pp. 609–615.

2916. Seliger, V. Effect of rivet pitch upon the fatigue strength of single row riveted joint of .025 to .025 inch 24S-T alclad. *Tech. Note nat. Adv. Comm. Aero., Wash.*, no. 900, July 1943, pp. 9.

2917. SHANNON, J. F. *and* LUCK, G. A. Preliminary fatigue experiments on W.2/500 turbine blades using pneumatic excitation. *Roy. Aircr. Estab. Tech. Note*, no. ENG. 93, Jan. 1943, pp. 3.

2918. STÄHLI, G. Das Verhalten von Leichtmetall-Kolbenlegierungen bei gleichzeitiger ruhender und schwingender Zug-Druck-Beanspruchung in der Wärme. [The behaviour of light metal piston alloys under simultaneous static and alternating tension-compression stressing at elevated temperatures.] *Zent. wiss. Ber., ForschBer.*, no. 1727, Jan. 1943, pp. 38.

2919. STUART, N. *and* EVANS, U. R. The effect of zinc on the corrosion-fatigue life of steel. *J. Iron St. Inst.*, vol. 147, no. 1, 1943, pp. 131–144. *Engineering, Lond.*, vol. 155, no. 4032, April 23, 1943, pp. 336–337. *Iron and Steel, Lond.*, vol. 16, no. 10, May 20, 1943, pp. 387–389.

2920. SWINDEN, T. Leaded manganese–molybdenum steel. *J. Iron St. Inst.*, vol. 148, no. 2, 1943, pp. 441p–474p.

2921. TAPSELL, H. J. *and* THORPE, P. L. Fatigue properties at 200°C of some light alloy impeller and compressor materials. *Rep. aero. Res. Comm., Lond.*, no. 6935, Aug. 3, 1943, pp. 5.

2922. TAYLOR, W. J. *and* BINNING, M. S. Endurance tests of flexible non-corrodible ordinary and preformed steel wire rope. *Roy. Aircr. Estab. Rep.* no. Mat/N/1/13875, Aug. 1943, pp. 16.

2923. THUM, A. Über den Einfluss der Schnittbedingungen auf die Dauerfestigkeit von Leichtmetallen. [The influence of machining conditions on the fatigue strength of light metals.] *Metallwirtschaft*, vol. 22, no. 15/17, April 20, 1943, pp. 239–241. Abstr.: *Z. Ver. dtsch. Ing.*, vol. 87, no. 25/26, June 26, 1943, p. 390. English abstr.: *Bull. Brit. non-ferr. Met. Res. Ass.*, no. 181, July 1944, p. 183.

2924. THUM, A. *and* KIRMSER, W. Überlagerte Wechselbeanspruchungen, ihre Erzeugung und ihr Einfluss auf die Dauerhaltbarkeit und Spannungsausbildung quergebohrter Wellen. [Combined alternating stresses, their origin and their influence on the fatigue durability and the stress distribution in transversely drilled shafts.] *Forschungsh. Ver. dtsch. Ing.*, no. 419, March–April 1943, pp. 33.

2925. THUM, A. *and* ZOEGE VON MANTEUFFEL, R. Das Sandstrahlen als Mittel zur einfachen und billigen mechanischen Oberflächenbearbeitung. [Sand blasting as a means of simple and economical surface treatment.] *Auto.-tech. Z.*, vol. 46, no. 13/14, July 25, 1943, pp. 304–313. English translation: *R.T.P. Trans.*, no. 2231, pp. 13.

2926. THUM, A. *and* PETERSEN, C. Zur Wechselfestigkeit von Gusseisen. [The fatigue strength of cast iron.] *Arch. Eisenhüttenw.*, vol. 16, no. 8, Feb. 1943, pp. 309–312.

2927. THUM, A. *and* PETERSEN, C. Die Vorgänge im zügig und wechselnd beanspruchten Metallgefüge. [Changes in the crystal structure of metals under tensile and alternating stress.] *Metallwirtschaft*, vol. 22, no. 39/41, Dec. 20, 1943, pp. 547–551.

2928. UNGER, A. M.; MATIS, H. H. *and* GRUCA, E. P. Behaviour of spot welds under fatigue stress. *Weld. J., Easton, Pa.*, vol. 22, no. 3, March 1943, pp. 135s–142s.

2929. WACHÉ, X. *and* CHEVENARD, P. Sur certaines causes de dispersion des résultats dans l'etude des phénomènes de fatigue. [Some causes of scatter of results in the study of the phenomena of fatigue.] *C. R. Acad. Sci., Paris*, vol. 216, no. 8, Feb. 22, 1943, pp. 264–266. Abstr.: *Génie civ.*, vol. 120, no. 18, Sept. 15, 1943, p. 210.

2930. WAHL, A. M. Recent developments in mechanical spring design and testing. *Wire and Wire Prod.*, vol. 18, no. 11, Nov. 1943, pp. 717–720, 735 and no. 12, Dec. 1943, pp. 778–780.

2931. WELLINGER, K. *and* STÄHLI, G. Verfestigungserscheinungen bei der Schwingungsbeanspruchung von Leichtmetall-Legierungen in der Wärme. [Strengthening phenomena by the alternating stressing of light metal alloys at elevated temperatures.] *T Z prakt. Metallbearb.*, vol. 53, no. 1/2, Jan. 1943, pp. 19–22.

2932. WELLINGER, K. *and* STÄHLI, G. Verhalten von Leichtmetall-Kolbenwerkstoffen bei betriebsähnlicher Beanspruchung. [Behaviour of light metal piston alloys under service type stresses.] *Z. Ver. dtsch. Ing.*, vol. 87, no. 41/42, Oct. 16, 1943, pp. 663–666. English translation: *Roy. Aircr. Estab., Lib. Trans.*, no. 56.

2933. WIEGAND, H. Verhalten harter Oberflächenschichten bei Betriebsbeanspruchung. [Behaviour of hard surface layers under service stresses.] *Z. Ver. dtsch. Ing.*, vol. 87, no. 9/10, March 6, 1943, pp. 137–138. English translation: *R.T.P. Trans.*, no. 2268, pp. 2.

2934. WILLIAMS, H. A. Fatigue tests of light weight aggregate concrete beams. *J. Amer. Concr. Inst.*, vol. 14, no. 5, April 1943, pp. 441–447.

2935. WILSON, W. M. Fatigue failure in its relationship to the strengthening and repair of steel bridge members. *Proc. Amer. Rly Engng Ass.*, vol. 45, no. 440, Sept.–Oct. 1943, pp. 25–32.

2936. WILSON, W. M. The fatigue strength of fillet-weld joints connecting steel structural members. *Weld. J., Easton, Pa.*, vol. 22, no. 12, Dec. 1943, pp. 605s–612s.

2937. WILSON, W. M.; BRUCKNER, W. H.; MCCRACKIN, T. H. *and* BEEDE, H. C. Fatigue tests of commercial butt welds in structural steel plates. *Uni. Ill. Engng Exp. Sta. Bull.*, no. 344, Oct. 12, 1943, pp. 140.

2938. WOOD, J. G. *and* SANDERS, R. F. Accelerated fatigue tests on N.E. steels. *Iron Age*, vol. 151, no. 9, March 4, 1943, pp. 68–70.

2939. YOST, F. L. Fatigue characteristics of rubber. *Trans. Amer. Soc. mech. Engrs*, vol. 65, no. 8, Nov. 1943, pp. 881–888.

2940. VON ZEERLEDER, A. Die Erzeugung und Verarbeitung des Aluminiums. [The fabrication and working of aluminium.] *Schweiz. Arch. angew. Wiss.*, vol. 9, no. 4, April 1943, pp. 127–132.

2941. Swiss fatigue tests of oxyacetylene welded aircraft tubing. *Weld. J., Easton, Pa.*, vol. 22, no. 2, Feb. 1943, p. 62s.

2942. Neue Prüfmaschine zur Bestimmung der Dauerfestigkeit. [New testing machine for determining fatigue strength.] *Flugw. u. Tech.* vol. 5, 1943, pp. 77–80.

2943. Creep strength. Its relation to fatigue strength. *Auto. Engr*, vol. 33, no. 436, May 1943, p. 202.

2944. Design of highly stressed studs to improve their fatigue strength. *Prod. Engng*, vol. 14, no. 5, May 1943, pp. 288–290. *Engrs' Dig* vol. 4, no. 8, Aug. 1943, pp. 239–241.

2945. Fatigue strength of butt welds in ordinary bridge steel. *Weld. J., Easton, Pa.*, vol. 22, no. 5, May 1943, pp. 189s–211s.

2946. Fatigue failure of metals. *Pract. Engng*, vol. 7, no. 179, June 25, 1943, pp. 584–585.

2947. Dynamic testing machines—ascertaining fatigue endurance limits of materials. *Elect. Rev., Lond.*, vol. 132, no. 3422, June 25, 1943, p. 858.

2948. Metal cleaners other than trichlorethylene—effect on the fatigue strength of duralumin. *Roy. Aircr. Estab. Rep.*, no. M.6227A, June 1943, pp. 13.

2949. Dynamic testing. Machines by W. & T. Avery Ltd., for applying alternating stresses. *Auto. Engr*, vol. 33, no. 438, July 1943, p. 276.

2950. Testing springs for endurance. *Mech. World*, vol. 114, no. 2953, Aug. 6, 1943, pp. 162–163, 166–167.

2951. Fatigue test fractures of light alloy turbine blades. *Roy. Aircr. Estab. Tech. Note*, no. M.8039, Aug. 7, 1943, pp. 3.

2952. Increasing the fatigue strength of machine parts. *Machinery, Lond.*, vol. 63, no. 1609, Aug. 12, 1943, p. 176.

2953. Prolonging the fatigue life of machine parts. *Machinery, Lond.*, vol. 63, no. 1610, Aug. 19, 1943, p. 204.

2954. Dynamic testing machines. *Pract. Engng*, vol. 8, no. 188, Aug. 27, 1943, p. 148.

2955. Screw threading and fatigue strength. *Machinery, Lond.*, vol. 63, no. 1614, Sept. 16, 1943, p. 316.

2956. Examination of two magnesium alloy propeller blades with particular reference to fatigue properties. *Roy. Aircr. Estab. Rep.*, no. M.6422, Nov. 1943, pp. 2.

2957. The Allison engine: developments in surface finishes. *Tool Engr*, vol. 12, Dec. 1943, pp. 92–94.

2958. Wöhler rotary fatigue tests on various sand cast magnesium alloys and a sand cast aluminium alloy to Specification D.T.D. 133B, before and after soaking at 140°C for 1000 hours. *Sci. Tech. Memor.*, no. 10/43, Dec. 1943, pp. 6.

1944

2959. AFANAS'EV, N. N. Formation of fatigue cracks from the point of view of microstructure. [R] *Zh. tekh. Fiz.*, vol. 14, no. 10/11, 1944, pp. 638–645.

2960. ALLEMAN, N. J. Second progress report of the investigation of fatigue failures in rail joint bars. *Uni. Ill. Engng Exp. Sta. Bull.*, vol. 41, no. 34, April 11, 1944, (*Reprint Series* no. 30, 1944), pp. 12. *Proc. Amer. Rly Engng Ass.*, vol. 45, 1944, pp. 434–449.

2961. ALMEN, J. O. Tightening is a vital factor in bolt endurance. *Mach. Design*, vol. 16, no. 2, Feb. 1944, pp. 158–162.

2962. ALMEN, J. O. On the strength of highly stressed dynamically loaded bolts and studs. *S.A.E. Jl.*, vol. 52, no. 4, April 1944, pp. 151–158.

2963. ALMEN, J. O. Some needed precautions when induction and flame hardening. *Metal Progr.*, vol. 46, no. 6, Dec. 1944, pp. 1263–1267.

2964. BENNETT, J. A. Early detection of cracks resulting from fatigue stressing. *Nat. Adv. Comm. Aero.*, *Wash.*, *Restricted Bulletin* no. 4I15 (*Wartime report* no. W-78), Sept. 1944, pp. 4.

2965. BLAND, R. B. *and* SANDORFF, P. E. The dynamic properties of flash-welded tubing. *Weld. J.*, *Easton*, *Pa.*, vol. 23, no. 6, June 1944, pp. 280s–302s.

2966. BOAS, W. *and* HONEYCOMBE, R. W. K. Thermal fatigue of metals. *Nature*, *Lond.*, vol. 153, no. 3886, April 22, 1944, pp. 494–495 and no. 3906, Sept. 9, 1944, p. 338.

2967. BOETCHER, H. N. Cracking and embrittlement in boilers. *Mech. Engng*, *N.Y.*, vol. 66, Sept. 1944, pp. 593–601.

2968. BOLLENRATH, F. *and* CORNELIUS, H. Eigenschaften von Widerstands-Abbrennschweissverbindungen aus Stahl mit hoher Festigkeit. [Properties of resistance flash welds in high tensile steels.] *Luftfahrtforsch.*, vol. 21, no. 1, Feb. 28, 1944, pp. 17–28. English abstr.: *Metallurgia*, *Manchr*, vol. 31, no. 181, Nov. 1944, p. 46.

2969. BOWEN, I. G. Repeated loading of single riveted joints. *Sci. Tech. Memor.*, no. 6/44, 1944, pp. 16.

2970. BOYVEY, H. O. Fatigue—the forgotten member of the design family. *Aero. Dig.*, vol. 44, no. 1, Jan. 1, 1944, pp. 74–76, 126, 128, 133, 135–136.

2971. BOYVEY, H. O. Fatigue tests of parts made basis for design. *Prod. Engng*, vol. 15, no. 7, July 1944, pp. 444–448.

2972. BRACE, P. H. Physical metallurgy of copper and copper-base alloys. *Elect. Engng*, *N.Y.*, vol. 63, no. 1, Jan. 1944, pp. 11–17.

2973. BRUCKNER, W. H. The metallurgical notch as a factor in fatigue failure. *Metal Progr.*, vol. 45, June 1944, pp. 1102–1103.

2974. BRUCKNER, W. H. *and* MUNSE, W. H. The effect of metallurgical changes due to heat treatment upon the fatigue strength of carbon-steel plates. *Weld. J.*, *Easton*, *Pa.*, vol. 23, no. 10, Oct. 1944, pp. 499s–510s.

2975. BRUEGGEMAN, W. C.; KRUPEN, P. *and* ROOP, F. C. Axial fatigue tests of ten airplane wing-beam specimens by the resonance method. *Tech. Note nat. Adv. Comm. Aero.*, *Wash.*, no. 959, Dec. 1944, pp. 14.

2976. BRUEGGEMAN, W. C. *and* MAYER, M. Guides for preventing buckling in axial fatigue tests of thin sheet-metal specimens. *Tech. Note nat. Adv. Comm. Aero.*, *Wash.*, no. 931, April 1944, pp. 6.

2977. BRUEGGEMAN, W. C.; MAYER, M. *and* SMITH, W. H. Axial fatigue tests at zero mean stress of 24S-T aluminium alloy sheet with and without a circular hole. *Tech. Note nat. Adv. Comm. Aero.*, *Wash.*, no. 955, Nov. 1944, pp. 10.

2978. BUCKINGHAM, E. Latest findings on surface fatigue. *Mach. Design*, vol. 16, no. 2, Feb. 1944, pp. 166–170, 248, 250.

2979. BUCKINGHAM, E. Surface fatigue of plastic materials. *Trans. Amer. Soc. mech. Engrs*, vol. 66, no. 4, May 1944, pp. 297–310.

2980. BUCKWALTER, T. V. *and* HORGER, O. J. Fatigue strength of welded aircraft joints. *Weld. J.*, *Easton*, *Pa.*, vol. 23, no. 1, Jan. 1944, pp. 50s–58s.

2981. CAMBOURNAC, M. Les rails de chemin de fer. [Railroad tracks.] *Rev. Métall.*, *Mem* vol. 41, no. 9, Sept. 1944, pp. 301–312.

2982. CAYLA, M.; VALLETTE, B. *and* L'HERMITE, R. Résistance des ponts-rails en béton armé. Essai de poutres en béton armé aux flexions répétées. [Strength of bridge rails of reinforced concrete. Tests on beams of reinforced concrete under repeated bending.] *C.R. Lab. Bâtim.*, 1944–1945, pp. 113–124.

2983. CHRISTENSEN, R. H. Fatigue analysis methods. *Douglas Aircraft Company Inc., Santa Monica*, Report no. 9262, Dec. 1944, pp. 31.

2984. CHRISTENSEN, R. H. Effect of damage in fatigue of 24ST aluminium alloy. *Douglas Aircraft Company Inc., Santa Monica*, Report no. SM.9564, Nov. 1, 1944, pp. 49.

2985. CLARK, H. H. Shot blasting prolongs life of leaf, torsion and helical springs. *Steel*, vol. 114, no. 9, Feb. 28, 1944, pp. 100–102, 137.

2986. COLLINS, W. L. *and* SMITH, J. O. Notch sensitivity of alloyed cast irons subjected to repeated and static loads. *Metallurgia, Manchr*, vol. 29, March 1944, pp. 281–282.

2987. CORNELIUS, H. Einfluss von Einspannungen auf das Dauerfestigkeitsverhalten einer kaltausgehärteten Aluminium–Kupfer–Magnesium-Knetlegierung. [Influence of clamping on the fatigue properties of cold age hardened wrought aluminium–copper–magnesium alloys.] *Z. Metallk.*, vol. 36, no. 5, May 1944, pp. 101–105.

2988. CORNELIUS, H. *and* SAMTLEBEN, W. Eigenschaften hochwertiger Stahlbänder und ihrer Punktschweissverbindungen. [Properties of high quality steel strips and their spotwelded joints.] *Luftfahrtforsch.*, vol. 20, no. 11, Jan. 6, 1944, pp. 311–322.

2989. CRAMER, R. E. Second progress report of the investigation of Shelly spots in railroad rails. *Uni. Ill. Engng Exp. Sta. Bull.*, vol. 41, no. 33, April 4, 1944, (*Reprint Series* no. 29, 1944), pp. 10. *Proc. Amer. Rly Engng Ass.*, vol. 45, 1944, pp. 462–469.

2990. DIETZ, A. G. H. *and* GRINSFELDER, H. Fatigue of urea assembly adhesives. *Mod. Plast.*, vol. 21, no. 8, April 1944, pp. 119–122, 160, 162.

2991. DIETZ, A. G. H. *and* GRINSFELDER, H. The behaviour of synthetic phenolic-resin adhesives in plywood under alternating stresses. *Trans. Amer. Soc. mech. Engrs*, vol. 66, no. 4, May 1944, pp. 319–328.

2992. DIETZ, A. G. H. *and* GRINSFELDER, H. Fatigue studies on urea assembly adhesives. *Trans. Amer. Soc. mech. Engrs*, vol. 66, July 1944, pp. 442–446.

2993. DIETZ, A. G. H. *and* GRINSFELDER, H. Fatigue tests on compressed and impregnated laminated wood. *Bull. Amer. Soc. Test. Mat.*, no. 129, Aug. 1944, pp. 31–34.

2994. FERGUSON, L. Early detection of fatigue cracks. *Metal Progr.*, vol. 45, March 1944, p. 512.

2995. FERGUSON, L. *and* BOUTON, G. M. The effect of a coating of Polybutene on the fatigue properties of lead alloys. *Symposium on stress-corrosion cracking of metals*, Amer. Soc. Test. Mater. — Amer. Inst. min. metall. Engrs, 1944, pp. 473–484. Abstr.: *Metal Ind., Lond.*, vol. 66, no. 22, June 1, 1945, pp. 345, 348.

2996. FORD, H. Machining fatigue test pieces. *Mach. Shop Mag.*, vol. 5, no. 4, April 1944, pp. 44–46.

2997. FOSTER, H. W. *and* SELIGER, V. Fatigue-testing methods and equipment. *Mech. Engng, N.Y.*, vol. 66, no. 11, Nov. 1944, pp. 719–725.

2998. Fox, F. A. *and* Walker, J. L. The fatigue strengths of four aluminium-containing magnesium-base casting alloys. *Magnes. Rev.*, vol. 4, no. 4, Oct. 1944, pp. 105–114. Abstr.: *Bull. Brit. non-ferr. Met. Res. Ass.*, no. 190, April 1945, p. 93.

2999. Frantz, F. Determining endurance limits of flexurally-stressed steel members. *Prod. Engng*, vol. 15, no. 2, Feb., 1944, pp. 97–98.

3000. Frocht, M. M. Studies in three-dimensional photoelasticity—stresses in bent circular shafts with transverse holes—correlation with results from fatigue and strain measurements. *Proc. Soc. exp. Stress Anal.*, vol. 2, no. 1, 1944, pp. 128–138.

3001. Gelman, A. S. Strength of spot welded joints under alternating loading. [R] *Vestn. mashinostr.*, vol. 24, no. 12, Dec. 1944, pp. 13–16.

3002. Godfrey, H. J.; Flury, A. H. *and* Lewis, D. Progress report on the corrosion fatigue failure of aircraft cables (N-101): The effect of metallic coatings and lubricants on the fatigue and internal friction properties of aircraft cables. *Rep. U.S. Off. Sci. Res. Developm.*, no. 3346, Ser. no. M-207, March 1, 1944, pp. 98.

3003. Godfrey, H. J.; Flury, A. H. *and* Lewis, D. Progress report on the corrosion fatigue failure of aircraft control cables (N-101): Fatigue tests under service loads. *Rep. U.S. Off. Sci. Res. Developm.*, no. 4543, Ser. no. M-439, Dec. 29, 1944, pp. 29.

3004. Hartmann, E. C.; Lyst, J. O. *and* Andrews, H. J. Fatigue tests of riveted joints —progress report of tests of 17S-T and 53S-T joints. *Nat. Adv. Comm. Aero., Wash., Restricted Bulletin* no. 4I15 (*Wartime report* no. W-55), Sept. 1944, pp. 20.

3005. L'Hermite, R. *and* Seferian, D. Les essais d'endurance en flexion répétée comme critère de soudabilité des tôles (1er rapport). [Repeated bending fatigue tests as a criterion for the weldability of sheet. (1st report).] *C.R. Lab. Bâtim.*, 1944–1945, pp. 125–148. *Ann. Inst. Bâtim.*, Circ. Ser. G, no. 9, Nov. 25, 1944, pp. 24.

3006. Herzog, A. Umlaufbiegefestigkeit von hartverchromten und inkromierten Stäben. [Rotating bending strength of hard chromed and chromed shafts.] *Zent. wiss. Ber. Untersuch. Mitt.*, no. 2121, Aug. 1944, pp. 8.

3007. Hess, W. F.; Wyant, R. A.; Winsor, F. J *and* Cook, H. C. An investigation of the fatigue strength of spot welds in the aluminium alloy alclad 24S-T. *Nat. Adv. Comm. Aero. Wash.*, O.C.R. no. 4H23, Aug. 1944, pp. 25

3008. Horger, O. J. *and* Buckwalter, T. V Endurance of NE steels in 1¾-in. specimens *Metal Progr.*, vol. 46, Oct. 1944, pp. 727–729

3009. Horger, O. J. *and* Buckwalter, T. V Fatigue resistance of NE steel shafts. *Iron Age*, vol. 154, no. 20, Nov. 16, 1944 pp. 60–63.

3010. Horger, O. J. *and* Neifert, H. R. Improving fatigue resistance by shot peening. *Proc Soc. exp. Stress Anal.*, vol. 2, no. 1, 1944 pp. 178–190.

3011. Huddle, A. U. *and* Evans, U. R. Som measurements of corrosion-fatigue made wit a new feeding arrangement. *J. Iron St. Inst* vol. 149, no. 1, 1944, pp. 109P–118P. Abstr. *Iron and Steel, Lond.*, vol. 17, no. 10, May 1£ 1944, pp. 405–407.

3012. Jackson, L. R.; Berry, J. M.; Jackson J. S.; Elsea, A. R. *and* Lorig, C. H. Progres report on investigation of boron in armor pla

(OD-87): Endurance properties of basic open hearth steel containing 0.4% carbon and 1.6% manganese without and with boron addition. *Rep. U.S. Off. Sci. Res. Developm.*, no. 4235, Ser. no. M-387, Oct. 10, 1944, pp. 28.

3013. JACKSON, L. R.; GROVER, H. J.; BEAVER W. W. *and* RUSSELL, H. W. Final report on fatigue properties of magnesium alloys and structures: Fatigue properties of magnesium alloy sheet—part II. *Rep. U.S. Off. Sci. Res. Developm.*, no. 3792, Ser. no. M-289, June 12, 1944, pp. 84.

3014. JACKSON, L. R.; GROVER, H. J.; BEAVER, W. W. *and* RUSSELL, H. W. Final report on fatigue properties of magnesium alloys and structures: Fatigue properties of magnesium alloy sheet—part III. *Rep. U.S. Off. Sci. Res. Developm.*, no. 4282, Ser. no. M-381, Oct. 20, 1944, pp. 56.

3015. JONES, J. The fatigue strength of welded joints. *Elect. Engng, N.Y.*, vol. 63, no. 8, Aug. 1944, pp. 288–290.

3016. KARIUS, A. Beitrag zur Frage der Werkstoffveränderungen bei Dauerbeanspruchung. [Contribution to the question of changes in materials under fatigue stressing.] *Metallwirtschaft*, vol. 23, no. 48/52, Dec. 20, 1944, pp. 419–434.

3017. KARIUS, A.; GEROLD, E. *and* SCHULZ, E. H. Werkstoffveränderungen bei der Dauerbeanspruchung. [Changes in materials subjected to fatigue stresses.] *Arch. Eisenhüttenw.*, vol. 18, no. 5/6, Nov.–Dec. 1944, pp. 113–124. English abstr.: *J. Iron St. Inst.*, vol. 152, no. 2, 1945, pp. 43A–44A.

3018. KELTON, E. H. Fatigue testing of zinc-base alloy die castings. *Metallurgia, Manchr*, vol. 29, no. 172, Feb. 1944, pp. 224–225.

3019. KINER, G. B. Early detection of fatigue cracks. *Metal Progr.*, vol. 45, no. 1, Jan. 1944, p. 89.

3020. KOMMERS, W. J. Supplement to the fatigue behaviour of wood and plywood subjected to repeated and reversed bending stresses. The fatigue behaviour of Douglas-fir and Sitka spruce subjected to reversed stresses superimposed on steady stresses. *Rep. For. Prod. Lab., Madison*, no. 1327-A, May 1944, pp. 5.

3021. LAZAN, B. J. *and* YORGIADIS, A. The behaviour of plastic materials under repeated stress. *Amer. Soc. Test. Mater., Symposium on Plastics*, S. T. P. no. 59, 1944, pp. 66–94. *Mod. Plast.*, vol. 21, no. 12, Aug. 1944, pp. 119–128, 164.

3022. LEHR, E. *and* RUEF, F. Fatigue strength of crankshafts of large diesels. *Engrs' Dig.*, vol. 5, no. 10, Oct. 1944, pp. 285–288.

3023. LLOYD, C. O. Fatigue resistance. *Pract. Engng*, vol. 10, no. 236, July 28, 1944, pp. 40–41.

3024. LUBIN, G. *and* WINANS, R. R. Preliminary studies on a drop ball impact machine. *Bull. Amer. Soc. Test. Mat.*, no. 128, May 1944, pp. 13–18.

3025. LUTHANDER, S. *and* WÅLLGREN, G. Statiska och dynamiska prov på koppelbultar av högvärdigt material. [Static and dynamic tests of attachment bolts of high alloy material.] *Flytek. Försöksanst., Stockholm, Rep.* no. 4 (HE-84), May 1944, pp. 16.

3026. LUTHANDER, S. *and* WÅLLGREN, G. Experimentell bestämning av utmattningsdiagrammet vid drag- och tryckbelastning av alcladplåt. [Experimental determination of the fatigue diagram for tension and compression in alclad sheet.] *Flytek. Försöksanst., Stockholm, Rep.* no. 5 (HE-106), June 1944, pp. 12.

3027. LUTHANDER, S. *and* WÅLLGREN, G. Statisk och dynamisk skjuvhållfasthet hos limförband i björk- och valnötsträ vid två temperaturer. [Static and dynamic shear strength of glued joints of Birch and Walnut at two different temperatures.] *Flytek. Försöksanst., Stockholm, Rep.* no. 2 (HE-55), Aug. 1944, pp. 25.

3028. McCLELLAND, A. E. The diagnosis and prevention of failures in colliery wire ropes. *Proc. S. Wales Inst. Engrs*, vol. 59, pt. 3, Jan. 1944, pp. 420–453.

3029. McQUAID, H. W. Needed better teamwork between designer and metallurgist. *Mach. Design*, vol. 16, no. 10, Oct. 1944, pp. 75–78.

3030. MADDEN, B. C. Investigation of R-301 Clad aluminium alloy sheet. *Tech. Rep. U.S. Army Air Force*, no. 5111, May 8, 1944, pp. 91.

3031. MAESER, M. The resistance of leather to flexural fatigue. *J. Amer. Leath. Chem. Ass.*, vol. 39, no. 2, Feb. 1944, pp. 35–57.

3032. MARCHANT, D. W. A study of the effects of surface recarburization on the torsion endurance limit. *Lab. Rep. Rock Island Arsenal*, no. 44–1595, March 31, 1944, pp. 8.

3033. MARTINAGLIA, L. The structural durability of crankshafts—its influence upon oil engine design. *Mot. Ship*, vol. 25, no. 292, May 1944, pp. 38–42.

3034. MATTSON, R. L. *and* ALMEN, J. O. Final report on effect of shot blasting on the mechan-ical properties of steel: Part I. *Rep. U.S. Off. Sci. Res. Developm.*, no. 3274, Ser. no. M-228, Feb. 19, 1944, pp. 109.

3035. MEHR, L.; OBERG, T. T. *and* TERES, J. Fatigue limit of chromium plated steel. *Tech. Rep. U.S. Army Air Force*, no. 5125, July 1944, pp. 77.

3036. MICKEL, E. *and* SOMMER, P. Einfluss des Stahles auf die Haltbarkeit von Kolbenbolzen. [Influence of type of steel on the service life of gudgeon pins.] *Arch. Eisenhüttenw.* vol. 17, no. 9/10, March–April 1944 pp. 227–234. English translation: *B.I.S.I Trans.*, no. 3087.

3037. MILLER, J. L. *and* WHATSON, H. Endurance testing and service failures. *Sheet Metal Ind.* vol. 19, no. 205, May 1944, pp. 854, 861–864

3038. MOORE, H. F. *Shot peening and the fatigue of metals.* Mishawaka, Ind., American Foundry Equipment Co., 1944, pp. 24.

3039. MOORE, H. F. Effect of shot peening on fatigue strength. *Mach. Design*, vol. 16, no. 11 Nov. 1944, pp. 145–150.

3040. MOORE, H. F. Shot peening and the fatigue of metals. *Iron Age*, vol. 154, no. 18 Nov. 2, 1944, pp. 67–71, 136, 138, 140, 142.

3041. MOORE, H. F. A study of residual stresses and size effect and a study of the effect of repeated stresses on residual stresses due to shot peening of two steels. *Proc. Soc. exp Stress Anal.*, vol. 2, no. 1, 1944, pp. 170–17

3042. MORKOVIN, D. *and* MOORE, H. F. Third progress report on the effect of size of specimen on fatigue strength of three types of steel *Proc. Amer. Soc. Test. Mater.*, vol. 44, 194 pp. 137–158.

3043. NISHIHARA, T. *and* KAWAMOTO, M. A new criterion for the strength of metals under combined alternating stresses. *Mem. Coll. Engng Kyoto*, vol. 11, no. 4, July 1944, pp. 65–83.

3044. NISHIHARA, T.; KAWAMOTO, M. *and* YAMADA, T. The fatigue strength of metallic materials under alternating stresses of varying amplitude. [J] *Trans. Soc. mech. Engrs, Japan*, vol. 10, no. 38, Feb. 1944, pp. I–23–I–27.

3045. NISHIHARA, T. *and* KOBAYASHI, A. Repeated stressing and crystalline state of annealed steel. *Mem. Coll. Engng Kyoto*, vol. 11, no. 4, July 1944, pp. 61–64.

3046. PERRIN, F. Considérations sur la limite de fatigue des aciers. [Notes on the fatigue limits of steels.] *Bull. Cerc. Étud. Métaux*, vol. 4, no. 5, Dec. 1944, pp. 125–156.

3047. PETERSON, R. E. *and* LESSELLS, J. M. Effect of surface strengthening on shafts having a fillet or a transverse hole. *Proc. Soc. exp. Stress Anal.*, vol. 2, no. 1, 1944, pp. 191–199.

3048. PETRUSEVICH, A. I. The design and construction of machine components. [R] *Vestn. mashinostr.*, no. 7/8, 1944, pp. 27–37 and no. 9, 1944, pp. 28–35.

3049. PHILLIPS, C. E. *and* THURSTON, R. C. A. A note on a rotating bending-fatigue machine for tests at 200°C. *Rep. Memor. aero. Res. Comm., Lond.*, no. 2674, Dec. 1944, pp. 4.

3050. PIWOWARSKY, E. Dauerschlagfestigkeit von Grauguss und Stahlguss bei Querschnitts-abmessungen von Flugmotoren-Kolbenringen. [The impact fatigue strength of grey iron and cast steel with sectional dimensions of aircraft engine piston rings.] *Stahl u. Eisen*, vol. 64, no. 25, June 22, 1944, pp. 405–406.

3051. POMEY, J. Contribution à l'étude de la fragilité. [Contribution to the study of brittleness.] *Rev. Métall. Mém.*, vol. 41, no. 1, Jan. 1944, pp. 17–25.

3052. PRETTYMAN, I. B. Tread cracking of natural and synthetic rubber stocks. *Industr. Engng Chem.*, vol. 36, no. 1, Jan. 1944, pp. 29–33.

3053. PUCHNER, O. Einfluss der Probengrösse auf die Dauerfestigkeit der Metalle insbesondere Gusseisen. [Influence of specimen size on the fatigue strength of metals particularly cast iron.] *Metallwirtschaft*, vol. 23, no. 48/52, Dec. 20, 1944, pp. 434–441.

3054. ROŠ, M. *and* ALBRECHT, A. Träger in Verbund-Bauweise. [Beams in composite structures.] *Ber. eidgenöss. MatPrüfAnst.*, no. 149, March 1944, pp. 94.

3055. ROŠ, M. *and* CERADINI, G. Statische und Ermüdungsversuche mit aufgeschweissten und aus dem vollen Stahlmaterial herausgearbeiten verschieden geformten Laschenkörpern sowie mit überlapptem Stoss. [Static and fatigue tests with welded and other types of steel cover plates machined from the solid, and also with lap joints.] *Application of arc welding*, Göteborg, Sweden, Elektriska Svetsningsaktiebolaget, 1944, pp. 311–327.

3056. RUSSELL, H. W.; JACKSON, L. R.; GROVER, H. J. *and* BEAVER, W. W. Fatigue strength and related characteristics of joints in 24S-T alclad sheet. *N.A.C.A. Advance Restricted Report* no. 4E30 (*Wartime Report* no. W-63), May 1944, pp. 28.

3057. RUSSELL, H. W.; JACKSON, L. R.; GROVER, H. J. *and* BEAVER, W. W. Fatigue strength and related characteristics of aircraft joints. I—Comparison of spot weld and rivet patterns in 24S-T alclad sheet. Comparison of

24S-T alclad and 75S-T alclad. *N.A.C.A. Advance Restricted Report* no. 4F01 (*Wartime Report* no. W-56), Dec. 1944, pp. 41.

3058. SACHS, G. *and* ESPEY, G. Fatigue strength properties of SAE X4130 tubing. *Iron Age*, vol. 153, no. 12, March 23, 1944, pp. 62–67.

3059. SCHALL, A. Spannungsverhalten der Oberflächenschichten bei statischer und dynamischer Zugdruck-Beanspruchung von legiertem Stahl. [Relation between strain in the surface layers and the static and dynamic tension-compression stress for alloy steel.] *Z. Metallk.*, vol. 36, no. 7, July 1944, pp. 153–163.

3060. SCHAUB, C. Värmebehandlingsstrukturens inverkan på utmattningshållfastheten hos svetsfogar i St 44. [The effect of the structure obtained by heat treatment on the fatigue strength of welds in steel St 44.] *Tekn. Tidskr., Stockh.*, vol. 74, May 20, 1944, pp. 617–623.

3061. SCHULTZ, G. Dynamische und statische Festigkeitsuntersuchungen an punktgeschweissten Verbindungen. [Dynamic and static strength investigations on spot welded joints.] *Elektroschweissung*, vol. 15, no. 8, Aug. 1944, pp. 101–105 and no. 9/10, Sept–Oct. 1944, pp. 125–132.

3062. SERENSEN, S. V. Conditions of failure and factors of safety in members subject to alternating stresses. *Phil. Mag.*, ser. 7, vol. 35, no. 246, July 1944, pp. 470–477.

3063. SERENSEN, S. V. On the evaluation of endurance under alternating stresses of variable amplitude. [R] *Vestn. mashinostr.*, no. 7/8, July–Aug. 1944, pp. 1–7.

3064. SÖRENSEN, P. G. Fatigue testing of metals. *Maskinteknikk*, no. 22, March 25, 1944, pp. 17–27. Abstr.: *Engrs' Dig.*, vol. 5, no. 7, July 1944, pp. 211–212.

3065. SPALDING, L. P. Endurance tests prove quality of gang-riveted joints. *Automot. Industr. N.Y.*, vol. 90, no. 7, April 1, 1944, pp. 43, 133.

3066. STAUDINGER, H. Biegewechselfestigkeit einsatzgehärteter und nitrierter Stähle mit Schleifrissen. [Bending fatigue strength of case hardened and nitrided steel with grinding cracks.] *Z. Ver. dtsch. Ing.*, vol. 88, no. 51/52, Dec. 23, 1944, pp. 681–686. English translation: *SLA Trans. Centre, John Crerar Lib.*, no. 59–20576.

3067. STEWART, W. C. *and* WILEY, R. E. Chemical and mechanical properties of some of the National Emergency steels. *J. Amer. Soc. nav. Engrs*, vol. 56, no. 3, Aug. 1944, pp. 396–411.

3068. THORNE, F. W.; HUNTER, A. *and* HIPPERSON, A. J. Fatigue tests on arc welded 3 % nickel steel. *Weld. J., Easton, Pa.*, vol. 23, no. 7, July 1944, pp. 357s–360s. *Trans. Inst Weld.*, vol. 7, no. 1, March 1944, pp. 38–41.

3069. THUM, A. Die Entwicklung der Lehre von der Gestaltfestigkeit. [The development of the study of design strength.] *Z. Ver. dtsch. Ing.* vol. 88, no. 45/46, Nov. 11, 1944, pp. 609–615

3070. THUM, A. *and* ZOEGE VON MANTEUFFEL, R. Geschweisste Magnesiumlegierungen und ihre Zug-Druck-Wechselfähigkeit. [Welded magnesium alloys and their tension-compression fatigue strength.] *Metallwirtschaft*, vol. 23, no. 22/26, June 20, 1944, pp. 215–220.

3071. TSUDA, T. Research on fatigue of metals 2nd report—X-ray study of fatigue of SDF material. [J] *J. Cent. Aero. Res. Inst., Mitaka* vol. 3, no. 2, Feb., 1944, pp. 51–66. English abstr.: *Metallurg. Abstr.*, vol. 16, 1948–1949, p. 540.

3072. TURNBULL, D. C. Fatigue life of stressed parts increased by shot peening. *Amer. Mach., N.Y.*, vol. 88, no. 18, Aug. 31, 1944, pp. 83–86. *Machinist*, vol. 88, no. 36, Dec. 16, 1944, pp. 83–86.

3073. ULRICH, M. Fatigue strength of gears increased by special shaping and lapping of the tooth base. *Automot. Industr., N.Y.*, vol. 91, no. 9, Nov. 1, 1944, pp. 35, 105–106.

3074. UZHIK, G. V. The determination of factors of safety in machine components with asymmetric cyclic changes of stress. [R] *Vestn. mashinostr.*, no. 5, May 1944, pp. 3–18.

3075. VALENTINE, K. B. Improving the impact stress endurance of a carburized gun-part. *Metal Progr.*, vol. 46, no. 3, Sept. 1944, pp. 467–472.

3076. VANAS, J. Service failure of forging die shanks. *Steel Process.*, vol. 30, Nov. 1944, pp. 718–720.

3077. VATER, M. *and* HENN, M. Die Lebensdauer wechselbeanspruchter Stähle bei gleichzeitiger Einwirkung von Süss- oder Seewasser. [The life of steels under alternating stress concurrent with the effects of fresh and salt water.] *Korros. Metallsch.*, vol. 20, no. 6, June 1944, pp. 179–185.

3078. WACHÉ, X. Contribution à l'etude de l'hystérésis mécanique et de la fatigue. [Contribution to the study of mechanical hysteresis and fatigue.] *Mètaux, Corros-Usure*, vol. 19, no. 227/228, July–Aug. 1944, pp. 79–82.

3079. WILLIAMSON, A. J. Fatigue studies of weld test triangular structures with NE8630 steel tubing. *Weld. J., Easton, Pa.*, vol. 23, no. 1, Jan. 1944, pp. 27s–32s, 49s and [discussion] no. 12, Dec. 1944, p. 635s.

3080. WILLIS, E. J. *and* ANDERSON, R. G. Operating temperatures and stresses of aluminium aircraft-engine parts. *S.A.E. Jl.*, vol. 52, no. 1, Jan. 1944, pp. 28–37.

3081. WILSON, W. M.; BRUCKNER, W. H.; DUBERG, J. E. *and* BEEDE, H. C. Fatigue strength of fillet-weld and plug-weld connections in steel structural members. *Uni. Ill. Engng Exp. Sta. Bull.*, no. 350, March 1944, pp. 94. Abstr.: *Iron Coal Tr. Rev.*, vol. 149, no. 3995, Sept. 22, 1944, pp. 417–418.

3082. WYMAN, L. L. Final report on high temperature properties of light alloys: Part I—aluminium. *Rep. U.S. Off. Sci. Res. Developm.*, no. 3607, Ser. no. M-251, April 15, 1944, pp. 149.

3083. WYMAN, L. L. Final report on high temperature properties of light alloys: Part II—magnesium. *Rep. U.S. Off. Sci. Res. Developm.*, no. 4150, Ser. no. M-292, Sept. 18, 1944, pp. 101.

3084. ZIMMERLI, F. P. Permissible stress range for small helical springs. *Metallurgia, Manchr*, vol. 30, no. 180, Oct. 1944, pp. 337–338.

3085. Verdreh-Dauerprüfmaschine für grosse Bauteile. [Torsion fatigue testing machine for large specimens.] *Z. Ver. dtsch. Ing.*, vol. 88, no. 3/4, Jan. 22, 1944, p. 54. English abstr.: *Aircr. Prod., Lond.*, vol. 6, no. 68, June 1944, p. 300.

3086. Fatigue strength of nitrided surfaces. *Iron Age*, vol. 153, no. 4, Jan. 27, 1944, p. 55.

3087. Strength of shafts—effect of transverse holes, splines and shoulders on torsional fatigue. *Auto. Engr*, vol. 34, no. 446, Feb. 1944, p. 68.

3088. Shot blasting gears to improve fatigue life. *Iron Age*, vol. 153, no. 11, March 16, 1944, p. 63.

3089. Progress report on fatigue and impact characteristics and notch effect in tension of artifically aged aluminium alloys. *Rep. U.S. Off. Sci. Res. Developm.*, no. 3579, Ser. no. M-216, April 18, 1944, pp. 25.

3090. Fatigue tests. The value of fatigue curves at high stress. *Metal Ind., Lond.*, vol. 64, no. 17, April 28, 1944, p. 262.

3091. New vibration fatigue tester. *Rev. sci. Instrum.*, vol. 15, no. 4, April 1944, p. 110.

3092. Fatigue bend strength. Experiments on Al–Mg alloys under corrosive conditions. *Metal Ind., Lond.*, vol. 65, no. 4, July 28, 1944, p. 60.

3093. Spannungsverteilung in Stabköpfen und deren Dauerhaltbarkeit. [Stress distribution in bar ends and their fatigue durability.] *Z. Ver. dtsch. Ing.*, vol. 88, no. 35/36, Sept. 2, 1944, pp. 492–493.

3094. The phenomenon of metal fatigue. *Engrs' Dig.*, vol. 5, no. 10, Oct. 1944, p. 312.

3095. Fatigue testing apparatus—a diversity of types of Avery machines. *Aircr. Engng*, vol. 16, no. 189, Nov. 1944, pp. 334–335.

3096. Resonant fatigue tests. *West. Metals*, vol. 2, no. 12, Dec. 1944, pp. 54–56.

3097. Fatigue tests on riveted joints in 22 gauge aluminium alloy sheet to BSS. 5L3. *Sci. Tech. Memor.*, no. C.1/44, 1944, pp. 3.

3098. The effect of surface finish and shot-blasting upon the fatigue strength of a light alloy and a steel. *Sci. Tech. Memor.*, no. C.5/44, 1944, pp. 5.

3099. The effect of aluminium cladding on the fatigue strength of riveted joints. *Sci. Tech. Memor.*, no. C.10/44, 1944, pp. 3.

1945

3100. ANDREWS, H. J. *and* HOLT, M. Fatigue tests on $\frac{1}{8}$ inch aluminium alloy rivets. *Tech Note nat. Adv. Comm. Aero., Wash.*, no. 971, Feb. 1945, pp. 8.

3101. BARKOW, A. G. An X-ray diffraction study of fatigue in metals at high stresses. *J. appl Phys.*, vol. 16, no. 2, Feb. 1945, pp. 111–120

3102. BENNETT, J. A. Effect of fatigue-stressing short of failure on some typical aircraft metals. *Tech. Note nat. Adv. Comm. Aero. Wash.*, no. 992, Oct. 1945, pp. 23.

3103. BERRY, J. M.; JACKSON, J. S.; ELSEA, A. R *and* LORIG, C. H. Final report on investigatio of boron in armorplate: Endurance and othe properties of some boron-containing carbo steels and NE 9400 type steels. *Rep. U.S. Of Sci. Res. Developm.*, no. 4995, Ser. no. M-50 April 25, 1945, pp. 31.

3104. BERTSCHINGER, R. Über die Ermüdungs festigkeit und die Kerbsicherheit von grauer Gusseisen. [On the fatigue strength and th notch safety of grey cast iron.] *Schweiz. tec Z.*, vol. 20, no. 48, Nov. 29, 1945, pp. 605–61

3105. BINKHORST, I. De sterkte van constructi delen tegenover wisselende belasting. [Th strength of structural elements under alternating loads.] *Nat. Luchtvaartlab., Amsterdam,* Rep. no. S. 301, 1945, pp. 34.

3106. BOLZ, H. A. Fatigue failures may be blamed on you. *Mod. Mach. Shop,* vol. 17, no. 8, Jan. 1945, pp. 142, 144, 146, 148, 150.

3107. BRADSHAW, W. H. Standard fatigue tester for use by the rayon manufacturers—report to Committee D-13. *Bull. Amer. Soc. Test. Mat.,* no. 136, Oct. 1945, pp. 13–16.

3108. BRUEGGEMAN, W. C.; MAYER, M. *and* SMITH, W. H. Axial fatigue tests at two stress amplitudes of 0.032 inch 24S-T sheet specimens with a circular hole. *Tech. Note nat. Adv. Comm. Aero., Wash.,* no. 983, July 1945, pp. 8.

3109. BRUMFIELD, R. C. A sulphur print method for study of crack growth in corrosion-fatigue of metals. *Proc. Amer. Soc. Test. Mater.,* vol. 45, 1945, pp. 544–553.

3110. BUDD, C. B. *and* LARRICK, L. Statistical comparison of rayon tire cord fatigue testing machines. *Bull. Amer. Soc. Test. Mat.,* no. 136, Oct. 1945, pp. 19–25.

3111. BURGHOFF, H. L. *and* BLANK, A. L. Fatigue tests on some copper alloys in wire form. *Metallurgia, Manchr,* vol. 32, no. 190, Aug. 1945, p. 189.

3112. LE CAMUS, B. Recherches sur le comportement du béton et du béton armé soumis à des efforts répétés. [Research on the behaviour of concrete and reinforced concrete subjected to repeated loads.] *C. R. Lab. Bâtim.,* 1945–1946, pp. 25–47.

113. CRAMER, R. E. How rail failures are being prevented. *Metal Progr.,* vol. 47, no. 3, March 1945, pp. 521–524.

3114. DAVIS, D. M. Fatigue failure of aircraft parts—their cause and cure. *Automot. Industr., N.Y.,* vol. 92, no. 9, May 1, 1945, pp. 34–37, 67–68, 72, 74.

3115. DRUCKER, D. C. *and* TACHAU, H. New design criteria for wire rope. *J. appl. Mech.,* vol. 12, no. 1, March 1945, pp. A33–A38.

3116. EAGAN, T. E. Effect of heat treatment on the endurance limit of alloyed grey cast iron. *Amer. Foundrym.,* vol. 5, no. 5, Dec. 1945, pp. 44–53.

3117. ENEMOTO, N. Study on fatigue and internal friction of metallic materials. 1st report—mechanism of fatigue of soft materials. [J] *Trans. Soc. mech. Engrs, Japan,* vol. 11, no. 41, Feb.–May 1945, pp. I-39–I-50.

3118. ENEMOTO, N. Study on fatigue and internal friction of metallic materials. 2nd report—mechanism of internal friction. [J] *Trans. Soc. mech. Engrs, Japan,* vol. 11, no. 41, Feb.–May 1945, pp. I-50–I-59.

3119. ERICKSON, J. L. Fatigue in light metals. *Light Metal Age,* vol. 3, no. 10, Oct. 1945, pp. 17–20, 31, 44 and no. 11, Nov. 1945, p. 39.

3120. FINDLEY, W. N. Fatigue tests of a laminated Mitscherlich paper plastic. *Proc. Amer. Soc. Test. Mater.,* vol. 45, 1945, pp. 878–908.

3121. FINDLEY, W. N.; WORLEY, W. J. *and* KACALIEFF, C. D. Effect of molding pressure and resin on static and fatigue properties of Compreg. *Mod. Plast.,* vol. 22, no. 12, Aug. 1945, pp. 143–148, 196.

3122. FORRESTER, P. G. *and* CHALMERS, B. Fatigue testing of bearing alloys. *Engineering,*

Lond., vol. 159, no. 4123, Jan. 19, 1945, pp. 41–43.

3123. Fowler, F. H. On fatigue failures under triaxial static and fluctuating stresses and a statistical explanation of size effect. *Trans. Amer. Soc. mech. Engrs*, vol. 67, no. 3, April 1945, pp. 213–215.

3124. Gadd, C. W. *and* Ochiltree, N. A. Full scale fatigue testing of crankshafts. *Proc. Soc. exp. Stress Anal.*, vol. 2, no. 2, 1945, pp. 150–158.

3125. Gadd, C. W.; Zmuda, A. *and* Ochiltree, N. A. Correlation of stress concentration with fatigue strength of engine components. *S.A.E. Jl.*, vol. 53, no. 11, Nov. 1945, pp. 640–647.

3126. Gassner, E. *and* Teichmann, A. Ansatz und Durchführung von Betriebsfestigkeits-Versuchen. [The planning and execution of service strength tests.] *Ber. dtsch. VersAnst. Luftf.*, Jan., 1945. *Laboratorium für Betriebsfestigkeit, Darmstadt*, Bericht No. V-24, pp. 16.

3127. George, P. F. Fracture characteristics of magnesium castings. *Alumin. Magnes.*, vol. 1, no. 9, June 1945, pp. 14–17.

3128. Godfrey, H. J.; Flury, A. H. *and* Lewis, D. Progress report on the corrosion fatigue failure of aircraft control cables: miscellaneous tests. *Rep. U.S. Off. Sci. Res. Developm.*, no. 4602, ser. no. M-452, Jan. 17, 1945, pp. 27.

3129. Grover, H. J.; Bennett, R. W. *and* Foley, G. M. Fatigue properties of flash welds. *Weld. J., Easton, Pa.*, vol. 24, no. 11, Nov. 1945, pp. 599s–617s.

3130. Harvey, G. W. *and* Campbell, H. Effect of bead contour on fatigue life of arc welds.

Prod. Engng, vol. 16, no. 5, May 1945, pp. 306–307.

3131. Herb, C. O. Shot peening now widely used for increasing fatigue resistance. *Machinery, New York*, vol. 51, no. 8, April 1945, pp. 170–179.

3132. L'Hermite, R.; Seferian, D. *and* Canac, F. Les essais d'endurance en flexion répétée comme critère de soudabilité des tôles (2ème rapport). Application à une tôle d'acier au chrome-molybdène. [Repeated bending fatigue tests as a criterion for the weldability of sheet (2nd report). Application to chromium-molybdenum steel sheet.] *Rapp. Gr. franc. Rech. aéro.*, no. 14, 1945, pp. 60. *C.R. Lab. Bâtim.*, 1945–1946, pp. 141–168.

3133. Hirst, G. W. C. The influence of radial pressure between members of a press fit in initiating fatigue failures. *J. Instn Engrs Aust.*, vol. 17, no. 6, June 1945, pp. 101–105.

3134. Horger, O. J. Improving the railroad axle. *Metal Progr.*, vol. 47, no. 3, March 1945, pp. 529–532.

3135. Horger, O. J. Mechanical and metallurgical advantages of shot peening. *Iron Age*, vol. 155, March 29, 1945, pp. 40–49, 100, 102, 104 and April 5, 1945, pp. 66–76, 146, 148–149.

3136. Horger, O. J.; Buckwalter, T. V. *and* Neifert, H. R. Fatigue strength of 5¼″ diameter shafts as related to design of large parts. *J. appl. Mech.*, vol. 12, no. 3, Sept. 1945, pp. A149–A155.

3137. Horger, O. J. *and* Neifert, H. R. Shot peening to improve fatigue resistance. *Proc. Soc. exp. Stress Anal.*, vol. 2, no. 2, 1945, pp. 1–10.

3138. HULL, F. C. *and* WELTON, H. R. Work hardened surfaces of fatigue specimens. *Metal Progr.*, vol. 48, no. 6, Dec. 1945, pp. 1287–1288.

3139. HYLER, J. E. Shot peening increases fatigue resistance. *Mill Fact. ill.*, vol. 36, no. 4, April 1945, pp. 102–105, 224, 228, 232.

3140. JACKSON, L. R. *and* GROVER, H. J. The application of data on strength under repeated stresses to the design of aircraft. *N.A.C.A. Advance Restricted Report* no. 5H27 (*Wartime Report* no. W-91), Oct. 1945, pp. 33.

3141. JEWETT, F. D. *and* GORDON, S. A. Repeated load tests–some experimental investigations on aircraft components. *Proc. Soc. exp. Stress Anal.*, vol. 3, no. 1, 1945, pp. 123–130.

3142. KARIUS, A.; GEROLD, E. *and* SCHULZ, E. H. Der Einfluss von Erholungspausen bei der Dauerbeanspruchung. [The influence of rest periods on the fatigue strength.] *Arch. Eisenhüttenw.*, vol. 18, no. 7/8, Jan.–Feb. 1945, pp. 155–159. English abstr.: *J. Iron St. Inst.*, vol. 152, no. 2, 1945, p. 44A.

3143. KENYON, J. N. Fatigue properties of some cold-drawn nickel-alloy wires. *Metallurgia, Manchr*, vol. 32, no. 187, May 1945, 42–43.

3144. KENYON, J. N. An automotive tire fatigue machine. *Bull. Amer. Soc. Test. Mat.*, no. 136, Oct. 1945, pp. 9–12.

3145. KOMMERS, J. B. The effect of overstress in fatigue on the endurance life of steel. *Proc. Amer. Soc. Test. Mater.*, vol. 45, 1945, pp. 532–543.

3146. LARRICK, L. Tension vibrator compares tire cord values. *Text. World*, vol. 95, no. 5, May 1945, pp. 107–109, 194, 196, 198, 200, 202.

3147. LEHR, E. *and* SKIBA, A. Torsional fatigue testing apparatus for large components. *Engrs' Dig.*, vol. 6, no. 4, April 1945, pp. 85–88.

3148. DE LEIRIS, H. L'analyse morphologique des cassures. [The examination of fractures after failure.] *Bull. Ass. tech. marit.*, vol. 44, mémoire 801, 1945, pp. 95–110.

3149. LEWIS, D.; FLURY, A. H. *and* GODFREY, H. J. Final report on the corrosion-fatigue failure of aircraft control cables. *Rep. U.S. Off. Sci. Res. Developm.*, no. 4819, ser. no. M-467, Feb. 26, 1945, pp. 26.

3150. LOVE, R. J. *and* MILLS, H. R. Bending fatigue strength of forged crankshafts: effect of lightening holes, shot peening and straightening. *Rep. Instn Auto. Engrs Res. Comm.*, no. 1945/R/7, July 1945, pp. 11.

3151. LUTHANDER, S. *and* WÅLLGREN, G. Utmattningsegenskaperna hos ett Cr-legerat konstruktionsstål vid pulserande dragbelastning. [The fatigue properties of a Cr-alloy constructional steel under pulsating tensile loads.] *Flytek. Försöksanst., Stockholm, Rep.* no. 13 (HE-85), Jan. 1945, pp. 19.

3152. MCFARLAND, F. R. Experiences with highly stressed aircraft engine parts. *Proc. Soc. exp. Stress Anal.*, vol. 3, no. 1, 1945, pp. 112–122.

3153. MACLEAN, J. A. Airplane landing-gear fatigue problems. *Aeronaut. Engng Rev.*, vol. 4, no. 11, Nov. 1945, pp. 12–21.

3154. MACK, D. J. The corrosion fatigue properties of some hard lead alloys in sulphuric acid. *Proc. Amer. Soc. Test. Mater.*, vol. 45, 1945, pp. 629–650.

3155. MANEY, G. A. *and* WYLY, L. T. Fatigue strength of flush-riveted joints for aircraft manufactured by various riveting methods. *N.A.C.A. Advance Restricted Report* no. 5H28 (*Wartime Report* no. W-82), Dec. 1945, pp. 10.

3156. MARIN, J. Applying fatigue test data in design. *Mach. Design*, vol. 17, no. 1, Jan. 1945, pp. 143–146, 200, 202.

3157. MATTSON, R. L. *and* ALMEN, J. O. Final report on the effect of shot blasting on the mechanical properties of steel: part II. *Rep. U.S. Off. Sci. Res. Developm.*, no. 4825, ser. no. M-476, March 16, 1945, pp. 127.

3158. MINER, M. A. Cumulative damage in fatigue. *J. appl. Mech.*, vol. 12, no. 3, Sept. 1945, pp. A159–A164 and [discussion] vol. 13, no. 2, June 1946, pp. A169–A171.

3159. MINER, M. A. How cumulative damage affects fatigue life. *Mach. Design*, vol. 17, no. 12, Dec. 1945, pp. 111–115.

3160. MINER, M. A. Experimental verification of cumulative fatigue damage. *Automot. Industr. N.Y.*, vol. 93, no. 11, Dec. 1, 1945, pp. 20–24, 56, 58.

3161. MOORE, H. F. Shot peening and fatigue of metals. *Canad. Mach.*, vol. 56, no. 5, May 1945, pp. 130, 132, 134, 136.

3162. MOORE, H. F. The fatigue strength of aluminium and magnesium alloys. *Alumin. Magnes.*, vol. 2, no. 2, Nov. 1945, pp. 14–17, 28–29.

3163. MOORE, H. F. A study of size effect and notch sensitivity in fatigue tests of steel. *Proc. Amer. Soc. Test. Mater.*, vol. 45, 1945, pp. 507–521.

3164. MOORE, H. F.; CRAMER, R. E.; ALLEMAN, N. J. *and* JENSEN, R. S. Progress reports of investigation of railroad rails and joint bars. *Uni. Ill. Engng Exp. Sta. Bull.*, vol. 42, no. 47, July 10, 1945 (Reprint Series no. 32), pp. 48.

3165. MOORE, R. L. *and* HILL, H. N. Comparative fatigue tests of riveted joints of alclad 24S-T, alclad 24S-T81, alclad 24S-RT, alclad 24S-T86 and alclad 75S-T sheet. *Nat. Adv. Comm. Aero., Wash., Restricted Bulletin* no. 5F11 (*Wartime report* no. W-76), Aug. 1945, pp. 4.

3166. MORETTO, O. La fatiga de aceros y su relacion con estructuras metalicas. [The fatigue of steel and its relation to metal structures.] *Cienc. y Téc.*, vol. 104, no. 511, Jan. 1945, pp. 7–15.

3167. MORIKAWA, G. K. *and* GRIFFIS, L. V The biaxial fatigue strength of low carbor steels. *Weld. J., Easton, Pa.*, vol. 24, no. 3 March 1945, pp. 167s–174s.

3168. MUHLENBRUCH, C. W. The effect of re peated loading on bond strength of concrete *Proc. Amer. Soc. Test. Mater.*, vol. 45, 1945 pp. 824–845.

3169. NISHIHARA, T. *and* KAWAMOTO, M. Th strength of metals under combined alternatin bending and torsion with phase difference *Mem. Coll. Engng Kyoto*, vol. 11, no. 5, Jun 1945, pp. 85–112.

3170. NISHIHARA, T. *and* KOBAYASHI, A. Zu Theorie der Wechselfestigkeit. [A theory c fatigue strength.] *Mem. Coll. Engng Kyot* vol. 11, no. 5, June 1945, pp. 113–120.

3171. OAKS, J. K. *and* TOWNSHEND, P. F Fluctuating load tests of Typhoon tail plane *Roy. Aircr. Estab., Rep.* no. SME 3354, De 1945, pp. 13.

3172. OLDBERG, S. *and* LIPSON, C. Structural evolution of a crankshaft. *Proc. Soc. exp. Stress Anal.*, vol. 2, no. 2, 1945, pp. 118–138.

3173. PARKINSON, D. *and* BLOXHAM, J. L. Behavior of natural rubber and GR-S under repeated stress cycles. *I. R. I. Trans.*, vol. 21, no. 2, Aug. 1945, pp. 67–77.

3174. PETERSON, R. E. Relation between life testing and conventional tests of materials. *Bull. Amer. Soc. Test. Mater.*, no. 133, March 1945, pp. 9–16.

3175. PHILLIPS, C. E. A note on rotating bending fatigue test on X76S alloy extended to endurances of over 500 million cycles. *Sci. Tech. Memor.*, no. 7/45, 1945, pp. 3.

3176. PHILLIPS, C. E. *and* RIDLEY, R. W. The effect on the fatigue resistance of shot blasting the machined surfaces of two cast aluminium alloys. *Sci. Tech. Memor.*, no. 8/45, 1945, pp. 6.

3177. PUGSLEY, A. G. The specification of test loads—loading conditions for strength tests on aeroplane structures. *Aircr. Engng*, vol. 17, no. 202, Dec. 1945, pp. 352–353.

3178. PUTNAM, A. A. An analysis of life expectancy of airplane wings in normal cruising flight. *N.A.C.A. Rep.*, no. 805, 1945, pp. 9.

3179. RAPPLEYEA, F. A.; PERRY, R. E. *and* ANSEL, G. Influence of prestressing and cyclic stressing on stress-strain characteristics of magnesium alloys. *J. aero. Sci.*, vol. 12, no. 4, Oct. 1945, pp. 448–454, 460.

3180. RAWLINS, R. E. Fatigue tests at resonant speed. *Metal Progr.*, vol. 47, no. 2, Feb. 1945, pp. 265–267.

3181. ROBERTS, A. M. *and* GRAYSHON, R. J. W. The effect of treatment on fatigue resistance of D.T.D. 331 steel. *Sci. Tech. Memor.*, no. 1/45, 1945, pp. 4.

3182. ROŠ, M. Influence de la surface sur la fatigue des métaux. [The influence of the surface on the fatigue of metals.] *Commission technique des états et propriétés de surface des métaux, Journées des états de surface, Paris,* Oct. 23–26, 1945, pp. 207–218.

3183. RUSSENBERGER, M. Eine dynamische Zug-Druck-Prüfmaschine zur Bestimmung von Wechselfestigkeit und Werkstoffdämpfung. [A dynamic tension-compression testing machine for the determination of the fatigue strength and the damping capacity of materials.] *Schweiz. Arch. angew. Wiss.*, vol. 11, no. 2, Feb. 1945, pp. 33–42.

3184. SCHAUB, C. Utmattningshållfasthet och hålkälskänslighet hos bågsvetsmaterial. [Fatigue strength and notch sensitivity of arc-welded material.] *Tekn. Tidskr. Stockh.*, vol. 75, Nov. 17, 1945, pp. 1263–1266.

3185. SCHENKEL, E. R. Fatigue failure in shear panels of aluminium and magnesium alloys. *Prod. Engng*, vol. 16, no. 4, April 1945, pp. 222–225.

3186. SELIGER, V. A suggested new parameter for fatigue strength analysis. *Bull. Amer. Soc. Test. Mat.*, no. 132, Jan. 1945, pp. 29–33.

3187. SERENSEN, S. V.; TETELBAUM, I. M. *and* PRIGOROVSKII, N. I. Dynamic strength in mechanical engineering. [R] Second edition. Moscow, Mashgiz, 1945, pp. 327.

3188. SHINN, D. A. *and* OBERG, T. T. Aluminium alloy aircraft extrusions. *Tech. Rep. U.S. Army Air Force* no. 5228, March 31, 1945, pp. 60.

3189. SIMS, C. E. Comparative properties of cast and forged steel. *Foundry, Cleveland*, vol. 73, no. 5, May 1945, pp. 90–93, 192, 195. Abstr.: *Metallurgia, Manchr*, vol. 33, no. 194, Dec. 1945, pp. 111–112.

3190. SOPWITH, D. G. *and* SETTLE, T. Research on fatigue strength of screw threads of different form. *Engineering, Lond.*, vol. 160, no. 4149, July 20, 1945, pp. 58–59.

3191. STONE, D. D. Fatigue life of parts as affected by grain direction. *Prod. Engng*, vol. 16, no. 7, July 1945, p. 442.

3192. STRAW, J. H. Endurance tests of high tensile steel wire under combined stresses. *Rep. Div. Aeronautics, Aust.*, no. SM.55, Aug. 1945, pp. 20.

3193. TURNER, T. H. Prevention of corrosion and corrosion fatigue. *Proc. Instn Loco. Engrs, Lewes*, vol. 35, no. 185, May–June 1945, pp. 153–219.

3194. VENABLE, C. S. Comparative fatigue test data. *Bull. Amer. Soc. Test. Mat.*, no. 136, Oct. 1945, pp. 17–19.

3195. WAHL, A. Effect of vibration on brackets, fastenings. *Mach. Design*, vol. 17, no. 3, March 1945, pp. 141–146.

3196. WAHL, A. M. Fatigue tests of airplane generator brackets with special reference to failure of screw fastenings. *J. appl. Mech.*, vol. 12, no. 2, June 1945, pp. A113–A122.

3197. WERNER, M.; DANGL, K. *and* KLOTZ, B. Einfluss der Schweissnahtform auf die Zugschwellfestigkeit von Rohrschweissungen. [Influence of the weld-bead shape on the tensile fatigue strength of tube welds.] *Stahl u. Eisen*, vol. 65, no. 5/6, Feb. 1, 1945, pp. 64–67.

3198. WILSON, J. B. Thermal fatigue in magnesium. *Magnes. Rev.*, vol. 5, no. 2, April 1945, pp. 47–50. Abstr.: *Bull. Brit. non-ferr. Met. Res. Ass.*, no. 194, Aug. 1945, p. 197.

3199. WILSON, W. K. Effect [of surface finish] on fatigue strength. *Proc. Instn mech. Engrs, Lond.*, vol. 153, no. 10, 1945, pp. 347–351. *Engineering, Lond.*, vol. 159, no. 4134, April 6, 1945, pp. 277–280.

3200. WYSS, TH. Untersuchungen an gekerbten Körpern, insbesondere am Kraftfeld der Schraube unter Berücksichtigung der Vergleichsspannung. [Investigations on notched solids, in particular a comparison of the stress fields of screws.] *Ber. eidgenöss. MatPrüfAnst.*, no. 151, Dec. 1945, pp. 73.

3201. ZHUKOV, S. L. On the question of the relationship between the endurance limit and the static characteristics of metals. [R] *Zav. Lab.* vol. 11, no. 11/12, 1945, pp. 1095–1104.

3202. Fatigue and tensile test specimens prepared by grinding. *Iron Age*, vol. 155, Jan. 18, 1945, p. 57.

3203. Fatigue strength of butt welds in ordinary bridge steel—maximum stress compressive. *Weld. J., Easton, Pa.*, vol. 24, no. 1, Jan. 1945, pp. 7s–9s.

3204. Fatigue testing machine for aircraft structural elements. *Metallurgia, Manchr*, vol. 31, no. 184, Feb. 1945, p. 78.

3205. Increasing fatigue life of heat treated gears. *Iron Age*, vol. 155, March 1, 1945, p. 6.

3206. Shot blasting—developments in the technique for increasing fatigue strength. *Auto. Engr*, vol. 35, no. 461, April 1945, pp. 163–164.

3207. Design of torsion rod springs used in M-18 tank destroyer. *Prod. Engng*, vol. 16, no. 6, June 1945, pp. 390–392.

3208. Fatigue strength of fillet, plug and slot welds in ordinary bridge steel. *Weld. J., Easton, Pa.*, vol. 24, no. 7, July 1945, pp. 378s–400s.

3209. Bibliography on the fatigue of materials. *Australia, Council for Scientific and Industrial Research. Div. Aeronautics*, July 1945, pp. 89.

3210. Shot blasting or shot-peening springs and the effect on fatigue life. *Mainspring*, Aug. 1945, pp. 3–7.

3211. Progress report on fatigue strength of structural welds. *Bull. Amer. Rly Engng Ass.*, vol. 47, no. 454, Sept.–Oct. 1945, pp. 55–57.

212. Increasing fatigue resistance by shot peening. *Machinery, Lond.*, vol. 67, no. 1724, Oct. 25, 1945, pp. 449–454.

1946

213. AFANAS'EV, N. N. A theory of fatigue under combined stress. [R] *Zh. tekh. Fiz.*, vol. 16, no. 4, 1946, pp. 443–454.

214. AFANAS'EV, N. N. The effect of shape and size factors on the fatigue strength. [R] *Sbornik dokladov po dinamicheskoi prochnostidetalei mashin* [*Collection of papers on the dynamic strength of machine parts*], Moscow, Academy of Sciences U.S.S.R., 1946, pp. 157–167. English abstr.: *Engrs' Dig.*, vol. 9, no. 3, March 1948, pp. 96–100.

3215. ALMEN, J. On the strength of highly stressed, dynamically loaded bolts and studs. *Diesel Pwr*, vol. 24, no. 8, Aug. 1946, pp. 961–965, 992, 994, 996, 998.

3216. ALMEN, J. O.; MATTSON, R. L. and FONDA, H. E. Final report on effect of shot blasting on the mechanical properties of steel—part III. *Rep. U.S. Off. Sci. Res. Developm.*, no. 6647, Ser. no. M-661, April 1, 1946, pp. 263.

3217. ANDERSON, A. R.; SWAN, E. F. and PALMER, E. W. Fatigue tests on some additional copper alloys. *Proc. Amer. Soc. Test. Mater.*, vol. 46, 1946, pp. 678–692.

3218. ARMSTRONG, J. A. W. and BROOKS, H. Static and fatigue strength of spot welds in MG5 and MG7. *Roy. Aircr. Estab. Tech. Note*, no. Met. 15, Jan. 1946, pp. 7.

3219. ARNSTEIN, K. Fatigue failures in aircraft. *Proc. Soc. exp. Stress Anal.*, vol. 3, no. 2, 1946, pp. 124–130.

3220. BENNETT, J. A. Early detection of fatigue cracks. *Mater. & Meth.*, vol. 23, no. 4, April 1946, p. 1057.

3221. BENNETT, J. A. A study of the damaging effect of fatigue stressing on X4130 steel. *J. Res. nat. Bur. Stand.*, vol. 37, no. 2, Aug. 1946, pp. 123–139. *Proc. Amer. Soc. Test. Mater.*, vol. 46, 1946, pp. 693–714.

3222. BOLZ, R. W. Production processes—their influence on design. Part XV—shotpeening. *Mach. Design*, vol. 18, no. 9, Sept. 1946, pp. 129–132.

3223. BOSCH, M. T. Ein neues Kriterium für die Berechnung von Drahtseilen. [A new criterion for the design of wire ropes.] *Schweiz. Bauztg.*, vol. 128, no. 19, Nov. 9, 1946, pp. 237–239. English abstr.: *Engrs' Dig.*, vol. 8, no. 3, March 1947, pp. 81–83.

3224. BUCHWALD, A. Détermination simplifiée de l'endurance de l'acier et des autres métaux. [A simplified method for determining the fatigue strength of steel and other metals.] *Génie civ.*, vol. 123, (no. 3) no. 3174, Feb. 1, 1946, pp. 37–38.

3225. LE CAMUS, B. Recherches sur le comportement de béton et du béton armé soumis à des efforts répétes. [Research on the behaviour of concrete and reinforced concrete subjected to repeated loads.] *Ann. Inst. Bâtim.*, Circ. Série F, no. 27, July 25, 1946, pp. 24.

3226. CASWELL, J. S. Internal stress and the fracture of metals. *Metallurgia, Manchr*, vol. 33, no. 197, March 1946, pp. 236–242.

3227. CHEVIGNY, R. Quelques résultats relatifs à la résistance a la fatigue sur les métaux légers. [Some results concerning the fatigue resistance of light metals.] *Rev. Métall. Mem.*, vol. 43, no. 11–12, Nov.–Dec. 1946, pp. 330–335.

3228. CRAMER, R. E.; ALLEMAN, N. J. and JENSEN, R. S. Progress reports of investigation of railroad rails and joint bars. *Uni. Ill. Engng Exp. Sta. Bull.*, vol. 43, no. 67, July 11, 1946, (*Reprint Series* no. 35, 1946), pp. 40. *Proc. Amer. Rly Engng Ass.*, vol. 47, no. 458, Feb. 1946, pp. 414–427, 443–448 and 464–466.

3229. CRATE, H.; OCHILTREE, D. W. and GRAVES, W. T. Effect of ratio of rivet pitch to rivet diameter on the fatigue strength of riveted joints of 24S-T aluminium alloy sheet. *Tech. Note nat. Adv. Comm. Aero., Wash.*, no. 1125, Sept. 1946, pp. 7.

3230. CRUSSARD, C. Le role des joints intergranulaires dans de déformation des métaux.

Application au fluage et à la fatigue. [The role of intergranular boundaries in the deformation of metals. Application to creep and to fatigue.] *Rev. Métall.*, vol. 43, no. 11/12, Nov.–Dec. 1946, pp. 307–317.

3231. DAVIS, D. M. Fatigue failure of aircraft parts—their cause and cure. *Aeronaut. Engng Rev.*, vol. 5, no. 1, Jan. 1946, pp. 15–23.

3232. DOLAN, T. J. An investigation of the behaviour of materials under repeated stress. *Office Naval Research, Contract N6-ori-71, Task Order IV, First Progr. Rep., Uni. Illinois*, Dec. 1946, pp. 125.

3233. DRAKE, D. W. The effect of notches on static and fatigue strength. *J. aero. Sci.*, vol. 13, no. 5, May 1946, pp. 259–269.

3234. EAGAN, T. E. Effects of heat treatment on the endurance limit of alloyed gray cast iron. *Trans. Amer. Foundrym. Ass.*, vol. 54, 1946, pp. 230–240.

3235. EICKNER, H. W.; MRAZ, E. A. and BRUCE H. D. Resistance to fatigue stressing of wood to metal joints glued with several types of adhesives. *Rep. For. Prod. Lab., Madison* no. 1545, Aug. 1946, pp. 6.

3236. FERGUSON, L. Fatigue cracking of coated lead alloys. *Bell Lab. Rec.*, vol. 24, no. 2, Feb. 1946, pp. 53–56.

3237. FISHER, M. S. Comments on the results of tensile tests and fatigue tests made at R.A.E on butt welded joints in alloy steel tubing *Roy. Aircr. Estab. Tech. Note* no. Met. 2 Feb. 1946, pp. 14.

3238. FORREST, G. Some experiments on the effects of residual stresses on the fatigue of aluminium alloys. *J. Inst. Met.*, vol. 72, pt.

1946, pp. 1–17 and [discussion] pp. 516–527. Abstr.: *Metallurgia, Manchr*, vol. 33, no. 197, March 1946, pp. 267–268.

3239. FOSTER, H. W. Plates in tension joints under repeated loading. *Prod. Engng*, vol. 17, no. 12, Dec. 1946, pp. 108–111.

3240. FOSTER, H. W. *and* SELIGER, V. An empirical approach to the fatigue strength analysis of structures. *J. appl. Mech.*, vol. 13, no. 3, Sept. 1946, pp. A-201–A-206.

3241. FOTIADI, A. Position de la résistance à l'endurance des aciers par rapport à leur limite élastique dynamique. [The fatigue resistance of steels in relation to their dynamic elastic limits.] *C.R. Acad. Sci., Paris*, vol. 222, no. 21, May 20, 1946, pp. 1229–1231.

3242. FOUND, G. H. The notch sensitivity in fatigue loading of some magnesium-base and aluminium-base alloys. *Proc. Amer. Soc. Test. Mater.*, vol. 46, 1946, pp. 715–740.

3243. FREUDENTHAL, A. M. The statistical aspect of fatigue of metals. *Proc. roy. Soc. (A)*, vol. 187, no. 1011, Dec. 13, 1946, pp. 416–429.

3244. GABER, E. Versuche und Betrachtungen über die Sicherheit von Stahlbrücken. [Investigation and consideration of the safety of steel bridges.] *Technik, Berl.*, vol. 1, no. 2, Aug. 1946, pp. 57–65.

3245. GASSNER, E. Strength investigations in aircraft construction under repeated application of load. *Tech. Memor. nat. Adv. Comm. Aero., Wash.*, no. 1087, Aug. 1946, pp. 8.

3246. GILLETT, H. W.; GROVER, H. J. *and* JACKSON, L. R. Les tendances Americaines dans les essais d'endurance. [American trends in fatigue testing.] *Rev. Métall.*, vol. 43, no. 9/10, Sept.–Oct. 1946, pp. 268–270.

3247. GOHN, G. R. *and* ARNOLD, S. M. Fatigue properties of beryllium copper strip and their relation to other physical properties. *Proc. Amer. Soc. Test. Mater.*, vol. 46, 1946, pp. 741–782.

3248. GOUGH, H. J. *and* SOPWITH, D. G. Inert atmospheres as fatigue environments. *J. Inst. Met.*, vol. 72, 1946, pp. 415–421 and [discussion] pp. 651–653. Abstr.: *Engineering, Lond.*, vol. 162, no. 4207, Aug. 30, 1946, pp. 197–198.

3249. GURNEY, C. *and* PEARSON, S. Cyclic fatigue of small duralumin tubes. *Roy. Aircr. Estab. Tech. Note* no. Mat. 53, Nov. 1946, pp. 5.

3250. HANSTOCK, R. F. *and* MURRAY, A. Damping capacity and fatigue of metals. *J. Inst. Met.*, vol. 72, Feb. 1946, pp. 97–132 and [discussion] pp. 511–515. *Engineering, Lond.*, vol. 161, no. 4187, April 12, 1946, pp. 358–360 and no. 4188, April 19, 1946, pp. 381–383. Abstr.: *Metallurgia, Manchr*, vol. 33, no. 197, March 1946, pp. 266–267.

3251. L'HERMITE, R.; SEFERIAN, D. *and* CANAC, F. Les essais d'endurance en flexion répétée comme critère de soudabilité des tôles. 2e Rapport — Application à une tôle d'acier au chrome-molybdène. [Repeated bending fatigue tests as a criterion for the weldability of sheet. 2nd report—application to chromium–molybdenum steel sheet.] *Ann. Inst. Bâtim.*, Circ. ser. G, no. 14, Oct. 15, 1946, pp. 28. English abstr.: *Mater. & Meth.*, vol. 25, no. 3, March 1947, pp. 128–129.

3252. L'HERMITE, R. *and* SÉFÉRIAN, D. Les essais d'endurance en flexion répétée comme critère de soudabilité des tôles. 3ème rapport — Application à une tôle d'un alliage aluminium-magnésium à 5% de magnesium. [Repeated bending fatigue tests as a criterion for the weldability of sheet. 3rd report —application to aluminium–magnesium alloy sheet with up to 5% of magnesium.] *Rapp. Gr. franc. Rech. aéro.*, no. 18, 1946, pp. 48.

3253. L'HERMITE, R. *and* SEFERIAN, D. The fatigue behaviour in reverse bending of welded steel sheet. *Weld. J., Easton, Pa.*, vol. 25, no. 9, Sept. 1946, pp. 512s–517s.

3254. HESS, W. F.; WYANT, R. A.; WINSOR, F. J. *and* COOK, H. C. An investigation of the fatigue strength of spot welds in the aluminium alloy Alclad 24S-T. *Weld. J., Easton, Pa.*, vol. 25, no. 6, June 1946, pp. 344s–358s.

3255. HOLLISTER, S. C. *and* GARCIA, J. Final report on fatigue tests of ship welds. *Rep. U.S. off. Sci. Res. Developm.*, no. 6544, ser. no. M-606, Jan. 17, 1946, pp. 38.

3256. HOLT, M. Fatigue failures of bolted, welded and riveted connections. *Proc. Soc. exp. Stress Anal.*, vol. 3, no. 2, 1946, pp. 131–132.

3257. HORGER, O. J. Fatigue tests of some manufactured parts. *Proc. Soc. exp. Stress Anal.*, vol. 3, no. 2, 1946, pp. 133–136.

3258. HORGER, O. J. *and* CANTLEY, W. I. Design of crankpins for locomotives. *J. appl. Mech.*, vol. 13, no. 1, March 1946, pp. A17–A33.

3259. HUBERT, P. Etude sur l'emploi du soudage autogène dans les pièces principales en acier à haute résistance. [Study of the use of autogeneous welding for high strength steel parts.] *Congr. nat. Aviat. franc.*, 2nd Congress, Report no. 43/193, 1946, pp. 26.

3260. HUGHES, E. Longitudinal grinding of fatigue test pieces. *Munit. Supp. Labs, Maribyrnong, Australia*, Circ. no. 7, Nov. 1946, pp. 2.

261. JACKSON, L. R.; BANTA, H. M. *and* MCMASTER, R. C. Progress report. Drill string research. *Drill. Contract.*, vol. 3, no. 1, Nov. 15, 1946, pp. 46–60.

3262. JACKSON, L. R. *and* GROVER, H. J. The fatigue strength of some magnesium sheet alloys. *Proc. Amer. Soc. Test. Mater.*, vol. 46, 1946, pp. 783–798.

3263. JACKSON, L. R.; GROVER, H. J. *and* MCMASTER, R. C. Advisory report on fatigue properties of aircraft materials and structures. *Rep. U.S. Off. Sci. Res. Developm.*, no. 6600, ser. no. M-653, March 1, 1946, pp. 291.

3264. JACKSON, L. R.; WILSON, W. M.; MOORE, H. F. *and* GROVER, H. J. The fatigue characteristics of bolted lap joints of 24S-T alclad sheet materials. *Tech. Note nat. Adv. Comm. Aero., Wash.*, no. 1030, Oct. 1946, pp. 67.

3265. JENSEN, R. S. *and* MOORE, H. F. Fatigue tests of rail steel under compressive stress. *Proc. Amer. Soc. Test. Mater.*, vol. 46, 1946, pp. 799–813.

3266. JOHNSTONE, W. W. Repeated load tests on a Mosquito wing. *Rep. Div. Aeronautics, Aust.*, no. SM.59, April 1946, pp. 9.

3267. JOHNSTONE, W. W. Review of current methods for strength testing aircraft wings. *Rep. A.C.A.*, no. ACA-28, Sept. 1946, pp. 34.

3268. JONES, A. B. Effect of structural changes in steel on fatigue life of bearings. *Steel*, vol. 119, no. 14, Sept. 30, 1946, pp. 68–70, 97, 99–100.

3269. KOMMERS, J. B. Effect of understressing and overstressing in fatigue. *Proc. Soc. exp. Stress Anal.*, vol. 3, no. 2, 1946, pp. 137–141.

3270. KONTOROVA, T. A. The statistical theory of the scale factor. [R] *Sbornik dokladov po dinamicheskoi prochnosti detalei mashin [Col-*

lection of papers on the dynamic strength of machine parts]. Moscow, Izd-vo Akademii nauk SSSR, 1946, pp. 178–184.

3271. KUNDRYAVTSEV, I. V. Fatigue at very low temperatures and critical temperature. [R] Zav. Lab., vol. 12, nos. 9/10, 1946, pp. 843–849. English abstr.: Engrs' Dig., vol. 8, no. 10, Oct. 1947, p. 354.

3272. DE LACOMBE, M. J. Difficultés dans l'exécution et l'interprétation des essais de fatigue à la flexion rotative. [Difficulties in carrying out and interpreting rotating bending fatigue tests.] Rev. Métall., vol. 43, no. 9/10, Sept.–Oct. 1946, pp. 271–285. English abstr.: Iron Coal Tr. Rev., vol. 155, no. 4160, Dec. 5, 1947, pp. 1087–1090.

3273. LAURENT, P. and FERRY, M. Fatigue et effet Bauschinger. [Fatigue and the Bauschinger effect]. Rev. Métall., vol. 43, no. 11/12, Nov.–Dec. 1946, pp. 327–329.

3274. LEWIS, W. C. Fatigue of sandwich constructions for aircraft—cellular cellulose acetate core material in shear. Rep. For. Prod. Lab., Madison, no. 1559, Dec. 1946, pp. 5.

3275. LEWIS, W. C. Fatigue of wood and glued-wood constructions. Proc. Amer. Soc. Test. Mater., vol. 46, 1946, pp. 814–835.

3276. LIPSON, C. and NOLL, G. Fatigue strength of steel forgings. Prod. Engng, vol. 17, no. 9, Sept. 1946, pp. 130–131.

3277. LLOYD, W. S. Surface preparation and drill pipe fatigue failure. Oil Wkly, vol. 124, no. 2, Dec. 9, 1946, p. 46.

3278. LUTHANDER, S. and WÅLLGREN, G. Bestämning av livslängden vid utmattningsprov med variabla lastgränser. [Determination of the fatigue life with stress cycles of varying amplitude.] Flytek. Försöksanst., Stockholm, Rep. no. 18, 1946, pp. 23. English translation: Roy. Aircr. Estab., Lib. Trans., no. 421, March 1953, pp. 19.

3279. McQUAID, H. W. The metallurgical phase of fatigue failures. Proc. Soc. exp. Stress Anal., vol. 3, no. 2, 1946, pp. 142–148.

3280. MANEY, G. A. Repeated loads on riveted joints. Fasteners, vol. 3, no. 2, 1946, pp. 14–15.

3281. MATTHIEU, P. Über transversale Stabschwingungen und Dauerbrüche. [Transverse vibrations of bars and fatigue fractures.] Schweiz. Arch. angew. Wiss., vol. 12, no. 11, Nov. 1946, pp. 329–338 and no. 12, Dec. 1946, pp. 361–372. English abstr.: Engrs' Dig., vol. 8, no. 6, June 1947, pp. 197–201 and no. 7, July 1947, pp. 218–222.

3282. MOORE, H. F. Strengthening metal parts by shot peening. Iron Age, vol. 158, no. 22, Nov. 28, 1946, pp. 70–76 and no. 23, Dec. 5, 1946, pp. 81–86.

3283. NISHIHARA, T. and KOBAYASHI, A. A method of deducing S-N curves. [J] Trans. Soc. mech. Engrs, Japan, vol. 12, no. 42, 1946, pp. 65–71.

3284. NOLL, G. C. and LIPSON, C. Allowable working stresses. Proc. Soc. exp. Stress Anal., vol. 3, no. 2, 1946, pp. 89–109.

3285. NORTON, J. T.; ROSENTHAL, D. and MALOOF, S. B. X-ray diffraction study of the effect of residual compression on the fatigue of notched specimens. Weld. J., Easton, Pa., vol. 25, no. 11, Nov. 1946, pp. 729s–735s.

3286. ODING, I. A. Influence of size and shape of specimens on their fatigue strength. [R] Sbornik dokladov po dinamicheskoi prochnosti *detalei mashin [Collection of papers on the dynamic strength of machine parts].* Moscow, Izd-vo Akademii nauk SSSR, 1946, pp. 141–156.

3287. OLSON, W. Z.; BENSEND, D. W. *and* BRUCE, H. D. Resistance of several types of glue in wood joints to fatigue stressing. *Rep. For. Prod. Lab. Madison,* no. 1539, March 1946, pp. 4.

3288. OWEN, J. B. B. Note on service failures in aircraft structures associated with fatigue, "repeated" or dynamic loads. *Roy. Aircr. Estab. Rep.* no. SME 3384, Aug. 1946, pp. 9.

3289. PARKINSON, D. *and* BLOXHAM, J. L. Behaviour of natural rubber and GR–S under repeated stress cycles. *Rubb. Chem. Technol.,* vol. 19, 1946, pp. 417–427.

3290. PETERSON, R. E. Amount gained in fatigue strength of machine parts by using a material of higher tensile strength. *Proc. Soc. exp. Stress Anal.,* vol. 3, no. 2, 1946, pp. 149–151.

3291. POOLE, S. W. *and* JOHNSON, R. J. A review of some mechanical failures of steel plant machine equipment. *Proc. Soc. exp. Stress Anal.,* vol. 3, no. 2, 1946, pp. 61–75.

3292. PUCHNER, O. Über die Erzegung synchroner überlagerter Wechselbiege- und Verdrehbeanspruchungen. [On the production of synchronous superimposed alternating bending and torsional loads.] *Schweiz. Arch. angew. Wiss.,* vol. 12, no. 9, Sept. 1946, pp. 289–293.

3293. PUTNAM, A. A. *and* REISERT, T. D. An analysis of the fatigue life of an airplane wing structure under overload conditions. *Nat. Adv. Comm. Aero., Wash., Restricted Bulletin*

no. L5K29 (*Wartime report* no. L-10), Feb. 1946, pp. 4.

3294. QUINLAN, F. B. Pneumatic fatigue machines. *Proc. Amer. Soc. Test. Mater.,* vol. 46, 1946, pp. 846–851.

3295. Roš, M. Die Festigkeit und Sicherheit der Schweissverbindungen. [The strength and safety of welded joints.] *Ber. eidgenöss. MatPrüfAnst.,* no. 156, Jan. 1946, pp. 51.

3296. Roš, M. G. Influence de la surface sur la fatigue des métaux. [Influence of surface conditions on the fatigue of metals.] *Journées des États de Surface,* Paris, Editions de l'Office Professionnel Général de la Transformation des Métaux, 1946, pp. 207–218.

3297. ROSEN, C. G. A. *and* KING, R. Some aspects of fatigue in diesel engine parts. *Proc. Soc. exp. Stress Anal.,* vol. 3, no. 2, 1946, pp. 152–160.

3298. ROTH, W. The effect of repeated loadings on the strength of aircraft components. *Roy. Aircr. Estab. Tech. Note* no. S.M.E. 350, Feb. 1946, pp. 6.

3299. SOPWITH, D. G. *and* SETTLE, T. Research on fatigue strength of screw threads of different form. *Proc. Instn mech. Engrs, Lond.,* vol. 155, 1946, pp. 156–158.

3300. STRAUB, J. C. Can compression stress cause fatigue failure? *Amer. Mach., N.Y.,* vol. 90, no. 22, Oct. 24, 1946, p. 120.

3301. STÜSSI, F. *and* KOLLBRUNNER, C. F. Schrumpspannungen und Dauerfestigkeit geschweisster Trägerstösse. [Shrinkage stresses and the fatigue strength of welded joints in girders.] *Mitt. Inst. Baustat. Zürich*, no. 18, 1946, pp. 47.

3302. SWERING, J. B. Fatigue cracks in retaining rings of turbine-generators. *Pwr Plant (Engng)*, vol. 50, July 1946, pp. 92–94.

3303. TAYLOR, W. J. *and* SAUNDERS, E. A. D. Endurance tests of flexible steel wire rope. *Roy. Aircr. Estab. Tech. Note* no. MET 37, June 1946, pp. 15.

3304. TOWNSEND, J. R. Fatigue failures in telephone apparatus parts. *Proc. Soc. exp. Stress Anal.*, vol. 3, no. 2, 1946, pp. 161–166.

3305. TRISHMAN, L. E. Plastic coating inside surface of drill pipe to combat corrosion fatigue failures. *Petrol. Engr*, vol. 18, no. 2, Nov. 1946, pp. 194, 196, 198, 200.

3306. UZHIK, G. V. Principles of metal fatigue testing methods. [R] *Zav. Lab.*, vol. 12, no. 11/12, 1946, pp. 949–954.

3307. DE WAARD, R. D. An expedient method for preparation of cantilever-beam fatigue specimens. *Bull. Amer. Soc. Test. Mat.*, no. 141, Aug. 1946, pp. 40–42.

3308. WALLER, R. C. *and* ROSEVEARE, W. E. Fatigue failure of rayon tire cord. *J. appl. Phys.*, vol. 17, no. 6, June 1946, pp. 482–491.

3309. WÅLLGREN, G. Experimentell bestämning av utmattningsdiagrammet för nithålsförsedda provstavar av alcladplåt vid drag-ock tryckbelastning. [Experimental determination of the fatigue diagram for alclad sheet specimens with rivet holes, subjected to tension and compression loading.] *Flytek. Försöksanst., Stockholm, Report* no. 14, 1946, pp. 10.

3310. WEIBULL, W. Inflytandet av volymen vid utmattnings-påkänningar. [Influence of specimen size in fatigue testing.] *I.V.A.*, vol. 17 no. 2, 1946, pp. 62–64.

3311. WELLINGER, K. *and* STÄHLI, G. Reversed stress testing of piston materials. *Automot. Industr. N.Y.*, vol. 94, no. 4, Feb. 15, 1946, pp. 20–23, 66.

3312. WIESCHHAUS, L. J. Shot peening and its importance in the spring industry. *Wire and Wire Prod.*, vol. 21, no. 9, Sept. 1946, pp. 665–667, 701–703.

3313. WILLIAMS, D. Strength of aircraft in relation to repeated loads (a review of the problem and suggestions for further research), *Roy. Aircr. Estab. Rep.* no. S.M.E. 3386. Sept. 1946, pp. 10.

3314. WILSON, W. M. Fatigue strength of weldments used to reinforce and repair steel bridge members. *Bull. Amer. Rly Engng Ass.*, vol. 48, no. 460, June–July 1946, pp. 19–33.

3315. WYLY, L. T. Fatigue testing machine built for Northwestern University. *Civ. Engng, Easton*, vol. 16, no. 10, Oct. 1946, p. 471.

3316. YAROVINSKII, L. M. *and* KUDRYAVTSEV, I. V. Fatigue resistance of cast steel and welded-on metal at 100 to –75°. [R] *Avtog. Delo,* nos. 5/6, 1946, pp. 1–5.

3317. Fatigue failures and measures to reduce their incidence. *Elect. Mfg*, vol. 37, no. 5, May 1946, pp. 121–124, 216, 218, 220, 222, 224.

3318. Effect of metallurgical changes due to welding upon fatigue strength of carbon steel plates. *Weld. J., Easton, Pa.*, vol. 25, no. 8, Aug. 1946, pp. 425s–450s.

3319. Pneumatic fatigue test. *Steel*, vol. 119, no. 11, Sept. 9, 1946, pp. 112–113, 150, 152.

3320. Effect of periods of rest on fatigue strength of welded joints. *Weld. J., Easton, Pa.*, vol. 25, no. 9, Sept. 1946, pp. 518s–521s.

3321. Fatigue tests on joints made in D.T.D. 390 using Redux cement. *Sci. Tech. Memor.*, no. C9/44, Nov. 1946, pp. 2.

3322. Prestressing of metals and metal parts, *Mater. & Meth.*, vol. 24, no. 5, Nov. 1946, pp. 1156–1159.

1947

3323. ALMEN, J. O. Fatigue of metals as influenced by design and internal stresses. *Surface stressing of metals*, Cleveland, Ohio, American Society for Metals, 1947, pp. 33–84.

3324. AMERICAN SOCIETY FOR TESTING MATERIALS. *Samposium on testing of parts and assemblies*. Philadelphia, Pa., American Society for Testing Materials, Special Technical Publication no. 72, April 1947, pp. 86.

3325. BARISH, T. Fatigue testing machines for ball and roller bearings. *Symposium on testing of bearings*, Philadelphia, Pa., American Society for Testing Materials, Special Technical Publication no. 70, 1947, pp. 19–34.

3326. BAXTER, A. M. The fatigue of bolts and studs under combined bending and tension. *Rep. Brit. Shipb. Res. Ass.*, no. 2, June 1947, pp. 6.

3327. BEALE, W. O. Types of fatigue failure in the steel industry. *The failure of metals by*

fatigue, Melbourne, Melbourne University Press, 1947, pp. 455–462.

3328. BEAN, W. T. Sound engine design thwarts parts fatigue. *S.A.E. Jl.*, vol. 55, Sept. 1947, pp. 44–45.

3329. BEAN, W. T. Endurance —a criterion of design. *Symposium on testing of parts and assemblies*, Philadelphia, Pa., American Society for Testing Materials, Special Technical Publication no. 72, April 1947, pp. 25–40.

3330. BERDAHL, E. O. Construction and operation of the Losenhausen 44,000 pound vibration generator. *David W. Taylor Model Basin, U.S. Navy, Report* no. 554, April 1947 pp. 30.

3331. BOAS, W. Theories of the mechanism of fatigue failure. *The failure of metals by fatigue* Melbourne, Melbourne University Press, 1947 pp. 28–39.

3332. BOLLENRATH, F. *and* CORNELIUS, H. Zugschwellfestigkeit gleichartig gekerbter Stahlstäbe mit verschiedener Querschnittsgrösse [Alternating tensile fatigue strength of similarly notched steel specimens of different diameter.] *Z. Metallk.*, vol. 38, no. 1, Jan. 1947 pp. 9–11.

3333. BOLLENRATH, F. *and* CORNELIUS, H. Einfluss der Gewindeherstellung auf die Dauerhaltbarkeit von Schrauben. [The effect of the method of manufacture of screw threads on the fatigue strength of bolts.] *Werkst. u. Betr.* vol. 80, no. 9, Sept. 1947, pp. 217–222. English translation: *Roy. Aircr. Estab., Lib. Trans.* no. 388.

3334. BRAITHWAITE, W. R The influence of protective coatings on the fatigue and corrosion fatigue of steel. *Coil Spring Jnl*, no. 6, March 1947, pp. 26–28.

3335. BROOKMAN, J. G. *and* KIDDLE, L. The prevention of fatigue failures in metal parts by shot peening. *The failure of metals by fatigue*, Melbourne, Melbourne University Press, 1947, pp. 395–415.

3336. BURGHOFF, H. L. *and* BLANK, A. I. Fatigue characteristics of some copper alloys. *Proc. Amer. Soc. Test. Mater.*, vol. 47, 1947, pp. 695–712.

3337. CAMPUS, F. *and* JACQUEMIN, R. Essais d'endurance sur traverses de voies ferrées en béton armé ou précontraint. [Fatigue tests on reinforced concrete and prestressed concrete railway sleepers.] *Bull. Cent. Rech. Constr.*, vol. 2, 1947, pp. 133–145.

3338. CASWELL, J. S. Design of parts for conditions of variable stress. *Prod. Engng*, vol. 18, no. 1, Jan. 1947, pp. 118–119.

3339. CAVANAGH, P. E. Some changes in physical properties of steels and wire rope during fatigue failure. *Trans. Canad. Min. Inst. (Inst. Min. Metall.)*, vol. 50, 1947, pp. 401–411.

3340. CAVANAGH, P. E. The progress of failure in metals as traced by changes in magnetic and electrical properties. *Proc. Amer. Soc. Test. Mater.*, vol. 47, 1947, pp. 639–650.

3341. CHIVERS, S. D. Fatigue failures of lead sheathing of telephone cables. *The failure of metals by fatigue*, Melbourne, Melbourne University Press, 1947, pp. 500–505.

3342. COLOMBIER, L. Influence de très petites quantités de soufre et d'inclusions oxydées sur la qualité des aciers. [Influence of small quantities of sulphur and oxide inclusions on the quality of steel.] *Rev. Métall.*, vol. 44, nos. 1/2, Jan.–Feb. 1947, pp. 47–57.

3343. CONNOR, E. Fatigue failure of axles of car and wagon railway rolling stock. *The failure of metals by fatigue*, Melbourne, Melbourne University Press, 1947, pp. 442–454.

3344. CRAMER, R. E.; ALLEMAN, N. J. *and* JENSEN, R. S. Progress reports of investigations of railroad rails and joint bars. *Proc. Amer. Rly Engng Ass.*, vol. 48, no. 465, Feb. 1947, pp. 714–729, 734–739, 756–766, 804–808. *Uni. Ill. Engng Exp. Sta. Bull.*, vol. 44, no. 46, March 28, 1947, (*Reprint Series* no. 37, 1947), pp. 47.

3345. CRUSSARD, C. Creep and fatigue as affected by grain boundaries. *Metal Treatm.*, vol. 14, no. 51, Autumn 1947, pp. 149–160.

3346. DAVIES, H. E. *and* McKEOWN, J. The significance of mechanical testing. *Metallurgia, Manchr*, vol. 37, no. 217, Nov. 1947, pp. 19–22.

3347. DOLAN, T. J. An investigation of the behaviour of materials under repeated stress. *Office Naval Research, Contract N6-ori-71, Task Order IV, Uni. Illinois, Second Progr. Rep.*, March 1947, pp. 7; *Third Progr. Rep.*, Aug. 1947, pp. 14.

3348. DUNKLEY, G. T. Corrosion fatigue. *Mech. World*, vol. 122, no. 3160, Aug. 8, 1947, pp. 137–141.

3349. EDWARDS, A. R. Fatigue problems in the gas turbine aero engine. *The failure of metals by fatigue*, Melbourne, Melbourne University Press, 1947, pp. 383–394.

3350. EGGINTON, H. H. Introduction to the fatigue of metals. *Bgham metall. Soc. J.*, vol. 27, pt. 1, March 1947, pp. 258–278.

3351. ENOMOTO, N. Study of fatigue and internal friction of metallic materials. 3rd report —mechanism of fatigue under pulsating torsional load. [J] *Trans. Soc. mech. Engrs, Japan,* vol. 13, no. 43, Jan. 1947, pp. 21–32.

3352. EVANS, U. R. The electrochemistry of corrosion fatigue. *The failure of metals by fatigue,* Melbourne, Melbourne University Press, 1947, pp. 84–94.

3353. EVANS, U. R. *and* SIMNAD, M. T. The mechanism of corrosion fatigue of mild steel. *Proc. roy. Soc. (A),* vol. 188, no. 1014, Feb. 11, 1947, pp. 372–392.

3354. FARDIN, R. Indications sommaires sur la résistance des assemblages aux efforts alternés. [Summary of results on the resistance of assemblies to alternating loads.] *Éspaces,* no. 15, July 1947, pp. 54–57.

3355. FINDLEY, W. N. New apparatus for axial-load fatigue testing. *Bull. Amer. Soc. Test. Mat.,* no. 147, Aug. 1947, pp. 54–56.

3356. FÖPPL, O. Drehstabfedern für verschiedene Verwendungszwecke bei Kraftfahrzeugen. [Torsion bars for various purposes in motor vehicles.] *Auto.-tech. Z.,* vol. 49, no. 4, July 1947, pp. 54–55. English abstr.: *Engineer, Lond.,* vol. 185, no. 4801, Jan. 30, 1948, p. 115.

3357. FORRESTER, P. G.; GREENFIELD, L. T. *and* DUCKETT, R. The fatigue strength of some tin–antimony–copper and other tin-base alloys. *Metallurgia, Manchr,* vol. 36, no. 213, July 1947, pp. 113–117.

3358. FOSTER, H. W. A method of detecting incipient fatigue failure. *Proc. Soc. exp. Stress Anal.,* vol. 4, no. 2, 1947, pp. 25–31.

3359. FOTIADI, A. Les propriétés dynamiques, les capacités d'endurance et la qualification des aciers pour pièces de fatigue des moteurs. [The dynamic properties, fatigue strength and classification of steels for engine components subjected to fatigue.] *Rev. Métall.,* vol. 44, no. 1–2, Jan.–Feb. 1947, pp. 12–38 and no. 3–4, March–April 1947, pp. 97–121.

3360. FOUND, G. H. Fatigue characteristics of magnesium castings. *Symposium on testing of parts and assemblies,* Philadelphia, Pa., American Society for Testing Materials, Special Technical Publication no. 72, April 1947, pp. 12–24.

3361. FULLER, F. B. *and* OBERG, T. T. Fatigue characteristics of rotating-beam versus rectangular cantilever specimens of steel and aluminium alloys. *Proc. Amer. Soc. Test. Mater.,* vol. 47, 1947, pp. 665–676.

3362. GADD, C.; OCHILTREE, N. A. *and* ZMUDA, A. Stress concentration and the fatigue strength of engine components. *Symposium on testing of parts and assemblies,* Philadelphia, Pa., American Society for Testing Materials, Special Technical Publication no. 72, April 1947, pp. 76–86.

3363. GEORGE, C. W.; GROVER, S. F. *and* CHALMERS, B. The factors contributing to fatigue failure in aircraft. *The failure of metals by fatigue,* Melbourne, Melbourne University Press, 1947, pp. 56–63.

3364. GEORGE, H. P. Lead-base babbitt alloys— II. Fatigue and wear properties. *Prod. Engng,* vol. 18, no. 6, June 1947, pp. 138–141.

3365. GIBBONS, H. R. Fatigue testing of rolling bearings. *Symposium on testing of bearings,* Philadelphia, Pa., American Society for Testing Materials, Special Technical Publication no. 70, 1947, pp. 53–61.

3366. GIBSON, W. H. H. Photo-elasticity and stress concentration. *The failure of metals by fatigue*, Melbourne, Melbourne University Press, 1947, pp. 196–211.

3367. GLAUBITZ, H. Oberflächenhärtung und Bauteilfestigkeit von Zahnrädern. [Surface hardening and strength of structural members and gear teeth.] *Werkst. u. Betr.*, vol. 80, no. 10, Oct. 1947, pp. 249–259 and no. 11, Nov. 1947, pp. 277–282.

3368. GLEASON, G. B. Influence of shot peening on fatigue strength of 14ST aluminium alloy. *Iron Age*, vol. 159, no. 2, Jan. 9, 1947, pp. 62–64.

3369. GOHN, G. R. *and* ELLIS, N. C. The fatigue characteristics of copper–nickel–zinc and phosphor bronze strip in bending under conditions of unsymmetrical loading. *Proc. Amer. Soc. Test. Mater.*, vol. 47, 1947, pp. 713–724.

3370. GORDON, S. A. Some repeated load investigations on aircraft components. *Proc. Soc. exp. Stress Anal.*, vol. 4, no. 2, 1947, pp. 39–41.

3371. GREEN, N. B. How processing affects bolt fatigue strength. *Mach. Design*, vol. 19, no. 12, Dec. 1947, pp. 138–140.

3372. GREENWOOD, J. N. The failure of metals by fatigue. *The failure of metals by fatigue*, Melbourne, Melbourne University Press, 1947, pp. 1–7.

3373. GROVER, H. J. *and* JACKSON, L. R. Fatigue tests on some spot welded joints in aluminium alloy sheet materials. *Weld. J.*, Easton, Pa., vol. 26, no. 4, April 1947, pp. 215s–232s.

3374. GROVER, H. J. *and* JACKSON, L. R. The fatigue strength of lap joints in some magnesium sheet alloys. *Symposium on testing of parts and assemblies*, Philadelphia, Pa., American Society for Testing Materials, Special Technical Publication no. 72, April 1947, pp. 2–11.

3375. GURNEY, C. *and* PEARSON, S. The scatter of fatigue test results on small duralumin tubes. *Roy. Aircr. Estab. Tech. Note* no. Met. 73, Dec. 1947 [Addendum, May, 1948], pp. 4.

3376. GUSTAFSON, J. R. Some of the effects of cadmium, zinc and tin platings on springs. *Proc. Amer. Soc. Test. Mater.*, vol. 47, 1947, pp. 782–802.

3377. HANSTOCK, R. F. Damping capacity, strain hardening and fatigue. *Proc. Phys. Soc.*, vol. 59, pt. 2, no. 332, March 1, 1947, pp. 275–287.

3378. HANSTOCK, R. F. *and* MURRAY, A. Relation entre la capacité d'amortissement et la fatigue dans les alliages d'aluminium. [Relation between damping capacity and fatigue of aluminium alloys.] *Rev. Métall.*, vol. 44, no. 1–2, Jan.–Feb. 1947, pp. 58–62.

3379. HARTMANN, E. C.; HOLT, M. *and* ZAMBOKY, A. N. Static and fatigue tests of arc-welded aluminium alloy 61S-T plate. *Weld. J.*, Easton, Pa., vol. 26, no. 3, March 1947, pp. 129s–138s.

3380. HERZOG, A. Six ton Schenck fatigue testing machine. *Tech. Rep. U.S. Army Air Force*, no. 5623, Aug. 15, 1947, pp. 24.

3381. HEYWOOD, R. B. The relationship between fatigue and stress concentration. *Aircr. Engng*, vol. 19, no. 217, March 1947, pp. 81–84.

3382. HIRST, G. W. C. The influence of radial pressure from a press fit on the endurance limit of axles and crank pins. *The failure of metals by fatigue*, Melbourne, Melbourne University Press, 1947, pp. 431–441. Summary: *Engineer, Lond.*, vol. 184, no. 4796, Dec. 26, 1947, pp. 598–599.

3383. HOLLOMAN, J. H.; JAFFE, L. D.; McCARTHY, D. E. *and* NORTON, M. R. The effects of microstructure on the mechanical properties of steel. *Trans. Amer. Soc. Metals*, vol. 38, 1947, pp. 807–847.

3384. HONEYCOMBE, R. W. K. Conditions leading to fatigue failure in sleeve bearings. *The failure of metals by fatigue*, Melbourne, Melbourne University Press, 1947, pp. 362–382.

3385. HOOKE, F. H. *and* WILLS, H. A. A procedure for the calculation of the endurance strength of aircraft structures. *Note Div. Aeronautics Aust.*, Structures and Materials no. 152, April 1947, pp. 10.

3386. HOOTON, F. W. The measurement of dynamic strain. *The failure of metals by fatigue*, Melbourne, Melbourne University Press, 1947, pp. 112–134.

3387. HORGER, O. J. Stressing axles and other railroad equipment by cold rolling. *Surface stressing of metals*, Cleveland, Ohio, American Society for Metals, 1947, pp. 85–142.

3388. HORGER, O. J. *and* LIPSON, C. H. Automotive rear axles and means of improving their fatigue resistance. *Symposium on testing of parts and assemblies*, Philadelphia, Pa., American Society for Testing Materials, Special Technical Publication no. 72, April 1947, pp. 47–75.

3389. HUYETT, R. B. Shot peening increases life of machinery parts. *Steel Process.*, vol. 33,

no. 9, Sept. 1947, pp. 553–557, 573 and no. 10, Oct. 1947, pp. 609–613, 638, 647.

3390. JACKSON, L. R. *and* GROVER, H. J. Structures liable to fatigue failure and some considerations in their design. *The failure of metal by fatigue*, Melbourne, Melbourne University Press, 1947, pp. 73–83.

3391. JACKSON, L. R. *and* POCHAPSKY, T. E. The effect of composition on the fatigue strength of decarburized steel. *Trans. Amer. Soc. Metals*, vol. 39, 1947, pp. 45–60.

3392. JOHNSON, E. T. Life testing of plain bearings for automotive engines. *Symposium o' testing of bearings*, Philadelphia, Pa., American Society for Testing Materials, Special Technical Publication no. 70, 1947, pp. 2–18

3393. JOHNSON, P. O. and LIPSON, C. Fatigue strength of steel parts. *Prod. Engng*, vol. 18 no. 10, Oct. 1947, pp. 144–146.

3394. JOHNSTONE, W. W. Methods of investigating the fatigue properties of material *The failure of metals by fatigue*, Melbourne Melbourne University Press, 1947, pp. 135 164.

3395. JOHNSTONE, W. W. Static and repeated load tests on "Mosquito" wings. *Rep. Div Aeronautics Aust.*, no. SM.104, Sept. 194 pp. 21.

3396. JONES, A. B. Metallographic observatio of ball bearing fatigue phenomena. *Symposium on testing of bearings*, Philadelphia, Pa American Society for Testing Materia Special Technical Publication no. 70, 194 pp. 35–52.

3397. JONES, G. M. *and* JONES, E. R. W. A statistical analysis of the results of Wöhler fatigue tests made on material to B.S.S. 6L1. *R*

Aircr. Estab. Tech. Note MET 53, May 1947, pp. 10.

3398. KEETH, J. A. Gage-glass condensate cracks metal by pocketing at drum counterbore. *Power*, vol. 91, no. 8, Aug. 1947, pp. 78–79.

3399. KOSTING, P. R. Progressive stress-damage. *Surface stressing of metals*, Cleveland, Ohio, American Society for Metals, 1947, pp. 143–190.

3400. KUKANOV, L. I. Fatigue test of leaf springs and torsion shafts. [R] *Zav. Lab.*, vol. 13, no. 8, Aug. 1947, pp. 997–1002.

3401. DE LACOMBE, J. Interpretation of results of rotary bending fatigue tests. *Iron Coal. Tr. Rev.*, vol. 155, no. 4160, Dec. 5, 1947, pp. 1087–1090.

3402. LASHKO, N. F. Repeated impact bending tests. [R] *Zav. Lab.*, vol. 13, no. 5, May 1947, pp. 600–606.

3403. LAURENT, P. Sur l'influence de la forme et des dimensions de éprouvette sur le résultat de l'essai de fatigue. [The influence of form and dimensions of specimen on fatigue test results.] *C. R. Acad. Sci., Paris.*, vol. 224, no. 10, March 10, 1947, pp. 719–721.

3404. LAZAN, B. J. Fatigue testing machine. *Mach. Design*, vol. 19, no. 5, May 1947, pp. 123–127.

405. DE LEIRIS, H. Sur quelques ruptures d'arbres extérieurs par fissuration progressive. [On the fracture of external shafts due to progressive cracking.] *Bull. Ass. tech. marit.*, vol. 46, mémoire 865, 1947, pp. 385–392.

3406. LEVEY, H. C. *and* BRETT, P. R. The vibration of telephone line wires. *The failure of metals by fatigue*, Melbourne, Melbourne University Press, 1947, pp. 477–499.

3407. LOVE, R. J. Cast crankshafts: a survey of published information. *Rep. Mot. Ind. Res. Ass.*, no. 1947/R/10, Dec. 1947, pp. 39. *J. Iron St. Inst.*, vol. 195, pt. 3, July 1948, pp. 247–274.

3408. LOVE, R. J. *and* MILLS, H. R. Fatigue strength of cast crankshafts: effect of proportions and size on bending fatigue strength. *Rep. Mot. Ind. Res. Ass.*, no. 1947/R/11, Dec. 1947, pp. 20.

3409. LUNDBERG, BO. "Bear-up" requirements for aircraft. *Aero Dig.*, vol. 55, no. 6, Dec. 1947, pp. 56–58, 120–122.

3410. LUNDBERG, B. Note on fatigue strength requirements for aircraft. *Flytek. Försöksanst.*, Stockholm, Rep. no. HE-206 : 1, March 13, 1947, pp. 23.

3411. LUNDBERG, G. *and* PALMGREN, A. Dynamic capacity of rolling bearings. *Acta polyt. Stockh.*, vol. 1, no. 3, 1947, pp. 50. *Ingen-VetenskAkad. Handl.*, no. 196, 1947, pp. 50.

3412. LUTHANDER, S. Det hållfasthetstekniska säkerhetsbegreppet ur statistisk synpunkt. [The safety concept in strength of materials from the statistical viewpoint.] *Tekn. Tidskr.*, Stockh., vol. 77, no. 42, Oct. 18, 1947, pp. 769–780.

3413. MCDONALD, G. G. The design of cylindrical shafts subjected to fluctuating loading. *The failure of metals by fatigue*, Melbourne, Melbourne University Press, 1947, pp. 248–259.

3414. MACKENZIE, B. Automobile suspension springs. *Proc. Instn mech. Engrs, Lond., (Automobile Division)*, pt. III., 1947–48, pp. 122–135.

3415. MCKEOWN, J. Fatigue testing with particular reference to tests at elevated temperatures. *Bgham metall. Soc. J.*, vol. 27, pt. 4, Dec. 1947, pp. 423–442.

3416. MCMASTER, R. C. and GROVER, H. J. Radiography and the fatigue strength of spot welds in aluminium alloys. *Weld. J., Easton, Pa.*, vol. 26, no. 3, March 1947, pp. 223–232.

3417. MCMASTER, R. C. and GROVER, H. J. Spot welded aluminium lap joints designed for repeated loads. *Prod. Engng*, vol. 18, no. 11, Nov. 1947, pp. 112–116.

3418. MCMASTER, R. C. and GROVER, H. J. Spot welded aluminium components designed for repeated loads. *Prod. Engng*, vol. 18, no. 12, Dec. 1947, pp. 158–161.

3419. MARIN, J. Utilizing mechanical properties in die casting design. Part 5—members subjected to repeated or fatigue loads. *Die Castings*, vol. 5, no. 8, Aug. 1947, pp. 22, 25–26, 28.

3420. MARIN, J. and STULEN, F. B. A new fatigue strength—damping criterion for the design of resonant members. *J. appl. Mech.*, vol. 14, no. 3, Sept. 1947, pp. A209–A212.

3421. MARKL, A. R. C. Fatigue tests of welding elbows and comparable double-mitre bends. *Trans. Amer. Soc. mech. Engrs*, vol. 69, no. 8, Nov. 1947, pp. 869–879.

3422. MARKOVETS, M. P. Relationship between fatigue limit and true fracture strength. [R] *Zav. Lab.*, vol. 13, no. 8, Aug. 1947, pp. 1003–1007.

3423. MATTHIEU, P. Transverse vibrations of bars and fatigue fractures. *Engrs' Dig.*, vol. 8, no. 6, June 1947, pp. 197–201 and no. 7, July 1947, pp. 218–222.

3424. MEHR, L.; OBERG, T. T. and TERES, J. Fatigue limit of chromium plated steel. *Mon. Rev. Amer. Electroplat. Soc.*, vol. 34, no. 12, Dec. 1947, pp. 1345–1360.

3425. MOORE, H. F. Effect of surface stressing metals on endurance under repeated loadings—the problem defined. *Surface stressing of metals*, Cleveland, Ohio, American Society for Metals, 1947, pp. 1–18.

3426. MOORE, H. F. Metallography, fatigue of metals and conventional stress analysis. *The failure of metals by fatigue*, Melbourne, Melbourne University Press, 1947, pp. 8–27.

3427. MORRIS, D. O. Composition and physical properties of steel in relation to fatigue. *The failure of metals by fatigue*, Melbourne, Melbourne University Press, 1947, pp. 336–361.

3428. NAIRN, R. A. Shot peening and fatigue failure. *Commonw. Engr*, vol. 35, no. 1, Aug. 1, 1947, pp. 10–18.

3429. NISHIHARA, T. and KOBAYASHI, A. New theory of fatigue of metals and its application. [J] *Trans. Soc. mech. Engrs, Japan*, vol. 13, no. 44, May 1947, pp. 56–67.

3430. NISHIHARA, T. and KOBAYASHI, A. New theory of fatigue of metals and various application. *Mem. Coll. Engng Kyoto*, vol. 11, no. 6, June 1947, pp. 121–134.

3431. O'DONNELL, D. *and* BUNDLE, A. S. Some practical aspects of wire fatigue in aerial telephone lines based on an analysis of wire breakages. *The failure of metals by fatigue*, Melbourne, Melbourne University Press, 1947, pp. 463–476.

3432. O'NEILL, H. Failures of railway materials by fatigue. *The failure of metals by fatigue*, Melbourne, Melbourne University Press, 1947, pp. 416–430.

3433. ODING, I. A. *The allowable stresses in machine construction and the fatigue strength of metals.* [R] Third Edition. Moscow, Gosudarstvennoe Nauchno-Tekhnicheskoe Izdatel'-stvo [GONTI] Mashinostroitel'noi Literatury, 1947, pp. 184.

3434. ORR, C. W. The detection of fatigue cracks. *The failure of metals by fatigue*, Melbourne, Melbourne University Press, 1947, pp. 95–111.

3435. OSBORN, C. J. The fatigue of welded steel tubing in aircraft structures. *The failure of metals by fatigue*, Melbourne, Melbourne University Press, 1947, pp. 290–308.

3436. PATERSON, M. S. Notch sensitivity of metals. *The failure of metals by fatigue*, Melbourne, Melbourne University Press, 1947, pp. 309–335.

3437. PATERSON, M. S. *and* WILLS, H. A. The strength of machine parts under fluctuating loads. *Rep. Div. Aeronautics Aust.*, no. SM.82, Nov. 1947, pp. 26.

3438. PERCIVAL, A. L. *and* WECK, R. Fatigue tests on four welded H-beams. *The failure of metals by fatigue*, Melbourne, Melbourne University Press, 1947, pp. 212–236.

3439. PERCIVAL, A. L. *and* WECK, R. Fatigue tests by the resonance vibration method on four welded H-beams. First interim report. *Trans. Inst. Weld.*, [Welding Research], vol. 10, no. 3, June 1947, pp. 6–22.

3440. PIERPONT, W. G. Fatigue tests of major aircraft structural components. *Proc. Soc. exp. Stress Anal.*, vol. 4, no. 2, 1947, pp. 1–15.

3441. POITMAN, I. M. *and* FRIDMAN, YA. B. Repeated-load tests on metals in the plastic zone. [R] *Zav. Lab.*, vol. 13, no. 4, April 1947, pp. 452–463.

3442. PROT, M. L'essai de fatigue sous charge progressive. [Fatigue test under progressive load.] *C. R. Acad. Sci., Paris*, vol. 225, no. 16, Oct. 20, 1947, p. 669.

3443. PÜCHNER, O. Einfluss der Querschnittgrösse auf die Dauerfestigkeit und Kerbempfindlichkeit von Gusseisen. [Influence of cross-section size on the fatigue strength and notch sensitivity of cast iron.] *Schweiz. tech. Z.*, vol. 44, no. 16, April 17, 1947, pp. 253–256.

3444. PUGSLEY, A. G. Repeated loading on structures. *The failure of metals by fatigue*, Melbourne, Melbourne University Press, 1947, pp. 64–72.

3445. PUGSLEY, A. G. The behaviour of structures under repeated loads. *J. R. aero. Soc.*, vol. 51, no. 441, Sept. 1947, pp. 715–720.

3446. QUINLAN, F. B. Pneumatic fatigue testing. *Automot. Industr. N.Y.*, vol. 96, no. 2, Jan. 15, 1947, pp. 30–31, 94, 96.

3447. QUINLAN, F. B. Pneumatic fatigue machines. *Symposium on testing of parts and assem-*

blies, Philadelphia, Pa., American Society for Testing Materials, Special Technical Publication no. 72, April 1947, pp. 41–46.

3448. RAUB, E. Der Einfluss der Hartverchromung auf die Dauerfestigkeit von Aluminium-legierungen. [The influence of hard chromium plating on the fatigue strength of aluminium alloys.] *Z. Metallk.*, vol. 38, no. 4, April 1947, pp. 121–126. English abstr.: *Metal Progr.*, vol. 59, no. 3, March 1951, pp. 412, 414.

3449. RITCHIE, J. G. Fatigue of bolts and studs. *The failure of metals by fatigue*, Melbourne, Melbourne University Press, 1947, pp. 260–289.

3450. Roš, M. La fatigue des metaux. [The fatigue of metals.] *Rev. Acad. Madr.*, Tomo XLI, 1947, pp. 351–407.

3451. Roš, M. Strength and safety of welded joints. *Welding, Lond.*, vol. 15, no. 3, March 1947, pp. 125–131.

3452. Roš, M. La fatigue des Métaux. [The fatigue of metals.] *Rev. Métall.*, vol. 44, nos. 5–6, May–June 1947, pp. 125–143.

3453. SACHS, G. Residual stresses, their measurement and their effects on structural parts. *The failure of metals by fatigue*, Melbourne, Melbourne University Press, 1947, pp. 237–247.

3454. SHAW, F. S. Determination of stress concentration factors. *The failure of metals by fatigue*, Melbourne, Melbourne University Press, 1947, pp. 165–195.

3455. SICHIKOV, M. F.; VISHNEVETSKII, S. D. and GINBERG, D. L. Fatigue of stainless steel

at elevated temperatures. [R] *Kotloturbostroenie*, no. 1, Feb. 1947, pp. 22–24. English abstr.: *Chemical Abstracts*, vol. 44, 1950, col. 7738g.

3456. SIEBEL, E. and PFENDER, M. Neue Enkenntnisse der Festigkeitsforschung. [New results of strength investigations.] *Technik, Berl.*, vol. 2, no. 3, March 1947, pp. 117–121.

3457. SIEBEL, E. and PFENDER, M. Weiterentwicklung der Festigkeitsrechnung bei Wechselbeanspruchung. [Further developments of strength calculations under alternating stress.] *Stahl u. Eisen*, vol. 66/67, no. 19/20, Sept. 11 1947, pp. 318–321.

3458. SILBERSTEIN, J. P. O. and WILLS, H. A. Analytical expressions for strength reduction factors. *Rep. Div. Aeronautics Aust.* no. SM.93, June 1947, pp. 17.

3459. SIMNAD, M. T. and EVANS, U. R. The mechanism of corrosion fatigue of steel in acid solution. *J. Iron St. Inst.*, vol. 156, pt. 4, Aug. 1947, pp. 531–539.

3460. SNIDERMAN, A. Criterion of static and fatigue failures. *Proc. Soc. exp. Stress Anal.* vol. 5, no. 1, 1947, pp. 26–30.

3461. SUTTON, H. Fatigue problems associated with aircraft materials. *The failure of metals by fatigue*, Melbourne, Melbourne University Press, 1947, pp. 40–55.

3462. TELFAIR, D.; ADAMS, C. H. and MOHMAN, H. W. Creep, long time tensile and flexural fatigue properties of Melamine and Phenolic-plastics materials. *Trans. Amer. Soc. mech. Engrs*, vol. 69, no. 7, Oct. 1947, pp. 789–793. *Mod. Plast.*, vol. 24, no. 9, May 1947, pp. 151–152, 236, 238, 240, 242, 244, 246, 248.

3463. TEMPLIN, R. L. *and* HOLT, M. Static and fatigue strengths of welded joints in aluminium–manganese alloy sheet and plates. *Weld. J., Easton, Pa.*, vol. 26, no. 12, Dec. 1947, pp. 705s–711s.

3464. TENOT, A. Nouvelle machine d'essai dynamique des textiles et du caoutchouc. [New machine for the dynamic testing of textiles and rubber.] *Génie civ.*, vol. 124, no. 18, Sept. 15, 1947, pp. 349–351. English abstr.: *Engrs' Dig.*, vol. 9, no. 2, Feb. 1948, p. 41.

3465. THRODAHL, M. C. Vibration fatigue of GR-S in the Goodrich Flexometer. *India Rubb. World*, vol. 116, no. 1, April 1947, pp. 69–70.

3466. TIMOSHENKO, S. Stress concentration and fatigue failure. *Proc. Instn mech. Engrs, Lond.*, vol. 157, no. 28, 1947, pp. 163–169. *Engineer, Lond.*, vol. 183, no. 4763, May 9, 1947, pp. 398–399 and no. 4764, May 16, 1947, pp. 421–422.

3467. TOOLIN, P. R. *and* MOCHEL, N. L. The high-temperature fatigue strength of several gas turbine alloys. *Proc. Amer. Soc. Test. Mater.*, vol. 47, 1947, pp. 677–694.

3468. UNDERWOOD, A. F. *and* GRIFFIN, C. B. A machine for fatigue testing full sized parts. *Proc. Soc. exp. Stress Anal.*, vol. 4, no. 2, 1947, pp. 32–38.

3469. VALETTE, R. Résistance des ponts sous rails en béton armé. Essais de poutres aux flexions répétées. [Strength of railroad bridges of reinforced concrete. Tests of beams under repeated bending.] *Mém. Ass. int. Ponts Charp.*, vol. 8, 1947, pp. 281–290.

3470. VOLDRICH, C. B. *and* ARMSTRONG, E. T. Effect of variables in welding technique on the strength of direct current metal–arc-welded joints in aircraft steel. I—static tension and bending fatigue tests of joints in S.A.E. 4130 steel sheet. *Tech. Note nat. Adv. Comm. Aero., Wash.*, no. 1261, July 1947, pp. 16.

3471. WÅLLGREN, G. Utmattningsprov På Flerradigt Nitförband av Material 75S. [Fatigue tests on multiple riveted joints of 75S material.] *Flytek. Försöksanst., Stockholm, Rep.* no. HU-220, Feb. 21, 1947, pp. 15.

3472. WÅLLGREN, G. Statiska prov och utmattningsprov på vinkelprofil av dural med brottanvisning i form av ett hål. [Static and fatigue strength of aluminium alloy angle extrusions notched by holes.] *Flytek. Försöksanst., Stockholm, Rep.* no. HE-100 : 1, Feb. 25, 1947, pp. 11.

3473. WÅLLGREN, G. Utmattningshållfasthet vid under drifttiden varierande lastgränser. [Endurance limit with time-variable limit stressing.] *Tekn. Tidskr., Stockh.*, vol. 77, no. 42, Oct. 18, 1947, pp. 781–783.

3474. WARNOCK, F. V. *and* POPE, J. A. The change in mechanical properties of mild steel under repeated impact. *Proc. Instn mech. Engrs, Lond.*, vol. 157, no. 26, 1947, pp. 33–51.

3475. WERREN, F. Fatigue of sandwich constructions for aircraft—aluminium face and paper honeycomb core sandwich material tested in shear. *Rep. For. Prod. Lab., Madison* no. 1559-A, Dec. 1947, pp. 6.

3476. WILLS, H. A. A method of obtaining fatigue endurance curves for any load range ratio. *Note Div. Aeronautics Aust.* no. Structures and Materials 155, April 1947, pp. 10.

3477. WILLS, H. A. The strength of components subjected to fluctuating loads—notes on some necessary investigations. *Note Div. Aeronautics Aust.* no. Structures and Materials 156, July 1947, pp. 11.

3478. WILSON, W. M. *and* BURKE, J. L. Rate of propagation of fatigue cracks in 12 inch ×¾ inch steel plates with severe geometrical stress raisers. *Uni. Ill. Engng Exp. Sta. Bull.*, no. 371, Sept. 1947, pp. 16.

3479. WOODWARD, G. R. Bibliography on the fatigue properties of cast iron. *Bull. Brit. cast Iron Ass.*, vol. 9, Nov. 1947, pp. 59–63.

3480. WYSS, T. Life of stranded steel-wire ropes. *Iron Coal Tr. Rev.*, vol. 155, no. 4146, Aug. 29, 1947, pp. 397–401 and no. 4147, Sept. 5, 1947, pp. 441–443.

3481. ZHUKOV, S. L. Determination of fatigue limit under repeated bending of steel or light alloys on the basis of their tensile strength. [R] *Zav. Lab.*, vol. 13, no. 10, Oct. 1947, pp. 1245–1252.

3482. *The failure of metals by fatigue*, Melbourne, Melbourne University Press, 1947, pp. 505.

3483. Fatigue of cadmium–copper trolley wires. Third report—grooved wires. *Rep. Brit. elect. Ind. Res. Ass.*, no. O/T1, 1947, pp. 13.

3484. Testing machines. Prüfmaschinen. *U.S. Army Air Force Air Material Command Translation*, Report no. F-TS-782-RE, Jan. 1947, pp. 9.

3485. Tuned air columns induce resonance for fatigue testing. *Prod. Engng*, vol. 18, no. 2, Feb. 1947, p. 98.

3486. American steel foundries coil-spring test machine. *Rly mech. Engr*, vol. 121, March 1947, pp. 120–122.

3487. The failure of metals by fatigue. *Metallurgia, Manchr*, vol. 35, no. 210, April 1947, pp. 289–293. *Iron Coal Tr. Rev.*, vol. 154, no. 4123, March 21, 1947, pp. 463–467; no. 4124, March 28, 1947, pp. 531–534 and no. 4125, April 4, 1947, pp. 581–585.

3488. Fatigue tester. Needs no attention during operation. *Sci. Amer.*, vol. 176, no. 5, May 1947, p. 224.

3489. Constant-force fatigue testing machine *Mach. Design*, vol. 19, no. 6, June 1947 p. 152.

3490. The Amsler high-frequency Vibrophore *Aircr. Engng*, vol. 19, no. 220, June 1947 pp. 206–207.

3491. Methods of increasing fatigue strength o gear teeth. *Machinery, Lond.*, vol. 70 no. 1806, June 5, 1947, pp. 601–602.

3492. Fatigue of ferrous materials: some factor that influence resistance to the damagin effect of fatigue stressing. *Metallurgia, Manch* vol. 36, no. 215, Sept. 1947, pp. 249–251.

3493. Fatigue tests on shot peened 64 ton ca steel. *Sci. Tech. Memor.*, no. 7/51, Dec. 194 pp. 5.

3494. Fatigue failure of press fitted member *Engineer, Lond.*, vol. 184, no. 4796, Dec. 2 1947, pp. 600–601.

1948

3495. AFANASEV, N. N. The effect of shape ar size factors on the fatigue strength. *Engr Dig.*, vol. 9, no. 3, March 1948, pp. 96–10

3496. AITCHISON, L. Regarding the cumulative rule. *Aeronautical Research Council, Structures Sub-Committee*, Struct. 1206, 11194, Jan. 28, 1948, pp. 3.

3497. ARTHUR, P. T. The fatigue of metals. *Pract. Engng*, vol. 16, no. 414, Jan. 9, 1948, p. 610.

3498. AUSTIN, C. R. Shot peening castings. *Steel*, vol. 122, April 5, 1948, pp. 79–82, 84.

3499. BASTIEN, P. *and* POPOFF, A. Sur l'existence de microfissures dans les depots de chrome electrolytique. Leur influence sur la limite de fatigue des pieces d'acier. [The existence of minute cracks in electrolytic chromium deposits. Their influence on the fatigue limit of steel parts.] *Métaux et Corros.*, vol. 23, no. 297, Sept. 1948, pp. 191–198.

3500. BERNSHTEIN, M. L. A small machine to test for prolonged endurance. [R] *Zav. Lab.*, vol. 14, no. 6, June 1948, pp. 760–761.

3501. BOCCON-GIBOD, R. L'endurance des métaux legers. [The fatigue of light metals.] *Métallurgie Constr. méc.*, vol. 80, no. 5, May 1948, pp. 33, 35, 37.

3502. BRANDENBERGER, H. Die Beziehungen der statischen und dynamischen Festigkeitswerte. [The relation between static and dynamic strength.] *Schweiz. Bauztg.*, vol. 66, no. 9, Feb. 28, 1948, pp. 121–123.

3503. BRUEGGEMAN, W. C. *and* MAYER, M. Axial fatigue tests at zero mean stress of 24S-T and 75S-T aluminium alloy strips with a central circular hole. *Tech. Note nat. Adv. Comm. Aero., Wash.*, no. 1611, Aug. 1948, pp. 23.

504. BUNGARDT, W. *and* OSSWALD, E. Presseffekt und Biegedauerfestigkeit einiger Aluminium-Kupfer-Magnesium-Knetlegierungen mit verschiedenen Mangangehalten. [Effect of forging of the bending fatigue strength of several aluminium–copper–magnesium alloys with varying manganese contents.] *Z. Metallk.*, vol. 39, no. 6, June 1948, pp. 185–189.

3505. BURGHOFF, H. L. *and* BLANK, A. I. Fatigue properties of some coppers and copper alloys in strip form. *Proc. Amer. Soc. Test. Mater.*, vol. 48, 1948, pp. 709–736.

3506. CAINE, J. B. What is strength? *Foundry, Cleveland*, vol. 76, no. 7, July 1948, pp. 80–81, 148–154 and no. 8, Aug. 1948, pp. 100–101, 166–170.

3507. CAMPUS, F. L'Equipement de la halle expérimentale (deux pulsateurs) et le pulsateur à efforts alternés de l'Université de Liège. [Equipment in the laboratory (two pulsators) and the pulsator for alternating loads at the University of Liege.] *Sci. et Tech.*, no. 9, 1948, pp. 203–208.

3508. CAZAUD, R. *La fatigue des métaux.* [*The fatigue of metals.*] Third edition. Paris, Dunod, 1948, pp. 318.

3509. CERADINI, G. La résistance à la fatigue des poutres en treillis soudées et rivées. [The fatigue resistance of welded and riveted truss girders.] *Mém. Ass. int. Ponts Charp.*, Final Report, 1948, pp. 205–213.

3510. CERARDINI, C. La résistance à l'endurance des poutres à treillis soudés et rivés. [The fatigue resistance of welded and riveted truss girders.] *J. Soud., Z. Schweiss-tech.*, vol. 38, no. 10, Oct. 1948, pp. 199–203 and no. 11, Nov. 1948, pp. 228–231.

3511. COLLINS, W. L. Fatigue and static load tests of an austenitic cast iron at elevated temperatures. *Proc. Amer. Soc. Test. Mater.*, vol. 48, 1948, pp. 696–708.

3512. CRAMER, R. E. *and* JENSEN, R. S. Progress reports of investigation of railroad rails and joint bars. *Uni. Ill. Engng Exp. Sta. Bull.*, vol. 45, no. 58, May 27, 1948, (*Reprint Series* no. 39, 1948), pp. 35.

3513. DOLAN, T. J. *and* HANLEY, B. C. The effect of size of specimen on the fatigue strength of S.A.E. 4340 steel. *Office Naval Research, Contract N6-ori-71, Task Order IV, Sixth Progr. Rep., Uni. Illinois*, May 1948, pp. 44.

3514. DOLAN, T. J.; McCLOW, J. H. *and* CRAIG, W. J. The influence of shape of cross-section on the flexural fatigue strength of steel. *Office Naval Research, Contract N6-ori-71, Task Order IV, Tenth Progr. Rep., Uni. Illionis*, Dec. 1948, pp. 21.

3515. DOLAN, T. J.; RICHART, F. L. *and* WORK, C. E. The influence of fluctuations in stress amplitude on the fatigue of metals (part 1). *Office Naval Research, Contract N6-ori-71, Task Order IV, Seventh Progr. Rep., Uni. Illinois*, July 1948, pp. 40.

3516. DOLAN, T. J. *and* YEN, C. S. Some aspects of the effect of metallurgical structure on fatigue strength and notch sensitivity of steel. *Office Naval Research, Contract N6-ori-71, Task Order IV, Fifth Progr. Rep., Uni. Illinois*, March 1948, pp. 54. *Proc. Amer. Soc. Test. Mater.*, vol. 48, 1948, pp. 664–695.

3517. DOREY, S. F. Large-scale torsional fatigue testing of marine shafting. *Proc. Instn mech. Engrs, Lond.*, vol. 159, W.E.P. no. 46, 1948, pp. 399–415. Abstr.: *Engineer, Lond.*, vol. 185, no. 4804, Feb. 20, 1948, pp. 183–185. *Engineering, Lond.*, vol. 165, no. 4284, March 5, 1948, pp. 229–230.

3518. DOREY, S. F. Torsional-fatigue testing of marine shafting. *Engineering, Lond.*, vol. 165, no. 4286, March 19, 1948, pp. 286–288 and no. 4287, March 26, 1948, pp. 310–312.

3519. DUKE, J. B. Some characteristics of residual stress fields during dynamic stressing above the endurance limit. *Proc. Amer. Soc. Test. Mater.*, vol. 48, 1948, pp. 755–766.

3520. DURHAM, R. J. Fatigue curves for some wrought aluminium alloys. *Sci. Tech. Memor.*, no. 8/50, Dec. 30, 1948, pp. 16.

3521. ENOMOTO, N. Fatigue and internal friction of metallic materials. 4th report—mechanism of fatigue in cast iron and super-duralumin. *Trans. Soc. mech. Engrs, Japan*, vol. 14, no. 46, 1948, pp. 90–95.

3522. EPPRECHT, J. *and* EBERHARDT, H. W. Selective shot peening extends fatigue life. *Amer. Mach., N.Y.*, vol. 92, no. 20, Sept. 23, 1948, pp. 95–98.

3523. EPREMIAN, E. *and* NIPPES, E. F. The fatigue strength of binary ferrites. *Trans. Amer. Soc. Metals*, vol. 40, 1948, pp. 870–896.

3524. FELTHAM, P. Fatigue in metals—a critical survey of recent research and theories. *Iron and Steel, Lond.*, vol. 21, no. 11, Oct. 1948, pp. 431–436.

3525. FINDLEY, W. N. *and* WORLEY, W. J. Mechanical properties of five laminated plastics. *Tech. Note nat. Adv. Comm. Aero., Wash.*, no. 1560, Aug. 1948, pp. 111.

3526. FISHER, W. A. P. Repeated loading and fatigue tests on a D.H. 104 (Dove) wing and fin. *Roy. Aircr. Estab. Tech. Note* no. Struct. 32, Dec. 1948, pp. 14.

3527. FÖPPL, O. Stress concentration and fatigue failures. *Engineer, Lond.*, vol. 185, no. 4801, Jan. 30, 1948, pp. 114–115.

3528. FOLKHARD,   . Die Dauerfestigkeit von Schweissverbindungen als Berechnungsgrundlage für geschweisste Brücken. [The fatigue strength of welded joints as a basis for the design of welded bridges.] *Schweisstechnik*, vol. 2, no. 8, Aug. 1948, pp. 93–104.

3529. FONTANA, M. G. Investigation of mechanical properties and physical metallurgy of aircraft alloys at very low temperatures. Part II—strength properties and hardness. *Tech. Rep. U.S. Air Force* no. 5662, Oct. 20, 1948, pp. 116.

3530. FORSYTH, A. C. *and* CARRECKER, R. P. Fatigue limit of S.A.E. 1095 steel after various heat treatments. *Metal Progr.*, vol. 54, no. 5, Nov. 1948, pp. 683–685.

3531. FOWLER, F. H. Theories of the fatigue of metals. *U.S. Office Naval Research, Contract N7-ONR-468, Tech. Report no. 1, Lessells and Assoc.*, May 28, 1948, pp. 37.

3532. FREUDENTHAL, A. M. *and* DOLAN, T. J. The character of fatigue of metals. *Office Naval Research, Contract N6-ori-71, Task Order IV, Fourth Progr. Rep., Uni. Illinois*, Feb. 1948, pp. 76.

3533. FREUDENTHAL, A. M.; YEN, C. S. *and* SINCLAIR, G. M. The effect of thermal activation on the fatigue life of metals. *Office Naval Research, Contract N6-ori-71, Task Order IV, Eighth Progr. Rep., Uni. Illinois*, Aug. 1948, pp. 59.

3534. FRITH, P. H. Fatigue tests on crankshaft steels. I—the effect of nitriding on the fatigue properties of a chromium–molybdenum steel. II—tests on nickel–chromium–molybdenum and chromium–molybdenum–vanadium steels. *J. Iron St. Inst.*, vol. 159, pt. 4, Aug. 1948, pp. 385–409 and [discussion] vol. 162, pt. 4, Aug. 1949, pp. 432–436.

3535. FRITH, P. H. Crankshaft steels: Part I—effect of nitriding and composition on fatigue properties. Part II—nickel–chromium–molybdenum and chromium–molybdenum–vanadium steels. *Iron and Steel, Lond.*, vol. 21, no. 13, Nov. 18, 1948, pp. 542–552.

3536. FUCHS, H. O. Trapped stresses—how they can be created to improve the performance of machine parts. *Mach. Design*, vol. 20, no. 7, July 1948, pp. 114–118, 178.

3537. GOULD, A. J. and EVANS, U. R. The effect of shot peening on the corrosion fatigue of a high carbon steel. *J. Iron St. Inst.*, vol. 160, pt. 2, Oct. 1948, pp. 164–168.

3538. GRIFFIN, C. B. Fatigue testing production parts. *Iron Age*, vol. 161, no. 2, Jan. 8, 1948, pp. 59–62.

3539. GURNEY, C. *and* PEARSON, S. Fatigue of mineral glass under static and cyclic loading. *Proc. roy. Soc. (A)*, vol. 192, no. 1031, March 18, 1948, pp. 537–544.

3540. HÄNCHEN, R. Berechnung der Maschinenteile unter oftmals wiederholter Belastung. [Design of machine parts under often repeated loading.] *Glasers Ann. Gew.*, vol. 72, no. 8, Aug. 1948, pp. 113–118 and no. 9, Sept. 1948, pp. 135–139.

3541. HALL, L. D. *and* PARKER, E. R. Effect of residual tension stress on the fatigue strength of mild steel. *Weld. J., Easton, Pa.*, vol. 27, no. 8, Aug. 1948, pp. 421s–425s.

3542. HANSTOCK, R. F. The effect of vibration on a precipitation hardening aluminium alloy. *J. Inst. Met.*, vol. 74, pt. 9, May 1948, pp. 469–492.

3543. HAWKINS, W. A. What do we know about rolled threads? *Machinist*, vol. 91, no. 36, Jan. 3, 1948, pp. 1207–1213.

3544. HEMPEL, M. Wechselfestigkeit und Kerbwirkungszahlen von unlegierten und legierten Stählen bei +20 und −78°. [Fatigue strength and notch effect of unalloyed and alloyed steels at +20 and −78°C.] *Stahl u. Eisen*, vol. 68, no. 1/2, Jan. 1, 1948, p. 25.

3545. HEMPEL, M. Low temperature fatigue test on steel. *U.S. Air Materiel Command Translation* no. F-TS-1855-RE, July 1948, pp. 73. Abstr.: *Prod. Engng*, vol. 20, no. 11, Nov. 1949, pp. 125–127.

3546. HEMPEL, M. *and* SANDER, H. R. Über den Einfluss der Formänderungsgeschwindigkeit auf die mechanischen Kennwerte, insbesondere bei wechselnder Belastung. [On the influence of rate of deformation on the characteristic mechanism of deformation, in particular under repeated loading.] *Glasers Ann. Gew.*, vol. 72, no. 10, Oct. 1948, pp. 145–149.

3547. L'HERMITE, R. *and* GRANGEON, M. Les essais d'endurance en flexion répétée comme critère de soudabilité des tôles. 4ème rapport: Influence du cordon de soudure sur les caractéristiques d'endurance en flexion alterné d'une tôle en acier au carbon. [Bending fatigue tests as a criterion of the weldability of sheet. 4th report: Influence of the welding rod on the fatigue characteristics of carbon steel sheet subjected to alternating bending.] *Rapp. Gr. franc. Rech. aéro.*, no. 35, 1948, pp. 43.

548. HOFFMANN, C. A. *and* AULT, G. M. Application of statistical methods to study of gas-turbine blade failures. *Tech. Note nat. Adv. Comm. Aero., Wash.*, no. 1603, June 1948, pp. 27.

3549. HOPPMANN, W. H. Effect of fatigue on tension-impact resistance. *Bull. Amer. Soc. Test. Mat.*, no. 155, Dec. 1948, pp. 36–38.

3550. HOWELL, F. M.; STICKLEY, G. W. *and* LYST, J. O. Effects of surface finish, of certain defects, and of repair of defects by welding, on fatigue strength of 355-T6 sand-castings and effects of prior fatigue stressing on tensile properties. *Tech. Note nat. Adv. Comm. Aero., Wash.*, no. 1464, April 1948, pp. 49.

3551. HUMFREY, J. C. W. Stresses induced by the shot-peening of leaf springs. *Symposium on internal stresses in metals and alloys*, London, Institute of Metals, 1948, pp. 189–193.

3552. ISIBASI, T. On fatigue limits of notched specimens. *Mem. Fac. Engng Kyushu*, vol. 11, no. 1, 1948, pp. 1–31. Abstr.: *Engrs' Dig.*, vol. 10, no. 5, May 1949, p. 176.

3553. JACKSON, L. R.; CROSS, H. C. *and* BERRY, J. M. Tensile, fatigue and creep properties of forged aluminium alloys at temperatures up to 800°F. *Tech. Note nat. Adv. Comm. Aero., Wash.*, no. 1469, March 1948, pp. 48.

3554. JENSEN, R. S. Fatigue tests of rail webs. *Repr. Ill. Engng Exp. Sta.*, no. 39, 1948, pp. 21–25. *Bull. Amer. Rly Engng Ass.*, vol. 49, no. 472, Feb. 1948, pp. 485–490.

3555. JENSEN, R. S. Sixth progress report of the rolling-load tests of joint bars. *Repr. Ill. Engng Exp. Sta.*, no. 39, 1948, pp. 26–35. *Bull. Amer. Rly Engng Ass.*, vol. 49, no. 472, Feb. 1948, pp. 416–425.

3556. KAMMERER, A. Le calcul des pièces soumises à des forces périodiques. [The design of members subjected to intermittent loads.] *Rev. gén. Chem.-de-Fer.*, vol. 67, no. 2, Feb. 1948, pp. 60–63.

3557. KAMMERER, A. Le module d'élasticité et la limit de fatigue. [The modulus of elasticity and fatigue limit.] *C.R. Acad. Sci., Paris,* vol. 227, Nov. 29, 1948, pp. 1144–1145.

3558. KHEIFETS, S. G. Additional stresses in fatigue test specimens. [R] *Zav. Lab.*, vol. 14, June 1948, pp. 742–748.

3559. KOCHETOV, A. I. *and* KROLEVETSKII, A. D. The problem of the determination of fatigue strength limits. [R] *Zav. Lab.*, vol. 14, June 1948, pp. 732–738.

3560. KOROVSKII, SH. YA. Influence of various surface-active substances on the fatigue strength of steel. [R] *Dokl. Akad. Nauk SSSR.*, vol. 59, no. 8, 1948, pp. 1449–1451.

3561. KUKANOV, L. I. Fatigue testing of spring steel and the influence of surface defects on the data obtained. [R] *Zav. Lab.*, vol. 14, Aug. 1948, pp. 977–984.

3562. LAMBERG, A. B. Détermination simplifiée de l'endurance de l'acier et des autres métaux. [Simplified determination of fatigue resistance of steel and other metals.] *Génie civ.* vol. 125, no. 22 (no. 3241), Nov. 15, 1948, pp. 434–435.

3563. LANDECKER, F. K. Shotpeening. *S.A.E. quart. Trans.*, vol. 2, no. 2, April 1948, pp. 191–194, 200.

564. LANGDON, H. H. *and* FRIED, B. Fatigue of gusseted joints. *Tech. Note nat. Adv. Comm. Aero., Wash.*, no. 1514, Sept. 1948, pp. 40.

3565. LIHL, F. Kristallographische Vorgänge an der Fleissgrenze von Stahl und ihre Bedeutung für die Dauerfestigkeit. [Crystallographic behaviour at the yield point of steel and its relationship to fatigue strength.] *Metall: Wirtsch. Wiss. Tech.*, vol. 2, no. 23/24, Dec. 1948, pp. 391–396 and vol. 3, no. 3/4, Feb. 1949, pp. 49–51.

3566. LIU, S. I.; LYNCH, J. J.; RIPLING, E. J. *and* SACHS, G. Low cycle fatigue of the aluminium alloy 24ST in direct stress. *Trans. Amer. Inst. min. (metall.) Engrs*, vol. 175, 1948, pp. 469–496. *Metals Tech.*, vol. 15, T. P. 2338, Feb. 1948, pp. 22.

3567. LIU, S. I. *and* SACHS, G. The flow and fracture characteristics of aluminium alloy 24ST after alternating tension and compression. *Metals Tech.*, vol. 15, no. 4, T. P. 2392, June 1948, pp. 12.

3568. LOVE, R. J. Cast crankshafts. A survey of published information. *J. Iron St. Inst.*, vol. 159, pt. 3, July 1948, pp. 247–274.

3569. LUNDBERG, BO. *and* WÅLLGREN, G. Jämförelse av utmattningsegenskaperna hos material 24ST och 75ST med speciell hänsyn till de under flygning uppträdande utmattningspåkänningarna. [Comparison of the fatigue strength of 24ST and 75ST materials with special regard to repeated loadings of aircraft.] *Flytek. Försöksanst., Stockholm, Rep.* no. HE-265:1, March 5, 1948, pp. 36.

3570. LUTHANDER, S. *and* WÅLLGREN, G. An experimental determination of the fatigue diagram for tension and compression in Alclad sheet. *Flytek. Försöksanst., Stockholm, Rep.* no. 5, 1948, pp. 11.

3571. MCDONALD, G. G. The graphics of pulsating stresses. *J. Instn Engrs Aust.*, vol. 20, no. 12, Dec. 1948, pp. 195–196.

3572. McDonald, J. C. Tensile, creep and fatigue properties at elevated temperatures of some magnesium-base alloys. *Proc. Amer. Soc. Test. Mater.*, vol. 48, 1948, pp. 737–754.

3573. Macgregor, C. W. *and* Grossman, N. Some new aspects of the fatigue of metals brought out by brittle transition temperature tests. *Weld. J., Easton, Pa.*, vol. 27, no. 3, March 1948, pp. 132s–144s and [discussion] no. 8, Aug. 1948, pp. 428s–432s.

3574. McKeown, J. *and* Back, L. H. A rotating-load, elevated temperature fatigue testing machine. *Metallurgia, Manchr*, vol. 38, no. 227, Sept. 1948, pp. 247–254.

3575. McMaster, R. C. An investigation of possibilities of organic coatings for prevention of premature corrosion-fatigue failures in steel. *Proc. Amer. Soc. Test. Mater.*, vol. 48, 1948, pp. 628–647.

3576. McMaster, R. C. Prevention of drill string failures in the Permian Basin. *World Oil*, vol. 127, no. 13, April 1948, pp. 75–80, 82 and vol. 128, no. 1, May 1948, pp. 134, 136, 138.

3577. Machlin, E. S. Dislocation theory of the fatigue of metals. *Tech. Note nat. Adv. Comm. Aero., Wash.*, no. 1489, Jan. 1948, pp. 33.

3578. Machlin, E. S. An application of dislocation theory to fracturing by fatigue. *Fracturing of metals*, Cleveland, Ohio, American Society for Metals, 1948, pp. 282–289.

3579. Majors, H. Change in hardness of a metal bar under low cycles of reversed and pulsating plastic bending. *Bull. Amer. Soc. Test. Mat.*, no. 155, Dec. 1948, pp. 39–43.

3580. Makhov, V. N. Étude des caractéristiques de fatigue des aciers employés en U.R.S.S. pour la fabrication des essieux de matériel roulant. [Study of the fatigue characteristics of steels used in the U.S.S.R. for the manufacture of rolling-stock axles.] *Circ. Cent. Docum. sidér.*, vol. 5, April 25, 1948, pp. 207–217.

3581. Marin, J. Some new testing machines for combined stress experiments. *Fracturing of metals*, Cleveland, Ohio, American Society for Metals, 1948, pp. 189–200.

3582. Markl, A. R. C. Fatigue tests of welding elbows and comparable double-mitre bends. *Weld. J., Easton, Pa.*, vol. 27, no. 6, June 1948, pp. 310s–320s.

3583. Melhardt, H. Stumpf- und Kehlnaht-verbindungen bei ruhender und bei wechselnder Beanspruchung. [Butt and fillet joints under static and alternating stresses.] *Schweisstechnik*, vol. 2, May 1948, pp. 55–57 and June 1948, pp. 72–75.

3584. Mills, H. R. *and* Love, R. J. Fatigue strength of cast crankshafts. *Proc. Instn mech Engrs, Lond. (Automobile Division)*, pt. 3 1948–1949, pp. 81–99.

3585. Mitinskii, A. N. *and* Reinberg, E. S Method of testing the protective value o corrosion resistant coatings on steel unde alternating stresses. [R] *Zav. Lab.*, vol. 14 Oct. 1948, pp. 1247–1250. English translation *Henry Brutcher Tech. Trans.*, no. 2262.

3586. Moore, H. F.; Dolan, T. J. *and* Hanley B. C. The effect of size and notch sensitivit on fatigue characteristics of two metalli materials. Part I—aluminium alloy 75S-T Part II—S.A.E. 4340 steel. *Tech. Rep. U.S Air Force*, no. 5726, Oct. 5, 1948, pp. 153

3587. MORTON, B. B. Corrosion fatigue of oil-well equipment. *World Oil*, vol. 127, no. 13, April 1948, pp. 156, 158, 160, 162, 168.

3588. MUHLENBRUCH, C. W. The effect of repeated loading on bond strength of concrete—II. *Proc. Amer. Soc. Test. Mater.*, vol. 48, 1948, pp. 977–987.

3589. NALESZKIEWICZ, J. On the computation of endurance limit stresses. *Proc. Int. Congr. app. Mech., Seventh Congr.*, 1948, vol. 4, pp. 190–193.

3590. NIKOLAEV, R. S. On the designation of the fatigue limit. [R] *Zav. Lab.*, vol. 14, no. 12, Dec. 1948, p. 1488.

3591. NISHIHARA, T.; TAIRA, S.; TSUNEDA, H. and OCHIAI, R. X-ray investigation of the change of residual stress due to repeated stress. [J] *Trans. Soc. mech. Engrs, Japan*, vol. 14, no. 48, 1948, pp. I107–I113.

3592. NISHIHARA, T. and YAMADA, T. The fatigue strength of metallic materials under alternating stresses of varying amplitude. [J] *Trans. Soc. mech. Engrs, Japan*, vol. 14, no. 47, Oct. 1948, pp. I6–I18. English abstr.: *Metallurg. Abstr.*, vol. 17, 1949–1950, p. 563.

3593. NOLL, G. C. and ERICKSON, M. A. Allowable stresses for steel members of finite life. *Proc. Soc. exp. Stress Anal.*, vol. 5, no. 2, 1948, pp. 132–143. Italian abstr., *Metallurg. ital.*, vol. 41, no. 1, Jan.–Feb. 1949, pp. 46–47.

3594. OBERG, T. T. and ROONEY, R. J. Static and fatigue tests of cast and forged I-beams of aluminium and magnesium alloys. *Tech. Rep. U.S. Air Force*, no. 5701, May 14, 1948, pp. 35.

3595. PAYSON, P. and NEHRENBERG, A. E. New steel features high strength and high toughness. *Iron Age*, vol. 162, no. 17, Oct. 21, 1948, pp. 64–71.

3596. PERLMUTTER, I. and ADENSTEDT, H. Service failures of turbine buckets. *Tech. Rep. U.S. Air Force*, no. 5716, July 19, 1948, pp. 50.

3597. POPE, J. A. and ANDREW, J. E. Mechanical factors affecting the strength of springs. Section II—fatigue strength. *Coil Spring Journal*, no. 12, Sept. 1948, pp. 33–41.

3598. PROT, M. Une nouvelle technique d'essai des matériaux—l'essai de fatigue sous charge progressive. [A new technique for material tests—fatigue test under progressive load.] *Ann. Ponts Chauss.*, vol. 118, no. 4, July–Aug. 1948, pp. 441–464.

3599. PROT, M. L'essai de fatigue sous charge progressive. Une nouvelle technique d'essai des matériaux. [Fatigue tests under progressive load. A new technique for testing materials.] *Rev. Métall.*, vol. 45, no. 12, Dec. 1948, pp. 481–489. English translation: *Wright Air Development Centre Tech. Report* no. 52-148, Sept. 1952. Italian abstr.: *Metallurg. ital.*, vol. 41, no. 4, July–Aug. 1949, pp. 211–212.

3600. PUCHNER, O. Zur Dauerhaltbarkeit von Formelementen der Welle bei überlagerter wechselnder Biege- und Verdrehbeanspruchung. [The fatigue resistance of various shapes of shaft under superimposed alternating bending and torsional loads.] *Schweiz. Arch. angew. Wiss.*, vol. 14, no. 8, Aug. 1948, pp. 217–229.

3601. RATNER, S. I. and ZAKHAROV, I. I. The problem of the increase in fatigue limit caused by surface cold working. [R] *Zav. Lab.*, vol. 14, Oct. 1948, pp. 1241–1246.

3602. RICHART, F. E. *and* NEWMARK, N. M. An hypothesis for the determination of cumulative damage in fatigue. *Proc. Amer. Soc. Test. Mater.*, vol. 48, 1948, pp. 767–800.

3603. ROBERTSON, T. S. *and* WINDER, R. H. B. Hysteretic heating in fatigue test pieces. *Naval Construction Research Establishment, Scotland*, Report no. R.77, Sept. 1948, pp. 2.

3604. ROŠ, M. G. Static failure and fatigue of steels. *Proc. Sh. Strip Metal Us. tech. Ass.*, vol. 3, 1948–1949, pp. 73–118.

3605. ROŠ, M. Qualité des matériaux et sécurité dans le batiment ainsi que dans la construction des machines. [Quality of materials for the safety of structures and in the construction of machines.] *Ann. Inst. Bâtim.*, no. 41, Sept. 1948, pp. 52.

3606. ROŠ, M. La fatigue des soudures. [The fatigue of welds.] *Rev. Métall.*, vol. 45, no. 11, Nov. 1948, pp. 421–446. *Ber. eidgenöss. MatPrüfAnst.*, no. 161, 1948, pp. 28.

3607. RUSSELL, H. W.; JACKSON, L. R.; GROVER, H. J. *and* BEAVER, W. W. Fatigue strength and related characteristics of aircraft joints. II—fatigue characteristics of sheet and riveted joints of 0.040″ 24S-T, 75S-T and R303–T275 aluminium alloys. *Tech. Note nat. Adv. Comm. Aero., Wash.*, no. 1485, Feb. 1948, pp. 97.

3608. RUSSENBERGER, M. Über neuere Schwingungsprüfmaschinen. [A new fatigue testing machine.] *Transactions of instruments and measurements Conference, Stockholm, 1947*, Norrköping, Norrköpings Tidn. A-B., 1948, pp. 105–111.

3609. SAUER, J. A. A study of fatigue phenomena under combined stress. *Proc. Int. Congr. app. Mech.*, Seventh Congr., 1948, vol. 4, pp. 150–164.

3610. SAUER, J. A. *and* ROOS, P. K. Obtaining fatigue-test data. *Mach. Design*, vol. 20, no. 10, Oct. 1948, pp. 115–117, 158–162.

3611. SCHOTTKY, H. *and* HILTENKAMP, H. How atmospheric nitrogen encourages galling and fatigue failures. *Steel*, vol. 123, no. 1, July 5, 1948, pp. 97, 110, 113–114.

3612. SICHIKOV, M. F. *and* VISHNEVETSKII, S. D. Method of testing fatigue of steel specimens at high temperatures. [R] *Zav. Lab.*, vol. 14, no. 1, Jan. 1948, pp. 86–91. English abstr.: *Engrs' Dig.*, vol. 10, no. 2, Feb. 1949, pp. 46–49. *Metal Progr.*, vol. 57, no. 1, Jan. 1950, pp. 96, 98.

3613. SIEBEL, E. Neue Wege der Festigkeitsrechnung. [New methods of stress analysis.] *Z. Ver. dtsch. Ing.*, vol. 90, no. 5, May 1948, pp. 135–139.

3614. SIEBEL, E. *and* BUSSMANN, K. H. Das Kerbproblem bei schwingender Beanspruchung. [The notch problem under alternating loads.] *Technik, Berl.*, vol. 3, no. 6, June 1948, pp. 249–252.

3615. STEWART, W. C. *and* WILLIAMS, W. L. Effects of inclusions on endurance properties of steels. *J. Amer. Soc. nav. Engrs*, vol. 60, no. 4, Nov. 1948, pp. 475–504.

3616. STRAUB, J. Why peening calls for uniform shot. *S.A.E. Jl.*, vol. 56, no. 11, Nov. 1948, pp. 37–38.

3617. TERMINASOV, YU. S. X-ray investigation of residual stresses of types II and III during fatigue testing of steel. [R] *Zh. tekh. Fiz.*, vol. 18, no. 4, April 1948, pp. 517–523.

3618. THURSTON, R. C. A. Dynamic calibration method uses modified proving ring. *Bull.*

Amer. Soc. Test. Mater., no. 154, Oct. 1948, pp. 50–52.

3619. TRAPEZIN, I. I. *The strength of metals under cyclic loads.* [R] Moscow, Gostekhizdat, 1948, pp. 107.

3620. UZHIK, G. V. *Methods of fatigue testing metals and machine parts.* [R] Moscow, Izd-vo Akademii nauk SSSR, 1948, pp. 262.

3621. VIDMAN, D. N. On the structure of the fracture faces of fatigue failures. [R] *Vestn. mashinostr.*, vol. 28, no. 9, Sept. 1948, pp. 18–24.

3622. VOLDRICH, C. B. *and* ARMSTRONG, E. T. Effect of variables in welding technique on the strength of direct-current metal arc-welded joints in aircraft steel. II—repeated stress tests of joints in S.A.E. 4130 seamless steel tubing. *Tech. Note nat. Adv. Comm. Aero., Wash.*, no. 1262, April 1948, pp. 85.

3623. VOTTA, F. A. New wire fatigue testing method. *Iron Age*, vol. 162, no. 7, Aug. 12, 1948, pp. 78–81. *Wire and Wire Prod.*, vol. 23, no. 12, Dec. 1948, pp. 1117–1123.

3624. VOŬTE, . Investigation of magnitude and frequency of loads for a fatigue test on wing. *N.V. Nederlandsche Vliegtuigenfabriek Fokker, Amsterdam*, Report no. P-11-527a, June 28, 1948, pp. 3.

3625. WALKER, H. L. *and* CRAIG, W. J. Effect of grain size on tensile strength, elongation and endurance limit of deep drawing brass. *Metals Tech.*, vol. 15, no. 6, T.P. 2478, Sept. 1948, pp. 10.

3626. WÅLLGREN, G. An experimental determination of the fatigue diagram for alclad sheet specimens with rivet holes, subjected to tension and compression. *Flytek. Försöksanst. Stockholm*, Rep. no. 14, (translation no. 9), 1948, pp. 12.

3627. WECK, R. The design and fabrication of welded structures subjected to repeated loading. *Welder, Lond.*, vol. 17, no. 98, Oct.–Dec. 1948, pp. 91–96; vol. 18, no. 99, Jan.–March 1949, pp. 15–19; no. 101, July–Sept. 1949, pp. 61–66; vol. 19, no. 103, Jan.–March 1950, pp. 15–24; no. 104, April–June 1950, pp. 43–46; no. 105, July–Sept. 1950, pp. 61–70; vol. 20, no. 107, Jan.–June 1951, pp. 12–22; no. 108, July–Dec. 1951, pp. 38–41.

3628. WELLINGER, K. *and* HOFMANN, A. Prüfung metallischer Werkstoffe in der Kälte. [Testing of metallic materials at low temperatures.] *Z. Metallk.*, vol. 39, no. 8, Aug. 1948, pp. 233–239.

3629. WELLINGER, K. *and* KEIL, E. Verbesserung der Haftfähigkeit von Nickel- und Hartchromschichten und der Wechselfestigkeit vernickelter und verchromter Teile durch Wärmebehandlung. [Improvement of the adhesion of nickel and hard chromium deposits and of the fatigue resistance of nickel and chromium plated parts by heat treatment.] *Metalloberfläche*, vol. 2, no. 11, Nov. 1948, pp. 233–236. English translation: *Henry Brutcher Tech. Trans.* no. 2477.

3630. WELTER, G. Fatigue tests of spot welds—improvement of their endurance limit by hydrostatic pressure. *Weld. J., Easton, Pa.*, vol. 27, no. 6, June 1948, pp. 285s–298s. *Engrs' Dig.*, vol. 10, no. 3, March 1949, pp. 76–79 and no. 4, April 1949, pp. 120–122.

3631. WERREN, F. Fatigue of sandwich constructions for aircraft: Aluminium face and end grain balsa core sandwich material tested in shear. *Rep. For. Prod. Lab., Madison*, no. 1559-B, April 1948, pp. 6.

3632. WERREN, F. Fatigue of sandwich constructions for aircraft: Fiberglas-honeycomb core material with fiberglas-laminate or aluminium facings, tested in shear. *Rep. For. Prod. Lab., Madison*, no. 1559-C, Aug. 1948, pp. 8.

3633. WERREN, F. Fatigue of sandwich constructions for aircraft: Fiberglas-laminate face and end-grain balsa core sandwich material tested in shear. *Rep. For. Prod. Lab., Madison*, no. 1559-D, Sept. 1948, pp. 7.

3634. WERREN, F. Fatigue of sandwich constructions for aircraft: Cellular-hard-rubber core material with aluminium or fiberglas-laminate facings, tested in shear. *Rep. For. Prod. Lab., Madison*, no. 1559-E, Oct. 1948, pp. 4.

3635. WERREN, F. Fatigue of sandwich constructions for aircraft: Cellular cellulose acetate core material with aluminium or fiberglas-laminate facings, tested in shear. *Rep. For. Prod. Lab., Madison*, no. 1559-F, Dec. 1948, pp. 10.

3636. WIEGAND, H. La Nitruration. [Nitriding.] *Rev. Métall.*, vol. 45, no. 3/4, March–April 1948, pp. 105–117.

3637. WIESCHHAUS, L. J. Shot peening for longer life. *Mod. Mach. Shop*, vol. 21, no. 4, Sept. 1948, pp. 112–114, 116, 118, 120, 122, 124, 126, 128, 130.

3638. WILLS, H. A. The life of aircraft structures. *J. Instn Engrs Aust.*, vol. 20, no. 10, Oct. 1948, pp. 145–156.

3639. WILSON, W. M. Flexural fatigue strength of steel beams. *Uni. Ill. Engng Exp. Sta. Bull.*, no. 377, Jan. 22, 1948, pp. 34. *Weld. J., Easton, Pa.*, vol. 27, no. 8, Aug. 1948, pp. 409s–417s.

3640. WILSON, W. M. *and* BURKE, J. L. Rate of propagation of fatigue cracks in 12 inch $\times \frac{3}{4}$ inch steel plates with severe geometrical stress-raisers. *Weld. J., Easton, Pa.*, vol. 27, no. 8, Aug. 1948, pp. 405s–408s.

3641. WINSON, J. The testing of rotors for fatigue life. *J. aero. Sci.*, vol. 15, no. 7, July 1948, pp. 392–402.

3642. WOLDMAN, N. E. Some notes on fatigue failures in aircraft parts. *Iron Age*, vol. 162, no. 24, Dec. 9, 1948, pp. 97–101.

3643. WORK, C. E. *and* DOLAN, T. J. The influence of fluctuations in stress amplitude on the fatigue of metals (part II). *Office Nav. Research, Contract N6-ori-71, Task Order I, Ninth Progr. Rep., Uni. Illinois*, Sept. 1948, pp. 44.

3644. WORLEY, W. J. Simplified dynamic strain equipment. *Instruments*, vol. 21, April 1948, pp. 330–332.

3645. WYSS, TH. Erfahrungen mit Schweissungen an Vorderachsen, Kurbelwellen und Differentialwellen von Motorfahrzeugen Expériences faites avec des soudures pratiquées sur des essieux avant, des arbres arrière de véhicules à moteur. [Experiences with welded front axles, crankshafts and half shafts in motor vehicles.] *Z. Schweisstech. J. Soudure*, vol. 38, no. 9, Sept. 1948, pp. 175–182; no. 10, Oct. 1948, pp. 204–213 and no. 11, Nov. 1948, pp. 223–228.

3646. YATSKEVICH, S. I. A new fatigue test machine. [R] *Zav. Lab.*, vol. 14, no. 6, June 1948, pp. 739–741.

3647. ZHITKOV, D. G. Methods of testing steel wire cables for endurance. [R] *Zav. Lab.*, vol. 14, no. 7, July 1948, pp. 858–867.

3648. ZIMMERLI, F. P. Shot quality: how it affects fatigue life. *Steel*, vol. 123, no. 16, Oct. 18, 1948, pp. 126–129.

3649. ZIMMERLI, F. P. Effect of shot type on spring fatigue life. *S.A.E. Jl.*, vol. 56, no. 11, Nov. 1948, pp. 36, 39.

3650. Report on physical properties influenced by as-quenched hardness. *S.A.E. Iron and Steel Technical Committee SP-53, Report no. MR-73*, 1948, pp. 42.

3651. Report on torsion tests of splined shafts. *Contributions to the metallurgy of steel—no. 19*, American Iron and Steel Institute, Feb. 1948, pp. 9.

3652. Fatigue cracking of retaining ring causes turbine accident. *Pwr Generat.*, vol. 52, no. 3, March 1948, pp. 104, 106, 108.

3653. Corrosion fatigue. *Allen Engng Rev.*, no. 18, March 1948, pp. 4–5.

3654. Fatigue testing heavy structures. *Iron Age*, vol. 161, no. 20, May 13, 1948, p. 77.

3655. Fatigue testing: a rapid machine employing unmachined specimens. *Auto. Engr*, vol. 38, no. 503, July 1948, p. 278.

3656. Symbols and nomenclature for fatigue testing. *Bull. Amer. Soc. Test. Mat.*, no. 153, Aug. 1948, pp. 36–37.

3657. Fatigue properties of 1″ diameter extruded rod to Specification D.T.D. 364B, as affected by periods of standing unstressed at intervals during the fatigue tests. *Sci. Tech. Memor.*, no. 9/48, Sept. 1948, pp. 6.

3658. Shot peened parts last longer. *Sth. Pwr Ind.*, vol. 66, no. 9, Sept. 1948, pp. 66–68, 124, 126, 128.

3659. Fatigue tests on crankshaft steels: nitriding effects and tests on Ni–Cr–Mo and Cr–Mo–V material. *Auto. Engr*, vol. 38, no. 506, Oct. 1948, p. 384.

3660. X-ray measuring strains in metal. *Steel*, vol. 123, no. 22, Nov. 29, 1948, pp. 79–80, 90.

3661. Fatigue testing machine for wire. *Rev. sci. Instrum.*, vol. 19, no. 12, Dec. 1948, pp. 930–931.

3662. Research report on fatigue tests on shot peened 74/78 (Cr–Mo) cast steel. *Sci. Tech. Memor.*, no. 11/51, Dec. 1948, pp. 5.

3663. From a technician's notebook—analysis of failed material. *J. auto. aero. Engrs*, vol. 8, no. 12, Dec. 1948, pp. ix–xiv and vol. 9, no. 1, Jan. 1949, pp. xii–xvi.

1949

3664. AMERICAN SOCIETY FOR TESTING MATERIALS. *Manual on fatigue testing*, Philadelphia, Pa., American Society for Testing Materials, Special Technical Publication no. 91, Dec. 1949, pp. 82.

3665. ANDREW, J. E. Strength of springs. *Coil Spring J.*, no. 15, June 1949, pp. 31–35.

3666. ANDREW, J. E. A new torsion-fatigue testing machine. *Coil Spring J.*, no. 17, Dec. 1949, pp. 12–17.

3667. ARCHER, S. Screwshaft casualties—the influence of torsional vibration and propeller immersion. *Trans. Instn nav. Archit., Lond.,* vol. 91, 1949, pp. J56–J104.

3668. BEAN, W. T. Analysis of stress in aircraft engines. *Properties of metals in materials engineering,* Cleveland, Ohio, American Society for Metals, 1949, pp. 100–123.

3669. BECKER, A. Elektrisches Mehrfach-Schalt-zählwerk zur Steuerung von Belastungsprogrammen bei Beitriebsfestigkeits-Versuchen. [Electric multiple cut-out counting mechanism for controlling load programmes in service strength tests.] *FeinwTech.,* vol. 53, no. 7, 1949, pp. 189–193.

3670. BEDESCHI, G. Una macchina per prove di fatica a flessione rotante; lavorazione e misura delle provette. [A rotating bending fatigue machine; manufacture and measurement of specimens.] *Alluminio,* vol. 18, no. 2, March–April 1949, pp. 139–146.

3671. BERNSHTEIN, M. L. Fatigue test machine for high temperature testing. [R] *Zav. Lab.,* vol. 15, no. 4, April 1949, pp. 497–500.

3672. BILLING, B. F. Equipment and technique for the measurement of fluctuating and mean loads in Haigh fatigue testing machines. *Roy. Aircr. Estab. Tech. Note* no. Met. 116, Dec. 1949, pp. 6.

3673. BOCCON-GIBOD, R. L'endurance des métaux légers. [The endurance of light metals.] *Rev. Alumin.,* vol. 26, no. 158, Sept. 1949, pp. 279–286.

3674. BUDD, R. T. *and* PARKER, R. J. Wire strain gauges—applications to tensile, compression and fatigue testing. *Metal Ind., Lond.,* vol. 75, no. 25, Dec. 16, 1949, pp. 511–514, 521.

3675. BÜRNHEIM, H. Neuere Erkenntnisse zur Dauerhaltbarkeit der Nietverbindungen von Blechen aus Leichtmetallen. [New information on the fatigue durability of riveted joints in light metal sheet.] *Schweiz. tech. Z.,* vol. 46, no. 10, March 10, 1949, pp. 151–156 and no. 11, March 17, 1949, pp. 167–175. German abstr.: *Z. Ver. dtsch. Ing.,* vol. 91, no. 19, Oct. 1, 1949, pp. 504–508.

3676. CAZAUD, R. La fatigue des métaux — son importance dans la construction aéronautique. [The fatigue of metals—its importance in aeronautical construction.] *Tech. et Sci. aéro.,* no. 3, 1949, pp. 147–159.

3677. CAZAUD, R. La résistance à fatigue des aciers. [The fatigue resistance of steels.] *Tech. mod.,* vol. 41, no. 23–24, Dec. 1–15, 1949 pp. 377–384.

3678. CERARDINI, C. Fatigue of welded and riveted trusses. *Weld. J., Easton. Pa.,* vol. 28 no. 6, June 1949, pp. 241s–245s.

3679. CHALMERS, B. *and* STENHOUSE, A. C. The effect of shot blasting on the fatigue life of leaf springs. *Ministry of Supply (Great Britain), Permanent Records of Research and Development, Monograph* no. 20.301, July 1949, pp. 7.

3680. CODE, C. J. *and* BILLSTEIN, A. E. F. Fatigue life of rail webs in service. *Proc. Soc. exp Stress Anal.,* vol. 7, no. 1, 1949, pp. 103–110.

3681. COLEGATE, G. T. Shot peening: a survey of modern methods and applications. *Sheet Metal Ind.,* vol. 26, no. 261, Jan. 1949 pp. 141–148, 152 and no. 262, Feb. 1949 pp. 371–380, 384.

3682. CORSON, M. G. The N : S relationship in endurance testing. *Iron Age,* vol. 16 no. 10, March 10, 1949, pp. 103–105.

3683. CORSON, M. G. Economy in fatigue testing. *Metal Progr.*, vol. 56, no. 4, Oct. 1949, pp. 518–519.

3684. CRAMER, R. E. Laboratory tests of two welded rails. *Bull. Amer. Rly Engng Ass.*, vol. 50, no. 479, Feb. 1949, pp. 510–512. *Repr. Ill. Engng Exp. Sta.*, no. 43, 1949, pp. 8–10.

3685. DAVIDENKOV, N. N. *Metal fatigue* [R]. Kiev, Izdatel'stvo Akademii Nauk Ukrainskoi SSR, 1949, pp. 60.

3686. DENKHAUS, G. Über Veränderungen des Werkstoffs bei Dauerbeanspruchung von gedrückten und ungedrückten Gewinden aus Stahl. [On changes in materials with pressed and unpressed steel threads under fatigue loading.] *Werkst. u. Betr.*, vol. 82, no. 10, Oct. 1949, pp. 355–363.

3687. DOLAN, T. J.; RICHART, F. E. and WORK, C. E. The influence of fluctuations in stress amplitude on the fatigue of metals. *Proc. Amer. Soc. Test. Mater.*, vol. 49, 1949, pp. 646–682.

3688. DOLAN, T. J. and YEN, C. S. A critical review of the criteria for notch sensitivity in fatigue of metals. *Office Naval Research, Contract N6-ori-71, Task Order IV, Uni. Illinois, 13th Progr. Rep.*, Nov. 1949, pp. 55.

3689. DOSOUDIL, A. Dauerfestigkeit der verdichteten Hölzer. [The fatigue strength of compressed wood.] *Z. Ver. dtsch. Ing.*, vol. 91, no. 4, Feb. 15, 1949, pp. 85–88.

3690. VAN DER EB, W. J. Onderzoek van staalconstructies en vermoeiingsverschijnselen. [Study of fatigue in steel structures.] *Ingenieur, 's Grav.*, vol. 61, no. 32, Aug. 12, 1949, pp. 0.51–0.57.

3691. EILENDER, W.; AREND, H. and SCHMIDTMANN, E. Schwingungsuntersuchungen an hartverchromten Stählen. [Fatigue tests on hard chromium plated steels.] *Metalloberfläche*, vol. 3, no. 18, Aug. 1949, pp. 161–163. English translation: *Henry Brutcher Tech. Trans.* no. 2488.

3692. EPPRECHT, J. and EBERHARDT, H. W. Selective shot peening extends fatigue life. *Machinist*, vol. 92, no. 41, Feb. 5, 1949, pp. 1315–1318.

3693. FENNER, A. J. Fatigue of metals. *J. Lond. Ass. Engrs*, March 1949, pp. 16–26.

3694. FIEK, G. Werkstoffprüfung und Festigkeitsberechnung (Gestaltfestigkeit). [Material testing and strength calculations (shape strength).] *Arch. Metallk.*, vol. 3, no. 8, Aug. 1949, pp. 271–273.

3695. FINK, K. and HEMPEL, M. Über magnetische Messungen an dauerbeanspruchten Stahlstäben. [Magnetic measurements on fatigued steel specimens.] *Arch. Eisenhüttenw.*, vol. 20, nos. 1/2, Jan.–Feb. 1949, pp. 75–78.

3696. FISHER, W. A. P. A comparison of the endurance of various aircraft structures under fluctuating loading. *Roy. Aircr. Estab. Rep.* no. Struct. 45, July 1949, pp. 15.

3697. FORSYTH, P. J. E. Microscopical examination of specimens under fatigue stresses at elevated temperatures. *Roy. Aircr. Estab. Rep.* no. Met. 39, Jan. 1949, pp. 7.

3698. FORSYTH, P. J. E. A method of examining metallographic specimens while subjected to fatigue stresses. *J. sci. Instrum.*, vol. 26, no. 6, May 1949, pp. 160–161.

3699. Fosberry, R. A. C. Bending fatigue strength of gear teeth: preliminary report. *Rep. Mot. Ind. Res. Ass.*, no. 1949/7, Dec. 1949, pp. 8.

3700. Found, G. H. Efficient magnesium castings—their design and production. *Metal Progr.*, vol. 56, no. 6, Dec. 1949, pp. 833–840, 892.

3701. Freudenthal, A. M. Cold work and fatigue. *Cold working of metals*, Cleveland, Ohio, American Society for Metals, 1949, pp. 248–261.

3702. Fukui, S. *and* Sato, S. The fatigue of work-hardened steel. [J] *Rep. Inst. Sci., Tokyo*, vol. 3, no. 11–12, 1949, pp. 311–316.

3703. Gardiner, G. C. I. Some aspects of propeller fatigue testing. *Aircr. Engng*, vol. 21, no. 243, May 1949, pp. 149–152.

3704. Gardner, E. R. *and* Williams, P. L. Fatigue tests on the rubber-cord bond. *I.R.I. Trans.*, vol. 24, no. 6, April 1949, pp. 284–295.

3705. Gartside, F. The rig testing of coil springs. *Coil Spring Journal*, no. 15, June 1949, pp. 48–53.

3706. Gerold, E. Korrosion und mechanische Beanspruchung. [Corrosion and mechanical stress.] *Metalloberfläche*, vol. 3, no. 2, Feb. 1949, pp. 29–32.

3707. Gillig, F. J. Short-time high-temperature bending fatigue properties of sheet materials. *Cornell Aeronautical Lab., New York, Tech. Memo.* no. CAL-30, Sept. 8, 1949, pp. 30.

3708. Glaubitz, H. Einfluss der Oberflächenrauhigkeit auf die Biege-Wechselfestigkeit von un-gehärtetem und vergütetem Stahl. [The effect of surface roughness on the alternating bending strength of unhardened and tempered steel.] *Institut für Maschinenelemente der Technische Hochschule, Braunschweig, Einzel-bericht* no. 115, Sept. 6, 1949. English translation: *Dept. sci. industr. Res., Lond., Sponsored Research (Germany)*, Report no. 1, 1949, pp. 10.

3709. Glikman, L. A.; Zhuravlev, V. A. *and* Snezhkova, T. W. Variation of damping under cyclic stresses below and above the fatigue limit. [R] *Zh. tekh. Fiz.*, vol. 19, no. 4, April 1949, pp. 448–464.

3710. Gohn, G. R. *and* Morton, E. R. A new high-speed sheet metal fatigue testing machine for unsymmetrical bending studies. *Proc. Amer. Soc. Test. Mater.*, vol. 49, 1949, pp. 702–716.

3711. Gough, H. J. Engineering steels under combined cyclic and static stresses. *Proc. Instn mech. Engrs, Lond.*, vol. 160, no. 4, 1949, pp. 417–440. *Engineer, Lond.*, vol. 188, no. 4892, Oct. 28, 1949, pp. 497–500; no. 4893, Nov. 4, 1949, pp. 510–514; no. 4894, Nov. 11, 1949, pp. 540–543 and no. 4895, Nov. 18, 1949, pp. 570–573.

3712. Gould, A. J. Corrosion-fatigue of steel under asymmetric stress in sea water. *J. Iron St. Inst.*, vol. 161, pt. 1, Jan. 1949, pp. 11–16.

3713. Graf, O. Versuche über die Widerstandsfähigkeit geschweisster Bleche aus Aluminium-Legierungen beim Zerreissversuch und bei oftmals wiederholter Zugbelastung. [Tests of the strength of welded aluminium alloy sheet in tension and under repeated tensile loading. *Schweiss. Schneid.*, vol. 1, no. 11, Nov. 1949, pp. 183–189.

714. HÄNCHEN, R. Berechnung von Laufkran-Fachwerkträgern auf Dauerhaltbarkeit. [Design of overhead crane truss girders for fatigue durability.] *Schweiss. Schneid.*, vol. 1, no. 9, 1949, pp. 139–153 and no. 10, 1949, pp. 170–174.

715. HARRISON, S. T. Techniques of metallurgical investigations: the life testing of engine components. *Metal Ind., Lond.*, vol. 75, no. 26, Dec. 23, 1949, pp. 535–538.

716. HARTMANN, E. C.; HOWELL, F. M. *and* TEMPLIN, R. L. How to use high strength aluminium alloys. *Aviat. Week*, vol. 51, no. 15, Oct. 10, 1949, pp. 21–27.

17. HAUK, V. Statische und dynamische Festigkeitsuntersuchungen an Punktschweissverbindungen aus hochfesten Stahlfeinblechen. [Static and fatigue strength of spot welded joints in high tensile thin steel sheets.] *Arch. Eisenhüttenw.*, vol. 20, nos. 1/2, Jan.–Feb. 1949, pp. 41–51.

18. HELBIG, FR. Die Grübchenbildung an Wälzflächen. [The pitting of rolling surfaces.] *Werkst. Tech. MaschBau*, vol. 39, no. 4, 1949, pp. 111–115.

19. HEMPEL, M. Dauerversuche an Schraubenfedern. [Fatigue tests on coiled springs.] *Stahl u. Eisen*, vol. 69, no. 20, Sept. 29, 1949, pp. 712–713.

20. HEMPEL, M. Dauerfestigkeit von Sintereisen-Werkstoffen. [Fatigue strength of sintered iron material.] *Stahl u. Eisen*, vol. 69, no. 23, Nov. 10, 1949, pp. 852–853.

1. HEMPEL, M. *and* MÖLLER, H. Die Auswirkung von Schweissfehlern in Proben aus Stahl St37 auf deren Zugschwellfestigkeit. [The effect of weld defects in specimens of steel

St37 on their tensile strength.] *Arch. Eisenhüttenw.*, vol. 20, no. 11/12, Nov.–Dec. 1949, pp. 375–383. English translation: *Associated Technical Services Inc.*, Trans. no. 01L34G.

3722. HEMPEL, M. *and* WIEMER, H. Dauerfestigkeit von Sintereisen-Werkstoffen. [Fatigue strength of sintered iron material.] *Arch. Metallk.*, vol. 3, no. 1, Jan. 1949, pp. 11–17.

3723. HERSCHMAN, H. K. *and* THOMAS, C. Fatigue characteristics of electroformed sheets with and without iron backing. *J. Res. nat. Bur. Stand.*, vol. 43, no. 5, Nov. 1949, pp. 477–486.

3724. HOLT, M. *and* CLARK, J. W. A study of end connections for struts. *Proc. Amer. Soc. civ. Engrs*, vol. 75, no. 10, Dec. 1949, pp. 1477–1499.

3725. HUSTIN, M. P. *and* SOETE, W. La soudure des rails et la résistance à la fatigue du joint soudé. [The welding of rails and the fatigue resistance of welded joints.] *Rev. Soud.*, vol. 5, no. 2, 1949, pp. 87–105.

3726. JACOBSON, J. M. Problems of aircraft life evaluation. *S.A.E. quart. Trans.*, vol. 3, no. 4, Oct. 1949, pp. 616–633.

3727. JACQUES, H. E. The effect of cyclic stress on the transition temperature of steel. *Committee on Ship Construction, Ship Structure Committee, Division of Engineering and Industrial Research, National Research Council*, Res. Rep. Ser. no. SSC-31, July 18, 1949, pp. 27.

3728. JACQUESSON, R. *and* LAURENT, P. Les renseignements fournis par des essais de fatigue sur l'état cristallin des toles. [The information on the crystalline state of sheet metal provided by fatigue tests.] *Rev. Métall.*, vol. 46, no. 2, Feb. 1949, pp. 88–101.

3729. JAFFE, L. D.; REED, E. L. and MANN, H. C. Discontinuous crack propagation. *J. Metals, N.Y.*, vol. 1, no. 8, Aug. 1949, p. 526.

3730. JAFFE, L. D.; REED, E. L. and MANN, H. C. Discontinuous crack propagation—further studies. *J. Metals, N.Y.*, vol. 1, no. 10, Oct. 1949, pp. 683–687.

3731. JENSEN, R. S. Seventh progress report of the rolling load tests of joint bars. *Bull. Amer. Rly Engng Ass.*, vol. 50, no. 479, Feb. 1949, pp. 517–532. *Repr. Ill. Engng Exp. Sta.*, no. 43, 1949, pp. 11–26.

3732. JENSEN, R. S. Fatigue tests of manganese steel. *Bull. Amer. Rly Engng Ass.*, vol. 50, no. 479, Feb. 1949, pp. 579–588. *Repr. Ill. Engng Exp. Sta.*, no. 43, 1949, pp. 33–42.

3733. JOHNSON, S. Utmattningsförsök under stigande belastning. [Fatigue tests under step loading.] *Tekn. Tidskr., Stockh.*, vol. 79, no. 43, Nov. 26, 1949, p. 900.

3734. JOHNSON, W. and MATTHEWS, E. Fatigue studies on some dental resins. *Brit. dent. J.*, vol. 86, no. 10, May 20, 1949, pp. 252–253.

3735. KARPENKO, G. V. The influence of surface-active substances on the fatigue strength of metals. [U] *Dokl. Akad. Nauk URSR*, no. 3, 1949, pp. 39–43.

3736. KARPENKO, G. V. The influence of lubricants on the fatigue strength of metals. [U] *Dokl. Akad. Nauk URSR*, no. 6, 1949, pp. 58–61.

3737. KAWADA, Y. On some experimental results for notched and bored bars under repeated loading. III—effect of surface finishing upon fatigue strength. IV—relationship be-tween diameter of holes perpendicular to the axis of bars and their fatigue limit. [J] *Nippon kink. Gakk.*, vol. 13, no. 3, March 1949 pp. 33–39.

3738. KENYON, J. N. The fatigue problem in metals with special reference to wire materials *Wire and Wire Prod.*, vol. 24, no. 4, April 1949, pp. 317–319 and no. 6, June 1949 pp. 498–500, 525–527.

3739. KHRUSHCHOV, M. M. and BABICHEV M. A. Determination of the fatigue limit of metals by means of a three ring test machine [R] *Zav. Lab.*, vol. 15, no. 8, Aug. 1949 pp. 962–967. English abstr.: *Engrs' Dig* vol. 12, no. 7, July 1951, p. 230.

3740. KOLLMANN, F. and DOSOUDIL, A. Holz faserplatten: Ihre Eigenschaften und Prüfung mit besonderer Berücksichtigung der Dauer festigkeit. [Types of fibreboard, their proper ties and their testing with special considera tion of the fatigue strength.] *V.D.I. Forsch* no. 426, 1949, pp. 32. English translation C.S.I.R.O. Trans. no. 1142, May 1950, pp. 6

3741. KUDRYAVTSEV, I. V. The influence of residual tensile stresses on the fatigue strength of notched and unnotched specimens. [R] *Trudy tsent. nauchno-issled. Inst. tekhno Mashinost.*, vol. 24, 1949, pp. 40–51.

3742. KUDRYAVTSEV, I. V.; SAVERIN, M. M and RYABCHENKOV, A. V. *Metody poverk nostnogo uprochneniia detalei mashin. [Meth ods of surface hardening of machine compo nents.]* Moscow, Mashgiz, 1949, pp. 220.

3743. LAURENT, P. Influence de la forme et de dimensions de l'éprouvette sur la limite de fatigue. [Influence of the shape and size of the test piece on the fatigue limit.] *Rev. Méta* vol. 46, no. 1, Jan. 1949, pp. 55–59.

3744. LAZAN, B. J. Dynamic creep and rupture properties of temperature resistant materials under tensile fatigue stress. *Proc. Amer. Soc. Test. Mater.*, vol. 49, 1949, pp. 757–787.

3745. LENZEN, K. H. Bolted joints under fatigue loads. *Fasteners*, vol. 6, no. 1, 1949, pp. 6–9.

3746. LENZEN, K. H. The effect of various fasteners on the fatigue strength of a structural joint. *Bull. Amer. Rly Engng Ass.*, vol. 51, no. 481, June–July 1949, pp. 1–28.

3747. LIHL, F. Kristallographische Vorgänge an der Fleissgrenze von Stahl und ihre Bedeutung für die Dauerfestigkeit. [Crystallographic behaviour at the yield point of steel and its relationship to the fatigue strength.] *Metall: Wirtsch. Wiss. Tech.*, vol. 2, no. 23/24, Dec. 1948, pp. 391–396 and vol. 3, no 3/4, Feb. 1949, pp. 49–51.

3748. LIPSON, C. Surface finish, hardness and life of steel parts. *Prod. Engng*, vol. 20, no. 6, June 1949, p. 179.

3749. LIPSON, C.; NOLL, G. C. and CLOCK, L. S. Significant strength of steels in the design of machine parts. *Prod. Engng*, vol. 20, no. 4, April 1949, pp. 142–146 and no. 5, May 1949, pp. 124–128.

3750. LIPSON, C.; NOLL, G. C. and CLOCK, L. S. Significant stress and failure in static and fatigue loading. *Prod. Engng*, vol. 20, no. 7, July 1949, pp. 130–135.

3751. LIPSON, C.; NOLL, G. C. and CLOCK, L. S. Equality of strength and stress in design of machine parts. *Prod. Engng*, vol. 20, no. 8, Aug. 1949, pp. 86–87.

3752. LOGAN, H. L. Effect of chromium plating on the endurance limit of steels used in aircraft. *J. Res. nat. Bur. Stand.*, vol. 43, no. 2, Aug. 1949, pp. 101–112. Abstr.: *Metal Finish.*, vol. 47, no. 11, 1949, pp. 60–61, 98. *Steel*, vol. 125, no. 19, Nov. 7, 1949, pp. 110–111, 142. *Iron Age*, vol. 164, Nov. 24, 1949, p. 82. *Aviat. Week*, vol. 51, no. 24, Dec. 12, 1949, pp. 25–26.

3753. LUNDBERG, Bo. K. and WÅLLGREN, G. A study of some factors affecting the fatigue life of aircraft parts with application to structural elements of 24S-T and 75S-T aluminium alloys. *Flytek. Försöksanst.*, Stockholm, Rep. no. 30, 1949, pp. 33.

3754. LUNDBERG, G. and PALMGREN, A. Dynamic capacity of rolling bearings. *J. appl. Mech.*, vol. 16, no. 2, June 1949, pp. 165–172 and [discussion] no. 4, Dec. 1949, pp. 415–417.

3755. LUTHANDER, S. and WALLGREN, G. Determination of fatigue life with stress cycles of varying amplitude. *Flytek. Försöksanst.*, Stockholm, Rep. no. 18 (Translation no. 10), 1949, pp. 20.

3756. McDONALD, G. G. The graphics of pulsating stress. *Engineer, Lond.*, vol. 187, no. 4855, Feb. 11, 1949, pp. 154–155.

3757. MACHLIN, E. S. Dislocation theory of the fatigue of metals. *N.A.C.A. Rep.*, no. 929, 1949, pp. 10.

3758. MAIER, H. J. Reducing cost of fatigue testing. *Mach. Design*, vol. 21, no. 9, Sept. 1949, pp. 137–139.

3759. MAJORS, H.; MILLS, B. D. and MACGREGOR, C. W. Fatigue under combined pulsating stresses. *J. appl. Mech.*, vol. 16, no. 3, Sept. 1949, pp. 269–276.

3760. MANJOINE, M. J. Effect of pulsating loads on the creep characteristics of aluminium alloy 14S-T. *Proc. Amer. Soc. Test. Mater.*, vol. 49, 1949, pp. 788–803.

3761. MANSION, H. D. A hydraulic fatigue testing machine for gear teeth. *Rep. Mot. Ind. Res. Ass.*, no. 1949/4, June 1949, pp. 9.

3762. MARIN, J. Biaxial tension–tension fatigue strengths of metals. *J. appl. Mech.*, vol. 16, no. 4, Dec., 1949, pp. 383–388 and [discussion] vol. 17, no. 2, June 1950, p. 222.

3763. MARIN, J. *and* SHELSON, W. Biaxial fatigue strength of 24ST aluminium alloy. *Tech. Note nat. Adv. Comm. Aero., Wash.*, no. 1889, May 1949, pp. 41.

3764. MARKOVETS, M. P.; SMIYAN, I. A. *and* MIKHEEV, N. I. Fatigue testing machine for use at high temperatures. [R] *Zav. Lab.*, vol. 15, no. 1, Jan. 1949, pp. 82–85.

3765. VAN MEER, H. P. *and* PLANTEMA, F. J. Vermoeiing van constructies en constructie-delen. [Fatigue of structures and structural components.] *Nat. Luchtvaartlab., Amsterdam*, Rep. no. S.357, Aug. 18, 1949, pp. 39. English translation:. *T. I. B. (England)*, Trans. no. T 4080, Feb. 1953, pp. 63.

3766. MELCHOR, J. L.; GOOD, W. B.; PAGE, W. A. *and* SHEARIN, P. E. Failure of metals subjected to large repeated strains, properties of curved tubes, and force reaction units. *Uni. North Carolina, Dept. Physics, Rep. no. VIII to Naval Research Lab.*, June 1949, pp. 30. Abstr.: *Bull. Brit. non-ferr. Met. Res. Ass.*, no. 282, Dec. 1952, p. 504.

3767. MILLS, H. R. *and* LOVE, R. J. Fatigue strength of cast crankshafts. *Mach. Mkt*, no. 2541, July 29, 1949, pp. 489–490 and no. 2542, Aug. 5, 1949, pp. 507–509.

3768. MITCHELL, G. R. Research on the strength of bridges. (c) Problems of impact and fatigue and their effect on permissible stresses in cast iron girder bridges. *Mém. Ass. int. Ponts Charp.*, vol. 9, Nov. 1949, pp. 61–68.

3769. MITINSKII, A. N. *and* BYKOV, V. A. Fatigue testing machine for plane bending. [R] *Zav. Lab.*, vol. 15, no. 1, Jan. 1949, pp. 89–91.

3770. MOROZOV, IU. N. Errors in the measurement of loads during fatigue tests in hydraulic pulsators. [R] *Inzh. Sb.*, vol. 5, no. 2, 1949, pp. 148–163.

3771. NEERFELD, H. *and* MÖLLER, H. Zur Frage des Spannungsabbaues durch Schwingungsbeanspruchung. [The problem of relieving stress by alternating stresses.] *Arch. Eisenhüttenw.*, vol. 20, no. 5/6, May–June 1949, pp. 205–210.

3772. NIJHAWAN, B. R. Failure of railway materials. Fracture in rails and locomotive parts. *Iron Coal Tr. Rev.*, vol. 158, no. 4217, Jan. 7, 1949, pp. 1–5.

3773. NIKOLAEV, R. S.; MIKHNENKO, E. F. *an* SHKOLNIK, L. M. Application of the pulsato to the testing of heavily loaded pinion trans missions of high modulus. [R] *Zav. Lab.* vol. 15, no. 1, Jan. 1949, pp. 124–125.

3774. NIKOLAEV, R. S. *and* SHKOLNIK, L. M Experimental utilization of a pulsator fc endurance testing of gear teeth. [R] *Za Lab.*, vol. 15, Oct. 1949, pp. 1264–1265.

3775. NOVOKRESHCHENOV, P. D.; MARKOVA, N. E. and REBINDER, P. A. Adsorption effect during alternating torsion in connection with the problem of fatigue of metals. [R] Dokl. Akad. Nauk SSSR, vol. 68, no. 3, Sept. 21, 1949, pp. 549–552. English abstr.: Chem. abstr., vol. 44, 1950, col. 507f.

3776. OBERG, T. T. and ROONEY, R. J. Fatigue characteristics of aluminium alloy 75S-T6 plate in reversed bending as affected by type of machine and specimen. Proc. Amer. Soc. Test. Mater., vol. 49, 1949, pp. 804–814.

3777. OBERG, T. T. and ROONEY, R. J. Reversed bending fatigue characteristics of steel and high strength aluminium alloys as affected by type of specimen. Part I—extruded aluminium alloys and rolled 4130 steel plate. Part II—aluminium alloy 75S-T6 plate. Tech. Rep. U. S. Air Force, no. 5775, July 1949, pp. 36.

3778. ODING, I. A. Structural signs of metal fatigue as a means for determining the causes of failure in machinery. [R] Moscow, Izdatel' stvo Akademii Nauk SSSR, 1949, pp. 80.

3779. OROWAN, E. Fracture and strength of solids. Rep. Progr. Phys., vol. 12, 1949, pp. 185–232.

3780. PARDUE, T. E.; MELCHOR, J. L. and GOOD, W. B. Energy losses and fracture of some metals resulting from a small number of cycles of strain. Proc. Soc. exp. Stress Anal., vol. 7, no. 2, 1949, pp. 27–39.

3781. PETERSON, R. E. Approximate statistical method for fatigue data. Bull. Amer. Soc. Test. Mat., no. 156, Jan. 1949, pp. 50–52.

3782. PETERSON, R. E. Application of fatigue data to machine design. Properties of metals in materials engineering, Cleveland, Ohio, American Society for Metals, 1949, pp. 60–78.

3783. PETRACCHI, G. Intorno all'interpretazione del processo di corrosione per cavitazione. [On the interpretation of the corrosion process by cavitation.] Metallurg. ital., vol. 41, no. 1, Jan. 1949, pp. 1–6.

3784. POGODIN-ALEKSEEV, G. I. On the influence of short periods of overloading on the fatigue limit of steel. [R] Zav. Lab., vol. 15, no. 1, Jan. 1949, pp. 91–95.

3785. POMP, A. and HEMPEL, M. Bruchhäufigkeit und Oberflächengüte von Schraubenfedern. [Frequency of fracture and surface quality of coil springs.] Arch. Eisenhüttenw., vol. 20, no. 11/12, Nov.–Dec. 1949, pp. 385–393.

3786. POPE, J. A. Surface effects in spring design. Coil Spring Journal, no. 15, June 1949, pp. 17–25.

3787. RAITSES, V. B. Fatigue testing of valve springs with surface defects. [R] Zav. Lab., vol. 15, Dec. 1949, pp. 1494–1496.

3788. RANSOM, J. T. and MEHL, R. F. The statistical nature of the endurance limit. J. Metals, N.Y., vol. 1, no. 6, June 1949, pp. 364–365.

3789. ROOS, P. K.; LEMMON, D. C. and RANSOM, J. T. Influence of type of machine, range of speed, and specimen shape on fatigue test data. Bull. Amer. Soc. Test. Mat., no. 158, May 1949, pp. 63–65.

3790. ROŠ, M. Les bases des contraintes admissibles dans les constructions métalliques. [The basis of permissible stresses in metallic structures.] Ann. Inst. Bâtim., Nouv. Ser. no. 78, June 1949, pp. 44.

3791. Roš, M. G. Static failure and fatigue of steels with particular reference to welded structures. *Sheet Metal Ind.*, vol. 26, no. 271, Nov. 1949, pp. 2417–2426, 2440 and no. 272, Dec. 1949, pp. 2625–2656, 2658.

3792. Roš, M. G. La fatigue des métaux. [The fatigue of metals.] *Ber. eidgenöss. MatPrüf-Anst.*, no. 160, 1949, pp. 19.

3793. Roš, M.; Bühler, F. *and* Ceradini, G. Fachwerkträger für Eisenbahnbrücken aus "St N" in völlig geschweisster Auführung. [All welded lattice girders for railway bridges constructed in steel "St N".] *Ber. eidgenöss. MatPrüfAnst.*, no. 168, June 1949, pp. 40.

3794. Roš, M. *and* Ceradini, G. Statische und Ermüdungsversuche mit aufgeschweissten und aus dem vollen Stahlmaterial herausgearbeiteten, verschieden geformten laschenkörpern sowie mit überlapptem Stoss. [Static and fatigue tests with different types of cover straps welded on and machined from the solid steel material, and also with lap joints.] *Ber. eidgenöss. MatPrüfAnst.*, no. 168, June 1949, pp. 17.

3795. Rosenthal, D.; Sines, G. *and* Zizicas, G. Effect of residual compression on fatigue. *Weld. J., Easton, Pa.*, vol. 28, no. 3, March 1949, pp. 98s–103s.

3796. Roseveare, W. E. *and* Waller, R. C. A dynamically balanced fatigue tester for rayon tire cord. *Text. Res. (J.)*, vol. 19, no. 10, Oct. 1949, pp. 633–637.

3797. Rozovskii, M. I. Thermal stresses associated with fatigue. [R] *Zh. tekh. Fiz.*, vol. 19, no. 6, June 1949, pp. 696–710.

3798. Sachs, G. Fretting fatigue — literature report no. 1. *J. sci. industr. Res.*, vol. 8, no. 8, Aug. 1949, pp. 329–333.

3799. Sato, H. On the internal stress and internal friction of metals. I—the mechanism of fatigue. *Sci. Rep. Res. Insts Tôkohu Univ. Ser. A*, vol. 1, no. 3, Oct. 1949, pp. 203–206.

3800. Saverin, M. M. Investigations into the process of shot peen hardening of components. [R] *Trudy tsent. nauchno-issled. Inst. tekhnol. Mashinost.*, vol. 24, 1949, pp. 7–39.

3801. Saxton, R. Cold-worked metals and fatigue. *Metallurgia, Manchr*, vol. 41, no. 241, Nov. 1949, pp. 32–33.

3802. Schaal, A. Röntgenographische Untersuchungen über das Verhalten der Werkstoffe bei Schwingungsbeanspruchung. [X-ray investigations on the behaviour of materials under alternating stress.] *Z. Metallk.*, vol. 40, no. 11, Nov. 1949, pp. 417–427. English translation: *Henry Brutcher Tech. Trans.*, no. 3102.

3803. Siebel, E. *and* Meuth, H. O. Die Wirkung von Kerben bei schwingender Beanspruchung. [The effect of notches under fatigue stresses.] *Z. Ver. dtsch. Ing.*, vol. 91, no. 13, July 1, 1949, pp. 319–323.

3804. Sinclair, G. M. *and* Dolan, T. J. The influence of austenitic grain size and metallurgical structure on the mechanical properties of steel. *Office Naval Research, Contract N6-ori-71, Task Order IV, 15th Progr. Rep., Uni. Illinois*, Dec. 1949, pp. 35.

3805. Soete, W. *and* Van Crombrugge, R. La résistance à la fatigue ondulée des fils utilisés en béton précontraint. [The fatigue resistance of wires used in prestressed concrete.] *Ann Trav. publ. Belg.*, vol. 102, no. 5, Oct. 1949 pp. 513–533. English translation: *Cement and Concrete Assoc. Library Trans.* no. 25.

3806. Sonneville, R. Essais d'endurance de quelques attaches de rails. [Fatigue tests on some rail attachments.] *Rev. gén. Chem.-de-Fer*, vol. 68, no. 12, Dec. 1949, pp. 550–557.

3807. SPERLING, E. Festigkeitsversuche an Eisenbahnwagen-Achsen als Grundlage für deren Berechnung. [Strength tests of railway wagon axles as a basis for their design.] *Z. Ver. dtsch. Ing.*, vol. 91, no. 6, March 15, 1949, pp. 134–136. English translation: *Brit. Transport Comm. Derby*, Trans. no. 303.

3808. STEINHARDT, O. *and* MÖHLER, K. Betrachtungen zu Dauerversuchen an grösseren Nietverbindungen. [Regarding the fatigue testing of large riveted joints.] *Bautechnik*, vol. 26, no. 9, 1949, pp. 265–269.

3809. STRAUB, J. C. *and* MAY, D. Stress peening. *Iron Age*, vol. 163, no. 16, April 21, 1949, pp. 66–70.

3810. TATNALL, F. G. Fatigue and mechanical properties. Is there a relation between them? *West. Mach.*, vol. 40, no. 7, July 1949, pp. 88–89, 100.

3811. TEED, P. L. Materials from the aircraft manufacturer's point of view. *Proceedings 2nd International Aeronautical Conference*, [Editor B. H. Jarck], New York, Institute of Aeronautical Sciences, Inc., 1949, pp. 242–311.

3812. TEICHMANN, A. Belastungs-Kollektive und Festigkeitsnachweis. [Load summation and strength determination.] *Konstruktion*, vol. 1, no. 4, 1949, pp. 103–112.

3813. THUM, A. *and* SVENSON, O. Beanspruchung bei mehrfacher Kerbwirkung. Enlastungs- und Überlastungskerben. [Applied load and multiple notch effect. Unloading and overloading notches.] *Schweiz. Arch. angew. Wiss.*, vol.15, no. 6, June 1949, pp. 161–174.

3814. ULRICH, M. *and* GLAUBITZ, H. Stand der Induktionshärtung von Zahnrädern. Festigkeits- und Verschleissverhalten. [Status of induction hardening of gear teeth. Strength and wear behaviour.] *Z. Ver. dtsch. Ing.*, vol. 91, no. 22, Nov. 15, 1949, pp. 577–583.

3815. UZHIK, G. V. The effect of stress concentrations under asymmetric loading cycles. [R] *Vestn. mashinostr.*, vol. 29, no. 4, April 1949, pp. 5–9.

3816. UZHIK, G. V. Endurance limit of steel under simultaneous action of constant and alternating stresses. [R] *Izv. Akad. Nauk SSSR. Otd. tekh. nauk*, no. 5, May 1949, pp. 657–665.

3817. WAKEFIELD, P. S. Fatigue. *Mach. Lloyd*, vol. 21, no. 17, Aug. 13, 1949, pp. 68–73.

3818. WALKER, H. L, *and* CRAIG, W. J. Effect of grain size on tensile strength, elongation and endurance limit of deep drawing brass. *Trans. Amer. Inst. min. (metall.) Engrs*, vol. 180, 1949, pp. 42–51.

3819. WALKER, P. B. Fatigue. *J. R. aero. Soc.*, vol. 53, no. 464, Aug. 1949, pp. 763–778. Abstr.: *Flight*, vol. 55, no. 2102, April 7, 1949, p. 400.

3820. WALLACE, W. P. *and* FRANKEL, J. P. Relief of residual stress by a single fatigue cycle. *Weld. J.*, Easton, Pa., vol. 28, no. 11, Nov. 1949, p. 565s.

3821. WALLGREN, G. Fatigue tests with stress cycles of varying amplitude. *Flytek. Försöksanst.*, Stockholm, Rep. no. 28, 1949, pp. 34.

3822. WECK, R. De invloed van na en door het lassen over blijvende spanningen in gelaste constructies. [The influence of residual stresses in welded structures.] *Polyt. Tijdschr. (A)*, vol. 4, no. 15/16, April 19, 1949, pp. 279a–285a.

3823. WEIBULL, W. A statistical representation of fatigue failure in solids. *K. tekn Högsk. Handl.*, no. 27, 1949, pp. 51. *Acta polyt. Stockh.*, no. 49, 1949, pp. 51.

3824. WELTER, G. Fatigue tests of spot welds. Improvement of their endurance limit by hydrostatic pressure. *Engrs' Dig.*, vol. 10, no. 3, March 1949, pp. 76–78 and no. 4, April 1949, pp. 120–122.

3825. WELTER, G. Fatigue tests of spot welded steel plates. *Weld. J., Easton, Pa.*, vol. 28, no. 9, Sept. 1949, pp. 414s–438s.

3826. WERREN, F. Fatigue of sandwich constructions for aircraft. Fiberglass laminate facing and paper honeycomb core sandwich material tested in shear. *Rep. For. Prod. Lab., Madison*, no. 1559-G, Feb. 1949, pp. 5.

3827. WERREN, F. Fatigue of sandwich constructions for aircraft. Aluminium facing and aluminium honeycomb core sandwich material tested in shear. *Rep. For. Prod. Lab., Madison*, no. 1559-H, Dec. 1949, pp. 5.

3828. WILKES, G. B. Changes in internal damping of gas-turbine materials due to continuous vibration. *Trans. Amer. Soc. mech. Engrs*, vol. 71, 1949, pp. 631–634.

3829. WILLIAMS, W. L. The effects of metallizing procedures on the fatigue properties of steel. *Proc. Amer. Soc. Test. Mater.*, vol. 49, 1949, pp. 683–701.

3830. WILLIAMS, W. L. *and* STEWART, W. C. Fatigue and corrosion of sintered and rolled

titanium. *Metal Progr.*, vol. 55, no. 3, March 1949, pp. 351–353.

3831. WILLS, H. A. The life of aircraft structures. *Proceedings 2nd International Aeronautical Conference*, 1949, [Editor, B. H. Jarck], New York, Institute of Aeronautical Sciences, Inc., pp. 361–403. Abstr.: *Aeroplane*, vol. 76, no. 1984, June 17, 1949, pp. 695–696.

3832. WILSON, W. M. *and* MUNSE, W. H. The fatigue strength of various details used for the repair of bridge members. *Uni. Ill. Engng Exp. Sta. Bull.*, no. 382, Dec. 1949, pp. 60.

3833. WILSON, W. M.; MUNSE, W. H. *and* BRUCKNER, W. H. Fatigue strength of fillet-weld, plug-weld, and slot-weld joints connecting steel structural members. *Uni. Ill. Engng Exp. Sta. Bull.*, no. 380, May 1949, pp. 104.

3834. WYSS, TH. Einfluss der sekundären Biegung und der inneren Pressungen auf die Lebensdauer von Stahldraht-Litzenseilen mit Hanfseele. [Influence of secondary bending and internal pressure on the life of steel wire stranded rope with hemp core.] *Schweiz. Bauztg*, vol. 67, no. 14, April 2, 1949, pp. 193–198; no. 15, April 9, 1949, pp. 212–215 and no. 16, April 16, 1949, pp. 225–228.

3835. YATSKEVICH, S. I. Machine for fatigue testing a stationary test specimen at elevated temperatures. [R] *Zav. Lab.*, vol. 15, no. 1, Jan. 1949, pp. 86–88.

3836. YEN, C. S. Stress distribution in a metal specimen as affected by plastic deformation during repeated loading. *Office Naval Research, Contract N6-ori-71, Task Order IV, 14th Progr. Rep., Uni. Illinois*, Nov. 1949, pp. 17.

3837. YEN, C. S. *and* DOLAN, T. J. An experimental study of the effect of thermal activation on the fatigue life of 75S-T aluminium alloy. *Office Naval Research, Contract N6-ori-71, Task Order IV, 12th Progr. Rep., Uni. Illinois,* Oct. 1949.

3838. YU, AI-TING *and* JOHNSTON, B. G. A method for vibration fatigue tests of stranded conductor. *Proc. Soc. exp. Stress Anal.,* vol. 6, no. 2, 1949, pp. 1–6.

3839. ZAMBROW, J. L. *and* FONTANA, M. G. Mechanical properties, including fatigue, of aircraft alloys at very low temperatures. *Trans. Amer. Soc. Metals,* vol. 41, 1949, pp. 480–518.

3840. ZHUKOV, S. L. Method of determination of fatigue strength under high stresses. [R] *Zav. Lab.,* vol. 15, Aug. 1949, pp. 971–976.

3841. Stroke of fatigue tester is varied automatically. *Prod. Engng,* vol. 20, no. 2, Feb. 1949, pp. 90–91.

3842. Research on factors affecting spring endurance. The radial fatigue test rig. *Coil Spring Journal,* no. 14, March 1949, pp. 20–21.

3843. Baldwin machine for testing rails. *Rly Age, N.Y.,* vol. 126, no. 18, April 30, 1949, p. 48.

3844. Variable stroke fatigue mechanism. *Mach. Design,* vol. 21, no. 5, May 1949, pp. 144–145.

3845. Corrosion fatigue cracking in steam pipe systems at power stations. *British Electricity Authority, Report of the Corrosion Fatigue Committee,* May 1949, pp. 19. Abstr.: *Beama J.,* vol. 56, no. 149, Nov. 1949, pp. 381–385. *Combustion int. Combust. Engng Corp.,* vol. 21, no. 6, Dec. 1949, pp. 57–58.

3846. Selective shot peening—stronger propeller hubs and operating units produced without undesirable weight increase. *Aust. Mfr,* vol. 34, no. 1737, July 16, 1949, pp. 28–29.

3847. Fatigue resistance of extruded 'Z' section stringers in aluminium alloy to Specification D.T.D. 364. *Sci. Tech. Memor.,* no. C2/49, Sept. 1949, pp. 5.

3848. Fatigue testing machine [for wheels]. *Mech. World,* vol. 126, no. 3269, Sept. 9, 1949, p. 299.

3849. Progress report of Research Council on riveted and bolted structural joints. *Bull. Amer. Rly Engng Ass.,* vol. 51, no. 482, Sept.–Oct. 1949, pp. 74–86.

3850. Tentative method of test for repeated flexural stress (fatigue) of plastics (D671-49T). *A.S.T.M. Standards,* Philadelphia, Pa., American Society for Testing Materials, Part 6, 1949, pp. 559–568.

1950

3851. ALMEN, J. O. Fatigue weakness of surfaces. *Prod. Engng,* vol. 21, no. 11, Nov. 1950, pp. 117–140.

3852. ALMEN, J. O. Surface deterioration of gear teeth. *Mechanical wear,* [Editor, J. T. Burwell], Cleveland, Ohio, American Society for Metals, 1950, pp. 229–288.

3853. ARBLASTER, H. E. Fatigue life of "75S". *Commonwealth Aircraft Corp. Pty. Ltd., Melbourne, Memorandum* Nov. 6, 1950, pp. 6.

3854. ARCHER, S. Screwshaft casualties—the influence of torsional vibration and propeller immersion. *J. Amer. Soc. nav. Engrs,* vol. 62, Aug. 1950, pp. 715–743.

3855. AVERY, H. S. and WILKS, C. R. Alloy Casting Institute thermal fatigue testing. *Alloy Cast. Bull.*, no. 14, May 1950, pp. 1–9.

3856. BACHMAN, W. S. Fatigue testing—and development of drill pipe-to-tool joint connections. *Oil Gas J.*, vol. 49, no. 24, Oct. 19, 1950, pp. 109–110, 112, 115–118.

3857. BACHMAN, W. S. Drill pipes and tool joints undergo rigid tests for fatigue. *Drilling*, vol. 12, Dec. 1950, pp. 20–21, 78.

3858. BAILEY-WATSON, C. B. Quest for efficiency. *Flight*, vol. 57, no. 2151, March 16, 1950, pp. 346–351.

3859. BALLETT, J. T. Tensile and fatigue properties of titanium at room and elevated temperatures (interim report). *Roy. Aircr. Estab. Rep.*, no. Met. 59, Nov. 1950, pp. 10.

3860. BANKS, L. B. Machine for testing rails in bending fatigue. *Engineering, Lond.*, vol. 169, no. 4400, May 26, 1950, pp. 585–587.

3861. BECKER, A. Last-Steurautomaten für Bauteil-Prüfmaschinen mit selbsttätigem Versuchsablauf. [Automatic load control equipment for component testing machines with automatic control of test programme.] *Z. Ver. dtsch. Ing.*, vol. 92, no. 11, April 11, 1950, pp. 266–271.

3862. BENNETT, J. A. and BAKER, J. L. Effect of prior static and dynamic stresses on the fatigue strength of aluminium alloys. *J. Res. nat. Bur. Stand.*, vol. 45, no. 6, Dec. 1950, pp. 449–457.

3863. BERRY, J. M. and GROVER, H. J. Surface hardening versus fatigue in steel. *Amer. Gas Ass. Mon.*, vol. 32, no. 9, Sept. 1950, pp. 20–21, 26. *Industr. Gas*, vol. 29, no. 5, Nov. 1950, pp. 10–11, 25.

3864. BILLING, B. F. and BALLETT, J. T. Fatigue tests at room and elevated temperatures on high purity aluminium alloys. *Roy. Aircr. Estab. Tech. Note*, no. Met. 134, Sept. 1950, pp. 14.

3865. BOAS, W. Théories sur le mechanisme de la rupture par fatigue. [Theories of the mechanism of fatigue fracture.] *Métaux et Corros.*, vol. 25, no. 296, April 1950, pp. 100–104.

3866. BOEGEHOLD, A. L. Test bar results compared with tests on components. *Metal Progr.*, vol. 57, no. 3, March 1950, pp. 349–357.

3867. BOLLENRATH, F.; BUNGARDT, W. and GRÖBER, H. Einfluss der Warmbehandlung auf die Festigkeit von Luftschrauben aus Aluminium – Zinc – Magnesium-Legierungen [Effect of heat treatment on the strength of propellers of aluminium–zinc–magnesium alloys.] *Z. Metallk.*, vol. 41, no. 12, Dec. 1950, pp. 463–469.

3868. BOLLENRATH, F. and TROOST, A. Wechselbeziehungen zwischen Spannungs- und Verformungsgradient. I—Behinderung der plastischen Verzerrung. [Relationship between alternating stress and deformation gradient I—Restraint of plastic deformation.] *Arch Eisenhüttenw.*, vol. 21, no. 11/12, 1950, pp. 431–436.

3869. BOONE, P. W. Caustic etching detects fatigue damage. *Iron Age*, vol. 165, no. 20, May 18, 1950, pp. 99–100.

3870. BOYER, H. E. Surface treatments and their effects on endurance limits. *Mod. Mach. Shop* vol. 23, no. 5, Oct. 1950, pp. 98–100, 102, 104, 106, 108, 110, 112, 114, 116, 118, 120, 122, 124, 126, 128, 130, 132.

3871. BRENNER, P. Einfluss einer Überhitzung auf die Wechselfestigkeit von Aluminium–Kupfer–Magnesium-Legierungen. [Influence of overheating on the fatigue strength of aluminium–copper–magnesium alloys.] *Metall: Wirtsch. Wiss. Tech.*, vol. 4, no. 23/24, Dec. 1950, pp. 502–504.

3872. BUCKINGHAM, E. *and* TALBOURDET, G. J. Recent roll tests on endurance limits of materials. *Mechanical wear*, [Editor, J. T. Burwell], Cleveland, Ohio, American Society for Metals, 1950, pp. 289–307.

3873. BUCKMAN, M. *and* RUDNICK, J. Repeated-load tests of Metalite stabilizers under vibration at sub-normal temperatures. *U.S. Naval Air Material Center Report*, no. ASL NAM DE-211.1, June 27, 1950, pp. 13.

3874. CAMPUS, F. Le béton précontraint. [Pre-stressed concrete.] *Ann. Trav. publ. Belg.*, vol. 103, no. 2, April 1950, pp. 295–331.

3875. CARPENTER, O. R.; JESSEN, N. C.; OBERG, J. L. *and* WYLIE, R. D. Some considerations in the joining of dissimilar metals for high-temperature high pressure service. *Proc. Amer. Soc. Test. Mater.*, vol. 50, 1950, pp. 809–860.

3876. CAZAUD, R. La forme des pièces de machines et la tenue à la fatigue. [The form of machine components and their behaviour in fatigue.] *Prat. Industr. méc.*, vol. 33, Feb. 1950, pp. 35–40 and March 1950, pp. 67–74.

3877. COOPER, W. E. Helicopter maintenance—materials and fatigue aspect. *J. Helicopter Ass. Great Brit.*, vol. 4, no. 3, Oct.–Dec. 1950, pp. 97–103.

3878. COX, H. L. Fracture by fatigue. *The fracture of metals*, London, The Institution of Metallurgists, Jan. 1950, pp. 42–67.

3879. COX, H. L. *and* COLEMAN, E. P. A note on repeated loading tests on components and complete structures. *J. R. aero. Soc.*, vol. 54, no. 469, Jan. 1950, pp. 1–10.

3880. CRAMER, R. E. *and* JENSEN, R. S. Progress reports of investigation of railroad rails, joint bars, and rail webs. *Repr. Ill. Engng Exp. Sta.*, no. 47, June 1950, pp. 37.

3881. DENKHAUS, G. Über eine Anwendung des oberflächendrückens bei der Einspannung von Probestäben bei Dauerbeanspruchung. [On the application of surface pressure for clamping fatigue test specimens.] *Metalloberfläche*, vol. 4, no. 2, Feb. 1950, p. A17.

3882. DOLAN, T. J. Past work on fatigue of metals in high temperature field. *Uni Ill. Dep. Theor. Appl. Mech. Tech. Rep.* no. 17 on *Behaviour of materials under repeated stress*, June, 1950, pp. 37. *Cornell Aero. Lab. Inc., Project Squid Tech. Rep.*, no. 21, May 24, 1950, pp. 5–35.

3883. DOLAN, T. J.; McCLOW, J. H. *and* CRAIG, W. J. The influence of shape of cross-section on the flexural fatigue strength of steel. *Trans. Amer. Soc. mech. Engrs*, vol. 72, no. 5, July 1950, pp. 469–477.

3884. DONANDT, H. Zur Dauerfestigkeit von Seildraht und Drahtseil. [The fatigue strength of cable wire and wire ropes.] *Arch. Eisenhüttenw.*, vol. 21, no. 9/10, Sept.–Oct. 1950, pp. 283–292.

3885. DOREY, S. F. Large scale torsional fatigue testing of marine shafting. *J. Amer. Soc. nav. Engrs*, vol. 62, no. 1, Feb. 1950, pp. 185–201.

3886. DROZD, A.; GEROLD, E. *and* SCHULZ, E. H. Der Einfluss von wechselnden und schlagartigen Überbelastungen auf die Lebensdauer von Stahl bei Biegewechselbeanspruchung. [The influence of alternating and impact overloading on the endurance of steel under alternating bending stress.] *Arch. Eisenhüttenw.*, vol. 21, no. 5/6, May–June 1950, pp. 181–189.

3887. EAGAN, T. E. Notch sensitivity of various cast materials. *Amer. Foundrym.*, vol. 18, no. 5, Nov. 1950, pp. 22–24.

3888. EICHINGER, A. Zur Frage der Wirkung des Oberflächendrückens auf die Dauerfestigkeit. [The question of the effectiveness of surface rolling on fatigue strength.] *Z. Ver. dtsch. Ing.*, vol. 92, no. 2, Jan. 11, 1950, pp. 35–39.

3889. FELTHAM, P. Fatigue of metals—consideration of Orowan and Dehlinger theories. *Iron Coal Tr. Rev.*, vol. 161, no. 4308, Nov. 3, 1950, pp. 599–604.

3890. FISHER, W. A. P. Repeated loading and fatigue tests on a D.H. 104 (Dove) wing and fin. *Aircr. Engng*, vol. 22, no. 256, June 1950, pp. 166–171.

3891. FISHER, W. A. P.; CROSS, R. H. *and* NORRIS, G. M. Pretensioning as a means of preventing fatigue in bolts. *Roy. Aircr. Estab. Rep.*, no. Structures 84, July 1950, pp. 16.

3892. FLUSIN, F. Mesure des efforts en marche normale sur un cadre de bicyclette. [Measurement of the loads in bicycle frames under normal operating conditions.] *Rev. Alumin.*, vol. 27, no. 164, March 1950, pp. 89–95. English translation: *Light Metals*, vol. 13, no. 147, April 1950, pp. 175–182.

3893. FORREST, G. *and* WOODWARD, A. R. Report on fatigue of complete extruded sec-

tions. *Sci. Tech. Memor.*, no. 10/50, March 24, 1950, pp. 4.

3894. FORSYTH, P. J. E. Some metallographic observations relating to the fatigue of metals. *Roy. Aircr. Estab. Rep.*, no. Met. 54, June 1950, pp. 12.

3895. FOSBERRY, R. A. C. *and* MANSION, H. D. Bending fatigue strength of gear teeth—a comparison of some typical gear steels. *Rep. Mot. Ind. Res. Ass.*, no. 1950/7, July 1950, pp. 13.

3896. FREUDENTHAL, A. M. Current theories of fatigue. *Cornell Aero. Lab. Inc., Project Squid Tech. Rep.*, no. 21, May 24, 1950, pp. 36–52.

3897. FUKUI, S. *and* SATO, S. On the fatigue of work hardened steel (2nd report). [J] *Rep. Inst. Sci., Tokyo*, vol. 4, no. 7/8, July–Aug. 1950, pp. 210–215.

3898. FULLER, F. B. Elevated temperature fatigue testing at Air Materiel Command. *Cornell Aero. Lab. Inc., Project Squid Tech. Rep.*, no. 21, May 24, 1950, pp. 53–67.

3899. GANDELOT, H. K. Fatigue and service testing. *S.A.E. Jl.*, vol. 58, no. 8, Aug. 1950, pp. 39–40.

3900. GARF, M. E. Dynamic arrangements of fatigue testing machines. [R] *Zav. Lab.*, vol. 16, no. 3, March 1950, pp. 331–338.

3901. GARF, M. E. Dynamic analysis of testing machines with mechanical load excitation. [R] *Zav. Lab.*, vol. 16, no. 6, June 1950, pp. 709–721.

3902. GARMAN, C. P. Vibration fatigue hit Boulder transmission lines. *Elect. West* vol. 104, no. 5, 1950, pp. 62–64.

3903. GARTSIDE, F. Effect of surface condition on the torsional fatigue properties of 0.63% carbon steel. *Coil Spring Journal*, no. 21, Dec. 1950, pp. 19–27.

3904. GASSNER, E. Preliminary results from fatigue tests with reference to operational statistics. *Tech. Memor. nat. Adv. Comm. Aero. Wash.*, no. 1266, May 1950, pp. 18.

3905. GATEWOOD, A. R. Some notes on propeller shaft failures. *Trans. Soc. nav. Archit., N.Y.*, vol. 58, 1950, pp. 753–787.

3906. GEORGE, C. W. Failure of light alloy nose wheel up lock jack on Hermes IV, GAL DB. *Roy. Aircr. Estab. Tech. Note*, no. Met. 132, Sept. 1950, pp. 10.

3907. GEROLD, E. *and* KARIUS, A. Abkürzungsverfahren zur Ermittlung der Wechselfestigkeit. [Rapid methods for the determination of fatigue strength.] *Arch. Eisenhüttenw.*, vol. 21, no. 5/6, May–June 1950, pp. 191–195.

3908. GEROLD, E. *and* TRACHTE, K. Das Verhalten von Stahl St37 im Gebiet der Zeitfestigkeit. [The behaviour of steel St37 in the region of the fatigue limit.] *Arch. Eisenhüttenw.*, vol. 21, no. 5/6, May–June 1950, pp. 175–179.

3909. GLAUBITZ, H. Biegefestigkeits- und Verschleisskennwerte für Zahnräder. [Bending strength and wear resistance of gear wheels.] *Arch. tech. Messen*, V. 8224-4, Nov. 1950, p. T126.

3910. GOUGH, H. J. Engineering steels under combined cyclic and static stresses. *J. appl. Mech.*, vol. 17, no. 2, June 1950, pp. 113–125 and [discussion] vol. 18, no. 2, June 1951, pp. 211–216. *J. Amer. Soc. nav. Engrs*, vol. 62, Aug. 1950, pp. 646–682.

3911. GOULD, A. J. *and* EVANS, U. R. Effect of shot-peening upon corrosion-fatigue of high carbon steel. *J. Iron St. Inst.*, vol. 165, pt. 3, July 1950, pp. 294–297.

3912. GRAF, O. Einfluss rostschützender Siluminüberzüge im Innern von Nietverbindungen aus St52 auf deren Dauerzugfestigkeit. [Effect of rust-preventive silumin coatings in the interior of riveted joints of St52 on their tensile fatigue strength.] *Z. Ver. dtsch. Ing.*, vol. 92, no. 26, Sept. 11, 1950, p. 747.

3913. GRANT, J. W. Notched and unnotched fatigue tests on flake and nodular cast irons. *J. Res. Brit. Cast Iron Ass.*, vol. 3, no. 5, April 1950, pp. 333–354. Abstr.: *Engrs' Dig.*, vol. 11, no. 6, June 1950, p. 198.

3914. GRANT, N. J. Some strain rate aspects of fatigue at high temperatures. *Cornell Aero. Lab. Inc., Project Squid Tech. Rep.*, no. 21, May 24, 1950, pp. 81–91.

3915. GRIGOR'EV, V. P. Machine for the fatigue testing of structural members. [R] *Zav. Lab.*, vol. 16, no. 7, July 1950, pp. 893–894.

3916. GROVER, H. J. Fatigue notch sensitivities of some aircraft materials. *Proc. Amer. Soc. Test. Mater.*, vol. 50, 1950, pp. 717–734.

3917. GUTFREUND, K. Verdrehdauerprüfmaschine für grosse Proben. [Torsional fatigue testing machine for large specimens.] *Industrie-Anzeiger, Essen*, vol. 72, no. 29, 1950, pp. 315–316.

3918. GUYOT, H. *and* SCHIMKAT, G. Les essais de fatigue à plusieurs étages. [Multiple stage fatigue tests.] *Rech. áero.*, no. 18, Nov.–Dec. 1950, pp. 3–9.

3919. HÄNCHEN, R. *Berechnung und Gestaltung der Maschinenteile auf Dauerhaltbarkeit. [Cal-*

culation and design of machine components for fatigue resistance.] Berlin–Hannover–Frankfurt(Main), Pädagog Verlag Berthold Schulz, 1950, pp. 232.

3920. HÄNCHEN, R. Grundlagen der Berechnung von Maschinenteilen auf Dauerhaltbarkeit. [Basis for the design of machine components for fatigue resistance.] *Konstruktion*, vol. 2, no. 1, 1950, pp. 7–13; no. 2, 1950, pp. 53–58 and no. 3, 1950, pp. 76–84.

3921. HANLEY, B. C. *and* DOLAN, T. J. Influence of surface finish on fatigue properties. *Uni. Ill. Dep. Theor. Appl. Mech. Tech. Rep.* no. 19 on *Behaviour of materials under repeated stress*, Sept. 1950, pp. 25.

3922. HEAD, A. K. Statistical properties of fatigue data on 24S-T aluminium alloy. *Note Aeronaut. Res. Labs Aust.*, no. SM. 180, Feb. 1950, pp. 9. *Bull. Amer. Soc. Test. Mater.*, no. 169, Oct. 1950, pp. 51–53.

3923. HENSHAW, R. C.; WALLERSTEIN, L. *and* ZAND, S. J. Fatigue life of aircraft engine mounting components. *S.A.E. Jl.*, vol. 58, no. 4, April 1950, pp. 42–45. Abstr.: *Automot. Industr. N.Y.*, vol. 102, no. 3, Feb. 1, 1950, pp. 52, 80.

3924. HERZOG, A. Elevated temperature fatigue testing of turbine buckets. Part I—calculations of natural frequencies and stresses, and proposed testing methods. *Tech. Rep. U.S. Air Force*, no. 5936, part I, May 1950, pp. 56.

3925. HEYES, J. *and* FISCHER, W. A. Über das elektrolytische Polieren von Stahl. [On the electropolishing of steel.] *Metalloberfläche*, vol. 4, no. 3, March 1950, pp. A38–A44.

3926. HOLT, M. Results of shear fatigue tests of joints with 3/16 inch diameter 24S–T31 rivets in 0.064 inch thick alclad sheet. *Tech. Note nat. Adv. Comm. Aero., Wash.*, no. 2012, Feb. 1950, pp. 45.

3927. VAN ITERSON, F. K. TH. De theoretische vermoeiingskrome. [The theoretical fatigue curve.] *Ingenieur, 's Grav.*, vol. 62, no. 24, June 16, 1950, pp. 55–59.

3928. IVANOV YU. M. Rupture of timber by repeated stresses above the elastic limit. [R Dokl. Akad. Nauk SSSR*, vol. 73, no. 5, Aug. 11, 1950, pp. 905–908.

3929. JACKSON, J. S. The effect of acid pickling on the endurance of hardened and tempered springs. *Coil Spring Journal*, no. 18, March 1950, pp. 5–7.

3930. JACQUESON, R. *and* LAURENT, P. Une nouvelle machine d'essais de fatigue des metaux a frequence elevee. [A new machine for fatigue testing metals at high frequency.] *Rev gén. Mécan.*, vol. 34, no. 13, Jan. 1950, pp. 31–36. English abstr.: *Engrs' Dig.*, vol. 11 no. 5, May 1950, p. 183.

3931. JACQUET, P. A. Electrolytic polishing c metallic surfaces—Part VI. *Metal Finish.* vol. 48, no. 1, Jan. 1950, pp. 56–62.

3932. JÄNICHE, W. Untersuchungen über Oberbaustoffe und Aufgaben auf diesem Gebiete. [Investigations on materials for rail construction and problems in this field.] *Stahl Eisen*, vol. 70, no. 5, March 2, 1950, pp. 174–186.

3933. JENSEN, R. S. Eighth progress report of the rolling load tests of joint bars. *Bull. Ame Rly Engng Ass.*, vol. 51, no. 486, Feb. 1950, pp. 585–593. *Repr. Ill. Engng Exp. Sta.*, no. 4 June 1950, pp. 10–18.

3934. JENSEN, R. S. Fatigue tests of rail webs. *Bull. Amer. Rly Engng Ass.*, vol. 51, no. 486, Feb. 1950, pp. 640–647. *Repr. Ill. Engng Exp. Sta.*, no. 47, June 1950, pp. 30–37.

3935. JOHNSTONE, W. W.; PATCHING, C. A. *and* PAYNE, A. O. An experimental determination of the fatigue strength of CA-12 "Boomerang" wings. *Rep. Aeronaut. Res. Labs Aust.*, no. SM.160, Sept. 1950, pp. 33.

3936. JONES, W. E. *and* WILKES, G. B. Effect of notches and shot peening on high temperature fatigue properties. *Cornell Aero. Lab. Inc., Project Squid Tech. Rep.*, no. 21, May 24, 1950, pp. 92–108.

3937. JONES, W. E. *and* WILKES, G. B. The effect of various treatments on the fatigue strength of notched S-816 and Timken 16–25–6 alloys at elevated temperatures. *Proc. Amer. Soc. Test. Mater.*, vol. 50, 1950, pp. 744–762.

3938. JOUKOFF, A. S. Recherches expérimentales sur les charpentes à points soudés. [Experimental research on spot welded trusses.] *Rev. Soud.*, vol. 6, no. 3, 1950, pp. 136–143. English abstr.: *Weld. J., Easton, Pa.*, vol. 30, no. 5, May 1951, pp. 264s–265s.

3939. KARPENKO, G. V. On the effect of surface active substances on the fatigue limit of metals. [R] *Zav. Lab.*, vol. 16, no. 8, Aug. 1950, pp. 984–985.

3940. KARPENKO, G. V. The influence of surface-active substances on fatigue strength of steel. [R] *Dokl. Akad. Nauk SSSR*, vol. 73, no. 6 Aug. 21, 1950, pp. 1225–1228.

3941. KARPENKO, G. V. On the problem of the formation of fatigue microcracks. [R] *Dokl. Akad. Nauk SSSR*, vol. 74, no. 1, Sept. 1, 1950, pp. 95–98.

3942. KEEL, C. G. Über die Entwicklung neuer Zusatzstäbe für die Autogenschweissung von Stahl. [On the development of new composition rods for the gas welding of steel.] *Ber. eidgenöss. MatPrüfAnst.*, no. 175, Feb. 1950, pp. 50.

3943. KENYON, J. N. The reverting of hard drawn copper to soft condition under variable stress. *Proc. Amer. Soc. Test. Mater.*, vol. 50, 1950, pp. 1073–1084.

3944. KIMMEL'MAN, D. N. *Stressing of machine components under cyclic stresses.* [R] Moscow, Gos. nauchn-tekhn. izd-vo mashinostroitel'noi lit-ry, 1950, pp. 128.

3945. KOBRIN, M. M. Fatigue fracture of machine components operating under static conditions. [R] *Vestn. mashinostr.*, vol. 30, no. 3, March 1950, pp. 18–20.

3946. KOOISTRA, L. F. *and* BLASER, R. V. Experimental technique in pressure-vessel testing. *Trans. Amer. Soc. mech. Engrs*, vol. 72, July 1950, pp. 579–589.

3947. KOSTING, P. R. Initiation and growth of cracks in gun sections of one size during hydraulic fatigue tests at one pressure. *U.S. Watertown Arsenal Lab., Memo. Rep.* no. 731/325, May 24, 1950, pp. 10.

3948. LANGEVIN, A.; PAUL, E. *and* REIMBERT, M. Prédetermination électromagnétique de la limite probable de fatigue. [Electromagnetic pre-determination of the probable fatigue limit.] *C.R. Acad. Sci., Paris*, vol. 230, no. 12, March 20, 1950, pp. 1138–1140.

3949. LAZAN, B. J. Dynamic creep and rupture properties of temperature resistant materials under tensile fatigue stress. *Tech. Rep. U.S. Air Force*, no. 5930, Feb. 1950, pp. 39.

3950. LAZAN, B. J. Dynamic creep, fatigue, damping and elasticity of temperature resistant materials. *Cornell Aero. Lab. Inc., Project Squid Tech. Rep.*, no. 21, May 24, 1950, pp. 131–162.

3951. LAZAN, B. J. A study with new equipment of the effects of fatigue stress on the damping capacity and elasticity of mild steel. *Trans. Amer. Soc. Metals*, vol. 42, 1950, pp. 499–558.

3952. LESSELLS, L. M. *and* JACQUES, H. E. Effect of fatigue on transition temperature of steel. *Weld. J., Easton, Pa.*, vol. 29, no. 2, Feb. 1950, pp. 74s–83s.

3953. LIGIER, A.-G. Le 'shot peening' et son application au traitment des ressorts en acier [Shot peening and its application to the treatment of steel springs.] *Rev. gén. Mécan.*, vol. 34, no. 22, Oct. 1950, pp. 365–369.

3954. LIHL, F. Eine neue Theorie der Korrosionsermüdung. [A new theory of corrosion-fatigue.] *Berg- u. hüttenm. Mh.*, vol. 95, no. 2, Feb. 1950, pp. 25–34.

3955. LIHL, F. Die Korrosionsermüdung als intrakristalliner Korrosioneffekt. [Corrosion-fatigue as an intra crystalline corrosion effect.] *Metall: Wirtsch. Wiss. Tech.*, vol. 4, no. 7/8, April 1950, pp. 130–132.

3956. LIPPERT, K. Vibration fatigue testing of TV antennas. *Telev. Engng*, vol. 1, no. 8, Aug. 1950, pp. 11, 31.

3957. LIPSON, C. Why machine parts fail. *Mach. Design*, vol. 22, no. 5, May 1950, pp. 95–100; no. 6, June 1950, pp. 111–116; no. 7, July 1950, pp. 141–145; no. 8, Aug. 1950, pp. 157–160; no. 9, Sept. 1950, pp. 147–150;

no. 10, Oct. 1950, pp. 97–100; no. 11, Nov. 1950, pp. 158–162 and no. 12, Dec. 1950 pp. 151–156.

3958. LIPSON, C. Materials data that are significant to the designer. *Automot. Industr. N.Y.* vol. 103, no. 8, Oct. 15, 1950, pp. 38–41, 94 96.

3959. LIPSON, C.; NOLL, G. C. *and* CLOCK, L. S *Stress and strength of manufactured parts* New York, McGraw-Hill Book Co. Inc. 1950, pp. 259.

3960. LITTLE, J. C.; MACMILLAN, D. G. an MAJERCAK, J. V. Vibration and fatigue lif of steel strand. *Trans. Amer. Inst. elect Engrs*, vol. 69, pt. 2, Dec. 1950, pp. 1473-1479.

3961. LOCATI, L. La fatica dei materiali metallici [The fatigue of metallic materials.] *Costruz metall.*, vol. 2, no. 5, 1950, pp. 35–38.

3962. LOCATI, L. Forma e resistenza alla fatic degli organi meccanici. [The influence o shape on the fatigue resistance of mechanica components.] *Riv. Mecc.*, vol. 1, no. 2 Sept. 15, 1950, pp. 11–14.

3963. LOCATI, L. Fatica durata e sovraccaric degli organi nelle costruzioni automobilis tiche. [Fatigue endurance under overloa of automobile components.] *ATA*, vol. 3 no. 12, Dec. 1950, pp. 41–45.

3964. LOCATI, L. *La fatica dei materiali metallic [The fatigue of metallic materials.]* Milar V. Hoepli, 1950, pp. 333.

3965. LOGAN, H. L. Effect of chromium platin on the endurance limit of steels used in ai craft. *Proc. Amer. Soc. Test. Mater.*, vol. 50, 1950, pp. 699–716.

3966. Love, R. J. Fatigue strength of cast crankshafts: First report on the influence of the crankshaft material on bending fatigue strength. *Rep. Mot. Ind. Res. Ass.*, no. 1950/2, June 1950, pp. 15.

3967. Love, R. J. The fatigue strength of steels—with particular reference to the influence of material condition and surface treatment. *Rep. Mot. Ind. Res. Ass.*, no. 1950/9, Oct. 1950, pp. 47.

3968. McBrian, R. Problem of decarburization in railroad materials. *Metal Progr.*, vol. 58, no. 1, July 1950, pp. 51–54.

3969. McKeown, J. The British Non-Ferrous Metals Research Association: creep and fatigue testing equipment. *Metallurgia, Manchr*, vol. 42, no. 251, Sept. 1950, pp. 189–196.

3970. McKeown, J.; Dineen, D. E. *and* Back, L. H. Fatigue properties of four cast aluminium alloys at elevated temperatures. *Metallurgia, Manchr*, vol. 41, no. 247, May 1950, pp. 393–396.

3971. McKeown, J. *and* Hopkin, L. M. T. Creep and fatigue tests on commercially extruded lead and lead alloy pipes. *Metallurgia, Manchr*, vol. 41, no. 243, Jan. 1950, pp. 135–143 and no. 244, Feb. 1950, pp. 219–223.

3972. McMurrich, R. New ideas in mechanical testing—(2) A note on high speed fatigue testing. *Aust. Engr*, July 7, 1950, pp. 49–51.

3973. McMurrich, R. P. Some effects of shot peening on fatigue resistance of a medium carbon steel. *Rep. Defence Res. Labs Aust.*, no. 178, Nov. 1950, pp. 29.

3974. Majors, H.; Mills, B. D. *and* Macgregor, C. W. Fatigue under combined pulsating stresses. *J. appl. Mech.*, vol. 17, no. 2, June 1950, pp. 218–221.

3975. Malcolm, V. T. *and* Low, S. Nitriding—effect on fatigue strength of stainless steel. *J. Metals, N.Y.*, vol. 2, no. 8, Sept. 1950, pp. 1094–1095.

3976. Malinin, N. N. Influence of residual stress caused by pre-loading on the fatigue strength of helical compression springs. [R] *Dinamika i prochnost' pruzhin [Dynamics and strength of springs]*, Moscow, Izd-vo Akademii nauk SSSR, 1950, pp. 214–237.

3977. Mann, J. Y. The effect of surface finish on the fatigue resistance of 24S-T aluminium alloy. *Rep. Aeronaut. Res. Labs Aust.*, no. SM. 147, June 1950, pp. 58.

3978. Markl, A. R. C. *and* George, H. H. Fatigue tests on flanged assemblies. *Trans. Amer. Soc. mech. Engrs*, vol. 72, no. 1, Jan. 1950, pp. 77–87.

3979. Minamiozi, K. *and* Okubo, H. A note on the notch effect of metals. *J. Franklin Inst.*, vol. 249, no. 1, Jan. 1950, pp. 49–55.

3980. Mitsuhashi, T. *and* Tsuya, K. On the fatigue strength and damping capacity of time piece springs. [J] *J. mech. Lab., Tokyo*, vol. 4, no. 5, Sept. 1950, pp. 180–185.

3981. Moore, H. F.; Gohn, G. R.; Howell, F. M. *and* Wilson, B. L. Report of the Task Group on effect of speed of testing on fatigue test results. *Proc. Amer. Soc. Test. Mater.*, vol. 50, 1950, pp. 421–424.

3982. Moore, S. C. Causes and prevention of drill collar failures. *Drilling*, vol. 12, Nov. 1950, pp. 28, 30, 85.

3983. MOWBRAY, A. Q. Effect of superposition of stress raisers on members subjected to static or repeated loads. *Uni. Ill. Dep. Theor. Appl. Mech. Tech. Rep.*, no. 18 on *Behaviour of materials under repeated stress,* Aug. 1950, pp. 28.

3984. NEWMARK, N. M. A review of cumulative damage in fatigue. *Office Naval Research, Contract N6-ori-71, Task Order V, Project NR-031-182, Tech. Rep., Uni. Illinois,* July 15, 1950, pp. 38.

3985. NIERHAUS, F. W. Verbesserung der Herstellung von Kurbelwellen für Grosskolbenmaschinen. [Improving the manufacture of crankshafts for large piston engines.] *Stahl u. Eisen,* vol. 70, no. 9, April 27, 1950, pp. 372–375.

3986. NISHIHARA, T. *and* YAMADA, T. The fatigue strength of metallic materials under alternating stresses of varying amplitude. *Japan Sci. Rev.,* vol. 1, no. 3, Sept. 1950, pp. 1–6.

3987. NOVIK, A. A. On the evaluation of fatigue strength. [R] *Zav. Lab.,* vol. 16, no. 3, March 1950, pp. 352–355.

3988. OAKS, J. K. Fatigue test on Meteor tailplane. *Roy. Aircr. Estab. Rep.,* no. Structures 59, Jan. 1950, pp. 10.

3989. OICLES, C. W. *and* LANDECKER, F. K. Multiply spring life without changing design. *Iron Age,* vol. 166, no. 25, Dec. 21, 1950, pp. 80–82.

3990. OKUBO, H. On the endurance of a round bar with longitudinal grooves. *J. appl. Phys.,* vol. 21, no. 11, Nov. 1950, pp. 1105–1108.

3991. ORLOV, D. P. *and* LOGGINOV, G. I. The endurance of crystals of gypsum and of mica

in periodic torsion. [R] *Dokl. Akad. Nauk SSSR,* vol. 70, no. 2, 1950, pp. 249–251.

3992. PARDUE, T. E.; MELCHOR, J. L. *and* GOOD, W. B. Energy losses and fracture of some metals resulting from a small number of cycles of strain. *Proc. Soc. exp. Stress Anal.,* vol. 7, no. 2, 1950, pp. 27–29.

3993. PEARSON, B. M. Fatigue properties of helical springs for I. C. engine valve springs and other duties. *Wire Ind.,* vol. 17, no. 204, Dec. 1950, pp. 987–989.

3994. PETERSON, R. E. Discussions of a century ago concerning the nature of fatigue and review of some of the subsequent researches concerning the mechanism of fatigue. *Bull. Amer. Soc. Test. Mater.,* no. 164, Feb. 1950, pp. 50–56.

3995. PETERSON, R. E. Nature of fatigue of metals—discussions of a century ago and review of some of the subsequent researches. *Mech. Engng, N.Y.,* vol. 72, no. 5, May 1950, pp. 371–375.

3996. PIPER, T. E.; FINLAY, K. F. *and* BINSACCA, A. P. Fatigue characteristics of aircraft materials and fastenings. *Bull. Amer. Soc. Test. Mater.,* no. 166, May 1950, pp. 60–64.

3997. POMP, A. *and* HEMPEL, M. Wechselfestigkeit und Kerbwirkung von unlegierten und legierten Baustählen bei +20° und −78°C [Fatigue strength and notch factors for unalloyed and alloyed constructional steels at +20° and −78°C.] *Arch. Eisenhüttenw.* vol. 21, no. 1/2, Jan.–Feb. 1950, pp. 53–66 English abstr.: *Engineer, Lond.,* vol. 190 no. 4938, Sept. 15, 1950, p. 266.

3998. POMP, A. *and* HEMPEL, M. Kerbschlag zähigkeit und Zeit- und Dauerfestigkeit zug schwellbeanspruchter Voll- und Kerbstäbe verschiedener Stähle. [Notch impact toughness, endurance and fatigue strength under pulsating tensile stress of various unnotched and notched steels.] *Arch. Eisenhüttenw.*, vol. 21, no. 1/2, Jan.–Feb. 1950, pp. 67–76.

3999. POMP, A. *and* HEMPEL, M. Dauerfestigkeit von Schraubenfedern unterschiedlicher Fertigungsart. [Fatigue strength of helical springs manufactured by different processes.] *Arch. Eisenhüttenw.*, vol. 21, no. 7/8, July–Aug. 1950, pp. 243–262. English abstr.: *Engineer, Lond.*, vol. 190, no. 4948, Nov. 24, 1950, pp. 496–497.

4000. POMP, A. *and* HEMPEL, M. Dauerfestigkeit von Schraubenfedern bei erhöhter Temperatur. [Fatigue strength of helical springs at elevated temperatures.] *Arch. Eisenhüttenw.*, vol. 21, no. 7/8, July–Aug. 1950, pp. 263–272.

4001. PRINGLE, C. N. S. The fatigue of simple riveted joints. 1.—Single rivet lap joints. *Roy. Aircr. Estab. Tech. Note*, no. Structures 62, Aug. 1950, pp. 8.

4002. PRINGLE, C. N. S. *and* BINNING, M. S. Fatigue of specimens from 'Z' section, D.T.D. 364 extrusions. *Roy. Aircr. Estab. Rep.*, no. Structures 88, Aug. 1950, pp. 11.

4003. PROMISEL, N. E. High temperature fatigue program of the Bureau of Aeronautics. *Cornell Aero. Lab. Inc., Project Squid Tech. Rep.*, no. 21, May 24, 1950, pp. 68–80.

4004. RATNER, S. I. *and* DANILOV, I. S. Change in the limits of proportionality and yield points on repeated loading. [R] *Zav. Lab.*, vol. 16, no. 4, April, 1950, pp. 468–475. English abstr.: *J. Iron St. Inst.*, vol. 174, pt. 2, July 1953, p. 291.

4005. REININGER, H. Die Dauerwechselfestigkeit poröser Silumin-Gamma-Sandgussteile. [The fatigue strength of porous silumin-gamma sand castings.] *Z. Metallk.*, vol. 41, no. 10, Oct. 1950, pp. 348–357.

4006. ROŠ, M. Mit Tor-Stahl "40" von 10, 20 und 30 mm Durchmesser bewehrte, durch statischen Bruch und Ermüdung erschöpfte Biegebalken Stahlbewehrung ohne Endhaken. [Static and fatigue tests on beams reinforced with 10, 20 and 30 mm Tor-Stahl "40" reinforcement without end hooks.] *Ber. eidgenöss. MatPrüfAnst.*, no. 176, Jan. 1950, pp. 46.

4007. ROŠ, M. La fatigue des métaux. [The fatigue of metals.] *Metallurg. ital.*, vol. 42, no. 1, Jan. 1950, pp. 7–21.

4008. ROŠ, M. Ermüdungsversuche mit Hohlstäben aus reinem Schweissgut, "Arcos-Stabilend-B" und "Arcos-Ductilend-55" bei mehrachsigen Spannungszuständen. [Fatigue tests on hollow bars of pure welding material, "Arcos-Stabilend-B" and "Arcos-Ductilend 55" under multiaxial stress conditions.] *Schweiz. Arch. angew. Wiss.*, vol. 16, no. 7, July 1950, pp. 193–199.

4009. ROŠ, M. Experiments for the determination of the influence of residual stresses on the fatigue strength of structures. *Trans. Inst. Weld.*, vol. 13, no. 5, Oct. 1950, pp. 83r–93r.

4010. ROŠ, M. *and* EICHINGER, A. Der Bruchgefahr fester Körper bei wiederholter Beanspruchung — Ermüdung. [The danger of failure of solid parts under repeated stress—fatigue.] *Ber. eidgenöss. MatPrüfAnst.*, no. 173, Sept. 1950, pp. 161.

4011. ROSENTHAL, D. *and* SINES, G. Residual stress and fatigue strength. *Metal Progr.*, vol. 58, no. 1, July 1950, pp. 76.

4012. RYABCHENKOV, A. V. The protection of steel from corrosion-fatigue failure. The endurance of metal subject to variable stress in a corrosive medium. [R] *Trudy tsent. nauchno-issled. Inst. tekhnol. Mashinost.*, vol. 31, 1950, pp. 5–25.

4013. SAUER, J. A. *and* LEMMON, D. C. Effect of steady stress on fatigue behaviour of aluminium. *Trans. Amer. Soc. Metals*, vol. 42, 1950, pp. 559–576. Abstr.: *Automot. Industr. N.Y.*, vol. 101, no. 10, Nov. 15, 1949, pp. 70, 72.

4014. SAUER, J. A.; LEMMON, D. C. *and* LYNN, E. K. Bolts—how to prevent their loosening. *Mach. Design*, vol. 22, no. 8, Aug. 1950, pp. 133–139.

4015. SCHAAL, A. Einflussfaktoren bei der Bestimmung der Schwingungsfestigkeit aus der statischen Fleissgrenze. [Factors which influence the determination of the fatigue strength from the static yield strength.] *Z. Metallk.*, vol. 41, no. 10, Oct. 1950, pp. 334–339.

4016. SCHWARTZ, A. New thread form reduces bolt breakage. *Steel*, vol. 127, no. 10, Sept. 4, 1950, pp. 86–87, 94.

4017. SCORTECCI, A. Über die Schweissung von Ermüdungsrissen in der Hauptwelle einer Umkehrwalzenzugmaschine von 10000PS. [On the welding of fatigue cracks in the main drive shaft of a reversing rolling mill.] *Berg-u. hüttenm. Mh.*, vol. 95, Dec. 1950, pp. 326–327.

4018. SIGOV, I. V. The starting point of fatigue failure in bending. [R] *Zav. Lab.*, vol. 16, no. 4, April 1950, pp. 479–482.

4019. SIMMONS, W. F. *and* CROSS, H. C. Some high temperature tensile fatigue data on gas turbine alloys. *Cornell Aero. Lab. Inc., Project Squid Tech. Rep.*, no. 21, May 24, 1950, pp. 163–174.

4020. SINCLAIR, G. M. *and* DOLAN, T. J. Some effects of austenitic grain size and metallurgical structure on the mechanical properties of steel. *Proc. Amer. Soc. Test. Mater.*, vol. 50, 1950, pp. 587–618.

4021. SMITH, E. W. P. Effect of residual stresses on fatigue of compressor valves. *Metal Progr.*, vol. 57, no. 4, April 1950, pp. 480–481.

4022. SMITH, F. C.; BRUGGEMAN, W. C. *and* HARWELL, R. H. Comparison of fatigue strengths of bare and alclad 24S-T3 aluminium alloy sheet specimens tested at 12 and 1000 cycles per minute. *Tech. Note nat. Adv. Comm. Aero., Wash.*, no. 2231, Dec. 1950, pp. 18.

4023. SMITH, L. W. High temperature fatigue program at Cornell Aeronautical Laboratory, Inc. *Cornell Aero. Lab. Inc., Project Squid Tech. Rep.*, no. 21, May 24, 1950, pp. 109–130.

4024. SMOLENSKII, S. I.; POLEZKAEV, A. A. *and* KARNEEV, L. I. Electromechanical machine for the fatigue testing of specimens in bending. [R] *Zav. Lab.*, vol. 16, no. 10, Oct. 1950, pp. 1272–1275.

4025. SOETE, W. *and* VAN CROMBRUGGE, R. Etude de la résistance à la fatigue des assemblages soudés. [Study of the fatigue resistance of welded joints.] *Rev. Soud.*, vol. 6, no. 2, 1950, pp. 72–82; no. 4, 1950, pp. 199–212 and vol. 7, no. 2, 1951, pp. 90–97.

4026. SOPWITH, D. G. The resistance of aluminium and beryllium bronzes to fatigue and corrosion-fatigue. *Rep. Memor. aero. Res. Comm., Lond.*, no. 2486, 1950, pp. 10.

4027. Späth, W. Die Auswertung von Dauerversuchen in der Werkstoffprüfung. [The value of fatigue tests in materials testing.] *Metall: Wirtsch. Wiss. Tech.*, vol. 4, no. 17/18, Sept. 1950, pp. 374–377. English abstr.: *Prod. Engng*, vol. 22, no. 7, July 1951, pp. 232, 234, 236.

4028. Spretnak, J. W. *and* Fontana, M. G. Investigation of mechanical properties and physical metallurgy of aircraft alloys at very low temperatures. Part 3—effect of notches on static tensile and fatigue strength at −320°F (−196°C). *Tech. Rep. U.S. Air Force*, no. 5662, April 1950, pp. 54.

4029. Stickley, G. W. *and* Howell, F. M. Effects of anodic coatings on the fatigue strength of aluminium alloys. *Proc. Amer. Soc. Test. Mater.*, vol. 50, 1950, pp. 735–743.

4030. Stirling, J. F. The science of steel. XVIII—steel in fatigue. *Pract. Engng*, vol. 22, no. 546, July 21, 1950, pp. 38–40.

4031. Stormont, D. H. Induflux: new method for testing fatigue in sucker rods. *Oil Gas J.*, vol. 48, no. 45, March 16, 1950, pp. 84, 86.

4032. Tapsell, H. J., Forrest, P. G. *and* Tremain, G. R. Creep due to fluctuating stresses at elevated temperatures. *Engineering, Lond.*, vol. 170, no. 4413, Aug. 25, 1950, pp. 188–191.

4033. Tarasov, L. P. *and* Grover, H. J. Effects of grinding and other finishing processes on the fatigue strength of hardened steel. *Proc. Amer. Soc. Test. Mater.*, vol. 50, 1950, pp. 668–698.

4034. Taylor, W. J. *and* Gunn, N. J. F. The effect of notches on the fatigue strength of three light alloys. *Roy. Aircr. Estab. Rep.*, no. Met. 42, Aug. 1950, pp. 41.

4035. Taylor, W. J. *and* Whillans, M. E. Endurance tests of flexible steel wire rope (non-corrodible). *Roy. Aircr. Estab. Tech. Note*, no. Met. 127, June 1950, pp. 10.

4036. Templin, R. L.; Howell, F. M. *and* Hartmann, E. C. Effect of grain direction on fatigue properties of aluminium alloys. *Prod. Engng*, vol. 21, no. 7, July 1950, pp. 126–130.

4037. Thorpe, P. L.; Tremain, G. R. *and* Ridley, R. W. The mechanical properties of some wrought and cast aluminium alloys at elevated temperatures. *J. Inst. Met.*, vol. 77, pt. 2, April 1950, pp. 111–140.

4038. Thum, A. *and* Svenson, O. Mehrfache Kerbwirkung. Enlastungskerben. Überlastungskerben. [Multiple notching effect. Stress relieving notches. Overloading notches.] *Z. Ver. dtsch. Ing.*, vol. 92, no. 10, April 1, 1950, pp. 225–230. English translation: *Aluminium Laboratories Ltd., Banbury, Rep.*, no. B-TM-141-54.

4039. Thurston, R. C. The fatigue of metals. *Canad. Metals*, vol. 13, no. 6, June 1950, pp. 8–11, 48–49.

4040. Trapp, W. J. Mechanical properties of extruded magnesium alloys: ZK60 and AZ80. *Tech. Rep. U.S. Air Force*, no. 5926, Nov. 1950, pp. 22.

4041. Trubin, G. K. Contact fatigue of the teeth of spur gears. [R] *Sbornik tsent. nauchno-issled. Inst. tekhnol. Mashinost.*, no. 37, 1950, pp. 1–150.

4042. Unckel, H. A. *and* Nyander, H. H. The effect of homogenising on the fatigue properties of an extruded Al–Cu–Mg alloy. *Metallurgia, Manchr*, vol. 41, no. 243, Jan. 1950, pp. 155–156.

4043. VOLKOV, S. D. and SOKOLOV, P. S. Machine for fatigue testing of metals in torsion. [R] *Zav. Lab.*, vol. 16, no. 7, July 1950, pp. 891–892.

4044. WALKER, P. B. Fatigue of aircraft structures. *Roy. Aircr. Estab. Rep.*, no. Structures 79, June 1950, pp. 22.

4045. WECK, R. The application of the resonance vibration method to the fatigue testing of spot welded light alloy structures. *Trans. Inst. Weld.*, vol. 13, no. 2, April 1950, pp. 33r–38r.

4046. WEIBULL, I. and DAVIDSON, P. Några iakttagelser beträffande utmattningshållfastheten i värme hos svetsat 18–8-stål. [Concerning the fatigue strength of welded 18–8 steel at high temperatures.] *Jernkontor. Ann.*, vol. 134, no. 12, Dec. 1950, pp. 559–571.

4047. WEIBULL, W. Statistika synpunkter pa utmattningshållfastheten. [Statistical viewpoints on fatigue strength.] *Tekn. Tidskr., Stockh.*, vol. 80, no. 42, Nov. 18, 1950, pp. 1059–1064. English translation: *Engrs' Dig.*, vol. 12, no. 2, Feb. 1951, pp. 57–60.

4048. WEISMAN, M. H. and KAPLAN, M. H. The fatigue strength of steel through the range from ½ to 30,000 cycles of stress. *Proc. Amer. Soc. Test. Mater.*, vol. 50, 1950, pp. 649–667.

4049. WELLINGER, K.; KEIL, E. and STÄHLI, G. Über die Temperaturabhängigkeit von Wechselfestigkeit und Dauerstandfestigkeit verschiedener Press- und Guss-legierungen aus Leichtmetall. [Influence of temperature on the fatigue strength and creep strength of different light metal forging and casting alloys.] *Z. Metallk.*, vol. 41, no. 9, Sept. 1950, pp. 309–313.

4050. WELTER, G. Stresses around a spot weld under static and cyclic loads. *Weld. J., Easton, Pa.*, vol. 29, no. 11, Nov. 1950, pp. 565s–576s.

4051. WERREN, F. Fatigue of sandwich constructions for aircraft. Glass fabric laminate facing and waffle type core sandwich material tested in shear. *Rep. For. Prod. Lab., Madison*, no. 1559–I, Oct. 1950, pp. 5.

4052. WHITWHAM, D. and EVANS, U. R. Corrosion fatigue—the influence of disarrayed metal. *J. Iron St., Inst.*, vol. 165, pt. 1, May 1950, pp. 72–79.

4053. WILLIAMS, D. Strength of aeroplanes in relation to repeated loads. *Aeronaut. Quart.*, vol. 1, pt. 4, Feb. 1950, pp. 291–304.

4054. WILLIAMS, W. L. The titanium program at U.S. Naval Experiment Station. *J. Amer. Soc. nav. Engrs*, vol. 62, no. 4, Nov. 1950, pp. 855–869.

4055. WILLS, H. A. Structural fatigue. *Aircraft, Melbourne*, vol. 28, no. 8, May 1950, pp. 12–14, 46, 48.

4056. WILSON, W. M. Fatigue of structural joints. *Weld. J., Easton, Pa.*, vol. 29, no. 3, March 1950, pp. 204–210.

4057. WILSON, W. M.; MUNSE, W. H. and SNYDER, I. S. Fatigue strength of various types of butt welds connecting steel plates. *Uni. Ill. Engng Exp. Sta. Bull.*, no. 384, March 1950 pp. 60.

4058. WOMELDORPH, R. C.; PLANKENHORN W. J. and BENNETT, D. G. The fatigue strength and fatigue life of uncoated and coated specimens of 18–8 stainless steel type 347. *Tech Rep. U.S. Air Force*, no. 6092, Oct. 1950 pp. 33.

4059. WOOD, W. A. and HEAD, A. K. Some new observations on the mechanism of fatigue c metals. *Rep. Aeronaut. Res. Labs Aust.*, no. SM.158, July 1950, pp. 14.

4060. WYLY, L. T. *and* CARTER, J. W. Stresses near bolt and rivet holes. *Fasteners*, vol. 6, no. 4, 1950, pp. 15–17.

4061. YEARIAN, H. J. *and* PORTER, P. K. Record of conference on fatigue of metals at high temperatures. *Cornell Aero. Lab. Inc.*, *Project Squid Tech. Rep.*, no. 21, May 24, 1950, pp. 178.

4062. YEN, C. S. An hypothesis for explanation of the effects of size of specimen and of stress raisers on the value of the endurance limit of metals. *Uni. Ill. Dep. Theor. Appl. Mech. Tech. Rep.* no. 16 *on Behaviour of materials under repeated stress*, Feb. 1950, pp. 43.

4063. Corrosion fatigue and zinc coatings. *Zinc Development Association, England, Tech. Memo.*, no. 13, Jan. 1950, pp. 8.

4064. Measurement of stresses in 132RE rail on tangent track—Santa Fe railway. *Bull. Amer. Rly Engng Ass.*, vol. 51, no. 486, Feb. 1950, pp. 626–640.

4065. Hydraulic cylinder varies spring load in fatigue tester. *Prod. Engng*, vol. 21, no. 2, Feb. 1950, pp. 102–103.

4066. Inductance unit regulates fatigue loads. *Prod. Engng*, vol. 21, no. 3, March 1950, pp. 100–101.

4067. Chromium plate affects fatigue limit in steels. *Prod. Engng*, vol. 21, no. 5, May 1950, p. 89.

4068. Onderzoek naar de wisselsterkte tegen buiging van vier soorten geplateerd lichtmetaalplatt.[Investigation of the alternating bend-ing strength of four types of clad light alloy sheet.] *Nat. Luchtvaartlab., Amsterdam Rep.*, no. M.1612, July 1950, pp. 8.

4069. Fatigue strength of welded butt joints. *Weld. J., Easton, Pa.*, vol. 29, no. 8, Aug. 1950, pp. 404s–408s.

4070. Fatigue tests of steels for hollow propeller blades. *Roy. Aircr. Estab. Tech. Note*, no. Met. 131, Aug. 1950, pp. 13.

4071. Impact value and fatigue strength. *Engineer, Lond.*, vol. 190, no. 4935, Aug. 25, 1950, p. 207.

4072. Vermoeiingsproeven op enkelvoudige lapnaden van 24S-T alclad gelijmd met Redux en 1 serie geklonken proefstukken uit 24S-T alclad plaat en 24S nagels. [Fatigue test on single lap joints of 24S-T alclad glued with Redux and one series of riveted test pieces from 24S-T alclad plates with 24S rivets.] *Nat. Luchtvaartlab., Amsterdam Rep.*, no. M.1627, Sept. 1950, pp. 23.

4073. Fatigue and vibration testing equipment. *Engineering, Lond.*, vol. 170, no. 4414, Sept. 1, 1950, pp. 203–204.

4074. Fatigue strength of steel at low temperatures. *Engineer, Lond.*, vol. 190, no. 4938, Sept. 15, 1950, p. 266.

4075. Pulsating pressure plant. *Chem. Age, delete, Lond.*, vol. 63, no. 1629, Sept. 30, 1950, pp. 467–468. *Mech. World*, vol. 128, no. 3329, Nov. 3, 1950, pp. 430–432.

4076. Fatigue tests of beams in flexure: advance report of Committee 15—iron and steel structures. *Bull. Amer. Rly Engng Ass.*, vol. 52, no. 489, Sept.–Oct. 1950, pp. 111–129.

4077. Oriënterende proeven ter bepaling van: 1. invloed van voorafgaande dynamische belasting van lichtmetaal op de mechanische eigenschappen bij statische belasting. 2. verband tussen de vermoeiingssterkte van lichtmetaal bij constante en variabele gemiddelde spanning ("accumulated damage"). [Preliminary investigation regarding: 1. The influence of preceding dynamic loading of light alloys on the mechanical behaviour under static loading. 2. Accumulated damage in rotating bending and in fluctuating tension.] *Nat. Luchtvaartlab., Amsterdam Rep.*, no. M.1654, Oct. 1950, pp. 6.

4078. Pulsating pressure test plant of the British Welding Research Association. *Engineer, Lond.*, vol. 190, no. 4942, Oct. 13, 1950, pp. 370–371.

4079. Increasing fatigue strength of stainless steels. *Engrs' Dig.*, vol. 11, no. 11, Nov. 1950, pp. 374–375.

4080. Pulsating-pressure testing plant for pressure vessels. *Engineering, Lond.*, vol. 170, no. 4425, Nov. 17, 1950, pp. 384–385.

Author index

(Numbers refer to items in the bibliography and *not* to page numbers)

Subject index

(Numbers refer to items in the bibliography and *not* to page numbers)

Abrasion 1746, 1884

Accelerated testing, *see* Testing; Testing methods; Tests, accelerated

Accelerators, corrosion fatigue 672, 770

Adhesive bonded joints, *see* Redux adhesive joints

Adhesives, *see* Phenolic resins; Urea

Adsorption effects 3775

Aeronautical Research Committee 433

Air compressor, breakdown 1560

Air Material Command 3898

Aircraft
   design requirements 3140, 3313, 3409, 3410
   failures 3219, 3363
   flight loads 2189, 2276, 2514, 2552, 2682, 3904
   industry, problems 2157

Aircraft components 504, 976, 3298, 3753
   failure 3114, 3196, 3231, 3642, 3906
   testing 1233, 3141, 3370, 3879

Aircraft engine components 355, 3152
   stresses, stress analysis 1619, 3080, 3668

Aircraft engine crankshafts 1480, 1899, 1900

Aircraft engine materials 1261, 1414, 1827, 1874, 1985

Aircraft engine mounts 2638, 3923

Aircraft engines, gas turbine (problems) 3349

Aircraft fastenings 3996

Aircraft fin, D.H. 104 *(Dove)* 3526, 3890

Aircraft joints, riveted and welded 2658, 2980, 3155

Aircraft landing-gear, problems 3153

Aircraft materials 420, 938, 1057, 1555, 3461, 3811
   fatigue strength 617, 1261, 1341, 1354, 1414, 1464, 1513, 1911, 2319, 3102, 3263, 3676, 3996
   notch sensitivity 3916, 4028
   temperature, low 1016, 1243, 3529, 3839, 4028

Aircraft propeller hubs 1480, 3846

Aircraft propeller materials 652, 2608, 2673, 2756, 2785, 2897, 2912, 2956, 3867, 4070

Aircraft propellers
   design 1832
   Elektron 2075
   failure, fracture, life 680, 1315, 2061
   surface rolling 2265
   testing 2845, 3703, 3858

Aircraft stabilizers, low temperature 3873

Aircraft structural members 355, 692, 3765
   beams, girders 1994, 2062, 2151, 2975
   design 2840, 3240, 3753
   spars (Hydronalium) 1333
   spars (wooden) 563
   testing 963, 1994, 2859, 3204, 3440, 3879, 3904

Aircraft structures
   design, stressing 2682, 3240, 3390
   fatigue 479, 1555, 2405, 2514, 2670, 3390, 3696, 3765, 4044, 4055
   fatigue strength 2095, 2249, 2652, 3263, 3445, 4053
   life, life estimation 2802, 3385, 3638, 3726, 3831
   service failures 355, 784, 2802, 3288
   testing 2190, 2514, 2652, 3177, 3245, 3444, 3879

Aircraft tailplanes
   Meteor 3988
   Typhoon 3171

Aircraft tubing, welded 1838, 1840, 2014, 2484, 2941

Aircraft wings 479, 2552, 3178, 3293
   CA. 12 *(Boomerang)* 3935
   D.H. 104 *(Dove)* 3526, 3890
   Mosquito 3266, 3395
   loads 3624
   testing 3267

Albert, W. A. J. (experiments of) 68, 1503

Albrac, corrosion fatigue 2607

Alclad, *see* Aluminium clad sheet

Aldrey wire 887

Alloy Casting Institute 3855

Almite anodic process 2366, 2890